Formulas for slopes and deflections for prismatic members

D0142712

Case 1

$$\theta_A = \frac{PL^2}{2EI} \qquad v_A = \frac{PL^3}{3EI}$$

Case 2

$$\theta_A = \frac{wL^3}{6EI} \qquad v_A = \frac{wL^4}{8EI}$$

Case 3

$$\theta_A = \frac{ML}{EI} \qquad v_A = \frac{ML^2}{2EI}$$

Case 4

$$\theta_A = \theta_B = \frac{PL^2}{16EI} \; ; \qquad v_C = \frac{PL^3}{48EI}$$

Case 5

$$\theta_A = \frac{P}{6EI}\left(bL - \frac{b^3}{L}\right) ; \quad v_C = \frac{Pa^2 b^2}{3EIL}$$

$$\theta_B = \frac{P}{6EI}\left(2bL + \frac{b^3}{L} - 3b^2\right)$$

$$v_{@x \leq a} = \frac{Pbx}{6EIL}(L^2 - b^2 - x^2)$$

Case 6

$$\theta_A = \theta_B = \frac{wL^3}{24EI} ; \qquad v_C = \frac{5wL^4}{384EI}$$

$$v_{@x} = \frac{wx(L^3 - 2Lx^2 + x^3)}{24EI}$$

Case 7

$$\theta_A = \frac{ML}{3EI} \; ; \quad \theta_B = \frac{ML}{6EI}$$

$$v_C = \frac{ML^2}{16EI}$$

$$v_{@x} = \frac{Mx(L^2 - x^2)}{6EIL}$$

Case 8

$$\frac{M}{2} \qquad \theta_A = \frac{ML}{4EI}$$

Case 9

$$\frac{wL^2}{8} \qquad \theta_A = \frac{wL^3}{48EI}$$

STRUCTURAL ANALYSIS

STRUCTURAL ANALYSIS

J. C. Smith
North Carolina State University

HARPER & ROW, PUBLISHERS, New York
Cambridge, Philadelphia, San Francisco, Washington
London, Mexico City, São Paulo, Singapore, Sydney

Sponsoring Editor: Cliff Robichaud
Text Design Adaptation: Barbara Bert/North 7 Atelier Ltd.
Cover Design: Michel Craig
Text Art: RDL Artset Ltd.
Production Manager: Jeanie Berke
Production Assistant: Beth Maglione
Compositor: TAPSCO, Inc.
Printer and Binder: R. R. Donnelly & Sons Company

STRUCTURAL ANALYSIS

Copyright © 1988 by Harper & Row, Publishers, Inc.

All rights reserved. Printed in the United States of America. No part of this book may be used
or reproduced in any manner whatsoever without written permission, except in the case of brief
quotations embodied in critical articles and reviews. For information address Harper & Row,
Publishers, Inc., 10 East 53d Street, New York, NY 10022.

Library of Congress Cataloging-in-Publication Data

Smith, J. C., 1933
   Structural analysis.

   Bibliography: p.
   Includes index.
   1. Structures, Theory of.   I. Title.
TA645.S495   1988   624.1'7   87-23609
ISBN 0-06-046317-1

87 88 89 90 9 8 7 6 5 4 3 2 1

To
My Teachers
and
My Students

CONTENTS

CHAPTER 16
DISPLACEMENT METHOD—STABILITY ANALYSIS   642

PREFACE

The ever-increasing availability and affordability of microcomputers and structural analysis software for them are having a profound influence on the contents of textbooks being prepared for undergraduate engineering students. Currently, classical methods of structural analysis are being taught to undergraduate students at most of the engineering schools. However, the microcomputer software is based on matrix methods of structural analysis. Therefore, it is imperative to teach the undergraduate students matrix methods of structural analysis, beginning with the first course in indeterminate structural analysis. However, the matrix methods should be taught in a manner that is very similar to the way classical methods are taught. That is, textbooks on computerized methods should emphasize the physical behavior of the structure instead of presenting the methods in an abstract, mathematical, rote memory manner.

This book is intended to serve as the textbook for the first two or three courses in structural analysis, assuming that microcomputers are to be gradually introduced and utilized in the educational process. I have fifteen years of experience teaching the introductory course in computer programming for civil engineering students using either FORTRAN or PL/1, and more than twenty years of experience teaching students how to use computers in structural analysis and design. Therefore, this text contains only those methods of structural analysis for beams, frames, and trusses that I believe are appropriate for a three-course sequence designed to prepare aspiring structural engineers for a successful computer-oriented career. Most schools probably will choose to teach the third course at the graduate level. A subsequent graduate level course in finite element analysis is also recommended.

When a potential textbook is examined the following questions should be asked. Does the textbook contain material on every topic that should be covered in an undergraduate course? Does it contain most of the material to be covered in a graduate course? Does the textual material supplement what will be presented in the classroom? Do the example problems supplement the ones to be presented in the classroom? Will it be a good textbook for more than one course? Will it be a good reference book for subsequent courses and/or in professional practice? The potential textbook that receives the most yes answers should be the correct choice.

Basically, this text parallels my approach in the classroom, which is to teach by showing examples. However, an experienced computer user knows that all of the pertinent definitions and details to be used must be carefully stated in order to successfully solve a problem by using computer software. Consequently, I first give a detailed set of procedural steps for each method of indeterminate analysis, but the teaching and learning of each method are conducted in the numerical examples. A matrix approach is used from the very beginning in the Force Method and the Displacement Method of indeterminate structural analysis since a matrix formulation is the most efficient approach to use in computerized solutions. However, the most efficient matrix approach for a computerized formulation is not the most efficient matrix approach for a noncomputerized solution. Consequently, in the early stages of the matrix methods of analysis, the most easily understood approach is used to explain all of the basic concepts of a matrix method of analysis. Also, the Moment Distribution Method is presented since it is still an appropriate method for micro-computer structural analysts to learn. Some instructors require their students to learn how to do moment distribution on a microcomputer using spreadsheet software.

In most of the examples in this book, realistic numerical values are used for two reasons: (1) computer analysts have to input numbers instead of symbols to obtain a computerized solution; and (2) the beginner needs to gain a feel for the order of magnitude of correct computer input and output. In the chapters dealing with matrix methods of analysis, the deformed structure is always drawn, used in presenting the theory, and used in formulating the example problem solutions. Such usage of the deformed structure emphasizes the structural behavior, enables the student to quickly acquire a "feel" for the structure, and helps the instructor (and the students) avoid the tendency to lapse into an abstract, mathematical approach, which is difficult for a beginner to follow. This is a very important issue. The inherent and most valuable ingredient of classical methods of structural analysis is retained and emphasized. A correct, rough sketch of the deflected structure takes very little time for the student to make, but the student gains many benefits in return. During the learning process, sketching the deflected structure helps to keep the student from relying on an abstract, rote memory approach and should help the student to think about the sign convention more clearly (the units, sign, and magnitude of each computed value should be examined for correctness). Every good structural designer knows how the structure should deform and designs the structure to deform that way. Thus, sketching the deformed structural shape is innate to the structural design process.

The textual material is sequentially arranged in the order that I teach it. Chapters 1 through 6 deal with determinate structures, keeping in mind that usage of a micro-computer is imminent. All of Chapters 1 through 6 and the introductory material in Chapters 7 through 10 are taught in the first structural analysis course. It must be noted that I teach more structural design courses per year than structural analysis courses. Consequently, I teach Chapters 11 and 12 (Influence Lines for Indeterminate Structures and Approximate Methods of Indeterminate Analysis) in a structural design course before or concurrently with the second structural analysis course which deals with Chapters 7 through 10 and 13. Also, I teach a graduate level structural analysis course which deals in detail with Chapters 13 through 16. If a school wishes to present all of the topics in this text in analysis courses only, I propose that Chapters 1 through 6, the introductory part of Chapter 7, Chapters 8 and 9, and the introductory part of Chapter 10 be presented in the first course. The advanced part of Chapters 7 through 10, Chapters 13, 11, and 12 should be presented in the second course. A review of Chapter 13 should precede the presentation of Chapters 14 through 16 in the third course (preferably at the graduate level).

A computer or programmable calculator should be used minimally in the first structural analysis course. Students should not be required to use software simply because it is available and obtain computerized solutions until they have demonstrated without a computer that they have learned the underlying structural theory. Otherwise, the students will not be able to decide for themselves when they have correctly modeled and (finally!—Thank You, God!) obtained an acceptable computerized solution for their structural problem. The students must be ingrained with the need for critical examination of computer output for acceptability at the outset.

Some instructors prefer to use only educationally oriented software (either their own packages, or CAL, or similar packages) to teach the concepts of the computerized matrix approach. If only commercially available software with all of the latest and nicest features is used, the instructor must require the students to solve some problems either indirectly or in a piecemeal fashion in order to demonstrate that they have learned the computerized analysis concepts embodied in the software. Somehow, the students should be required to demonstrate a working knowledge of all the steps in the computerized approach before they are required to show the ability to read and understand a user's manual and to routinely zap out computer solutions.

ACKNOWLEDGMENTS

I would like to thank the following people for their contributions to this textbook: my students and some of my colleagues' students for identifying where the textual material needed to be improved and for finding previously undetected errors; J. F. Ely, V. C. Matzen, and C. C. Tung for providing appropriate problems with answers or/and making constructive criticisms after using the manuscript as a trial textbook; L. T. LaSalle, Jr. and H. D. Penny for their assistance in writing the educationally oriented microcomputer software and user manuals which are available upon request from the publisher to the schools that adopt this book as a textbook for a structural analysis course; Achintya Haldar, S. M. Holzer, Pierre Leger, and P. C. Perdikaris for their careful, critical reviews and valuable suggestions; and Cliff Robichaud, engineering editor of Harper & Row, for persuading the author to write this book.

Since I am a teacher, I would be remiss if I did not acknowledge that my teachers have had a profound influence on my life. From the first grade through the PhD degree, almost all of my teachers were either good or excellent and some were superb. Those teachers who had the most influence on my academic and personal development were:

Julia Fox, Third Grade, Hudson, NC
Claudia Kincaid, Senior High English, Hudson, NC
Karl P. Throneburg, Jr., Senior High Math, Hudson, NC
Charles R. Bramer, N. C. State University in Structures
Adolphus Mitchell, N. C. State University in Mechanics
John E. Goldberg, Purdue University in Structural Mechanics
Martin J. Gutzwiller, Purdue University in Structures
Ervin O. Stitz, Purdue University in Experimental Mechanics

In addition to being superb teachers, they influenced me in ways that show up frequently in my classroom teaching approach. Today only four of them are still living and I am unable to thank all of them now, but I will always remember each of them in a special way. Teachers like these cannot be replaced by a computer.

J. C. Smith

Chapter 1

Introduction

1.1 STRUCTURAL BEHAVIOR, ANALYSIS, AND DESIGN

A *structure* is an assembly of members interconnected by joints.

Structural behavior is the response of a structure to applied loads and environmental effects (wind, earthquakes, temperature changes, snow, ice, rain, etc.).

Structural analysis is the determination of the reactions, member forces, and deformations of the structure due to applied loads and environmental effects.

Structural design involves:

1. Arranging the general layout of the structure to satisfy the owner's requirements (for nonindustrial-type buildings, an architect usually does this part).
2. Preliminary cost studies of alternative structural framing schemes or/and materials of construction.
3. Preliminary analyses and designs for one or more of the possible alternatives studied in step 2.
4. Choosing the alternative to be used in the final design.
5. Performing the final design which involves:
   a. Choosing the analytical model to use in the analyses.
   b. Determination of the loads.
   c. Performing the analyses using assumed member sizes which were obtained in the preliminary design phase.
   d. Using the analysis results to determine if the trial member sizes satisfy the design code requirements.
   e. Resizing the members, if necessary, and repeating items b through e of step 5, if necessary.

Structural engineers deal with the analysis and design of buildings, bridges, conveyor support structures, cranes, dams, offshore oil platforms, pipelines, stadiums, transmission towers, storage tanks, tunnels, pavement slabs for airports and highways, and structural components of airplanes, spacecraft, automobiles, buses, and ships. The same basic principles of analysis are applicable to each of these structures.

The engineer in charge of the structural design must (1) decide how it is desired for the structure to behave when the structure is subjected to applied loads and environmental effects, and (2) ensure that the structure is designed to behave that way. Otherwise a designed structure must be studied to determine how it responds to applied loads and environmental effects. These studies may involve making and testing a small-scale model of the actual structure to determine the structural behavior (this approach is warranted for a uniquely designed structure—no one has ever designed one like it before). Full scale tests to collapse are not economically feasible for one of a kind structures. For mass produced structures such as airplanes, automobiles, and multiple-unit (repetitive) construction, the optimum design is needed and full scale tests are routinely made to gather valuable data which are used in defining the analytical model employed in computerized solutions.

Analytical models (some analysts prefer to call them *mathematical models*) are studied to determine which analytical model best predicts the desired behavior of the structure due to applied loads and environmental effects. Determination of the applied loads and the effects due to the environment is a function of the structural behavior, any available experimental data, and the designer's judgment based on experience.

A properly designed structure must have adequate strength, stiffness, stability, and durability. The applicable structural design code is used to determine if a structural component has adequate strength to resist the forces required of it based on the results obtained from structural analyses. Adequate stiffness is required, for example, to prevent excessive deflections and undesirable structural vibrations. There are two types of possible instability: (1) A structure may not be adequately configured either externally or internally to resist a completely general set of applied loads; and (2) a structure may buckle due to excessive compressive axial forces in one or more members. A skateboard, for example, is externally unstable in its length direction. If a very small force is applied to a skateboard in its length direction, the skateboard begins to roll in that direction. Overall internal structural stability of determinate frames may be achieved by designing either truss-type bracing schemes or shear walls to resist the applied lateral loads. In the truss-type bracing schemes, members that are required to resist axial compression forces must be adequately designed to prevent buckling, otherwise the integrity of the bracing scheme is destroyed. Indeterminate structural frames do not need shear walls or truss-type bracing schemes to provide the lateral stability resistance required to resist the applied lateral loads. However, indeterminate frames can become unstable due to sidesway buckling of the structure as shown in Chapters 15 and 16.

In the coursework that an aspiring structural engineer must master, the traditional approach has been to teach at least one course in structural analysis and to require that course as a prerequisite for the first course in structural member behavior and design. This traditional approach of separately teaching analysis and design is the proper one in the author's opinion, but in this approach the student is not exposed to the true role of a structural engineer unless the student takes a structural design course that deals with the design of an entire structure. In the design of an entire structure, it becomes obvious that structural behavior, analysis, and design are inter-related. The most bothersome thing to the student in the first design of an entire structure using plane frame analyses is the determination of the loads and how they are transferred from floor slab to beams, from beams to girders, from girders to columns, and from columns to supports. Transferral of the loads is dependent on the analytical models that are deemed to best represent the behavior of the structure. Consequently, in structural analysis courses the analytical model and the applied loads are given information, and the focus of the analysis courses is on the applicable analysis tech-niques.

1.2 IDEALIZED ANALYTICAL MODELS

Structural analyses are conducted on an *analytical model* which is an *idealization* of the actual structure. Engineering judgment must be used in defining the idealized structure such that it represents the actual structural behavior as accurately as is prac-tically possible. Certain assumptions have to be made for practical reasons: Idealized material properties are used, estimations of the effects of boundary conditions must be considered, and complex structural details that have little effect on the overall structural behavior can be ignored (or studied later as a localized effect after the overall structural analysis is obtained).

All structures are three-dimensional, but in many cases it is possible to analyze the structure as being two-dimensional in two mutually perpendicular directions. This text deals only with either truss-type or frame-type structures. If a structure must be treated as being three-dimensional, in this text it is classified as being either a space truss or a space frame.

The points at which two or more members of a structure are connected are called *joints*. A *truss* is a structural system of members that are assumed to be pin connected at their ends. Truss members are designed to resist only axial forces and truss joints are designed to simulate a no moment resistance capacity.

A *frame* is a structural system of members that are connected at their ends to joints that are capable of receiving member-end moments and capable of transferring member-end moments between two or more member ends at a common point.

If all of the members of a structure lie in the same plane, the structure is a two-dimensional or planar structure. Examples of planar structures shown in Figure 1.1 are beams, plane grids, plane frames, and plane trusses. Note that all members of a plane grid lie in the same plane, but all loads are applied perpendicular to that plane. For all of the other planar structures, all applied loads and all members of the structure lie in the same plane. In Figure 1.1 each member is represented by only one straight line between two joints. Each joint is assumed to be a point that has no size. Members have dimensions of depth and width, but a single line (the longitudinal centroidal axis) is chosen for graphical convenience to represent the member spanning between two joints. Thus, the *idealized structure is a line diagram configuration.* The length of each line defines the span length of a member and usually each line is the trace

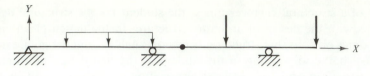

(a) Beam lying in XY plane

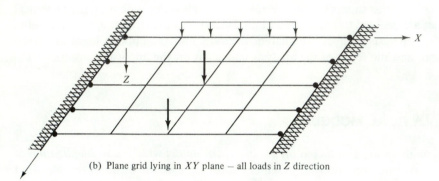

(b) Plane grid lying in XY plane – all loads in Z direction

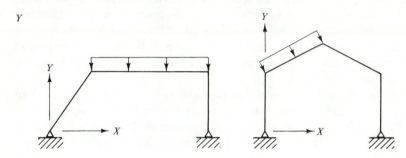

(c) One–story plane frames lying in XY plane

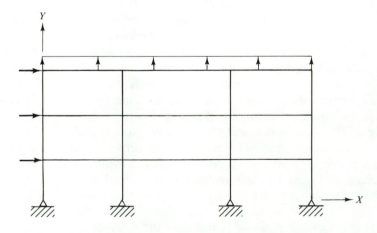

(d) Multistory, multibay, plane frame lying in XY plane

Figure 1.1 Examples of planar structures.

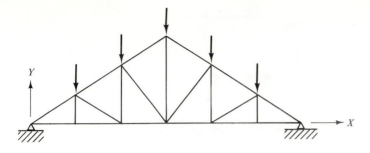

(e) Plane truss lying in XY plane

Figure 1.1 (*continued*)

along the member's length of the intersecting point of the centroidal axes of the member's cross section.

1.3 BOUNDARY CONDITIONS

For simplicity purposes in the following discussion, the structure is assumed to be a plane frame. At one or more points on the structure, the structure must be connected either to a foundation or to another structure. These points are called *support joints* (or *boundary joints,* or *exterior joints*). The manner in which the structure is connected to the foundation and the design of the foundation influence the number and type of restraints provided by the support joints. Since the support joints are on the boundary of a structure and since special conditions can exist at the support joint locations, the term *boundary conditions* is used for brevity purposes to embody the special conditions controlling displacements at the support joints. The various idealized boundary condition symbols for the line diagram structure are shown in Figure 1.2 and discussed in the following paragraphs.

A *hinge,* Figure 1.2a, represents that a structural part is pin connected to a foundation that does not allow translational movements in two mutually perpendicular directions. The pin connection is assumed to be frictionless. Therefore, the attached structural part is completely free to rotate with respect to the foundation. Since many of the applied loads on the structure are caused by and act in the direction of gravity, one of the two mutually perpendicular support directions is chosen to be parallel to the gravity direction. In conducting a structural analysis, the analyst assumes that the correct direction of this support force component is either opposite to the direction of the forces caused by gravity or in the same direction as the forces caused by gravity. In Figure 1.2, the reaction components are shown as vectors with a slash on them and the arrows indicate the author's choice for the assumed directions of each vector. It must be noted that the author could have assumed the opposite direction for each vector.

A *roller,* Figure 1.2b, represents a foundation that permits the attached structural part to rotate freely with respect to the foundation and to translate freely in the direction

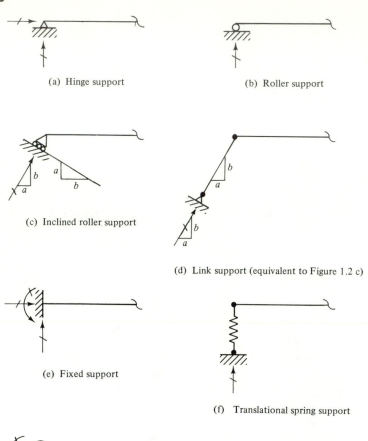

(a) Hinge support

(b) Roller support

(c) Inclined roller support

(d) Link support (equivalent to Figure 1.2 c)

(e) Fixed support

(f) Translational spring support

(g) Rotational spring support

Figure 1.2 Boundary condition symbols and their reaction components.

parallel to the foundation surface, but does not permit any translational movement in any other direction. To avoid any ambiguity for a roller on an inclined surface, Figure 1.2c, the author prefers to use a different roller symbol than he uses on a horizontal surface. Some analysts prefer to use a link, Figure 1.2d, instead of a roller to represent the boundary condition described at the beginning of this paragraph. A *link* is defined as being a fictitious, weightless, nondeformable, pinned-ended member which never has any loads applied to it except at the ends of the member.

A *fixed support,* Figure 1.2e, represents a bed-rock-type of foundation that does not deform in any manner whatsoever and the structural part is attached to the foundation in such a manner that no relative movements can occur between the foundation and the attached structural part.

A *translational spring,* Figure 1.2f, is a link that can deform only along its length direction. This symbol is used to represent either a joint in another structure or a foundation resting on a deformable soil.

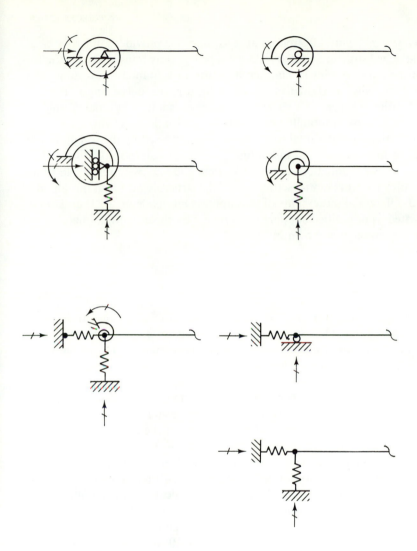

(h) Various possible combinations of Figures 1.2 a through g

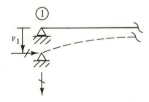

(i) Prescribed Y–direction support displacement

Figure 1.2 (*continued*)

A *rotational spring,* Figure 1.2g, represents a support that provides some rotational restraint for the attached structural part, but does not provide any translational restraint. The support can be either a joint in another structure or a foundation resting on a deformable soil. Generally, as shown in Figure 1.2h, a rotational spring is used in conjunction with either a hinge, or a roller, or a roller plus a translational spring, or a translational spring, or two mutually perpendicular translational springs.

The soil beneath each individual foundation is compressed by the weight of the structure. Soil conditions beneath all individual foundations are not identical. Also, the weights acting on the foundations are not identical and vary with respect to time. Therefore, nonuniform or differential settlement of the structure occurs at the support joints. Estimated differential settlements of the supports are made by the foundations engineer and treated as prescribed support movements by the structural engineer. A *prescribed support movement* is shown in Figure 1.2i.

1.4 INTERIOR JOINTS

For simplicity and generality purposes in the following discussion, the structure is assumed to be a plane frame. On a line diagram structure, an *interior joint* is a point on the structure at which two or more member length axes intersect. For example, in Figure 1.3 points 2, 4, 5, 7, 8, and 10 are interior joints, whereas points 1, 3, 6, and 9 are support joints (or exterior joints, or boundary joints).

The manner in which the member ends are connected at an interior joint must be accounted for on the line diagram. The types of connections for a structure composed of steel members can be broadly categorized as being one of the following types:

1. *Shear connection* (or a no moment connection)—*an internal hinge.* If the connection at joint 10 of Figure 1.3 is as shown in Figure 1.4, it is classified by designers as being a shear connection (or a no moment connection). Thus, an internal hinge is shown on the line diagram at joint 10 of Figure 1.4 to indicate that no moment can be transferred between the ends of members 2 and 10 at joint 10. However, the internal hinge is capable of transferring translational-type member-end forces (axial forces and shears) between the ends of members 2 and 10 at joint 10. It should be noted that

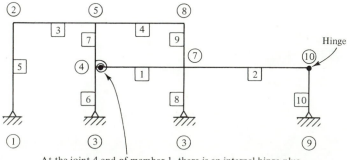

At the joint 4 end of member 1, there is an internal hinge plus a rotational spring spanning across the hinge

Figure 1.3   Idealized interior joint conditions.

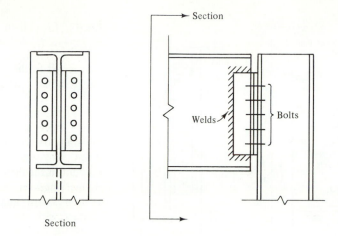

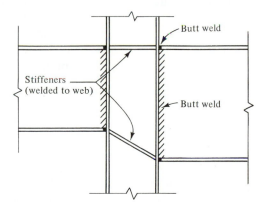

Figure 1.4 Web connection (shear connection).

this type of connection can transfer a small amount of moment, but the moment is so small that it can be ignored in design.

2. *Rigid connection* (fully restrained type of construction). If the connection at joint 7 of Figure 1.3 is as shown in Figure 1.5, it is classified by designers as a joint that behaves like a rigid body (or a nondeformable body). Thus, if joint 7 of Figure 1.3 rotates 5 degrees in the counterclockwise direction, the ends of members 1, 2, and 9 at joint 7 also rotate 5 degrees in the counterclockwise direction.

Figure 1.5 Rigid connection—fully welded plus stiffeners.

3. *Semirigid connection* (partially restrained type of construction). If the beam to column connection at joint 4 of Figure 1.3 is as shown in Figure 1.6, it is classified by designers as being a semirigid connection. (Webster's dictionary definition of semirigid is: rigid to some degree or in some parts.) The web angles fully ensure that the Y-direction displacement at the end of member 1 is identical to the Y-direction displacement of joint 4. (On the line diagram structure in Figure 1.3, joints 3, 4, and 5 lie on the same straight line which is the longitudinal axis of members 6 and 7. Thus, joint 4 is located at the point where the longitudinal axes of members 1, 6, and 7 intersect.) Consequently, joint 4 is treated as being rigid in the Y direction. However, the top and bottom flange angles in Figure 1.6 are not flexurally stiff enough to ensure

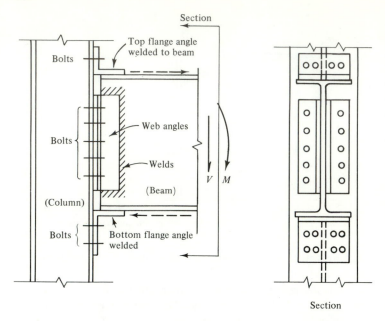

Section

Top flange angle
welded to beam

Bolts

Web angles

Bolts

Welds

V M

(Beam)

(Column)

Bolts

Bottom flange angle
welded

Section

Figure 1.6 Semirigid connection—angles welded to beam and bolted to column. M effectively is transferred to the top and bottom flange angles as shown by the dashed lined vectors. Consequently, due to the action of M, the top flange angle and the web angles flexurally deform allowing the top beam flange to translate a finite amount in the direction of the top dashed lined vector. However, the bottom flange angle remains in contact with the column flange. Thus, the gap between the end of the beam end and the column flange is trapezoidal after the angle deformations occur. The bolts resist V and ensure that the beam end does not translate in the Y direction.

that the flanges of member 1 always remain completely in contact with the flanges of members 6 and 7. Thus, joint 4 cannot be treated as being completely rigid. Therefore, at the left end of member 1 in Figure 1.3, a rotational (or spiral) spring is shown to denote that a rotational deformation occurs between joint 4 and the end of member 1. It should be obvious to the reader that a semirigid connection is capable of developing more moment than a web connection can develop, but not as much moment as a rigid connection can develop.

1.5 LOADS AND ENVIRONMENTAL EFFECTS

In structural analysis courses the analytical model and the applied loads are given information, and the focus of the analysis courses is on the applicable analysis techniques. In structural design the loads that are to be applied to the analytical model of the structure must be established by the structural designer.

Each state in the United States of America has a building code which is mandated by law to be used in the design of an engineered structure. The state building code gives minimum design loads that must be used in the design of a building to ensure a desired level of public safety unless in the structural engineer's judgment higher design loads should be used. Since coping with building codes and determining the applied loads are more appropriately a part of a structural design course, the author chooses to give only a brief description of loads and environmental effects in this text. However, he chooses to use the same terminology in the discussion as is used in the building code definitions for the loads and environmental effects.

All loads are treated as being statically applied to the structure and are classified

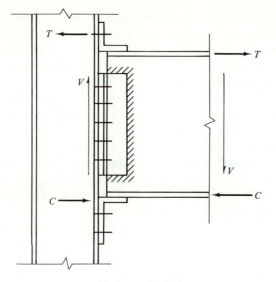

(a) Assumed behavior

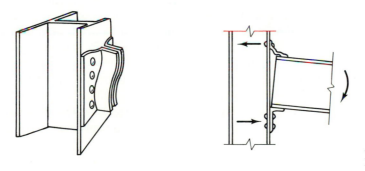

(b) Deformation of the connection (separated for clarity)

Figure 1.6 (*continued*) Behavior of semirigid connection.

as being one of the following types: *dead loads; live loads;* and *impact loads. Environmental effects* due to wind, earthquakes, snow, rain, ice, temperature changes, differential settlement of supports, soil pressures, and hydrostatic pressures are converted into equivalent statically applied live loads.

Examples of *dead loads* are the weight of the structure; heating, air-conditioning, plumbing, and electrical conduits and fixtures; floor and roof covers; and suspended ceilings. Dead loads do not vary with time in regard to position and weight. Thus, they are not moved once they are in place and, therefore, are called dead loads. A worn floor or roof cover is removed and replaced with a new one in a matter of days. A load that is not there for only a few days in the, say, 50-year life of a structure can be considered a permanent load and classified as a dead load.

Gravity loads that vary with time in regard to magnitude or/and position are called *live loads.* Examples of live loads are people, furniture, movable equipment, partition walls, file cabinets, and stored goods in general. Forklifts and other types of slow moving vehicles (in a parking garage, for example) may be treated as live loads.

Impact loads are live loads that have dynamic effects that cannot be ignored. Examples of impact loads are cranes, elevators, reciprocating machinery, and vehicular traffic on highway or railroad bridges.

The *effect of an earthquake* on a building is similar to the effect of a football player being clipped (for our purposes, say a clip is a hit around or below the knees and from the blind side). The football player is unaware that he is going to be hit. Consequently, his feet must go in the direction of the person who hits him, but his upper body does not want to move in that direction until the momentum of his lower body tends to drag the upper body in that direction. An earthquake consists of horizontal and vertical ground motions. The horizontal ground motion effect on a structure is similar to the football player being clipped. It is this type of motion that is converted into an equivalent static loading to simulate the effect of an earthquake on a building. An equivalent static loading is applied at all story levels and in the opposite direction of the gound motion since the foundation of the structure remains stationary in a static analysis. It should be noted that *all dynamic loads cannot always be replaced by equivalent static loads.* The *effects due to wind* are converted into an equivalent static pressure acting on the structure. There is a *basic wind pressure* which is a function of the mass density of air and the wind velocity. This basic wind pressure is given in the building code either as a formula or in tabular form (in pounds per square foot along the height direction of the building). Wind velocity is least at ground level and increases along the height direction of the building. The basic wind pressure is multiplied by a *shape factor* and possibly a *gust factor* to obtain a modified wind pressure which is applied to the structure.

Numerical examples are given in the text to show how *temperature changes* and *differential settlement of supports* are converted into equivalent static loadings.

1.6 BEHAVIORAL ASSUMPTIONS

In this text the idealized stress–strain relation for the structural *material* is assumed to be *linearly elastic.* Also, it is assumed that the material is never stressed beyond its elastic limit. Except for Chapter 15, the *change in geometry* of the structure due to applied loads is assumed to be *negligible* in writing the equilibrium equations of the structural system. Therefore, except for Chapter 15, the geometry of the structure before any loads are placed on it is used in writing the equilibrium equations.

Due to the above assumptions, except for Chapter 15, a *linear load-deformation relation exists* for the structure and *superposition is valid.* This means that if the solution for a 1^k concentrated load located at a specific point has been obtained, and we want the solution for a 60^k load located at the same point, we simply multiply all results obtained for the unit load by a factor of 60. Also, it means that if we have a solution for loading 1 (say due to dead load) and another solution for loading 2 (say for a particular live load configuration), the solution for a loading that consists of all of the loads in loading 1 and loading 2 is the algebraic sum of the results obtained for loadings 1 and 2.

1.7 METHODS OF STRUCTURAL ANALYSIS

The methods for analyzing the analytical structural model can be categorized as being either a *Force (or Flexibility) Method* or a *Displacement (or Stiffness) Method* of analysis. In the Force Method of analysis, the *internal forces in the members or/and the reactions* are directly obtained from the deformation compatibility equations, but the joint displacements are not obtained by this method in this text. In the Displacement Method, the *chosen unknown joint displacements* are directly obtained from the system equilibrium equations, and the member-end forces are obtained by substituting the appropriate system displacements into each member-end force-deformation relation. After the member-end forces are known, the reactions can be obtained by superposition of the appropriate member-end forces at each reaction.

1.8 SIGNIFICANT DIGITS AND COMPUTATIONAL PRECISION

Dead loads can be estimated very well after all of the member sizes are finalized. However, estimated live loads and equivalent static loads for the effects due to wind and earthquakes are much more uncertain than the estimated dead loads. Perfection is impossible to achieve in the fabrication and erection procedures of the structure. Also, certain simplifying assumptions have to be made by the analyst to obtain practical solutions. For example, joint sizes are usually assumed to be infinitesimal, whereas they really are finite. Interior joints and boundary joints are assumed to be either rigid or pinned, whereas they really are somewhere between rigid and pinned. Thus, the final structure is never identical to the one that the structural engineer designed, but the differences between the final structure and the designed structure are within certain tolerable limits.

A digit in a measurement is a significant digit if the uncertainty in the digit is less than 10 units. Standard steel mill tolerance for areas and weights is 2.5% variation. Consider a piece of steel listed in a steel handbook as weighing 100 pounds per foot and having a cross-sectional area of 29.4 square inches. The weight variation tolerance is $0.025 * 100 = 2.5$ and the actual weight lies between 97.5 and 102.5 pounds per foot. Since the third digit in 100 is uncertain by only 5 units, the 100 pounds per foot value is valid to three significant digits. However, the area variation tolerance is $0.025 * 29.4 = 0.735$ and the actual area lies between 28.665 and 30.135 square inches. Since there are 14.7 units of uncertainty in the third digit of 29.4, only the first two digits in the value 29.4 are significant and a recorded value of 29.0 would be more appropriate in the steel handbook.

Most computers will accept arithmetic constants having an absolute value in the range 1×10^{-35} to 1×10^{35}. Many computers will accept a much wider range. A computer holds numeric values only to a fixed number of digits, usually the equivalent of between 6 and 16 decimal (or base 10) digits. The number of decimal digits held is called the precision of the arithmetic constant.

For discussion purposes, suppose that the loads and structural properties are accurate to only 3 significant digits. Most commercial structural analysis programs use at least 16-digit precision in the solution of the set of simultaneous equations. In a multistory building, there may be hundreds or thousands of equations in a set. If only 3-digit precision were used in computerized solutions, the truncation and round-off errors in the mathematics would in some cases contribute much more uncertainty in the computed results than the structural engineer has in the loads and structural properties. Reconciliation of the actual number of significant digits in the computed results should be made by the structural designer after all of the computed results are available. The author always makes his electronic calculator computations using the maximum available precision in the same manner that a computer would make them in floating point form. If the loads have units of either kips or kilonewtons, the author records the computed reactions and final member-end forces rounded in the first digit to the right of the decimal in fixed decimal form.

Chapter 2

Reactions of Determinate Structures

2.1 INTRODUCTION

In this chapter, the discussion is limited to structures composed of one or more members whose longitudinal axes lie in a common plane. Also, these structures are assumed to be subjected to applied loads lying only in the common plane. Such structures are called planar structures and they are the ones most commonly seen in introductory structural analysis textbooks. Calculating the reactions of three-dimensional or space structures does not involve any new fundamental principles, but the computations are greatly complicated by the additional geometry of the third dimension. Fortunately, many three-dimensional structures can be decomposed into planar structures in two mutually perpendicular directions.

2.2 EQUATIONS OF STATIC EQUILIBRIUM

A structure that is initially at rest and remains at rest when acted upon by applied loads is said to be in a state of static equilibrium. Otherwise the structure is in a state of motion (for example, an in-flight airplane). This text considers only structures that are restrained by their supports such that they are not in motion relative to the earth's surface. *Reactions* are the restraining forces provided by the supports.

If the supports are replaced by their reactions, the structure is acted upon by the known applied loads and the unknown reactions. A sketch of the entire structure subjected to the known applied loads and the unknown reactions is called a Free Body Diagram (FBD) of the entire structure. To simplify the discussion of conditions defining static equilibrium, the applied loads and reactions on a FBD are collectively called forces. A force is a vector that has a magnitude, a direction, and a point of application.

15

Consider a planar FBD lying in the XY plane. X and Y are any two mutually perpendicular directions in the plane of the FBD. However, most analysts think of the X axis as being the horizontal (tangential to the earth's surface) axis and the Y-axis as being the vertical (perpendicular to the earth's surface) axis. This is natural and convenient since the weight of the structure acts in the direction of gravity. The axis perpendicular to the XY plane is called the Z axis. In this text X, Y, and Z are a right-handed set of orthogonally intersecting reference axes, that is, in a vector mathematics sense, X crossed into Y gives Z; Y crossed into Z gives X; Z crossed into X gives Y.

The conditions that ensure static equilibrium of a planar FBD lying in the XY plane are:

1. The algebraic sum of all force components in each of the two independent directions X and Y must be zero.
2. At any point i in the XY plane, the algebraic sum of the moments of all forces about an axis parallel to the Z axis must be zero.

These conditions stated in mathematical terminology are called the equations of static equilibrium:

$$\sum F_X = 0$$
$$\sum F_Y = 0$$
$$\sum M_Z = 0 \text{ at } i$$

In each of the following example problems, the given structure is stable and determinate. Therefore, the reactions can be calculated from the equations of statics. After some example problems have been solved for the reactions, conditions whereby a structure can be either unstable or indeterminate will be discussed. If a structure is *unstable,* all of the equations of statics are not applicable. For example, it is possible that only two equations of statics and an equation of motion are applicable to a skateboard. If a structure is *indeterminate,* the equations of statics are insufficient to obtain the reactions.

2.3 REACTIONS OF INDIVIDUAL MEMBERS

In each of the following example problems, the author first chooses a convenient set of orthogonal reference axes X, Y, and Z to be used in writing the equations of statics. The Z axis is not shown, but it is located at the origin of the X and Y axes and points out of the page toward the reader. Next the author individually examines each reaction symbol on the figure given in the problem statement, identifies, and labels the X and Y components of the unknown reactions. Each reaction component is depicted by a vector whose arrow is shown pointing in the direction assumed by the author. For convenience in the discussion, a slash is placed on each unknown vector. The analyst must be careful not to introduce invalid reaction components.

After the FBD has been completed, do not blindly start writing the equations of statics in the order shown in Section 2.2 and treating them as a set of simultaneous equations. Study the FBD to determine which reaction component is easiest to find by using only one of the equations of statics and find that unknown first. Then decide which of the two remaining equations of statics directly gives another of the reaction components, and so on.

EXAMPLE 2.1

Identify and calculate the reaction components for the structure shown in Figure 2.1a.

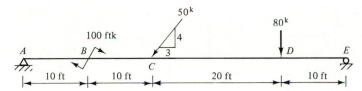

Figure 2.1a

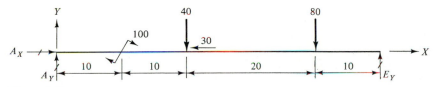

Figure 2.1b

Solution. The immovable hinge at support point A has X- and Y-direction reaction components, but the roller at support point E only has a Y-direction reaction component (the roller is free to roll on the "frictionless" surface beneath the roller) (Fig. 2.1b). The applied force at point B is a clockwise moment. For mathematical convenience, the inclined applied force at point C is decomposed into its X and Y components. In writing the equations of statics, the analyst always has the option of choosing and showing the direction of a positive force in each statical equation.

Since there is only one known applied force component in the X direction, first find A_X from the following equation of statics:

$$\overset{+}{\longrightarrow} \quad \Sigma F_X = 0: \qquad A_X - 30 = 0; \qquad A_X = 30^k$$

Since there are two unknown Y-direction reactions, $\Sigma F_Y = 0$ would give one equation that contains two unknowns. However, either $\Sigma M_Z = 0$ at A or $\Sigma M_Z = 0$ at F involves only one unknown.

$$\overset{\curvearrowleft +}{\Sigma M_Z = 0 \text{ at } A:} \qquad 50*E_Y - 100 - 20*40 - 40*80 = 0$$
$$E_Y = 82^k$$

$$\overset{+\uparrow}{\Sigma F_Y = 0:} \qquad A_Y - 40 - 80 + 82 = 0; \qquad A_Y = 38^k$$

Check the arithmetic by determining if $\Sigma M_Z = 0$ at E:

$$10*80 + 30*40 - 100 - 50*38 = 0, \quad \text{okay}$$

Alternate Solution Using Superposition. Reactions for each applied load one at a time can be calculated as shown in Figure 2.1c. These reactions are algebraically summed to obtain the total reactions.

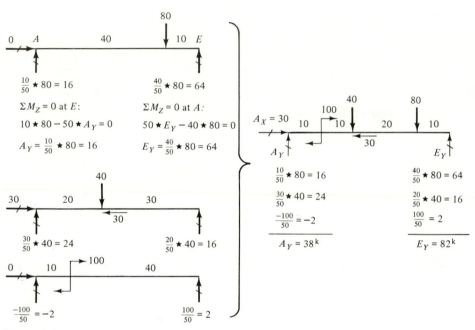

Figure 2.1c

EXAMPLE 2.2

Identify and calculate the reactions for the structure shown in Figure 2.2a.

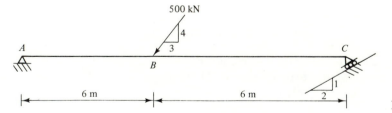

Figure 2.2a

Solution. The reaction at support C must act perpendicular to the surface on which the rollers are constrained to roll upon (Fig. 2.2b). Since the X- and Y-direction components of the reaction at C are needed in the equations of statics, C_Y is chosen as the unknown at C and C_X is expressed as a function of C_Y.

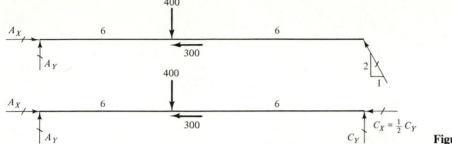

Alternatively, C_X could have been chosen as the unknown and C_Y expressed as a function of C_X.

$\circlearrowleft +$
$\sum M_Z = 0$ at C: $\qquad 12*400 - 18*A_Y = 0$; $\qquad A_Y = 240\,\text{kN}$

$+\uparrow$
$\sum F_Y = 0$: $\qquad 240 + C_Y - 400 = 0$; $\qquad C_Y = 160\,\text{kN}$

$\qquad\qquad C_X = \tfrac{1}{2}C_Y = 80\,\text{kN}$

$\qquad\qquad C = \sqrt{5}\,C_X = 80\sqrt{5} = 179\,\text{kN}$

$+$
$\overrightarrow{\sum F_X} = 0$: $\qquad A_X - 300 - 80 = 0$; $\qquad A_X = 380\,\text{kN}$

EXAMPLE 2.3 _____

Identify and find the reactions for the structure shown in Figure 2.3a.

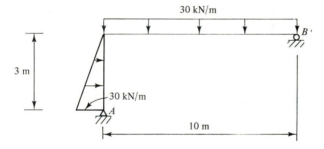

Figure 2.3a

Solution. For purposes of calculating the reactions, the distributed loads can be replaced by their resultants (areas under the load diagram) located at the centroids of the load diagrams (Fig. 2.3b).

$+$
$\overrightarrow{\sum F_X} = 0$: $\qquad A_X + 45 = 0$; $\qquad A_X = -45\,\text{kN}$ (the correct direction is 45 kN $\leftarrow$)

$\circlearrowleft +$
$\sum M_Z = 0$ at A: $\qquad 10*B_Y - 1*45 - 5*300 = 0$; $\qquad B_Y = 154.5\,\text{kN}$

$+\uparrow$
$\sum F_Y = 0$: $\qquad A_Y + 154.5 - 300 = 0$; $\qquad A_Y = 145.5\,\text{kN}$

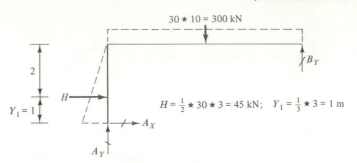

$$H = \tfrac{1}{2} \star 30 \star 3 = 45 \text{ kN}; \quad Y_1 = \tfrac{1}{3} \star 3 = 1 \text{ m}$$

Figure 2.3b

2.4 REACTIONS OF PIN-CONNECTED STRUCTURES

Cantilevered construction, three-hinged construction, and interconnected simple structures are the types of pin-connected structures considered in the following examples.

EXAMPLE 2.4 (CANTILEVERED CONSTRUCTION) ────────────────────

Find the reactions of the structure shown in Figure 2.4a.

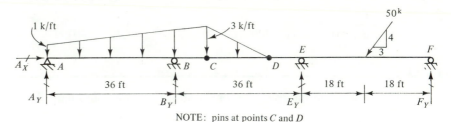

NOTE: pins at points C and D

Figure 2.4a

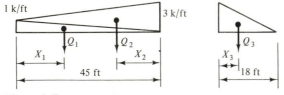

Figure 2.4b

Solution. The trapezoidal distributed area is subdivided into two triangles (Fig. 2.4b).

$$Q_1 = \tfrac{1}{2} * 1 * 45 = 22.5^k \qquad Q_3 = \tfrac{1}{2} * 3 * 18 = 27^k$$
$$X_1 = \tfrac{1}{3} * 45 = 15 \text{ ft} \qquad X_3 = \tfrac{1}{3} * 18 = 6 \text{ ft}$$
$$Q_2 = \tfrac{1}{2} * 3 * 45 = 67.5^k$$
$$X_2 = \tfrac{1}{3} * 45 = 15 \text{ ft}$$

In Figure 2.4c, there are four Y-direction unknowns and only two equations of statics remaining for this FBD. However, there are two internal hinges at points

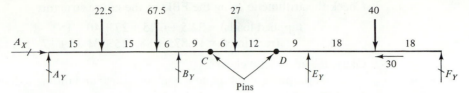

Figure 2.4c

$$\overset{+}{\underset{\longrightarrow}{\sum F_X}} = 0: \qquad A_X - 30 = 0; \qquad A_X = 30^k$$

C and D which enable us to disassemble the structure into three separate structures called substructures (Fig. 2.4d) to avoid ambiguity. The pins at points C and D are removed and substructure CD is lifted out; this reveals that substructure CD is supported in the Y direction by the cantilevered ends of substructures AC and DF. Only substructure AC has a true X-direction reaction. D_X of substructure DF is calculated, reversed, and applied to substructure CD. Similarly, C_X of substructure CD is calculated, reversed, and applied to substructure AC. The Y-direction reactions of substructure CD are calculated, reversed, and applied to the supporting substructures (Fig. 2.4d).

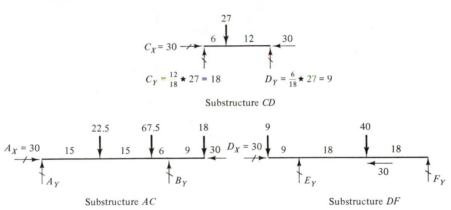

Figure 2.4d

For substructure AC:

$$\sum M_Z = 0 \text{ at } A: \qquad 36*B_Y - 15*22.5 - 30*67.5 - 45*18 = 0$$
$$B_Y = 88.125^k$$
$$\sum F_Y = 0: \quad A_Y + 88.125 - 22.5 - 67.5 - 18 = 0$$
$$A_Y = 19.875^k$$

For substructure DF:

$$\sum M_Z = 0 \text{ at } F: \qquad 18*40 - 36*E_Y + 45*9 = 0$$
$$E_Y = 31.25^k$$
$$\sum F_Y = 0: \qquad F_Y + 31.25 - 9 - 40 = 0$$
$$F_Y = 17.75^k$$

Check the arithmetic using the FBD of the entire structure:

$$(\text{applied loads}) = 22.5 + 67.5 + 27 + 40 = 157^k$$
$$(\text{reactions}) = 19.875 + 88.125 + 31.25 + 17.75 = 157^k$$

Okay, the arithmetic checks.

For purposes of subsequent analyses (shear and moment diagrams—Chapter 3), the reactions are rounded to the nearest 0.1^k:

$$A_X = 30.0^k \rightarrow; \quad A_Y = 19.9^k \uparrow; \quad B_Y = 88.1^k \uparrow; \quad E_Y = 31.2^k \uparrow; \quad F_Y = 17.8^k \uparrow$$

EXAMPLE 2.5 (THREE-HINGED CONSTRUCTION)

Find the reactions for the structure shown in Figure 2.5a; note the three hinges located at points A, B, and C.

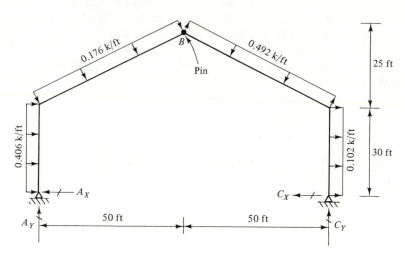

Figure 2.5a

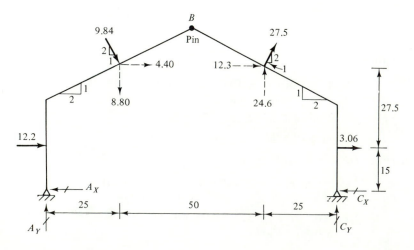

Figure 2.5b

Solution. Shown in Figure 2.5b are the resultants of the distributed loads. Also shown, as dashed vectors, are the X and Y components of the resultants acting on the rafters (the inclined roof members); these components are easier to use in the equations of statics (Fig. 2.5b). Note that the resultants and components were shown with a three-digit precision since the components and resultants are used in subsequent calculations.

$$\sum M_Z = 0 \text{ at } A: \quad 100*C_Y + 75*24.6 - 15*(12.2 + 3.06) - 25*8.80$$
$$- 42.5*(4.40 + 12.3) = 0$$
$$C_Y = -6.86^k \text{ (correct direction of } C_Y \text{ is down)}$$

$$\sum F_Y = 0: \quad A_Y + 24.8 - 8.80 - 6.86 = 0$$
$$A_Y = -8.94^k \text{ (correct direction of } A_Y \text{ is down)}$$

Since $\sum F_X = 0$ involves A_X and C_X, disassemble the structure into substructures by removing the pins at A, B, and C (Fig. 2.5c).

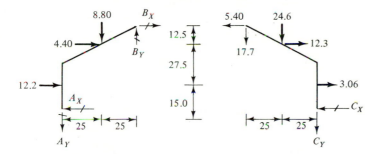

Substructure AB Substructure BC **Figure 2.5c**

For substructure AB:

$$\sum M_Z = 0 \text{ at } B: \quad 25*8.80 + 12.5*4.40 + 40*12.2 + 50*8.94 - 55*A_X = 0$$
$$A_X = 22.0^k$$

$$\sum F_X = 0: \quad 12.2 + 4.40 - 22.0 + B_X = 0$$
$$B_X = 5.40^k$$
$$\sum F_Y = 0: \quad B_Y = 8.80 + 8.94 = 17.74^k$$

For substructure BC: Reverse B_X and B_Y found on substructure AB and apply them at point B on substructure BC.

$$\sum F_X = 0: \quad 12.3 + 3.06 - 5.40 - C_X = 0$$
$$C_X = 9.96^k$$

$$\sum F_Y = 0: \quad 24.6 - 17.7 - 6.86 = 0.04^k \approx 0, \text{ okay (the arithmetic checks)}$$

For the entire structure (to check arithmetic):

$$\overset{+}{\overrightarrow{\sum F_X}} = 0: \qquad 4.40 + 12.3 + 12.2 + 3.06 - 22.0 - 9.96 = 0, \text{ checks}$$

EXAMPLE 2.6 (THREE-HINGED CONSTRUCTION)

In Figure 2.6a, two simple trusses are pin connected at point B. Actually all joints of trusses are considered to be pins, but in this problem solution only the pins at points A, B, and C are involved. Find the reactions.

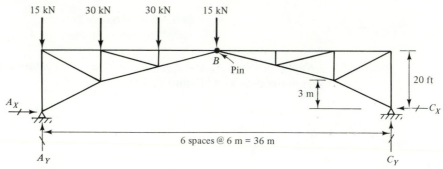

15 kN   30 kN   30 kN   15 kN

B

Pin

20 ft

3 m

A_X

C_X

6 spaces @ 6 m = 36 m

A_Y

C_Y

Figure 2.6a

Solution

$$\sum M_Z = 0 \text{ at } C: \qquad 18*15 + 24*30 + 300*30 + 36*15 - 36*A_Y = 0$$
$$A_Y = 67.5 \text{ kN}$$
$$\sum F_Y = 0: \qquad C_Y + 67.5 - (15 + 30 + 30 + 15) = 0$$
$$C_Y = 22.5 \text{ kN}$$

Since A_X and C_X are unknown and both appear in $\sum F_X = 0$, disassemble the structure into two substructures by removing the pins at A, B, and C. For purposes of calculating the reactions, the two trusses can be drawn as rigid bodies since the structure is determinate.

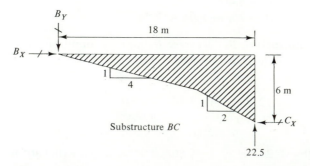

B_Y

18 m

B_X

1
4

6 m

1
2

C_X

Substructure BC

22.5   **Figure 2.6b**

For substructure BC (Fig. 2.6b):

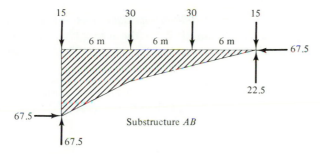

$\sum M_Z = 0$ at B: $18 * 22.5 - 6 * C_X = 0$

$C_X = 67.5\,\text{kN}$

$C = \sqrt{(67.5)^2 + (22.5)^2} = 22.5\sqrt{10} = 71.15\,\text{kN}$

Since there are no loads applied to substructure BC between points B and C, substructure BC acts like a link (straight line) connecting points B and C. The link has the same slope as the slope of resultant reaction at C. Using the other two equations of statics, $B_X = 67.5$ kN and $B_Y = 22.5$ kN.

For the entire structure as a FBD:

$\overrightarrow{\sum F_X} = 0$: $A_X - C_X = 0$;   therefore, $A_X = C_X = 67.5\,\text{kN}$

Check the arithmetic using substructure AB (Fig. 2.6c): Downward loads sum to 90 kN. Upward loads sum to 90 kN. Arithmetic checks out.

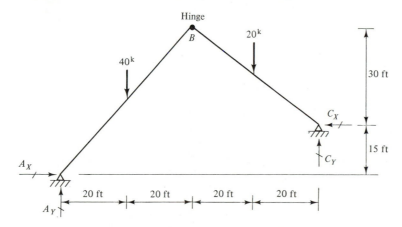

Substructure AB

Figure 2.6c

EXAMPLE 2.7 (THREE-HINGED CONSTRUCTION)

Find the reactions for the structure shown in Figure 2.7a.

Figure 2.7a

Solution (Involving Simultaneous Equations)

$\sum M_Z = 0$ at A: $80 * C_Y + 15 * C_X - 20 * 40 - 60 * 20 = 0$

For member BC as a FBD:

$$\overset{\curvearrowleft +}{\sum M_Z} = 0 \text{ at } B: \qquad 40*C_Y - 30*C_X - 20*20 = 0$$

The simultaneous equations in matrix form are:

$$\begin{bmatrix} 15 & 80 \\ -30 & 40 \end{bmatrix} \begin{Bmatrix} C_X \\ C_Y \end{Bmatrix} = \begin{Bmatrix} 2000 \\ 400 \end{Bmatrix}$$

The solution of the simultaneous equations is:

$$\begin{Bmatrix} C_X \\ C_Y \end{Bmatrix} = \begin{Bmatrix} 16^k \\ 22^k \end{Bmatrix}$$

For the entire structure:

$$\sum F_X = 0: \qquad A_X = C_X = 16^k$$
$$\sum F_Y = 0: \qquad A_Y = 40 + 20 - 22 = 38^k$$

Alternative Solutions (Avoiding Simultaneous Equations).   There are several ways to solve for the reactions in Example 2.7 without involving simultaneous equations. However, none of the ways is really attractive. Therefore, only two ways are described.

Alternative Solution 1.   Instead of choosing X and Y axes oriented as shown in Figure 2.7a, choose X' and Y' axes as shown in Figure 2.7b which necessitates first finding L.

$$\overset{\curvearrowleft +}{\sum M_{Z'}} = 0 \text{ at } A: \qquad L*C'_Y - 20*20 = 0; \qquad \text{solve for } C'_Y$$

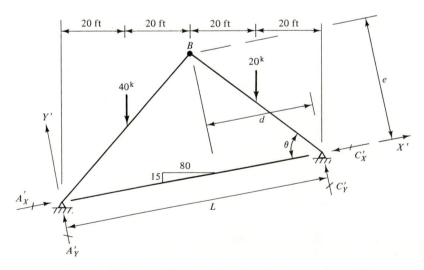

Figure 2.7b

Next, calculate θ, d, and e which are needed in the following equation. For FBD *BC*:

$$\sum M_{Z'} = 0 \text{ at } B: \qquad d*C_{Y'} - e*C_{X'} - 20*20 = 0; \qquad \text{solve for } C_{X'}$$

Now, find the *X*- and *Y*-direction components of $C_{X'}$ and $C_{Y'}$; combine these *X*-direction components to obtain C_X and combine these *Y*-direction components to obtain C_Y. Finally, using the C_X and C_Y values and Figure 2.7a, calculate A_X and A_Y by summing forces in the *X* and *Y* directions, respectively.

Alternative Solution 2.   Since superposition is valid, solve each of the problems shown in Figure 2.7c for the *X*- and *Y*-direction reactions. In Figure 2.7c(i), member *BC* is a link (since there is no load on member *BC*); sum moments at *A* to find the reactions at *C*. Then, sum forces in the *X* and *Y* directions to find the reactions at *A*. Similarly, in Figure 2.7c(ii), member *AB* is a link; sum moments at *C* to find the reactions at *A*. Then, sum forces in the *X* and *Y* directions to find the reactions at *C*.

Finally, superimpose the reactions of Figures 2.7c(i) and 2.7c(ii) to obtain the reactions for Figure 2.7b.

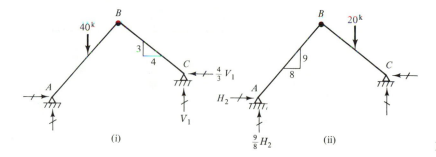

Figure 2.7c

EXAMPLE 2.8 (ALL JOINTS ARE PINS)

In Figure 2.8a members *BC* and *CE* are cables (tension only members). Find the reactions. Draw FBD for members *DE* and *AC*.

Solution

$$\sum M_Z = 0 \text{ at } A: \qquad 30*(3B_Y) - 15*B_Y - 7.5*2 - 15*30 = 0$$
$$B_Y = 4.43^k$$
$$B_X = 3B_Y = 13.3^k$$
$$B = \sqrt{10}\, B_Y = 14.0^k$$
$$\sum F_X = 0: \qquad A_X = 13.3^k$$
$$\sum F_Y = 0: \qquad A_Y = 30 + 2 - 4.43 = 27.6^k$$

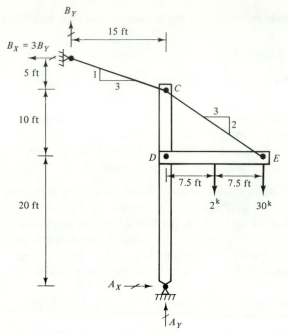

Figure 2.8a

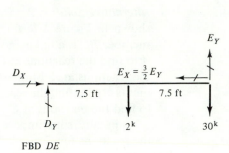

FBD *DE*

Figure 2.8b

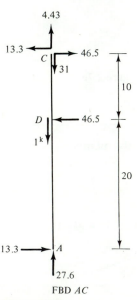

FBD *AC*

Figure 2.8c

For FBD *DE* (Fig. 2.8b):

$$\overset{\curvearrowleft +}{\sum M_Z} = 0 \text{ at } D: \qquad 15 * E_Y - 7.5 * 2 - 15 * 30 = 0$$

$$E_Y = 31^k$$

$$E_X = \frac{3}{2} E_Y = 46.5^k$$

$$E = 31 * \frac{\sqrt{13}}{2} = 55.9^k$$

$$\sum F_X = 0: \qquad D_X = 46.5^k$$

$$\sum F_Y = 0: \qquad D_Y = 32 - 31 = 1^k$$

Check arithmetic using FBD *AC* (Fig. 2.8c):

$$\overset{+\uparrow}{\sum F_Y} = 4.43 + 27.6 - (31 + 1) = 0.03 \approx 0, \quad \text{okay}$$

$$\overset{+}{\overrightarrow{\sum F_X}} = 46.5 + 13.3 - (13.3 + 46.5) = 0, \quad \text{checks}$$

2.5 STABILITY AND DETERMINACY

As shown in the example problems, for purposes of computing the reactions, a planar determinate structure can be considered as either a rigid body or pin-connected rigid bodies. In order to describe reaction configurations for which a structure is either not stable or not determinate, it must be imagined that the structure is subjected to a general applied loading (the resultant of the applied loads must have nonzero *X*- and *Y*-direction components). It must be noted that if the structure is stable and statically indeterminate, the structure *cannot* be treated as either a rigid body or pin-connected rigid bodies.

A planar structure composed of pin-connected rigid bodies which is initially at rest and remains at rest when subjected to a general loading is said to be in a state of static equilibrium. The general loading consists of the known applied loads and the unknown reactions provided by the supports. If the resultant of the general loading is either a force or a couple, the structure is statically unstable (is in motion). There are three independent equations of statics available for a FBD of the entire structure. If the structure has less than three independent reaction components, there are not enough unknowns to simultaneously satisfy all of the equations of statics and the structure is statically unstable. If there are a sufficient number of reaction components for statical equilibrium, but their geometrical arrangement renders the structure unstable, the structure is said to be geometrically unstable. That is, the statical stability of the structure is determined by the number and the arrangement of the reaction components.

The structures shown in Figure 2.9 are unstable. In Figure 2.9a, the structure (a skateboard) does not have a reaction component in the X direction. In Figure 2.9b, the structure does not have any resistance to moment about the Z axis passing through point i since all of the reactions intersect at point i. Rigid body B in Figure 2.9c does not have a reaction component in the X direction. In Figures 2.9d and 2.9e, the structures are geometrically unstable; that is, the interior joints in these structures must translate a large, finite amount in the X direction before any X-direction reaction components are developed.

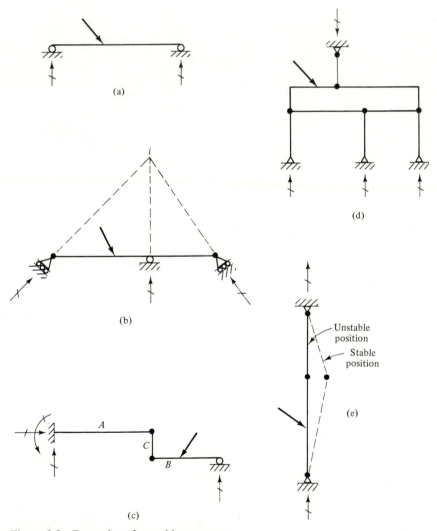

Figure 2.9 Examples of unstable structures.

The structures shown in Figure 2.10 are statically indeterminate. For Figure 2.10a, there are only three equations of statics available for computing the reactions and there are five unknown reactions.

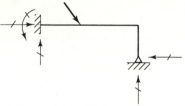

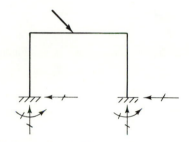

(a) Degree of statical indeterminancy is 2

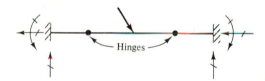

(b) Degree of statical inderminancy is 3

(c) Degree of statical indeterminancy is 1

Figure 2.10 Examples of statically indeterminate structures.

For purposes of the following discussion, let NR be the total number of reaction components and NIC be the total number of known, internal conditions (*internal hinges* can not resist any moment, for example). If a structure is stable and if (NR − NIC − 3) = 0, the structure is determinate. The 3 value in the preceding set of parentheses is the number of available equations of statics for the entire structure treated as a FBD. If a structure is stable and if (NR − NIC − 3) > 0, the structure is indeterminate and the degree of statical indeterminacy is the value of the expression (NR − NIC − 3).

For the structure in Figure 2.10a, there are not any known, internal conditions and (NR − NIC − 3) = (5 − 0 − 3) = 2; this structure is statically indeterminate to the second degree. There are no internal conditions for the structure in Figure 2.10b and the degree of statical indeterminacy is (6 − 0 − 3) = 3. In Figure 2.10c an internal condition of $M = 0$ is known at each pin connection, the X-direction reactions cannot be computed via statics only, and the degree of statical indeterminacy of this structure is (6 − 2 − 3) = 1. However, if this structure is subjected to a specialized loading having no X-direction components, from statics the X-direction reactions are equal to zero (they must be equal in magnitude, opposite in direction; therefore, they are equal to zero by inspection due to this special loading condition and the structure is statically determinate for this special loading condition).

PROBLEMS

2.1–2.18. Identify and compute the reaction components of the structures. Answers are not given for problems preceded by an asterisk.

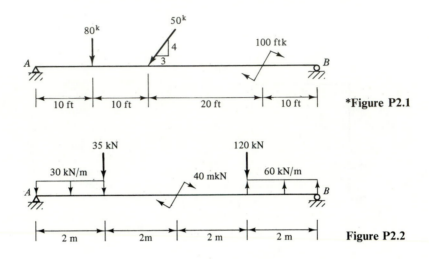

*Figure P2.1

Figure P2.2

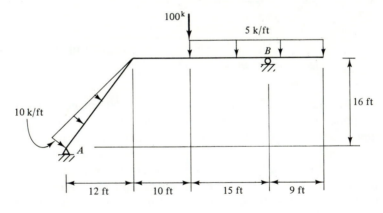

Figure P2.3

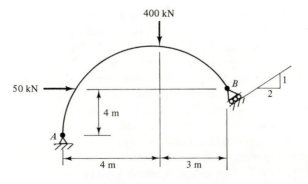

Figure P2.4

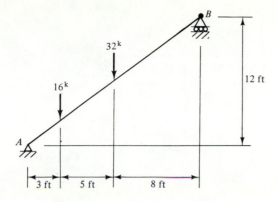

Figure P2.5

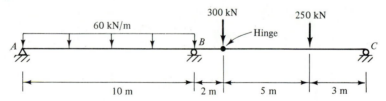

*Figure P2.6

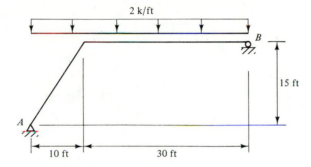

Figure P2.7

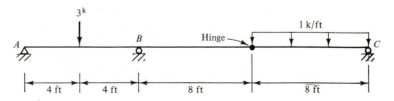

Figure P2.8

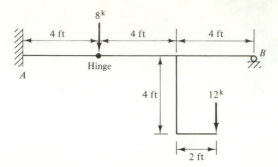

Figure P2.9

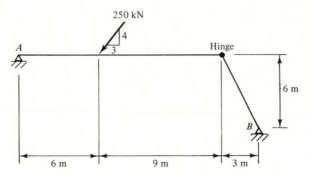

Figure P2.10

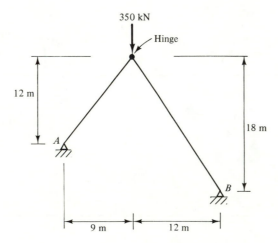

Figure P2.11

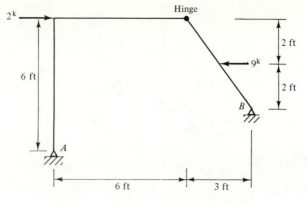

Figure P2.12

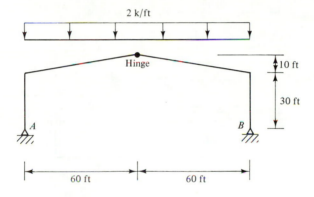

Figure P2.13

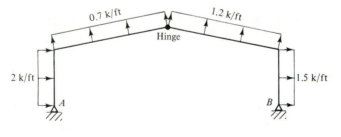

Figure P2.14 See Figure P2.13 for dimensions.

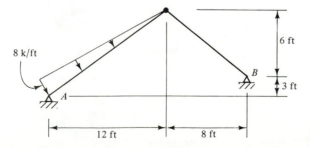

Figure P2.15

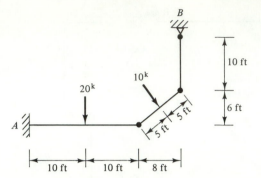

Figure P2.16

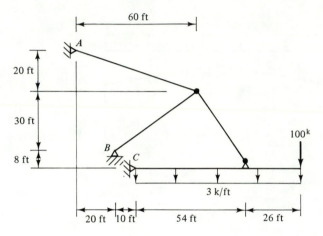

Figure P2.17

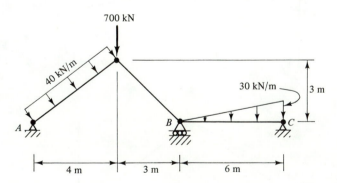

Figure P2.18

2.19–2.21. A rigid body weighs 300^k and is supported by three cables as shown in each problem. Ignore the weight of the cables. The rigid body is cross-hatched to distinguish it as a rigid body. If it is possible to do so, find the reactions in each problem. (Partial answer: One problem is indeterminate, one problem is unstable, and $B_Y = 124.9^k$ in the other problem.)

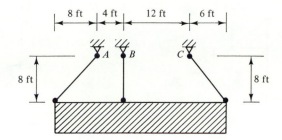

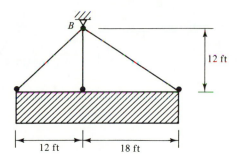

*Figure P2.19

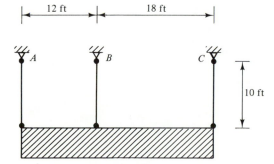

*Figure P2.20

*Figure P2.21

* **2.22.** Classify the following structures as being either determinate, indeterminate, or unstable. The cross-hatched structural parts are rigid bodies. All solid dots are hinges ($M = 0$ points).

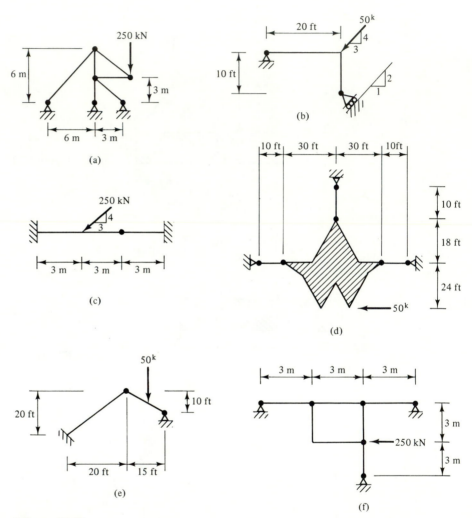

*Figure P2.22

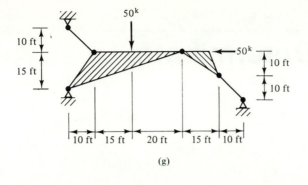

(g)

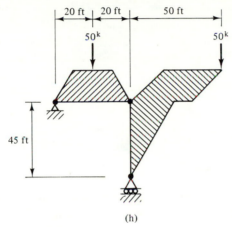

(h)

*Figure P2.22 (*continued*)

Chapter 3

Determinate Truss Analysis

3.1 DEFINITIONS AND ASSUMPTIONS

A *truss* is a structural frame that is designed to behave as though it were composed of pinned ended members. Each truss member is designed to resist only an axially directed tensile or compressive force. Since all truss members are treated as being pinned ended, all truss joints are assumed to be frictionless pins. Consequently, loads can be applied to trusses only at the joints. The self-weight of each truss member is halved and applied to the joints at the end of the member.

Figure 3.1 shows an actual joint configuration for a plane truss. The member ends are attached to a connector plate which is traditionally called a gusset plate. If the truss members are made of steel, the members are attached to the gusset plate by either fillet welds or bolts. Whenever the overall dimensions of the entire truss are not too large, the entire truss is completely prefabricated in a shop by using fillet welds and shipped to the construction site. Otherwise, shippable segments of the truss are fabricated by fillet welding, shipped to the construction site, and either bolted or fillet welded together to form the entire truss. For trusses composed of 2 by 4 wood members and nailed joints, a pair of prepunched steel gusset plates are used at each joint. The nominally 2-in.-thick members are placed between the two gusset plates and the gusset plates are nailed to the members. At each joint the steel or wood members are situated such that the centroidal axes of the members intersect at a common point as shown in Figure 3.1b. Consequently, the axial forces in the truss members intersect at a common point which is the pin joint of the analytical truss model. Therefore, truss joints are not real pins, but properly constructed truss joints essentially behave like pins.

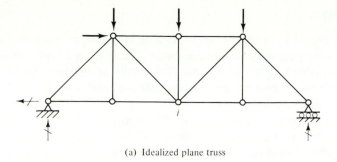

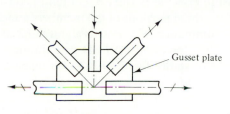

(a) Idealized plane truss

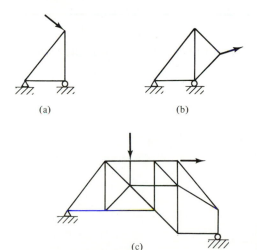

Gusset plate

Figure 3.1 Details for a plane truss.

(b) Actual configuration of joint *i* in (a)

3.2 ARRANGEMENT OF TRUSS MEMBERS

As shown in Figure 3.2a, the simplest plane truss is composed of only three members and three joints, and the shape of the void space formed by the member centerlines must be a triangle in order to ensure that the truss is stable. To this stable triangular configuration, two more members can be attached and subsequently connected to form another triangle (see Figure 3.2b). The triangle is the fundamental shape of the void spaces in a planar truss. However, as shown in Figure 3.2c, all void spaces need not be triangular.

(a)   (b)

(c)

Figure 3.2 Examples of stable and determinate plane trusses.

3.3 METHOD OF JOINTS

If a truss subjected to applied joint loads is in equilibrium, each member and each joint in the truss also must be in equilibrium. Each truss member is assumed to be capable of resisting only an axially directed tensile or compressive force. Therefore, all truss members must be straight, otherwise the members would bend when the joint loads were applied.

In order to keep the discussion as simple as possible, we begin by considering plane trusses. In a plane truss, the members all lie in a common plane. As shown in Figure 3.1b, the centroidal axes of all members attached to any joint i are assumed to intersect at a common point i. Therefore, the forces exerted on i by the members attached to i must intersect at i. The FBD of joint i consists of the pin at i, the forces acting on the infinitesimal portions of the members attached to joint i, and any loads applied at joint i. Summation of moments about the Z axis at i is automatically equal to zero since all of the forces on the FBD pass through point i. Therefore, the only available equations of statics for the FBD of a plane truss joint are $\Sigma F_X = 0$ and $\Sigma F_Y = 0$. In the *Method of Joints,* these two equations of statics are satisfied for the FBD of one joint at a time. Since there are only two available equations of statics at any joint, the analysis of a joint is possible if and only if there are no more than two unknown member forces at the joint being analyzed. Therefore, the analyst must successively choose to analyze joints which have no more than two unknown member forces. After an unknown member force has been determined, it is a known force at the joints on both ends of the member.

Since the objective of this text is to gradually progress toward computerized solutions, the joints and members of a truss are numbered for this reason in the example problems. Each joint number is encircled and each member number is enclosed within a rectangular box. In a computerized truss analysis, it is convenient to assume that all unknown member forces are tensile and that a tension force is positive. Then, if a computed member force is negative, the member is in compression.

EXAMPLE 3.1 ───

For the plane truss shown in Figure 3.3a, the reactions were computed by using the three equations of statics for the FBD of the entire structure. Find all of the member forces by using the Method of Joints.

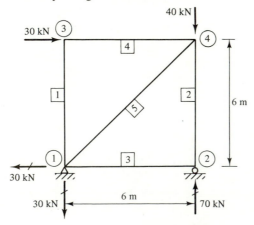

Figure 3.3a

Solution. As described below (Fig. 3.3b), this truss can be analyzed by the Method of Joints without first computing the reactions. It is shown in subsequent example problems that for most trusses only a few, if any, member forces can be found without first finding the reactions.

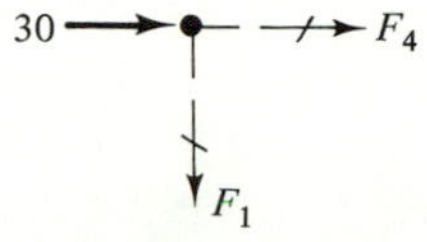

$$\sum F_X = 0: \qquad 30 + F_4 = 0; \qquad F_4 = -30^k$$
$$\sum F_Y = 0: \qquad F_1 = 0$$

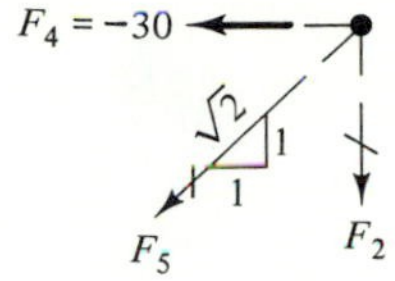

Joint 3 FBD

$$\sum F_X = 0: \qquad -\left(\frac{1}{\sqrt{2}} F_5 + F_4\right) = 0; \qquad F_5 = -\sqrt{2}\, F_4 = 30\sqrt{2}$$

$$\sum F_Y = 0: \qquad -\left(F_2 + \frac{1}{\sqrt{2}} F_5 + 40\right) = 0$$

$$F_2 = -\left(\frac{1}{\sqrt{2}} F_5 + 40\right) = -(30 + 40) = -70$$

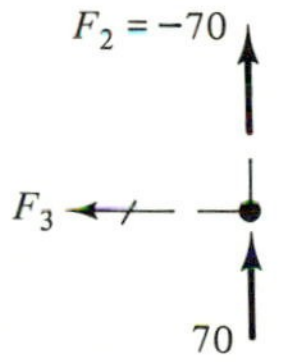

Joint 4 FBD

$$\sum F_X = 0: \qquad F_3 = 0$$
$$\sum F_Y = 0: \qquad 70 + (-70) = 0 \text{ (arithmetic checks)}$$

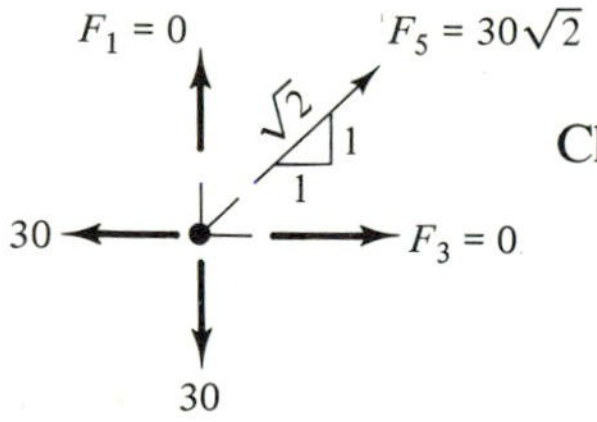

Joint 2 FBD

Check the arithmetic:

$$\sum F_X = \frac{1}{\sqrt{2}} * (30\sqrt{2}) - 30 = 0 \text{ (arithmetic checks)}$$

$$\sum F_Y = \frac{1}{\sqrt{2}} * (30\sqrt{2}) - 30 = 0 \text{ (arithmetic checks)}$$

Joint 1 FBD

Figure 3.3b

Alternative Solution. For a noncomputerized solution, it is more convenient to find the X and Y components of the member forces. Draw a reasonably large sketch of the structure to allow all details to be shown with clarity. For comparison to the previous solution, the solution order is joint 3, joint 4, joint 2, and joint

1 (Fig. 3.3c). At joint 4, $\Sigma F_X = 0$ gives the unknown horizontal component of the force in the diagonal member. Then, since the force in the diagonal member has the same slope as the member itself, the vertical component of the force in the diagonal member is found by using similar triangles; that is, the Y to X ratios of the member slope and member force components are equated. A slash was placed on each unknown force that was computed at each joint.

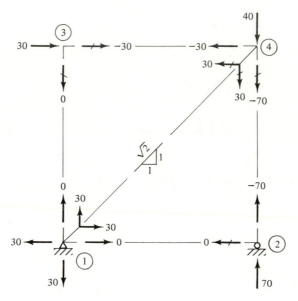

Figure 3.3c

EXAMPLE 3.2 ——————————————————————————————

For the plane truss shown in Figure 3.4a, find all of the member forces by using the Method of Joints.

Solution. The solution sequence in Figure 3.4b is joint 1; joint 2; joint 6—a separate FBD is shown for clarity (Fig. 3.4c); joint 5—a separate FBD is shown for clarity (Fig. 3.4c); joint 3; and joint 4 (to check the arithmetic).

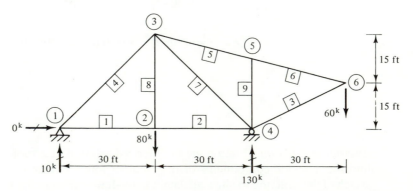

Figure 3.4a

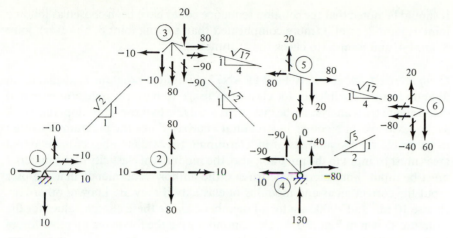

Figure 3.4b

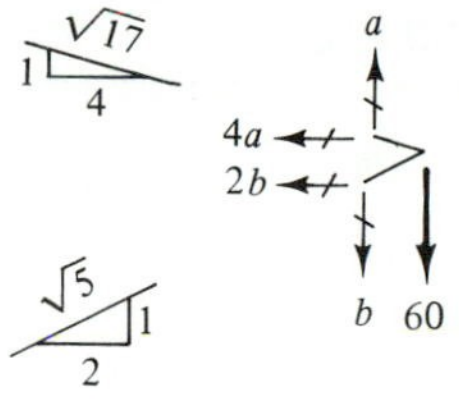

Joint 6 FBD

$$\sum F_X = 0: \qquad 4a + 2b = 0; \qquad b = -2a$$
$$\sum F_Y = 0: \qquad a = 60 + b = 60 + (-2a)$$
$$3a = 60$$
$$a = 20; \qquad b = -2a = -40$$
$$\sum F_3 = -40\sqrt{5}; \qquad F_6 = 20\sqrt{17}$$

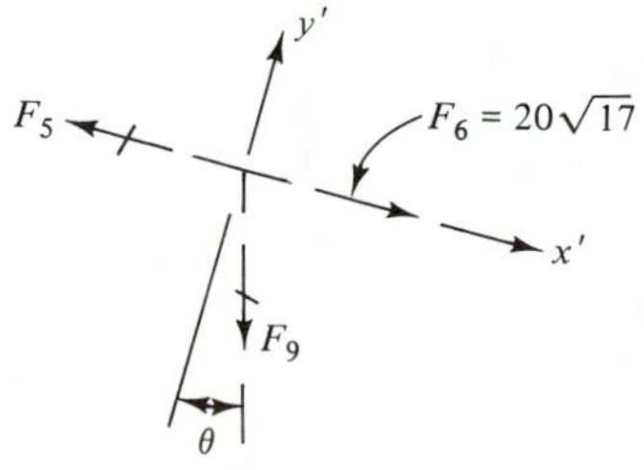

$$\sum F_{y'} = 0: \qquad F_9 * \cos\theta = 0; \qquad F_9 = 0$$
$$\sum F_{x'} = 0: \qquad F_5 = F_6 = 20\sqrt{17}$$

F_5 $F_6 = 20\sqrt{17}$ y' x' F_9 θ

NOTE: y' is perpendicular to x'

Joint 5 FBD

Figure 3.4c

It should be noted that the solution sequence could have been chosen as follows: joint 1; joint 2; joint 3 (more complicated than solving joint 6, however); joint 5; joint 4; and joint 6 (to check the arithmetic).

Computerized Solution. A FORTRAN77 computer program is available on diskette from the publisher for classroom usage to perform indeterminate and determinate truss analyses. The diskette is available to those who adopt this book as the textbook for a course in structural analysis. Since the program is written to be usable for the analysis of indeterminate trusses, member cross-sectional areas must be input to the program; also, the modulus of elasticity of the material must be input. For determinate trusses, the author recommends that the user input the correct areas and modulus of elasticity if they are known; otherwise, choose 10 in.^{2} and 29000 ksi for all members. Using these chosen values for the structure shown in Figure 3.4a, the computer gave the following member forces (and the joint displacements shown later in Example 6.15):

REACTIONS (kips)

JOINT	LOAD NO.	X	Y	Z
1				
	1	.0	10.0	.0
4				
	1	.0	130.0	.0

MEMBER FORCES (kips)

    + DENOTES TENSION
    − DENOTES COMPRESSION

MEMBER NUMBER	INCIDENCES JOINT 1	JOINT 2	AREA (in.^{2})	LENGTH (in.)	LOADING	FORCE
1	1	2	10.0000	360.000	1	10.0
2	2	4	10.0000	360.000	1	10.0
3	4	6	10.0000	402.492	1	−89.4
4	1	3	10.0000	509.117	1	−14.1
5	3	5	10.0000	371.079	1	82.5
6	5	6	10.0000	371.079	1	82.5
7	3	4	10.0000	509.117	1	−127.3
8	2	3	10.0000	360.000	1	80.0
9	4	5	10.0000	270.000	1	.0

3.4 METHOD OF SECTIONS

To keep the following discussion as simple as possible, only plane trusses are considered in the discussion. In the Method of Sections, an imaginary cutting line is drawn through a stable and determinate truss. For brevity purposes, this imaginary cutting line is called a section. Thus, a section subdivides the truss into two separate parts. Since the entire truss is in equilibrium, any part of it must also be in equilibrium. Either of the two parts of the truss can be considered as an FBD for which there are three equations of statics: $\Sigma F_X = 0$, $\Sigma F_Y = 0$, and $\Sigma M_Z = 0$ at i where i is any chosen point in the XY plane.

If only a few member forces of a truss are needed, generally the quickest way to find these forces is by the Method of Sections. Wherever it is possible to do so, the analyst should independently find each of the needed member forces. Otherwise, if an error is made in computing a member force and this erroneous value is used in finding another member force, the error is propagated in the calculations. Also, the analyst should choose the simpler of the two FBD defined by the section. The simpler FBD is the one that takes the least amount of time to draw. For notational convenience, each of the unknown member forces in the members cut by the section is assumed to be a tensile force. Some of the reasons for needing to find only a few member forces are (1) to verify a computerized solution; (2) to expedite a solution by the Method of Joints; and (3) to sharpen one's analytical skills.

EXAMPLE 3.3 ──

Independently find the forces in members 4, 7, 5, and 9 by the Method of Sections (Fig. 3.5a).

Solution. In many cases when the Method of Sections is being used, the analyst can quickly decide which section to choose by using a piece of blank paper to cover up the part of the truss on one side of the section. This suggestion enhances the analyst's visualization of the FBD being considered. The edge of the blank piece of paper is the imaginary cutting line or section. To find the force in member 4, consider section A–A shown in Figure 3.5a: this section cuts the member for which the force is being sought and it also cuts member 1, but the force in member 1 has no Y-direction component. Therefore, $\Sigma F_Y = 0$ for the FBD on the left side of section A–A (Fig. 3.5b) involves only the known reactions and the unknown Y-direction component of F_4.

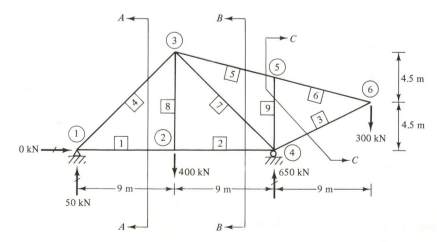

Figure 3.5a

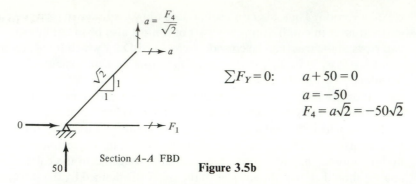

$$\sum F_Y = 0: \qquad a + 50 = 0$$
$$a = -50$$
$$F_4 = a\sqrt{2} = -50\sqrt{2}$$

Section A–A FBD

Figure 3.5b

The force in member 7 can be found directly from the FBD of section B–B (Fig. 3.5c) by summing moments at the point where F_2 and F_5 intersect. Since rigid body mechanics are applicable, the force in member 7 can be translated along its line of action to point 4 and decomposed into its X and Y components at point 4.

$$\sum M_Z = 0 \text{ at } C: \qquad 27b + 36*400 - 45*50 = 0$$
$$b = -450; \qquad F_7 = b\sqrt{2} = -450\sqrt{2}$$

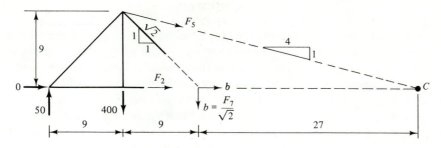

Section B–B FBD

Figure 3.5c

Either section B–B or section C–C can be used to find F_5 directly; in either case, the equation of statics to use is $\sum M_Z = 0$ at 4, and F_5 should be resolved into its X and Y components at joint 5 to facilitate the calculations. Since section C–C is needed to find F_9 directly (Fig. 3.5d), section C–C is used to find F_5 also (Fig. 3.5e); F_9 is found by summing moments at joint 6.

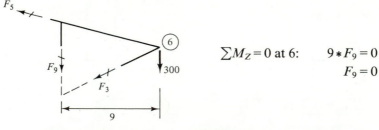

$$\sum M_Z = 0 \text{ at } 6: \qquad 9*F_9 = 0$$
$$F_9 = 0$$

Section C–C FBD to find F_9 **Figure 3.5d**

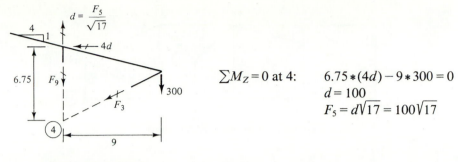

$$\sum M_Z = 0 \text{ at } 4: \quad 6.75*(4d) - 9*300 = 0$$
$$d = 100$$
$$F_5 = d\sqrt{17} = 100\sqrt{17}$$

Section C–C FBD to find F_5

Figure 3.5e

EXAMPLE 3.4 ——

The truss shown in Figure 3.6a contains most of the unusual features found in simple trusses. The truss features shown in Figure 3.6a are:

1. Member 9 is an inclined top chord member. In some trusses, all top chord members are inclined; such an example would exist in Figure 3.6a if members 6 and 11 were deleted and joint 8 were raised 5 ft.
2. Joints 9 and 10 plus members 20 and 21 is an example of a subdivided truss panel. Joint 9 subdivides the panel spanning from joint 7 to joint 11 since a Y-direction load is to be applied at the middle of panels 7 to 11.
3. Examples of K bracing are shown at joints 2 and 5. The purpose of braces is to reduce the length of a member which has to provide a compressive axial force due to some load configuration. Compression members are susceptible to buckling and the member length squared appears in the denominator of the buckling load equation. Therefore, halving a member's length increases its buckling load by a factor of 4.

Use the Method of Sections and independently find the forces in members 7, 9, 17, 20, 22, 23, and 24 of Figure 3.6a.

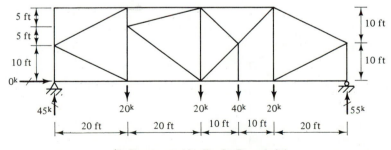

(i) Structure and loading for Example 3.4

Figure 3.6a

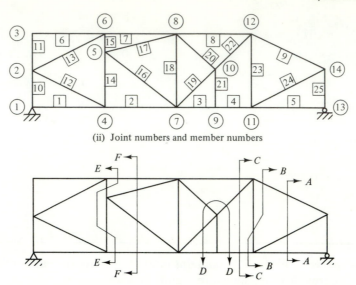

(ii) Joint numbers and member numbers

(iii) Sections to be used to solve Example 3.4

Figure 3.6a (*continued*)

Solution

To find F_{17}, $\Sigma F_Y = 0$ for the FBD of section *F–F* will be used (Fig. 3.6e). However, this requires the knowledge of the behavior of the K joint at joint ⑤. Consider Figure 3.6e FBD of joint ⑤: this joint is called a K joint because it has the shape of a K and because it does not have any applied joint load in the direction perpendicular to the two straight legs of the K. Consequently, one of

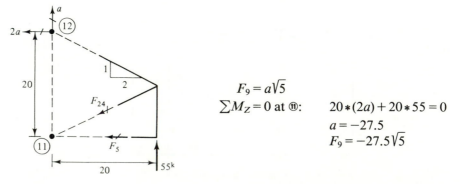

$$F_9 = a\sqrt{5}$$
$$\Sigma M_Z = 0 \text{ at ⑪:} \quad 20*(2a) + 20*55 = 0$$
$$a = -27.5$$
$$F_9 = -27.5\sqrt{5}$$

Section *A–A* FBD to find F_9

Figure 3.6b

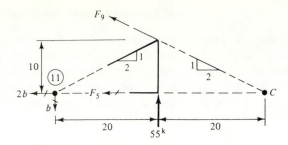

$$F_{24} = b\sqrt{5}$$
$$\sum M_Z = 0 \text{ at } C: \qquad 40*b - 20*55 = 0$$
$$b = 27.5$$
$$F_{24} = 27.5\sqrt{5}$$

Section A–A FBD to find F_{24}

Figure 3.6b (*continued*)

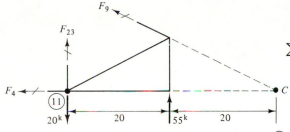

$$\sum M_Z = 0 \text{ at } C: \qquad 40*20 - 20*55 - 40*F_{23} = 0$$
$$F_{23} = -7.5$$

Note: F_9 can be found from this FBD by summing moments at ⑪

Section B–B FBD to find F_{23}

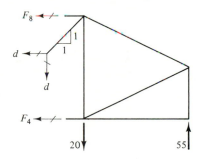

$$F_{22} = d\sqrt{2}$$
$$\sum F_Y = 0: \qquad 55 - 20 - d = 0$$
$$d = 35$$
$$F_{22} = 35\sqrt{2}$$

Section C–C FBD to find F_{22}

Figure 3.6c

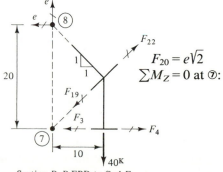

$$F_{20} = e\sqrt{2}$$
$$\sum M_Z = 0 \text{ at } ⑦: \qquad 20*e - 10*40 = 0$$
$$e = 20$$
$$F_{20} = 20\sqrt{2}$$

Section D–D FBD to find F_{20}

Figure 3.6d

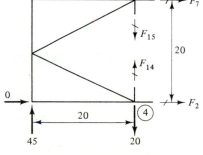

Section E–E FBD to find F_7

$$\sum M_Z = 0 \text{ at } ④: \qquad -(20*45 + 20*F_7) = 0$$
$$F_7 = -45$$

51

the inclined legs of the K is in tension and the other one is in compression since their components perpendicular to the straight legs of the K must be equal and opposite in direction. Using the knowledge that member 16 must be in compression if member 17 is assumed to be in tension (their X-direction components must cancel each other) and $\sum F_Y = 0$ gives the values shown in Figure 3.6e.

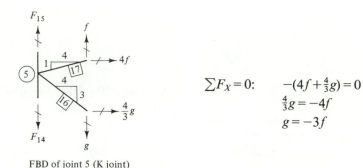

$$\sum F_X = 0: \quad -(4f + \tfrac{4}{3}g) = 0$$
$$\tfrac{4}{3}g = -4f$$
$$g = -3f$$

FBD of joint 5 (K joint)

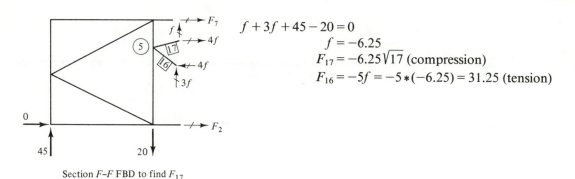

$$f + 3f + 45 - 20 = 0$$
$$f = -6.25$$
$$F_{17} = -6.25\sqrt{17} \text{ (compression)}$$
$$F_{16} = -5f = -5*(-6.25) = 31.25 \text{ (tension)}$$

Section F–F FBD to find F_{17}

Figure 3.6e

3.5 COMPOUND TRUSSES

The following discussion is restricted to plane trusses to simplify the discussion. A *compound truss* is formed by interconnecting two or more simple trusses. It may be possible to ship the fabricated simple trusses individually and to interconnect the simple trusses on the construction site to form a long-span compound truss. Some examples of compound trusses are shown in Figure 3.7; interconnection occurs at the joints where the hinges are shown.

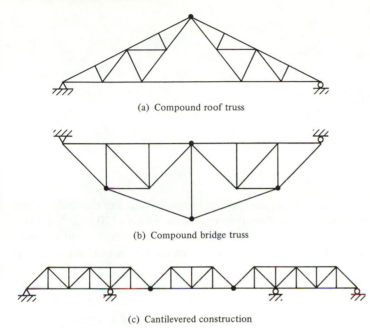

(a) Compound roof truss

(b) Compound bridge truss

(c) Cantilevered construction

Figure 3.7 Examples of compound trusses.

Figure 3.7a shows a compound roof truss composed of two simple trusses interconnected by a single member and a common joint. A compound bridge truss is shown in Figure 3.7b; the bridge deck (roadway surface) would be placed on the flat surface of the truss. This compound bridge truss is composed of two simple trusses interconnected by three members and a common joint. An example of cantilevered truss construction is shown in Figure 3.7c; in this case, the three simple trusses are interconnected at their common joints.

How does the analysis of a compound truss differ from the analysis of a simple truss? Basically, the Method of Sections is used to find the forces in the interconnecting member(s) in Figures 3.7a and 3.7b. For Figure 3.7c, the middle simple truss is isolated as a FBD to find its reactions; these reactions are reversed and applied to the interconnecting joints of the other two simple trusses. After the interconnecting forces between the simple trusses are found, the simple trusses are analyzed by the Method of Joints or the Method of Sections. These analysis procedures are illustrated in the following example problems.

EXAMPLE 3.5 ————————————————————————————————

In Figure 3.8a the roof truss is shown subjected to a snow load. Building codes specify the minimum snow load in pounds per square foot acting vertically on the horizontal projection of the roof surface. Suppose the minimum snow load specified is 20 psf and the trusses along the building length direction are spaced at 20 ft on centers; the uniformly distributed snow load is (20 ft)*20 psf = 400 lb/ft = 0.4 k/ft. Find the reactions and the forces in members *d, e,* and *f* of the compound truss.

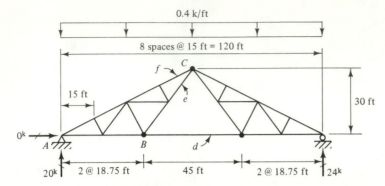

Figure 3.8a

Solution. Since the structure and loading are symmetric, half of the total vertical load is resisted by each of the *Y*-direction reactions: $\frac{1}{2}*0.4*120 = 24^k$. To find the force in member *d*, the pin at *C* is removed and member *d* is cut to arrive at the FBD shown in Figure 3.8b; summation of moments at *C* on this FBD gives $F_d = 24^k$.

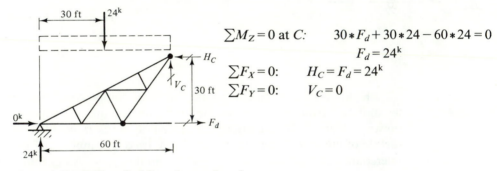

$$\sum M_Z = 0 \text{ at } C: \qquad 30*F_d + 30*24 - 60*24 = 0$$
$$F_d = 24^k$$
$$\sum F_X = 0: \qquad H_C = F_d = 24^k$$
$$\sum F_Y = 0: \qquad V_C = 0$$

Figure 3.8b   FBD to find force in member *d*.

To find the forces in members *e* and *f*, consider Figure 3.8c FBD. Since the simple truss is not intact in Figure 3.8c (it was intact in Figure 3.8b), the distributed snow load must be allocated to the truss joints along the roof top. At each interior joint along the roof top, the applied truss load is $(15 \text{ ft})*0.4 \text{ k/ft} = 6^k$ and at joint *A* the applied truss load is 3^k. Translate F_e to *B*, decompose F_e into its *X* and *Y* components, and $\sum M_Z = 0$ at *A* to find: $37.5*F_{eY} - 6*(15 +$

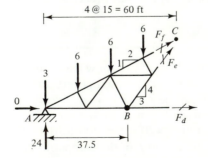

Figure 3.8c   FBD used to find F_e and F_f.

$30 + 45) = 0$; $F_{eY} = 14.4$; $F_e = \frac{5}{4} * 14.4 = 18^k$. Translate F_f to A, decompose F_f into its X and Y components, and

$$\sum M_Z = 0 \text{ at } B: \qquad 6*(7.5 + 22.5) + 37.5*(3 - 24 - F_{fY}) = 0$$
$$F_{fY} = -16.2$$
$$F_f = -16.2\sqrt{5}$$

Figure 3.8d shows the applied loads in the form which is convenient for the purpose of using the Method of Joints to find the other member forces. Since the structure and loading are symmetric, only half of the structure needs to be solved to find the other member forces. The applied loads normal to the roof surface are $n = 6*(2/\sqrt{5}) = 5.367^k$, whereas the applied loads parallel to the roof surface are $p = 6*(1/\sqrt{5}) = 2.683^k$.

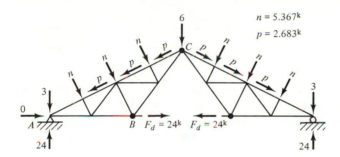

Figure 3.8d Joint loads in the form convenient for Method of Joints solution.

EXAMPLE 3.6 ───

Find the forces in members d, e, and f of the compound bridge truss shown in Figure 3.9a.

Solution. Remove the pin at B and cut member d to obtain the FBD shown in Figure 3.9b. Translate F_d to C, decompose F_d into its X and Y components, and $\sum M_Z = 0$ at B: $9*150 + 15*F_{dX} - 27*200 = 0$; $F_{dX} = 270$ kN; $F_d = 270\sqrt{10}$. Use the Method of Joints at joint C (FBD shown in Figure 3.9c) to obtain F_e and F_f. With F_d, F_e, and F_f known, the Method of Joints can be used to find all the other member forces in the compound truss (Figure 3.9a).

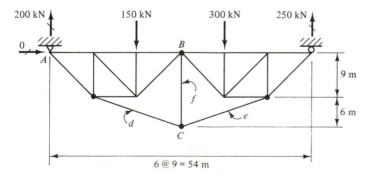

Figure 3.9a

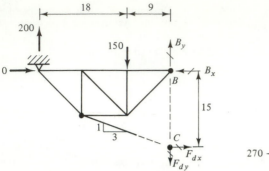

Figure 3.9b   **Figure 3.9c**

EXAMPLE 3.7 ————————————————————————————————————

Find the interconnecting forces at points A and B in the structure shown in Figure 3.10a which depicts cantilevered truss construction.

Solution. Remove the pins at points A and B; lift out the simple truss located between points A and B to produce the FBD shown below (Fig. 3.10b). Since there are not any X-direction applied loads on any of the three simple trusses, $A_X = B_X = 0$. For simple truss AB, $\sum M_Z = 0$ at A: $4 * 20 * B_Y - 3 * 20 * 40 = 0$ $B_Y = 30^k$; $\sum F_Y = 0$: $A_Y = 40 - 30 = 10^k$; Reverse these computed values of A_Y and B_Y, and apply them to the tips of the cantilevered trusses. Now, the reactions

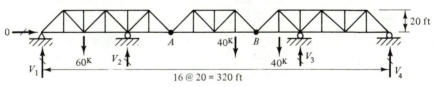

Figure 3.10a

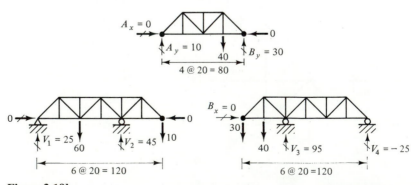

Figure 3.10b

of the cantilevered trusses can be obtained. Then, the Method of Joints can be used to obtain all of the forces in the members.

3.6 SPACE TRUSSES

Although most engineered structures are three-dimensional, fortunately many of them can be decomposed into planar structures and each planar structure is analyzed for loads lying in its plane. Figure 3.11 shows a building that has a rectangular plan view

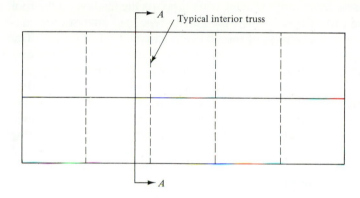

(a) Plan view of roof

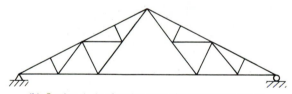

(b) Section A–A: elevation view of typical interior truss

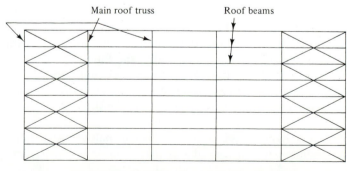

(c) Plan view of roof framing and wind bracing schemes

Figure 3.11 Roof trusses and wind bracing trusses for a building.

and a roof that is not flat. Suppose the roof structure consists of several, parallel, plane trusses spanning across the short direction of the plan view; these vertically oriented plane trusses resist gravity-type loads. Perpendicular to and spanning between these plane trusses, there must be roof beams located on and attached to the top truss joints. These roof beams are needed to support the roof. Also, between two or more of the plane trusses there must be diagonal members lying in the planes defined by the top chords of the trusses; these diagonal members must be attached to some of the top truss joints. These diagonal members, roof beams, and the top chords of a pair of vertically oriented plane trusses form another set of plane trusses which are needed to resist, for example, wind load components acting perpendicular to the vertically oriented set of plane trusses. Gravity direction roof loads are resisted by the vertically oriented set of plane trusses; each interior plane truss supports the loading on the roof halfway each side of the truss. Consequently, the gravity-type load analyses are conducted for a typical interior plane truss of the vertically oriented set of trusses. The other set of plane trusses must be designed to resist wind acting at the roof top and perpendicular to the end of the building. Similarly, a highway bridge truss (see Figure 3.12) consists of two vertically oriented plane trusses (one on each side of the bridge) and two horizontally oriented plane trusses (one beneath the roadway surface and the other forming the roof of the bridge). The vertically oriented pair of trusses resist gravity-type loads and the horizontally oriented pair of trusses resist wind loads acting on the side of the bridge. At the entrance and exit ends of the bridge, a frame attached to the bridge supports and the ends of the roof truss must be designed to carry the roof truss reactions to the bridge supports. Note that the top and bottom chords of the vertically oriented bridge trusses are common to the horizontally oriented trusses. Therefore, those truss members common to both sets of trusses must be designed to resist the algebraic sum of the forces obtained from the gravity-type and wind-type analyses.

Some trusses cannot be decomposed into plane trusses for analysis purposes. Electrical transmission towers, guyed radio or TV transmitter towers, spherically domed roof trusses, and two-way flat roof trusses are a few examples of such structures that must be analyzed as space trusses. It was shown previously that a triangle is the basic shape used in forming a plane truss. A tetrahedron is the basic shape used in forming a space truss.

Each member in a space truss is assumed to behave as though it were pinned ended and capable of resisting only an axially directed tensile or compressive force. Therefore, each joint of a space truss is treated as being a ball-shaped frictionless pin that does not resist any rotational motion. The members attached to an individual pin are assumed to intersect at a common point (the center of the pin). Loads can only be applied at the joints of a space truss. Consequently, the forces acting on a space truss joint intersect at the center of the pin joint and moment equilibrium about each of the X, Y, and Z axes at the center of the pin is automatically satisfied. The only available equations of statics for a space truss joint are $\sum F_X = 0$, $\sum F_Y = 0$, and $\sum F_Z = 0$.

Only one example of space truss analysis is shown since such analyses are lengthy and tedious. The space truss chosen to illustrate the analysis concepts is the space

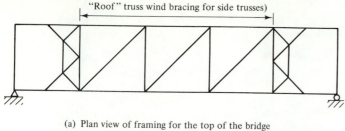

(a) Plan view of framing for the top of the bridge

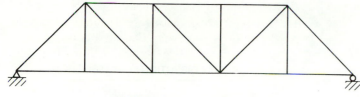

(b) Side elevation view of bridge

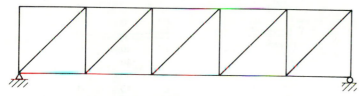

(c) Plan view of framing beneath the roadway of the bridge

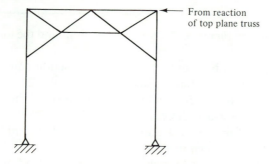

(d) End portal frame – members designed to resist moment, shear, axial force   **Figure 3.12** Bridge trusses.

truss in Example 4 of the user's manual for the TRUSS computer program available on diskette from the publisher to those who adopt this book as the textbook for a course in structural analysis. The purpose of the following example is to demonstrate how to compute the reactions and a few member forces such that an analyst can verify that a computerized solution is correct.

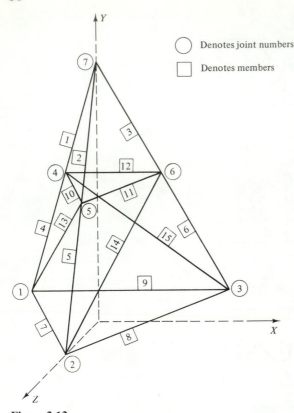

○ Denotes joint numbers

□ Denotes members

GRAVITY DIRECTION IS −Y DIRECTION

Joint number	Joint coordinates (ft)		
	X	Y	Z
1	−15.	0.	−12.5
2	0.	0.	12.5
3	15.	0.	−12.5
4	−7.5	40.	−6.25
5	0.	40.	6.25
6	7.5	40.	−6.25
7	0.	80.	0.

Figure 3.13a

EXAMPLE 3.8 ───

For the space truss shown in Figure 3.13a, find the reactions and the forces in members 1, 2, and 3.

The truss is designed to have only the following reaction components: force Y and force Z at joints 1 and 3, and force X and force Y at joint 2.

Loading 1 consists of only a 50^k load which is applied at joint 7 and is parallel to and acting in the positive Z-axis direction.

Loading 2 consists of only a 20^k load which is applied at joint 7 and is parallel to and acting in the negative Y-axis direction.

Loading 3 consists of the dead weight of all members. Half of each member's weight is applied in the $-Y$ direction at the joints to which the member's ends are attached. The members are made of steel (unit weight of material = 490 pcf and modulus of elasticity = 29000 ksi = 4176000 ksf) and their cross-sectional areas are 3 in.² for members 1 to 3, 4 in.² for members 4 to 10, 2.25 in.² for members 10 to 12, and 5 in.² for members 13 to 15.

Loading 4 is a dependent loading which is the algebraic sum of the independent loadings 1, 2, and 3. Loadings 3 and 4 are considered only in the computerized solution which is shown after the noncomputerized solutions for the reactions and forces in members 1, 2, and 3 of loadings 1 and 2 are obtained.

The following table of data was computed to facilitate the Method of Joints solutions:

Member number	Length L (ft)	Length components			L_X/L	L_Y/L	L_Z/L
		L_X	L_Y	L_Z			
1	41.1742	7.5	40	6.25	0.182153	0.971482	0.151794
2	40.4853	0	40	6.25	0	0.988013	0.154377
3	41.1742	7.5	40	6.25	0.182153	0.971482	0.151794
4	41.1742	7.5	40	6.25	0.182153	0.971482	0.151794
5	40.4853	0	40	6.25	0	0.988013	0.154377
6	41.1742	7.5	40	6.25	0.182153	0.971482	0.151794
7	29.1548	15	0	25	0.514495	0	0.857492
8	29.1548	15	0	25	0.514495	0	0.857492
9	30	30	0	0	1	0	0
10	14.5774	7.5	0	12.5	0.514495	0	0.857492
11	14.5774	7.5	0	12.5	0.514495	0	0.857492
12	15	15	0	0	1	0	0
13	46.6536	15	40	18.75	0.321519	0.857383	0.401898
14	44.8086	7.5	40	18.75	0.167379	0.892686	0.418447
15	46.3175	22.5	40	6.25	0.485778	0.863605	0.134938

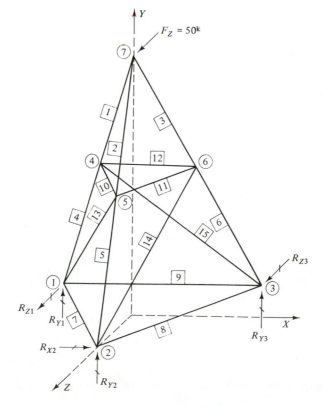

Figure 3.13b Loading 1.

Solution for Loading 1. Solution for the reactions using Figure 3.13b as FBD:

$\sum M_X = 0$ through ① and ③: $80 * 50 - 25 * R_{Y2} = 0$; $R_{Y2} = 160^k$

$\sum M_Z = 0$ at ①: $15 * R_{Y2} + 30 * R_{Y3} = 0$; $R_{Y3} = -0.5 * R_{Y2} = -80^k$

$\sum F_Y = 0$: $R_{Y1} + R_{Y2} + R_{Y3} = 0$; $R_{Y1} = -(160 - 80) = -80^k$

$\sum F_X = 0$: $R_{X2} = 0$

$\sum M_Y = 0$ at ③: $30 * R_{Z1} + 25 * R_{X2} + 15 * 50 = 0$; $R_{Z1} = -25^k$

$\sum F_Z = 0$: $R_{Z1} + R_{Z3} = 0$; $R_{Z3} = -R_{Z1} = 25^k$

Method of Joints at joint 7:

$\sum F_X = 0$: $-0.182153 * F_1 + 0 * F_2 + 0.182153 * F_3 = 0$

$\sum F_Y = 0$: $-0.971482 * F_1 - 0.988013 * F_2 - 0.971482 * F_3 = 0$

$\sum F_Z = 0$: $-0.151794 * F_1 + 0.154377 * F_2 - 0.151794 * F_3 + 50 = 0$

These simultaneous equations written in matrix form are

$$\begin{bmatrix} -0.182153 & 0 & 0.182153 \\ -0.971482 & -0.988013 & -0.971482 \\ -0.151794 & 0.154377 & -0.151794 \end{bmatrix} \begin{Bmatrix} F_1 \\ F_2 \\ F_3 \end{Bmatrix} = \begin{Bmatrix} 0 \\ 0 \\ -50 \end{Bmatrix}$$

$$\begin{Bmatrix} F_1 \\ F_2 \\ F_3 \end{Bmatrix} = \begin{Bmatrix} 82.3484 \\ -161.9414 \\ 82.3484 \end{Bmatrix}$$

Each of the remaining joints, joints 1 through 6, are indeterminate since there are four unknown member forces and only three available equations of statics at these joints.

Solution for Loading 2. Solution for the reactions using Figure 3.13c as FBD:

$\sum F_X = 0$: $R_{X2} = 0$

$\sum M_Y = 0$ at ③: $30 * R_{Z1} + 25 * R_{X2} = 0$; $R_{Z1} = 0$

$\sum F_Z = 0$: $R_{Z1} + R_{Z3} = 0$; $R_{Z3} = 0$

$\sum M_X = 0$ through ① and ③: $12.5 * 20 - 25 * R_{Y2} = 0$; $R_{Y2} = 10^k$

$\sum M_Z = 0$ at ①: $30 * R_{Y3} + 15 * R_{Y2} - 15 * 20 = 0$; $R_{Y3} = 5^k$

$\sum F_Y = 0$: $R_{Y1} + R_{Y2} + R_{Y3} - 20 = 0$; $R_{Y1} = 20 - (10 + 5) = 5^k$

Method of Joints at joint 7. The equations of statics in matrix form are:

$$\begin{bmatrix} -0.182153 & 0 & 0.182153 \\ -0.971482 & -0.988013 & -0.971482 \\ -0.151794 & 0.154377 & -0.151794 \end{bmatrix} \begin{Bmatrix} F_1 \\ F_2 \\ F_3 \end{Bmatrix} = \begin{Bmatrix} 0 \\ 20 \\ 0 \end{Bmatrix}$$

$$\begin{Bmatrix} F_1 \\ F_2 \\ F_3 \end{Bmatrix} = \begin{Bmatrix} -5.1468 \\ -10.1213 \\ -5.1468 \end{Bmatrix}$$

The remainder of the truss is indeterminate as it was in loading 1.

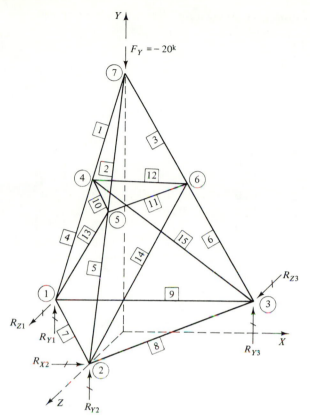

Figure 3.13c Loading 2.

Loading 3—dead weight of the truss members only. The weight of each member is: area $*$ length $*$ unit weight of material. Half of each member weight is applied at the member ends. For example, at joint 7 the applied joint load is

$$F_Y = -\tfrac{1}{2}*(A_1*L_1+A_2*L_2+A_3*L_3)*UW = -0.627^k$$

and at Joint 6 the applied joint load is

$$F_Y = -\tfrac{1}{2}*(A_3*L_3+A_6*L_6+A_{11}*L_{11}+A_{12}*L_{12}+A_{14}*L_{14})*UW = -0.985^k$$

Loading 3 summary is:

Joint	Force Y (kips)
1	−1.0797
2	−1.0535
3	−1.0768
4	−0.9976
5	−0.9907
6	−0.9848
7	−0.6270

The computerized solution is:

REACTIONS (kips)

JOINT	LOAD NO.	X	Y	Z
1				
	1	.0	−80.0	−25.0
	2	.0	5.0	.0
	3	.0	2.1	.0
	4	.0	−72.9	−25.0
2				
	1	.0	160.0	.0
	2	.0	10.0	.0
	3	.0	2.6	.0
	4	.0	172.6	.0
3				
	1	.0	−80.0	−25.0
	2	.0	5.0	.0
	3	.0	2.1	.0
	4	.0	−72.9	−25.0

MEMBER FORCES (kips)

+ DENOTES TENSION

− DENOTES COMPRESSION

MEMBER NUMBER	INCIDENCES JOINT 1	JOINT 2	AREA (ft^2)	LENGTH (ft)	LOADING	FORCE
1	4	7	.0208	41.174		
					1	82.3
					2	−5.1
					3	−.2
					4	77.0
2	5	7	.0208	40.485		
					1	−161.9
					2	−10.1
					3	−.3
					4	−172.4
3	6	7	.0208	41.174		
					1	82.3
					2	−5.1
					3	−.2
					4	77.0
4	1	4	.0278	41.174		
					1	82.3
					2	−5.1
					3	−1.2
					4	76.0

MEMBER FORCES (kips) (*continued*)

+ DENOTES TENSION
− DENOTES COMPRESSION

MEMBER NUMBER	INCIDENCES JOINT 1	JOINT 2	AREA (ft²)	LENGTH (ft)	LOADING	FORCE
5	2	5	.0278	40.485		
					1	−161.9
					2	−10.1
					3	−1.4
					4	−173.5
6	3	6	.0278	41.174		
					1	82.3
					2	−5.1
					3	−1.0
					4	76.2
7	1	2	.0278	29.155		
					1	14.6
					2	.9
					3	.1
					4	15.6
8	2	3	.0278	29.155		
					1	14.6
					2	.9
					3	.2
					4	15.7
9	1	3	.0278	30.000		
					1	−22.5
					2	.5
					3	.1
					4	−21.9
10	4	5	.0156	14.577		
					1	.0
					2	.0
					3	−.2
					4	−.2
11	5	6	.0156	14.577		
					1	.0
					2	.0
					3	−.1
					4	−.1
12	4	6	.0156	15.000		
					1	.0
					2	.0
					3	−.1
					4	−.1

MEMBER FORCES (kips) (*continued*)

+ DENOTES TENSION
− DENOTES COMPRESSION

MEMBER NUMBER	INCIDENCES JOINT 1	JOINT 2	AREA (ft²)	LENGTH (ft)	LOADING	FORCE
13	1	5	.0347	46.654		
					1	.0
					2	.0
					3	.1
					4	.1
14	2	6	.0347	44.809		
					1	.0
					2	.0
					3	−.1
					4	−.1
15	3	4	.0347	46.318		
					1	.0
					2	.0
					3	.0
					4	.0

3.7 STABILITY AND DETERMINACY

All of the plane trusses in the preceding example problems are stable and determinate. In the example problem for the space truss, it was shown that a portion of the stable space truss was indeterminate. In Figure 3.2 it was shown that:

1. The basic shape used in forming a stable plane truss is a triangle.
2. Two additional members are attached to the triangle and their ends are connected to form a fourth joint in another stable configuration.
3. Two more members are attached to the stable configuration and their ends are connected to form another joint in another stable configuration.
4. Item 3 is repeatedly performed to generate the entire plane truss configuration.

Therefore, the simplest stable and determinate plane truss has three members, three joints, and three reaction components. To form a stable and determinate plane truss of n joints, the three members of the original triangle plus two additional members for each of the remaining $(n - 3)$ joints are required. Thus, the minimum total number of members, m, required to form an internally stable plane truss is

$$m = 3 + 2*(n - 3) = 2n - 3$$

If a stable, simple, plane truss of n joints and $(2n - 3)$ members is supported by three independent reaction components, the structure is stable and determinate when subjected to a general loading. If the stable, simple, plane truss has more than three reaction components, the structure is externally indeterminate (all of the reaction components cannot be determined from the three available equations of statics). If the stable, simple, plane truss has more than $(2n - 3)$ members, the structure is internally indeterminate (all of the member forces cannot be determined from the $2n$ available equations of statics in the Method of Joints). Unfortunately, an equation cannot be given for the analyst to determine whether or not a simple plane truss is stable. The analyst must examine the arrangement of the truss members and the reaction components to determine if the simple plane truss is stable. Some examples are shown in Figure 3.14 to illustrate that simple plane trusses having $(2n - 3)$ members are not necessarily stable. For Figures 3.14b and d, $\Sigma F_Y = 0$ on the FBD of the indicated section reveals that the cut members have no resistance to Y-direction loads; therefore, these two truss configurations are internally unstable.

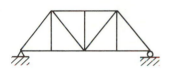

(a) Internally stable and determinate
     externally stable and determinate

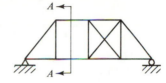

(b) Internally unstable
     externally stable and determinate

(c) Internally stable and determinate
     externally unstable

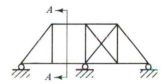

(d) Internally unstable
     externally unstable

Figure 3.14

The conditions of construction enable an analyst to decompose a stable, determinate, compound plane truss into $i \geq 2$ stable, simple, plane trusses. Therefore, the preceding discussion of internal stability and determinacy for a simple, plane truss also applies to a compound plane truss. However, in regard to external stability and determinacy of a compound plane truss, the following comments are applicable. A compound plane truss composed of i simple plane trusses has $k \geq (i - 1)$ conditions of construction and is externally indeterminate if there are more than $3 + k$ independent reaction components. The analyst must examine the arrangement of the provided reaction components of a compound plane truss to determine if it is externally stable. Sometimes, the arrangement appears to be stable, but when the analyst attempts to compute the reactions and member forces they are either inconsistent, infinite, or indeterminate. As shown in Figure 3.15, such structures are actually unstable and they can be made stable as shown in Figure 3.16 by rearrangement of the members.

A stable space truss is formed by starting with a tetrahedron which has four joints and six members. By attaching three members to the tetrahedron and connecting the other ends of the members at a common joint, another stable configuration is obtained. To form a stable and determinate space truss of n joints, the six members

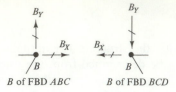

B of FBD *ABC* B of FBD *BCD*

Assumed directions of interconnective forces at B

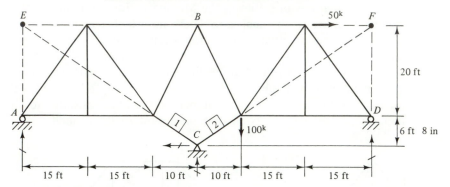

Figure 3.15 FBD of *ABC:* Reaction at A and force in member 1 intersect at point E. $\Sigma M_Z =$ 0 at E: $40*B_Y = 0$; $B_Y = 0$. FBD of *BCD:* Reaction at D and force in member 2 intersect at point F. $\Sigma M_Z = 0$ at F: $40*B_Y + 30*100 = 0$; $B_Y = -75^k$. Since there is no applied Y-direction load at point B, B_Y on FBD of *ABC* must be equal and opposite to B_Y on FBD of *BCD*. However, the above analyses give ($B_Y = 0$) and ($B_Y = -75$) which are inconsistent; this truss is internally unstable.

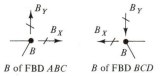

B of FBD *ABC* B of FBD *BCD*

Assumed direction of interconnective forces at B

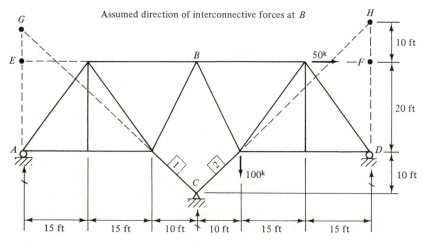

Figure 3.16 Changed geometry of Figure 3.15. FBD of *ABC:* Reaction A and force in member 1 intersect at point G. $\Sigma M_Z = 0$ at G: $40*B_Y + 10*B_X = 0$; $B_X = -4*B_Y$. FBD of *BCD:* Reaction D and force in member 2 intersect at point H. $\Sigma M_Z = 0$ at H: $40*B_Y - 10*B_X + 30*100 + 10*50 = 0$. Simultaneous equations give:

$$\begin{bmatrix} 10 & 40 \\ -10 & 40 \end{bmatrix} \begin{Bmatrix} B_X \\ B_Y \end{Bmatrix} = \begin{Bmatrix} 0 \\ -3500 \end{Bmatrix} \qquad \begin{Bmatrix} B_X \\ B_Y \end{Bmatrix} = \begin{Bmatrix} 175 \\ -43.75 \end{Bmatrix}$$

of the original tetrahedron plus three additional members for each of the remaining $(n - 4)$ joints are required. Therefore, the minimum total number of members required to form a stable space truss of n joints is $m = 6 + 3*(n - 4) = 3n - 6$. If an internally stable space truss has six independent reaction components that do not intersect a single straight line, the space truss is externally stable and determinate.

PROBLEMS

3.1–3.8. Use the Method of Joints and find the axial force in all members of the truss. Answers are not given for problems preceded by an asterisk.

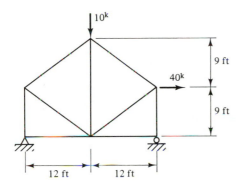

*Figure P3.1

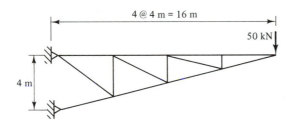

*Figure P3.2

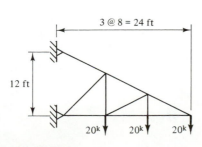

*Figure P3.3

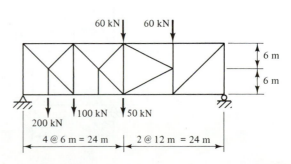

*Figure P3.4

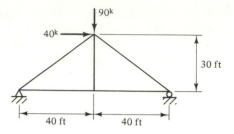

*Figure P3.5

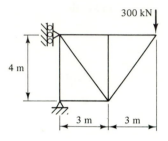

*Figure P3.6

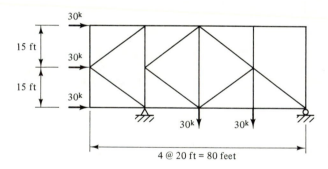

*Figure P3.7

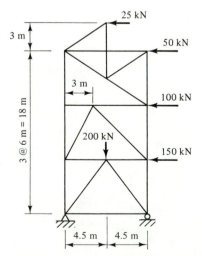

*Figure P3.8

3.9–3.22. Use the Method of Sections to find the axial force in the members that have a member number on them.

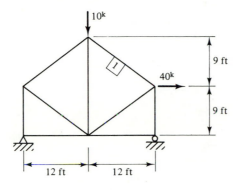

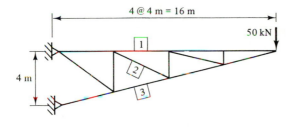

*Figure P3.9

*Figure P3.10

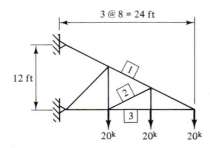

Figure P3.11

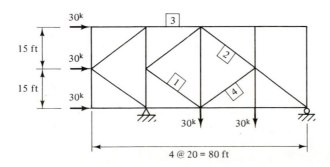

Figure P3.12 First, identify any zero force members.

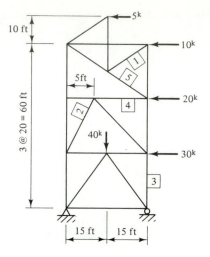

Figure P3.13

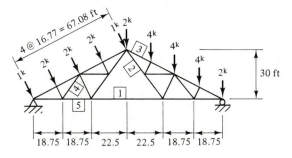

*Figure P3.14 First, identify any zero force members.

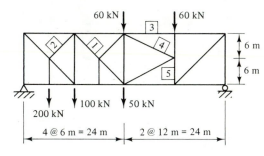

*Figure P3.15

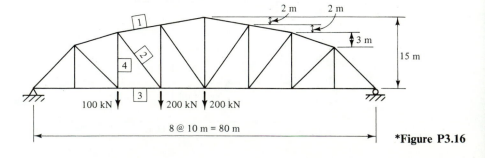

*Figure P3.16

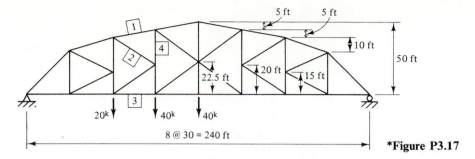

*Figure P3.17

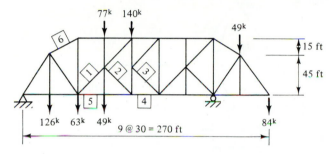

*Figure P3.18

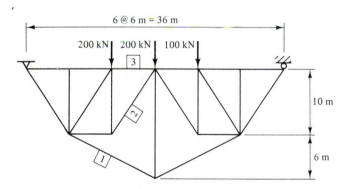

*Figure P3.19

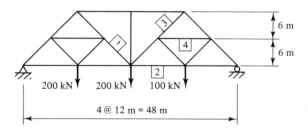

*Figure P3.20

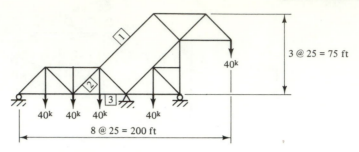

3 @ 25 = 75 ft

40ᵏ

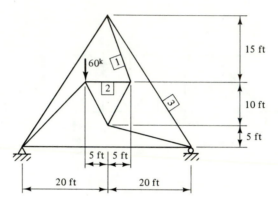

40ᵏ   40ᵏ   40ᵏ   40ᵏ

8 @ 25 = 200 ft

*Figure P3.21

60ᵏ

15 ft

10 ft

5 ft

5 ft | 5 ft

20 ft   20 ft

*Figure P3.22

Chapter 4

Determinate Beam and Frame Analysis

4.1 DEFINITIONS AND SIGN CONVENTION

In the following discussion it is assumed that:

1. The longitudinal centroidal axis (x axis) of each member in a structure is a straight line.
2. All member x axes lie in a common plane (XY plane).
3. The loads applied to a member lie in the XY plane.
4. The shear center and the centroid of a member's cross section are coincident (each member bends in the XY plane without twisting).
5. X, Y, Z are the global (or system) reference axes of the structural system.
6. x, y, z are the member (or local) reference axes of an individual member; also, y and z are the principal axes of inertia on the cross section of the member.

See Figure 4.1b for an example of global axes (X, Y, Z) and Figure 4.1c for examples of member axes (x, y, z). The analyst chooses convenient locations of the origins and directions of the global and member axes in each problem. To be compatible with the axes assumed in the author's FRAME computer program which will be used later in the text, all axes must be right handed and the member z axes must be parallel to the global Z axis and positive in the same direction as the global Z axis.

Eventually, the results obtained from the type of analyses shown in this chapter must be used to determine if a structure is properly designed to resist the worst loading(s) assumed to occur on the structure during its life time. Therefore, the following definitions and terminology are chosen to be compatible with those found in structural design codes.

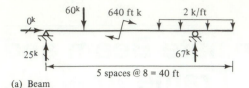

(a) Beam

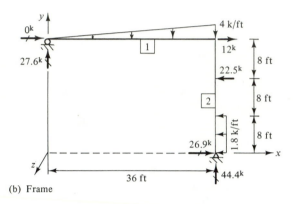

(b) Frame

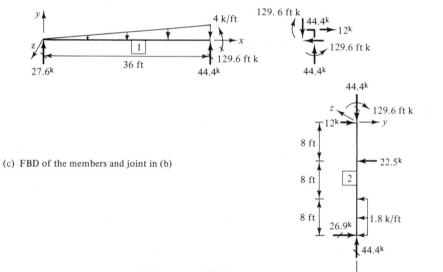

(c) FBD of the members and joint in (b)

Figure 4.1

A *beam* (see Figure 4.1a) is a structural member subjected to bending (or flexure) by loads applied transversely to its x axis. The beam in Figure 4.1a is subjected to two concentrated loads (the 60^k in the negative y direction and the 640 ftk couple) and a 2 k/ft uniformly distributed load on a portion of the member's length. A concentrated load acts on such a small portion of the member's length that it can be assumed to act at a point. Distributed loads act on a considerable length of the member and they may be uniformly varying (a triangular load) as shown in Figure 4.1b on member 1 or they may vary in some other fashion (for example, parabolically or sinusoidally). A triangular load may be due to water pressure or a pile of sand, for example. A *column* is a structural member subjected only to axial compression or tension-type loads which may vary along the x axis of the member. Truss members are examples of columns subjected to a constant axial load. A *beam-column* (see member 2 in Figure 4.1b) is a structural member that is subjected to combined bending plus axial load.

The most general type of member to be considered in this chapter is a beam-column. At the origin of member 2 in Figure 4.1c, the complete set of *member-end force types* for a beam-column are shown, namely:

1. An *axial* (x-direction) *force.*
2. A *shear* (y-direction) *force.*
3. A *bending moment* is a rotational force about the member's z axis (for three-dimensional structural analysis, a bending moment is depicted as a straight-line-type vector in the y and z directions with a double-headed arrow to denote the rotational direction using the right-hand screw rule).

In Figure 4.2a an FBD of a portion of member 2 in Figure 4.1c is shown. For Figure 4.2a to be in equilibrium, at the cut section of Figure 4.2a there must be an *axial force, P,* a *shear force, V,* and a *bending moment, M.* These section or *internal forces* are unknown and each unknown internal force is shown in its defined positive direction. The positive directions of the internal forces are as shown *on a differential element* in Figure 4.2b, namely:

1. *An axial tension force is positive.*
2. A *clockwise shear couple is positive* (that is, at the origin end of the differential element, V is positive in the y direction; at the other end of the differential element, V is positive in the negative y direction).
3. A *positive bending moment induces compression in the top fiber* (positive y-direction surface) of the differential element.

4.2 EQUATIONS FOR AXIAL FORCE, SHEAR, AND BENDING MOMENT

To check the design of a member, the designer needs detailed information on the variations of the internal forces along the x axis of the member. The variations of the internal forces as a function of x can be obtained by cutting sections through the member in the y direction at pertinent points and by satisfying the three equations of statics on the FBD involving a cut section. After the equations (functions) for the internal forces are obtained, the functions can be plotted along the length of the member. This approach for obtaining the variation of the internal forces is illustrated in the following example problems.

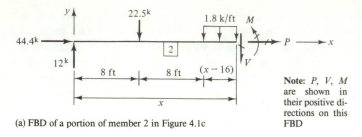

Note: *P*, *V*, *M* are shown in their positive directions on this FBD

(a) FBD of a portion of member 2 in Figure 4.1c

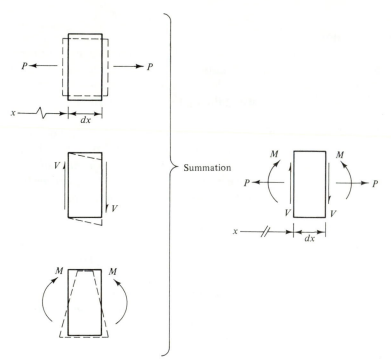

Summation

(b) Positive directions of the internal forces acting on a differential element

Figure 4.2 Sign convention for the internal forces.

It should be noted that the author computed the reactions in the following example problems using the approach shown in Chapter 2. That is, in computing the reactions, each distributed load was replaced by a concentrated load equal to the area under the load diagram and located at the centroid of the load diagram. After the reactions have been computed, the analyst *must* revert to the distributed load configuration in obtaining the equations for the internal forces.

EXAMPLE 4.1 ——————————————————————————————————

Find the equations for the variation of the internal forces of the beam in Figure 4.3a and plot the obtained functions.

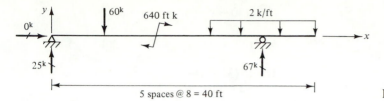

5 spaces @ 8 = 40 ft   **Figure 4.3a**

Solution

For $8 > x > 0$:

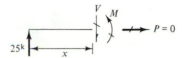

$$\sum F_y = 0: \qquad 25 - V = 0; \qquad \underline{V = 25}$$
$$\text{at } x, \sum M_z = 0: \qquad M - x*25 = 0; \qquad \underline{M = 25x}$$

For $16 > x > 8$:

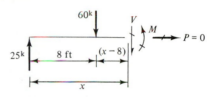

$$\sum F_y = 0: \qquad 25 - 60 - V = 0; \qquad \underline{V = -35}$$
$$\text{at } x, \sum M_z = 0: \qquad M + (x-8)*60 - x*25 = 0$$
$$\underline{M = 25x - 60*(x-8)}$$
$$\text{or} \qquad M = 480 - 35x$$

For $24 > x > 16$:

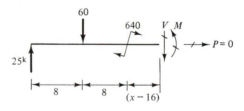

$$\sum F_y = 0: \qquad V = 25 - 60 = -35$$
$$\text{at } x, \sum M_z = 0: \qquad M = 25x + 640 - 60(x-8) \quad \text{or} \quad M = 1120 - 35x$$

The FBD on either side of the cut section can be used; therefore, use the simpler of the two FBD possibilities.

For $32 > x > 24$:

$$\sum F_y = 0: \qquad V = 2(40 - x) - 67 = 13 - 2x$$

$$\text{at } x, \ \sum M_z = 0: \qquad M = (32 - x) * 67 - \frac{(40 - x)}{2} * 2(40 - x)$$

$$M = 67(32 - x) - (40 - x)^2$$
$$M = 544 + 13x - x^2$$

For $40 > x > 32$:

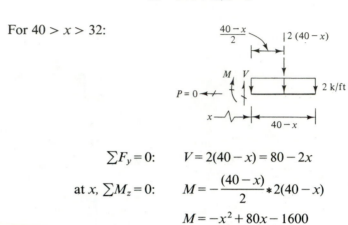

$$\sum F_y = 0: \qquad V = 2(40 - x) = 80 - 2x$$

$$\text{at } x, \ \sum M_z = 0: \qquad M = -\frac{(40 - x)}{2} * 2(40 - x)$$

$$M = -x^2 + 80x - 1600$$

On the M diagram (Fig. 4.3b), the direction that each segment curls is shown to aid in sketching the qualitative deflected shape. A positive moment causes a segment to curl such that it holds water. A negative moment causes a segment to curl such that it sheds water.

EXAMPLE 4.2 ——

For the members in the frame of Figure 4.4a, find the equations for the variation of the internal member forces and plot the obtained functions.

Solution.   Member 1. For $36 > x > 0$:

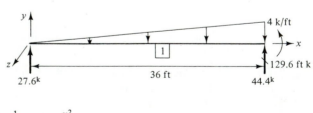

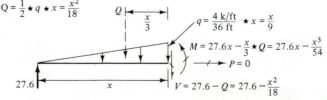

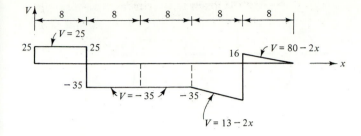

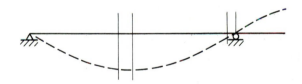

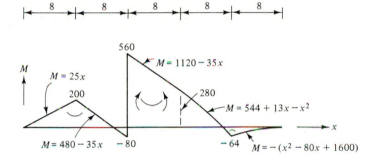

Figure 4.3b *Deflected shape.* On the M diagram, the direction that each segment curls is shown to aid in sketching the qualitative deflected shape. A positive moment causes a segment to curl such that it holds water. A negative moment causes a segment to curl such that it sheds water.

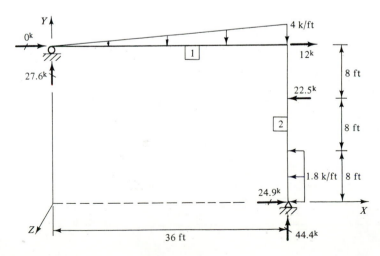

Figure 4.4a

Member 2:

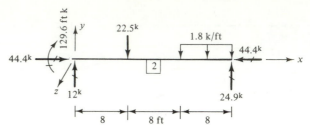

For $8 > x > 0$:

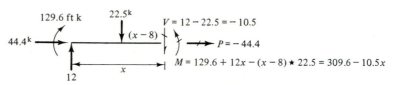

$$M = 129.6 + 12x$$
$$P = -44.4$$
$$V = 12$$

For $16 > x > 8$:

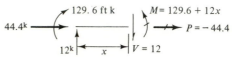

$$V = 12 - 22.5 = -10.5$$
$$P = -44.4$$
$$M = 129.6 + 12x - (x - 8) \star 22.5 = 309.6 - 10.5x$$

For $24 > x > 16$:

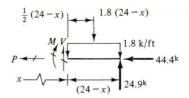

$$P = -44.4$$
$$V = 1.8(24 - x) - 24.9 = 18.3 - 1.8x$$
$$M = (24 - x) * 24.9 - \frac{(24 - x)}{2} * 1.8(24 - x) = 79.2 + 18.3x - 0.9x^2$$

For member 1 (Fig. 4.4b):

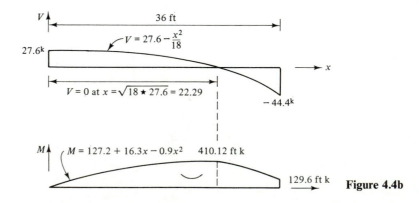

Figure 4.4b

For member 2 (Fig. 4.4c):

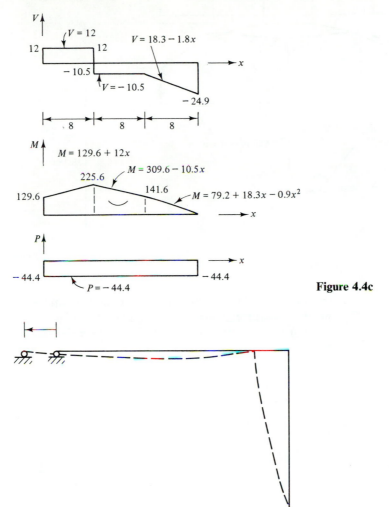

Figure 4.4c

Deflected shape   **Figure 4.4d**

EXAMPLE 4.3

Find the equations for the variation of the internal forces of the circular arch in Figure 4.5a.

Solution. It should be noted that the *Y*-direction reactions were obtained from the equations of statics, but the *X*-direction reactions were obtained from an indeterminate analysis. The purposes of this example are to emphasize that (1) the definition of a shear force is an internal force which is perpendicular to the member's length direction axis, and (2) if the function for the member's length

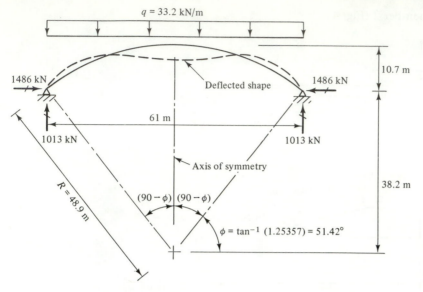

Figure 4.5a

direction axis is known, the equations of statics can be used to find the internal force functions for a curved axis member. See Figure 4.5b.

For $(180° - \phi) \geq \theta \geq \phi$, where $\phi = 51.42°$

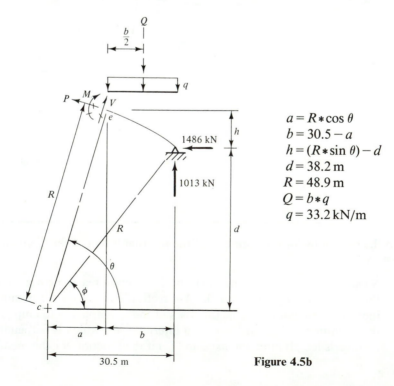

$a = R * \cos \theta$
$b = 30.5 - a$
$h = (R * \sin \theta) - d$
$d = 38.2 \, \text{m}$
$R = 48.9 \, \text{m}$
$Q = b * q$
$q = 33.2 \, \text{kN/m}$

Figure 4.5b

$$\text{At } e, \ \sum M_z = 0: \qquad M = b * 1013 - h * 1486 - \frac{b}{2} * Q$$

$$M = 1013 * (30.5 - R \cos \theta) + 1486 * (38.2 - R \sin \theta)$$

$$-\frac{q}{2} * (30.5 - R \cos \theta)^2$$

$$\text{At } c, \ \sum M_z = 0: \qquad R * P + d * 1486 + 30.5 * 1013 - M - \left(a + \frac{b}{2}\right) * Q = 0$$

$$P = \frac{M + q * b * \left(a + \frac{b}{2}\right) - 1486 * d - 30897}{R}$$

$$\sum F_X = 0: \qquad V \cos \theta - P \sin \theta - 334 = 0$$

$$V = \frac{1486 + P \sin \theta}{\cos \theta}$$

At $\theta = \phi$: $\sum M = 0$

$$P = -\frac{(1486 * 38.2 + 30897)}{48.9}$$

$$= -1793 \, \text{kN}$$

$$V = \frac{1486 - 1793 * 0.7817368}{0.6236085}$$

$$= 135.3 \, \text{kN}$$

At $\theta = 90°$: $V = 0$

$$M = -452 \, \text{mkN}$$

$$P = \ -1486 \, \text{kN}$$

4.3 RELATION BETWEEN TRANSVERSE LOADS, SHEAR, AND BENDING MOMENT

Except for curved members as in Example 4.3, construction of the Shear and Bending Moment Diagrams can be expedited by observing that certain relations exist between transverse load, shear, and bending moment. Consider the beam shown in Figure 4.6; all applied loads are shown to be acting in their positive directions and the correct directions of the unknown reactions are assumed to be as shown.

Figure 4.7 is an enlargement of the differential element in Figure 4.6; note that the element is taken from the region of x where the distributed load is a continuous function.

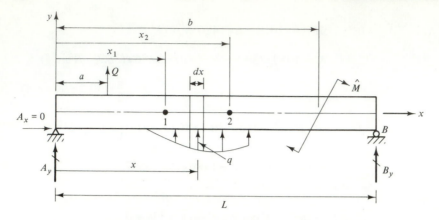

Note that the intensity of the distributed load is q at x

Figure 4.6

Consider the FBD shown in Figure 4.7:

$$\sum F_y = 0: \qquad V + q*dx - (V + dV) = 0$$

$$dV = q*dx$$

$$\frac{dV}{dx} = q$$

at x, slope of Shear Diagram = load intensity

*

$$\int_{V_1}^{V_2} dV = \int_{x_2}^{x_1} q\, dx$$

change in shear between points 1 and 2 = area under the load curve
between points 1 and 2

*

* *Reminder:* These relations are valid only in a region where the load q is a continuous function
of x.

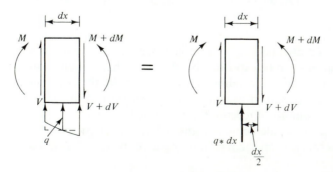

Figure 4.7

At the right-hand side of Figure 4.7,

$$\sum M_z = 0: \qquad M + V*dx + (q*dx)*\frac{dx}{2} - (M + dM) = 0$$

Since dx is very small, $(dx)^2 \approx 0$ and the preceding equation reduces to $dM = V*dx$.

$$\frac{dM}{dx} = V$$

at x, slope of Moment Diagram = shear

*

* *Reminder:* These relations are valid only where V is a continuous function of x.

$$\int_{M_1}^{M_2} dM = \int_{x_1}^{x_2} V\,dx$$

change in moment between points 1 and 2 = area under shear curve
between points 1 and 2

*

At a point on the shear curve where $V = 0$, $\dfrac{dM}{dx} = 0$

and the moment value at this point is either a maximum or minimum value.

The following relations apply at a point where there is a concentrated load:

$$\Delta V = Q$$

change in shear = load

$$\Delta M = \widehat{M}$$

change in moment = concentrated moment value

4.4 SHEAR AND BENDING MOMENT DIAGRAMS

In the process of satisfying a structural design code for shear and flexure of a beam, the structural designer needs to know the following:

1. Maximum magnitude of shear.
2. Maximum magnitudes of positive and negative moments; they occur at points of zero shear.
3. Locations along the length of the member at which:
   (a) The maximum magnitudes of $+M$ and $-M$ occur.
   (b) There are points of inflection on the Moment Diagram; at a point of

inflection, $M = 0$ and the sign of M changes from either $+$ to $-$ or from $-$ to $+$.

The needed Shear and Bending Moment information can be obtained from the Shear and Bending Moment Diagrams. In most beam and frame problems, the Shear and Bending Moment Diagrams can be plotted by using the relations derived in Section 4.3 without actually knowing the equations of shear and bending moment. This procedure for obtaining the Shear and Bending Moment Diagrams is illustrated in the following example problems.

EXAMPLE 4.4

Using the relations derived in Section 4.3, draw the Shear and Bending Moment Diagrams for the beam shown in Figure 4.8a.

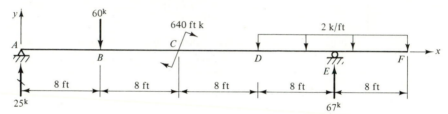

Figure 4.8a

Solution.   Comments: The Shear Diagram is created from the relations to the Load Diagram (Fig. 4.8b):

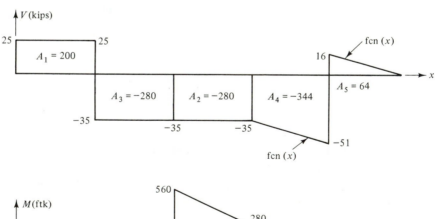

Figure 4.8b

1. At A, the known reaction is an applied concentrated load; $\Delta V = (Q = 25)$ at A.
2. Between A and B, there is not any applied load; $dV/dx = (q = 0)$ from A to B.
3. At B, $\Delta V = (Q = -60)$; $(V_{B+dx} - V_{B-dx}) = -60$; $V_{B+dx} = 25 + (-60) = -35$.
4. The concentrated moment at C is equivalent to a pair of equal but opposite y-direction forces separated by a small (compared to the member length) finite distance: $\overset{\uparrow Q}{\underset{d\ \downarrow Q}{\longmapsto}}$. Since the 640 ftk is concentrated at point C, at C, $dV = (Q - Q = 0)$; therefore, a concentrated moment does not cause any *net change in shear* at the point where the concentrated moment is applied. Therefore, from B to D, $dV/dx = (q = 0)$.
5. From D to E, $dV/dx = (q = -2)$ and $(V_E - V_D) =$ (area under the distributed load curve from D to E) $= 8\ \text{ft} * (-2\ \text{k/ft}) = -16^k$; $V_E = V_D + (-16) = -51$.
6. At E, $\Delta V = (Q = 67)$; $(V_{E+dx} - V_{E-dx}) = 67$; $V_{E+dx} = -51 + 67 = 16$.
7. From E to F, $dV/dx = -2$ and $(V_F - V_{E+dx}) =$ (area under the distributed load curve from E to F) $= 8 * (-2) = -16$; $V_F = V_{E+dx} + (-16) = 0$.

The Moment Diagram is created from the relations to the Shear Diagram plus the relation at the concentrated moment on the Load Diagram:

1. $M = 0$ at A since the boundary condition at A is a hinge.
2. $(M_B - M_A) = A_1$ (area under the Shear Diagram from A to B); from A to B, $dM/dx = (V = 25)$; $M_B = M_A + A_1 = 0 + 200 = 200$.
3. $M_{C-dx} = M_B + A_2 = 200 + (-280) = -80$; from B to C, $dM/dx = (V = -35)$.
4. $M_{C+dx} = M_{C-dx} + M_C = -80 + 640 = 560$.
5. $M_D = M_{C+dx} + A_3 = 560 + (-280) = 280$; from C to D, $dM/dx = (V = -35)$.
6. $M_E = M_D + A_4 = 280 + (-344) = -64$; from D to E, $dM/dx = (-35$ at D to -51 at E).
7. $M_F = M_E + A_5 = -64 + 64 = 0$; from E to F, $dM/dx = (16$ at E to 0 at F).

EXAMPLE 4.5

Using the relations derived in Section 4.3, draw the Shear and Bending Moment Diagrams for the beam shown in Figure 4.9a.

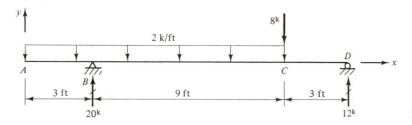

Figure 4.9a

Solution. Comments: The Shear Diagram is created from the relations to the Load Diagram (Fig. 4.9b):

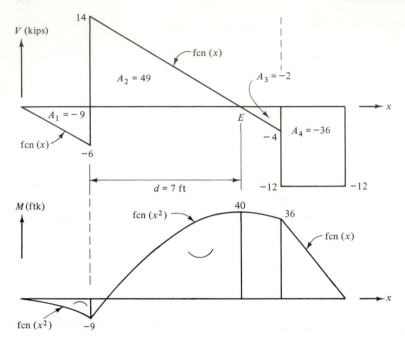

Figure 4.9b

1. $V_{B-dx} = V_A +$ (load area from A to B) $= 0 + (-2*3) = -6$. From A to B, $dV/dx = (q = -2)$.
2. $V_{B+dx} = V_{B-dx} + (Q = 20) = -6 + 20 = 14$.
3. $V_C = V_{B+dx} +$ (load area from B to C) $= 14 + (-2*9) = -4$. From B to C, $dV/dx = (q = -2)$ and $V = 0$ at $d = 14^k/2$ k/ft $= 7$ ft from B.
4. $V_{C+dx} = V_{C-dx} + (Q = -8) = -4 + (-8) = -12$.
5. From C to D, $dV/dx = (q = 0)$; $V_D = V_C +$ (load area from C to D) $= -12 + 0 = -12$.

The Moment Diagram is obtained from the relations to the Shear Diagram:

1. $M_B = M_A + A_1 = 0 + (-9) = -9$. From A to B, $dM/dx = (V = 0$ at A and -6 at B).
2. $M_E = M_B + A_2 = -9 + 49 = 40$. From B to E, $dM/dx = (V = 14$ at B and 0 at E).
3. $M_C = M_E + A_3 = 40 + (-4) = 36$. From E to C, $dM/dx = (V = 0$ at E and -4 at C).
4. $V_D = V_C + A_4 = 36 + (-36) = 0$ (checks, hinge at D). From C to D, $dM/dx = (V = -12)$.

EXAMPLE 4.6 ———

Using the relations derived in Section 4.3, draw the Shear and Bending Moment Diagrams for the members in the frame of Figure 4.10a.

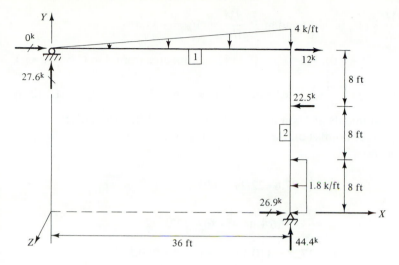

Figure 4.10a

Solution for Member 1. Comments (Fig. 4.10b):

1. $V_B = V_A +$ (Load Diagram area from A to B) $= 27.6 + \dfrac{1}{2}*36*(-4) = -44.4$.

2. At A, $\dfrac{dV}{dx} = (q = 0)$; load curve from A to B is a linear function of x. Since, $V_B = V_A + \int_0^{36} q\, dx$ and $q = \dfrac{-4\,k/ft}{36\,ft}*x$, the shear curve from A to B is a function of x^2.

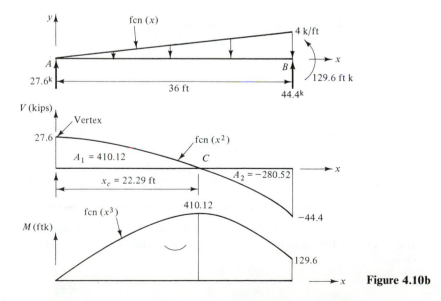

Figure 4.10b

3. At A, $\dfrac{dM}{dx} = (V_A = 27.6)$ and at B, $\dfrac{dM}{dx} = (V_B = -44.4)$.

4. At C, $\dfrac{dM}{dx} = (V_C = 0)$; $V_C = V_A + $ (Load Diagram area from A to $C = \dfrac{1}{2}*q*x_c$).

$V_C = 27.6 + \dfrac{1}{2}*\left(\dfrac{-4}{36}*x_c\right)*x_c = 0$; $x_c = 18*27.6 = 22.29$ ft. $M_C = 410.12$ by summing moments at C on a FBD of length 22.29 ft. Alternatively, $M_C = M_A + A_1$. Since point A on the Shear Diagram is a vertex of curve of fcn (x^2), from Appendix A,

$$A_1 = \frac{2}{3}*27.6*22.29 = 410.12$$

$$A_{AB} = \frac{2}{3}*(27.6 + 44.4)*36 = 1728$$

$$A_2 = 1728 - 410.12 - 36*44.4 = 280.52$$

Solution for Member 2

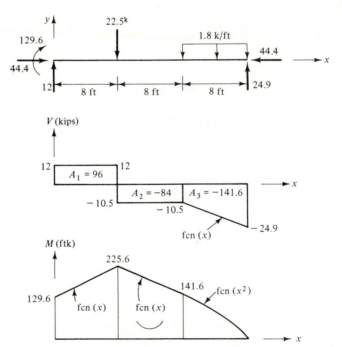

Figure 4.10c

EXAMPLE 4.7 ———

Using the relations derived in Section 4.3, draw the Shear and Bending Moment Diagrams for the members in the frame of Figure 4.11a; also, sketch the deflected structure.

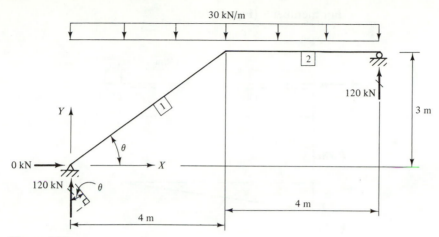

Figure 4.11a

Solution. To find the Shear Diagram for member 1, the reactions at the left support and the member loading need to be resolved into components parallel to and perpendicular to the axis of member 1 as shown in Figure 4.11b. The *Y*-direction loading distributed along the length of member 1 is $30*4/5 = 24$ kN/m; components of this loading are $24*\cos\theta = 24*0.8 = 19.2$ kN/m perpendicular to member 1, and $24*\sin\theta = 24*0.6 = 14.4$ kN/m parallel to member 1. Components of the reactions are $120*\cos\theta = 96$ kN perpendicular to member 1 and $120*\sin\theta = 12$ kN parallel to member 1. Also, the member-end moment for member 1 can be found from a FBD of member 1 in Figure 4.11a to be 240 mkN.

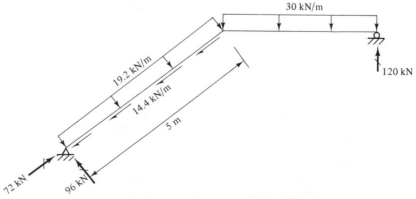

Figure 4.11b

For member 1 (Fig. 4.11c):

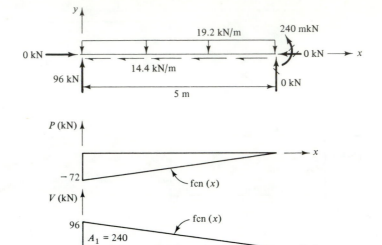

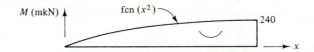

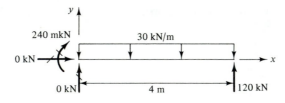

Figure 4.11c

For member 2 (Fig. 4.11d):

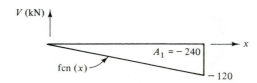

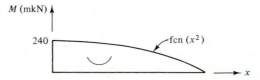

Figure 4.11d

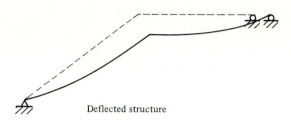

Deflected structure

Figure 4.11e

PROBLEMS

4.1–4.26. Draw the Shear and Moment Diagrams. Answers are not given for problems preceded by an asterisk.

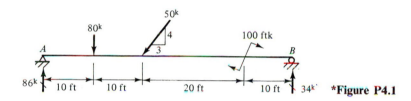

*Figure P4.1

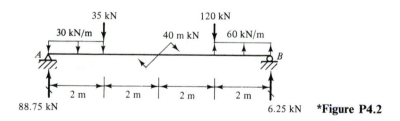

*Figure P4.2

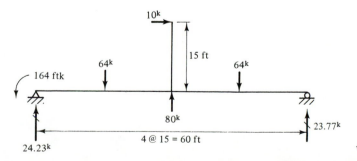

*Figure P4.3

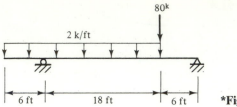

*Figure P4.4

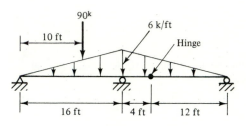

*Figure P4.5

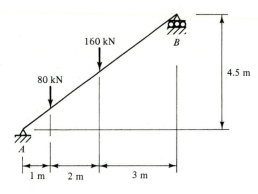

*Figure P4.6

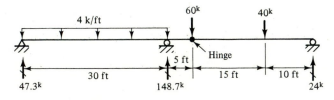

Figure P4.7

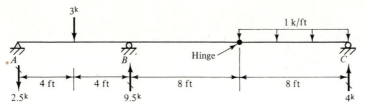

Figure P4.8

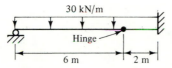

*Figure P4.9

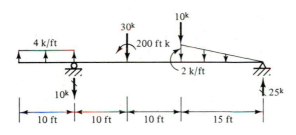

Figure 4.10

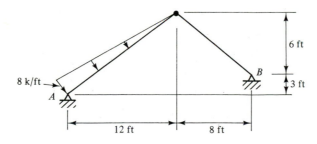

*Figure P4.11

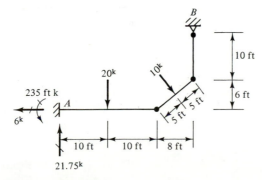

*Figure P4.12

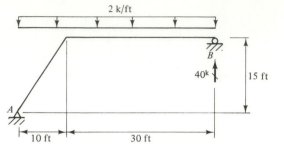

*Figure P4.13

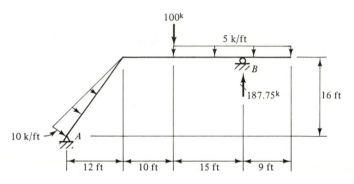

Figure P4.14

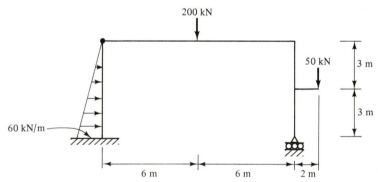

*Figure P4.15

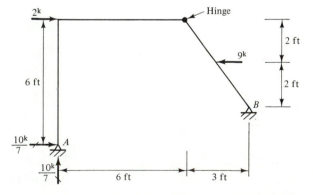

*Figure P4.16

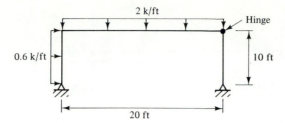

*Figure P4.17

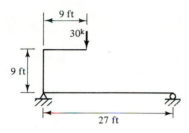

*Figure P4.18

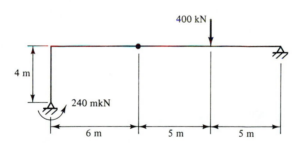

*Figure P4.19

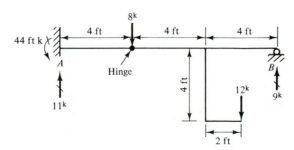

*Figure P4.20

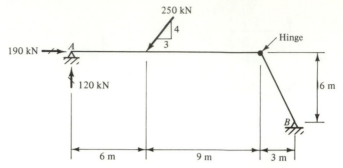

*Figure P4.21

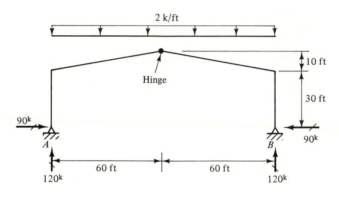

*Figure P4.22

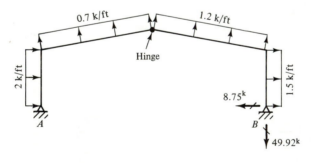

*Figure P4.23 See Figure P3.22 for the dimensions.

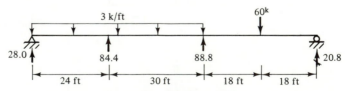

*Figure P4.24

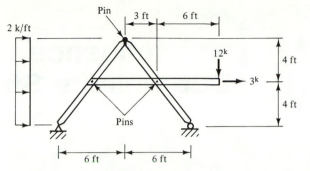

*Figure P4.25

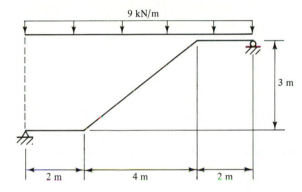

*Figure P4.26

Chapter 5

Influence Lines for Determinate Structures

5.1 INTRODUCTION

Live loads vary in position on the structure and they may vary with respect to time. In the design of a highway truss bridge, the structural designer may be required to position the traffic vehicle wheel loads to cause the maximum axial tension or compression force to occur separately in each truss member. Potentially, there are as many different live load configurations as there are truss members. Fortunately, a particular live load configuration may cause the maximum axial tension or compression force to occur simultaneously in several members. Similarly, to properly design each member in a warehouse, a school building, or an office building, the live loads must be positioned to cause the maximum magnitude of axial force, shear, and bending moment to occur separately in each member. Also, the maximum magnitude of each reaction is needed in the design of a structure.

Sometimes it is possible to determine by inspection where to position the live loads to maximize each structural response function. Usually, however, the designer must develop diagrams to be used in finding the live load configuration that maximizes a particular response function. The Influence Line (IL) is such a diagram.

5.2 DEFINITION OF AN INFLUENCE LINE

An influence line graphically shows how the movement of a single-wheel load along the platform of the structure influences some structural response function. Reactions, shears, bending moments, and axial forces are the structural response functions of interest in this chapter. For convenience in the subsequent design calculations, the magnitude of the single-wheel load is chosen as unity in constructing the influence line for a structural response function.

An IL is the graph of a structural response function due to a unit wheel load. The abscissa axis of the graph is named x, and x defines the position of the wheel load with respect to the origin of the platform on which the wheel load rolls. The name of the ordinate axis of the graph is the name of the structural response function being graphed. For example, suppose the structural response function being graphed is the Y-direction reaction at A for the structure in Figure 5.1a; then, the name of the ordinate axis of the graph is A_Y. The ordinate at x on the graph gives the value of the structural response function when the unit load is placed at x on the platform of the structure. For determinate structures, the graph is either a straight line or a series of straight lines with possible discontinuities in the ordinates at some positions of x. Consequently, to draw the influence line the analyst only needs to calculate the ordinates at each end of each straight line segment.

Since a graph shows quantities for the ordinate and abscissa axes, an influence line is a quantitative diagram. A rough sketch of a graph without quantities is called a qualitative diagram. The *Muller-Breslau principle* [5.1] can be used to quickly draw qualitative influence lines and in some cases to draw quantitative influence lines. This principle was developed in 1886 by Heinrich Muller-Breslau, a German professor, and can be stated as follows: If *a unit change in displacement* is induced at and in the direction of a particular structural response function, the deflection curve of the platform portion of the structure is the influence line for this structural response function. This principle is applicable to any linearly elastic determinate or indeterminate beam, frame, or truss structure. For indeterminate beams and frames, the influence line is a curved line or a series of curved lines with possible discontinuities in the ordinates at some positions of x.

In applying the Muller-Breslau principle, the structure must behave in a linearly elastic manner and:

1. The structure is treated as being weightless (the weight of the structure is accounted for in the dead load case).
2. Only the capability of the structure to provide the structural response function being studied is destroyed and a unit change in displacement is induced at and in the positive definition direction(s) of this structural response function.
3. The induced unit change in displacement is treated as being a small displacement.

The sign convention for the structural response functions are:

1. Y-direction components of the reactions are positive in the Y direction.
2. Tensile member-end axial forces are positive.

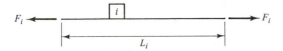

3. Positive member-end shears are (d is the relative displacement of the member ends.)

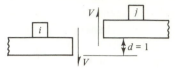

4. Positive member-end moments are (α and β are the member-end rotations; $\alpha + \beta = 1$ rad.)

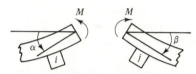

In each of the following example problems, the Muller-Breslau principle is used to obtain the qualitative influence line. In some of the example problems, the Muller-Breslau principle directly gives the quantitative influence line. If only the qualitative influence line can be obtained, the unit load is positioned at the key point or points and the ordinates at those points are obtained by statics.

5.3 EXAMPLES FOR BEAMS

In the following examples, the reader may find it convenient to think of the rolling unit load as being a wheel load of an overhead crane, for example. Therefore, the platform on which the unit load rolls is the beam shown in each of the following example problems.

EXAMPLE 5.1 ——

For the structure shown in Figure 5.1a, the platform on which the wheel rolls is the beam shown. Draw the influence line for each of the following structural response functions:

   1. A_Y
   2. C_Y
   3. M_A

Note: There is an internal hinge at point B.

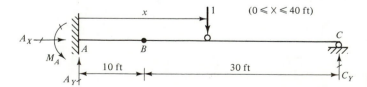

Figure 5.1a

Solution. First, we obtain the qualitative influence lines by the Muller-Breslau principle:

1. Qualitative IL for A_Y (Fig. 5.1b). Note that the assumed positive direction of A_Y shown in Figure 5.1a is upward. According to the Muller-Breslau principle, only the resistance to a Y-direction force is destroyed at A and a unit change in displacement is induced at and in the direction of A_Y to obtain the qualitative IL for A_Y. Since the resistance to A_X and M_A was not destroyed at A, segment AB must be constrained at A such that segment AB neither translates in the X direction nor rotates about the Z axis at A (Fig. 5.1c).

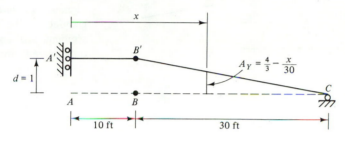

Figure 5.1b Qualitative IL for A_Y. Line segments $A'B'$ and $B'C$ are the qualitative IL for A_Y. The equation for line segment $B'C$ can be obtained from geometry or by the equations of statics as shown in Figure 5.1c.

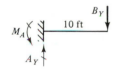

$$A_Y = B_Y = \frac{40-x}{30} = \frac{4}{3} - \frac{x}{30} \qquad B_Y = \frac{(40-x) \star 1}{30} = \frac{40-x}{30}$$

$$M_A = 10^* B_Y = \frac{40-x}{3} \qquad C_Y = \frac{(x-10) \star 1}{30} = \frac{x-10}{30}$$

Figure 5.1c FBD of Figure 5.1a.

2. Qualitative IL for C_Y (Fig. 5.1d). To destroy the resistance to a Y-direction force at C, the structure is disconnected from the roller support at C. Since point B is a hinge, no force is required at C to raise point C by an unit amount.

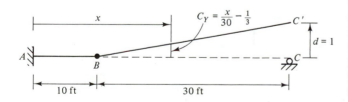

Figure 5.1d Qualitative IL for C_Y. Line segments AB and BC' are the qualitative IL for C_Y. The equation for line segment BC' was obtained in Figure 5.1c by statics.

3. Qualitative IL for M_A (Fig. 5.1e). The assumed positive direction of M_A is counterclockwise as shown in Figure 5.1a; only the external resistance provided by M_A is destroyed at A, and a counterclockwise unit angle change is induced at A as shown in Figure 5.1e.

Two observations in regard to implementation of the Muller-Breslau principle in obtaining an influence line for a determinate structure are:

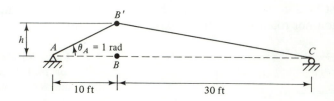

Figure 5.1e Qualitative IL for M_A. The unit change in displacement, $\theta_A = 1$ rad, must be treated as being a small displacement ($\theta_A = \tan \theta_A = h/10$ ft $= 1$ rad). Therefore, $h = (10$ ft$)*(1$ rad$) = 10$ ft. The equation of line segment $B'C$ was obtained in Figure 5.1c: $M_A = 40/3 - x/3$; and the equation of line segment AB' is obtained from geometry to be $M_A = x$. Line segments AB' and $B'C$ are the qualitative IL for M_A.

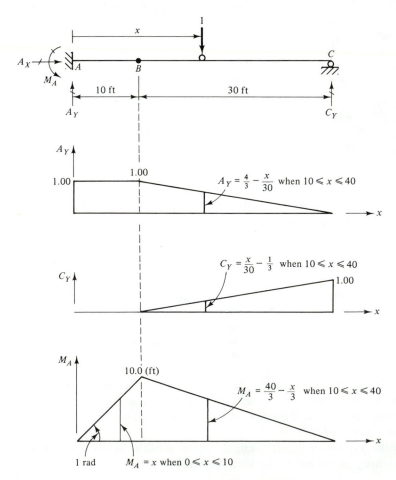

Figure 5.1f Influence Lines for A_Y, C_Y, and M_A of Figure 5.1a.

$A_Y = \frac{4}{3} - \frac{x}{30}$ when $10 \leqslant x \leqslant 40$

$C_Y = \frac{x}{30} - \frac{1}{3}$ when $10 \leqslant x \leqslant 40$

$M_A = \frac{40}{3} - \frac{x}{3}$ when $10 \leqslant x \leqslant 40$

$M_A = x$ when $0 \leqslant x \leqslant 10$

1. The modified structure on which the unit change in displacement is induced is unstable. This may be alarming to the reader. The modified structure is a mathematical model created for the special purpose of implementing the Muller-Breslau principle. That is, the modified structure is not real and is treated as a special, mathematical model; the modified structure serves only a mathematical purpose and does not have to be stable.
2. No force is necessary to induce the unit change in displacement since the modified structure is unstable. This observation is emphasized in subsequent example problems.

Any ordinate of a particular influence line for a determinate structure can be obtained by placing the unit load at a specific x location and using only the equations of statics. Consequently, the IL ordinates at the ends of each straight line segment can be obtained by statics for a determinate structure. However, this approach cannot be used to find an ordinate of an influence line for an indeterminate structure since the structure is indeterminate. In the implementation of the Muller-Breslau principle to obtain the qualitative influence line for an indeterminate structure, unknown forces are necessary to induce the unit change in displacement as shown in Figure 5.1g:

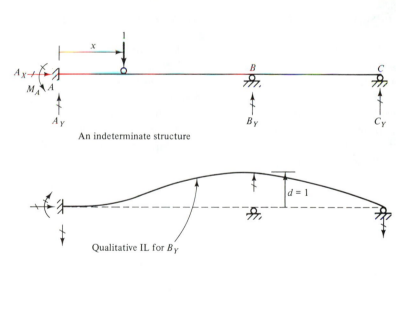

An indeterminate structure

Qualitative IL for B_Y

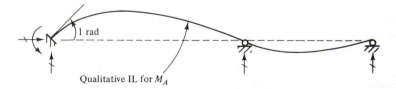

Qualitative IL for M_A

Figure 5.1g

EXAMPLE 5.2

The platform on which the wheel rolls is the beam shown in Figure 5.2a. Draw the influence line for each of the following structural response functions:

1. Shear to the left of B.
2. Shear to the right of B.
3. Bending moment at D.
4. Bending moment at E.

Note: There is an internal hinge at point B.

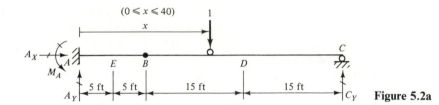

Figure 5.2a

Solution. First, we use the Muller-Breslau principle to obtain the qualitative IL:

1. Qualitative IL for $V_{B(\text{left})}$—(IL for shear to left of B). Only the internal resistance to shear is destroyed at dx to the left of point B. Therefore, when the unit change in the Y-direction displacement is induced at B, segments AB and BC must not be allowed to separate in the X direction at B and they must have the same final slope at point B. Since no force is required to displace segment BC at point B and V_B is positive in the Y direction on the left end of a section for segment BC, all of the unit change in Y-direction displacement at B occurs on segment BC as shown in Figure 5.2b.

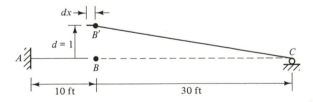

Figure 5.2b Qualitative IL for $V_{B(\text{left})}$.

2. Qualitative IL for $V_{B(\text{right})}$. Only the shear resistance is destroyed at dx to the right of point B. After the unit change in Y-direction displacement is induced at B, segments AB and BC must have the same final slope at B and they must not separate in the X direction. No force is needed to displace segment BC in the positive direction of V_B on segment BC; therefore, all of the unit change in Y-direction displacement at B occurs on segment BC as shown in Figure 5.2c.

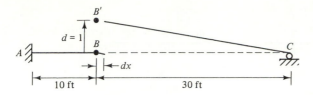

Figure 5.2c Qualitative IL for $V_{B(\text{right})}$.

3. Qualitative IL for M_D. Only the bending moment is destroyed at D; an internal hinge must be inserted at D to prevent separation from occurring in the X direction and Y direction at D. A unit angle change must be induced at D: $\alpha + \beta = 1$ rad; α and β must be treated as being small angles. The positive definition of M_D is counterclockwise on segment BD at D and clockwise on segment DC at D; therefore, α must be counterclockwise and β must be clockwise. See Figure 5.2d.

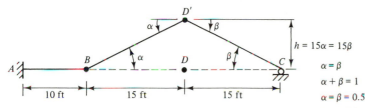

$h = 15\alpha = 15\beta$

$\alpha = \beta$

$\alpha + \beta = 1$

$\alpha = \beta = 0.5$

$h = 7.5$ ft **Figure 5.2d** Qualitative IL for M_D.

4. Qualitative IL for M_E. An internal hinge is inserted at point E to destroy the bending moment resistance at E and to prevent segments AE and EB from separating in the X and Y directions at E. A unit angle change must be induced at E. Since no moment is required to rotate segment EB, all of the unit angle change occurs on segment EB at E. The direction of the unit angle change must be clockwise since the positive direction of M_E on segment EB at E is clockwise. See Figure 5.2e.

The equations of statics can be used to obtain any ordinate of an IL for a determinate structure. Verify that $h = 5$ ft in Figure 5.2e by using the equations of statics. By summation of moments at B on FBD of BC in Figure 5.2f: $30 * C_Y = 0$; $C_Y = 0$.

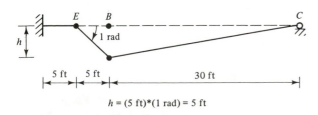

$h = (5 \text{ ft})*(1 \text{ rad}) = 5 \text{ ft}$

Figure 5.2e Qualitative IL for M_E.

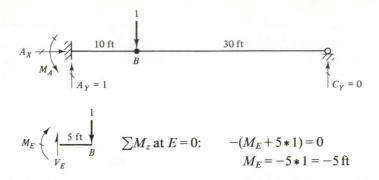

$$\sum M_z \text{ at } E = 0: \qquad -(M_E + 5*1) = 0$$
$$M_E = -5*1 = -5 \text{ ft}$$

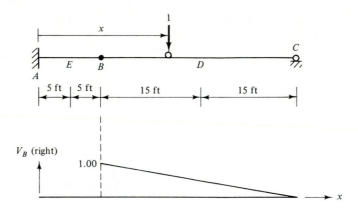

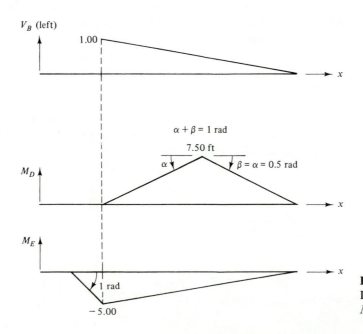

Figure 5.2f Influence Lines for $V_{B(\text{right})}$, $V_{B(\text{left})}$, M_D, M_E of Figure 5.2a.

EXAMPLE 5.3 —————

The platform on which the wheel rolls is the beam shown in Figure 5.3a. Draw the influence line for each of the following structural response functions:

1. Y-direction reactions.
2. Shear to the right of C.
3. Shear to the left of C.
4. Bending moment at C.
5. Shear at B.
6. Bending moment at B.

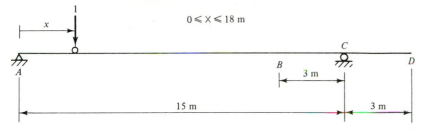

$0 \leqslant X \leqslant 18$ m

Figure 5.3a

Solution. Except for the last two IL in Figure 5.3b, the key ordinates were directly obtained from the Muller-Breslau principle. The shapes of the last two IL in Figure 5.3b were obtained by the Muller-Breslau principle and the values of the key ordinates were obtained by statics as shown below.

Verify by statics the ordinates at $x = 18$ m in Figure 5.3b (see Fig. 5.3c):

$$\sum M_Z = 0 \text{ at } A: \qquad 15 * C_Y - 18 * 1 = 0; \qquad C_Y = \tfrac{18}{15} = 1.20$$
$$\sum F_Y = 0: \qquad A_Y + C_Y - 1 = 0; \qquad A_Y = 1 - C_Y = 1 - 1.20 = -0.20$$

$$\sum F_Y = 0: \qquad V_{C(\text{right})} = 1$$
$$\sum M_Z = 0 \text{ at } C: \qquad M_C = -(3 \text{ m}) * 1 = -3 \text{ m}$$

$$\sum F_Y = 0: \qquad V_{C(\text{left})} = 1 - 1.20 = -0.2$$

$$\sum F_Y = 0: \qquad V_B = 1 - 1.20 = -0.20$$
$$\sum M_Z = 0 \text{ at } B: \qquad M_B = 3 * 1.20 - 6 * 1$$
$$M_B = 3.6 - 6 = -2.4 \text{ m}$$

Obtain the ordinates at $x = 12$ m in Figure 5.3b of IL for V_B and M_B (see Fig. 5.3d). Since there are two values of V_B at $x = 12$ m, the unit load must be positioned at (1) $x = 12$ m $- dx$, and (2) $x = 12$ m $+ dx$ (see Fig. 5.3e).

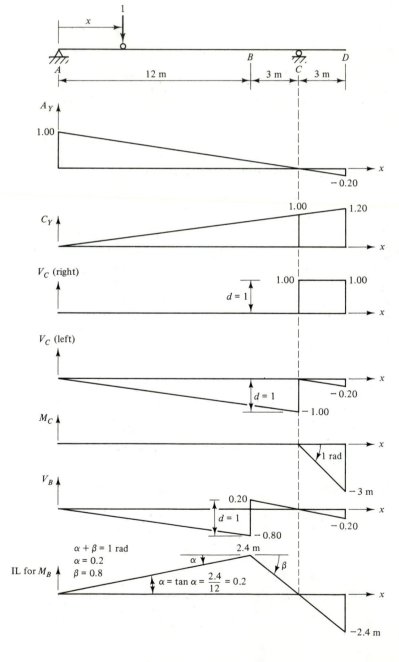

Figure 5.3b

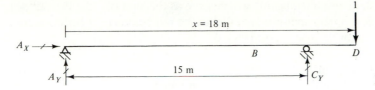

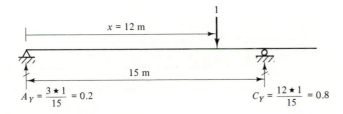

Figure 5.3c

Figure 5.3d

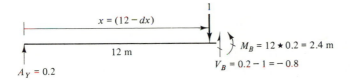

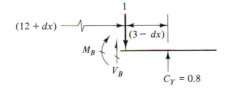

Figure 5.3e

$$\sum F_Y = 0: \qquad V_B = 1 - 0.8 = 0.2$$

Note: In Section 5.6 it will be shown that the equations of the influence lines are not needed. Therefore, the author does not show the equations of the influence lines hereafter.

5.4 EXAMPLES FOR FLOOR SYSTEMS

Sometimes the deck of a loading platform may consist of the following:

1. The deck on which a fork lift can roll.
2. Floor beams to support the deck.
3. Girders which serve as the supports for the floor beams.

Also, the structural framing scheme for an overhead crane may need to be as follows:

1. The rails are attached to the top flange of hinged ended rail beams spanning parallel to the rail direction.

2. Transverse beams which serve as the supports for the rail beams.
3. Girders spanning parallel to the rail beams (the girders serve as the supports for the transverse beams).

In obtaining influence lines for the structural response functions of the girders, as shown in Figure 5.4a the platform on which the unit load rolls is the set of hinged ended rail or deck beams. Therefore, the unit load does not roll directly on the girder for which the influence lines are desired. Such framing arrangements cause the unit load to be delivered to the girder at discretely located points along the girder. The influence lines for the girders in these framing schemes are quite different from those shown for the beams in the example problems of Section 5.3.

EXAMPLE 5.4 ——

In Figure 5.4a the unit load rolls on the top set of hinged ended members. Obtain the influence lines for the following structural response functions in the girder:

1. The shear in panel AB.
2. The shear in panel BC.
3. The shear in panel CD.
4. The moment at point B.
5. The moment at point D.

It should be noted that the IL for the reactions are identical to those in Figure 5.3b.

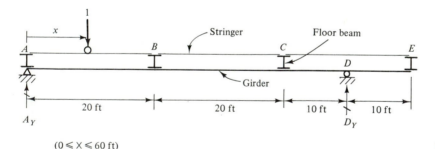

$(0 \leqslant X \leqslant 60 \text{ ft})$

Figure 5.4a

Solution. Since the stringers are hinged ended, for graphical convenience in the implementation of the Muller-Breslau principle the floor beams are shown as

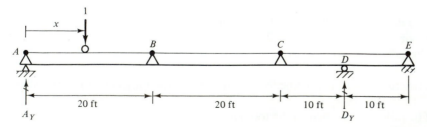

Figure 5.4b

hinges (Fig. 5.4b). First, we obtain the qualitative IL by the Muller-Breslau principle.

1. Qualitative IL for V_{AB}—(shear in panel A to B). Only the shear resistance is destroyed at any point between A and B; a unit change in Y-direction displacement is induced at the cut section where the shear resistance is destroyed. The deflection curve of the stringers (platform on which the unit load rolls) is the qualitative IL for V_{AB}. It should be noted that the cut section was chosen to be at $x = 17$ ft simply for graphical convenience; any point in the range of $0 < x < 20$ ft could have been chosen (Fig. 5.4c). To draw the quantitative IL for V_{AB}, we only need to compute h_B since line segments AB and BE are the IL for V_{AB}. To compute h_B by the equations of statics, position the unit load at B and compute the shear in panel AB as shown in Figure 5.4d. It should be noted that the stringers are not shown since the unit load was positioned on the hinge at B; therefore, the unit load is entirely transmitted through the hinge to the girder at point B.

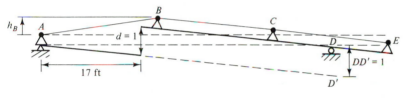

Figure 5.4c

Figure 5.4d

The following FBD (Fig. 5.4e) is taken from the preceding figure. Note that the value of a is not needed to find V on this FBD; that is, V is the same at all values of x in the range of $0 < x < 20$ on this FBD. However, to find M on this FBD a specific value of a must be known.

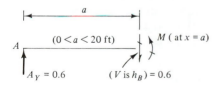

Figure 5.4e

Now, we have three known ordinate values on the IL for V_{AB}; namely, the zero values at points A and D as well as the 0.6 value at point B. This information enables us to plot Figure 5.4f and to compute any other ordinates on it by similar triangles.

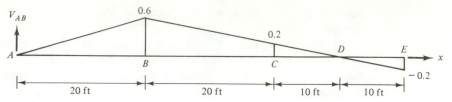

Figure 5.4f

Any ordinate on an IL for a determinate structure can be obtained by statics. It is of interest to verify that the ordinate at $x = 10$ ft is 0.3 on the IL for V_{AB}. As shown in Figure 5.4g, this requires that the reactions of stringer AB be determined, reversed, and applied to the girder.

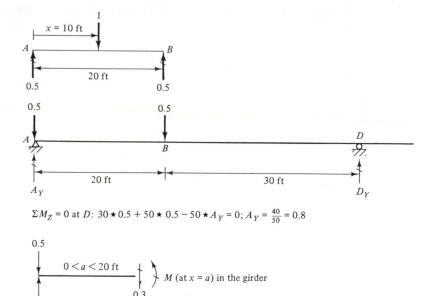

$\Sigma M_Z = 0$ at D: $30 \star 0.5 + 50 \star 0.5 - 50 \star A_Y = 0$; $A_Y = \frac{40}{50} = 0.8$

M (at $x = a$) in the girder

Figure 5.4g

2. Qualitative IL for V_{BC} (Figs. 5.4h and 5.4i). Since stringers AB and CE are parallel and have a slope of $-1/50 = -0.02$, we can use geometry instead of the equations of statics to find $h_B = 20 \ast (-0.02) = -0.4$ and the ordinates of IL for V_{BC} at C and E: $h_C = 10 \ast 0.02 = 0.2$; $h_E = -h_C = -0.2$.

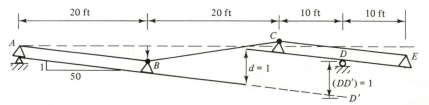

Figure 5.4h

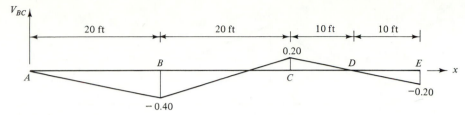

Figure 5.4i

3. Qualitative IL for V_{CD} (Figs. 5.4j and 5.4k). From geometry, $h_C = 40*(-0.02) = -0.8$ and $h_E = 10*(-0.02) = -0.1$ since both segments of the girder are parallel and their slope is -0.02.

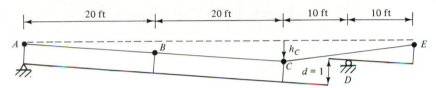

Figure 5.4j

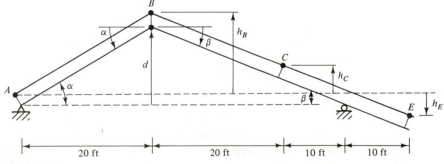

Figure 5.4k

4. Qualitative IL for M_B (Figs. 5.4l and 5.4m).

$$d = 20\alpha = 30\beta; \qquad \alpha = 1.5\beta$$
$$\alpha + \beta = 1; \qquad 2.5\beta = 1; \qquad \beta = 0.4$$
$$\alpha = 0.6$$

$$d = 20*0.6 = 12\,\text{ft}; \quad h_B = 20*0.6 = 12\,\text{ft}; \quad h_C = 10*0.4 = 4\,\text{ft}; \quad h_E = -10*0.4 = -4\,\text{ft}$$

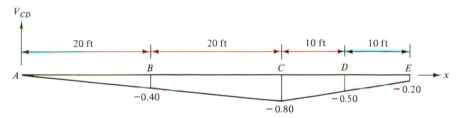

Figure 5.4l

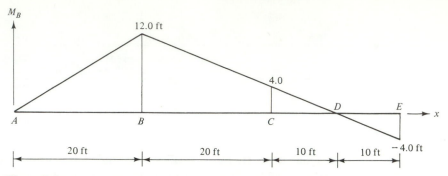

Figure 5.4m

5. Qualitative IL for M_D (Figs. 5.4n and 5.4o). A hinge is inserted at D in the girder and a unit angle change is induced. The downward displacement of the girder end at E is 10 ft $*$ (1 rad) = 10 ft. Stringer CE pivots about point C, but h_E must be identical to the downward displacement of the girder end at E; therefore, h_E = 10 ft.

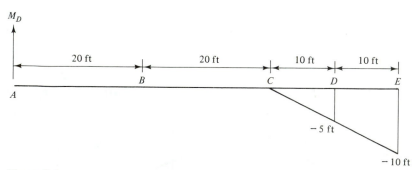

Figure 5.4n

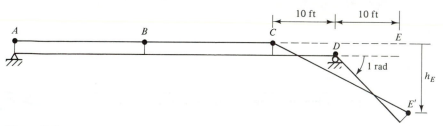

Figure 5.4o

EXAMPLE 5.5 ——————————————————————————

In Figure 5.5a the unit load rolls on the top set of cantilevered construction members. Obtain the influence lines for the following structural response functions in the girder: Shear in panel AB; shear in panel BE; M_B; and M_E.

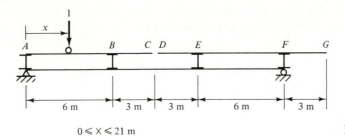

$0 \leqslant X \leqslant 21$ m

Figure 5.5a

Note that there is a gap between the ends of stringer AC and stringer DG (gap is between points C and D).

Solution

1. Qualitative IL for V_{AB} (Figs. 5.5b and 5.5c). The two girder segments are parallel and their slope is $-1/18$; also the slope of stringer DG is $-1/18$. Stringer DG and the right side segment of the girder pivot about point F; therefore, the Y-direction displacements of points B, D, E, and G are

$$h_B = 12*(1/18) = 2/3 = 0.667$$
$$h_D = 9*(1/18) = 0.5$$
$$h_E = 6*(1/18) = 1/3 = 0.333$$
$$h_G = -3*(1/18) = -1/6 = -0.167$$

The slope of stringer AC is $h_B/6 = (2/3)/6 = 1/9$; therefore, since stringer AC pivots about point A, the Y-direction displacement of point C is $h_C = 9*(1/9) = 1$.

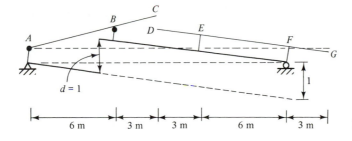

Figure 5.5b

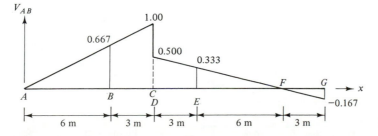

Figure 5.5c

2. Qualitative IL for V_{BE} (Figs. 5.5d and 5.5e). The clockwise angle of rotation for the girder segments is 1/18 m.

$$h_B = -6*(1/18) = -0.333$$
$$h_C = -9*(1/18) = -0.500$$

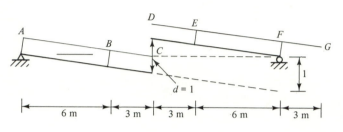

Figure 5.5d

$$h_D = 9*(1/18) = 0.500$$
$$h_E = 6*(1/18) = 0.333$$
$$h_G = -3*(1/18) = -0.167$$

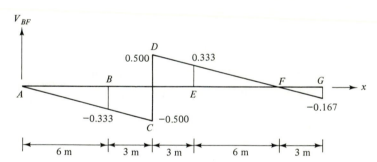

Figure 5.5e

3. Qualitative IL for M_B (Figs. 5.5f and 5.5g).

$$d = 6\alpha = 12\beta; \qquad \alpha = 2\beta$$
$$\alpha + \beta = 1; \qquad 3\beta = 1; \qquad \beta = 0.333$$
$$\alpha = 0.667$$
$$h_C = 9*0.667 = 6\,\text{m}$$
$$h_D = 9*0.333 = 3\,\text{m}$$

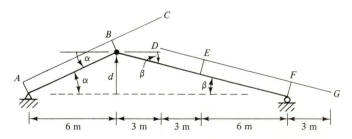

Figure 5.5f

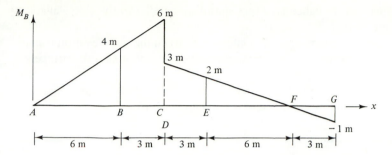

Figure 5.5g

4. Qualitative IL for M_E (Figs. 5.5h and 5.5i).

$$d = 12\alpha = 6\beta; \qquad \beta = 2\alpha$$
$$\alpha + \beta = 1; \qquad 3\alpha = 1; \qquad \alpha = 0.333$$
$$\beta = 0.667$$
$$h_C = 9*0.333 = 3\,\text{m}$$
$$h_D = 9*0.667 = 6\,\text{m}$$

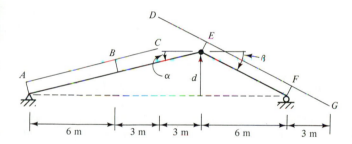

Figure 5.5h

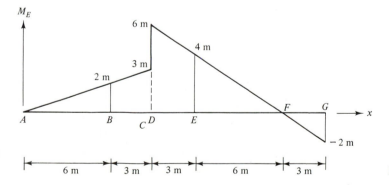

Figure 5.5i

5.5 EXAMPLES FOR TRUSSES

In a highway bridge truss, the traffic wheels roll on the reinforced concrete roadway deck slab. The floor framing beneath the concrete deck consists of:

1. Hinged ended beams called stringers spanning parallel to the side elevation trusses of the bridge.
2. Hinged ended floor beams spanning laterally between the side elevation trusses at the truss panel points (the floor beams serve as the supports for the stringers).

Thus, in obtaining the influence line for a truss member the platform on which the unit load rolls is the stringers in the plane of a side elevation truss.

EXAMPLE 5.6 ──

For the truss shown in Figure 5.6a, obtain the influence lines for the axial force in the members labeled 1, 2, and 3.

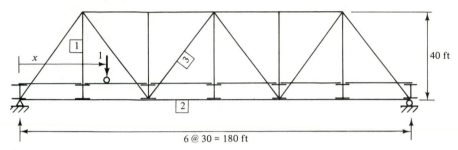

Figure 5.6a

Solution

1. IL for F_1 (axial tension force in member 1). By the Muller-Breslau principle, the capability of member 1 to resist an axial force is destroyed and a unit change in axial displacement is induced. Since all joints are pins, the platform can be displaced as shown in Figure 5.6b. Now that it has been established that the platform sits on the panel points of the bottom chords, the platform is omitted for graphical convenience in Figure 5.6b.

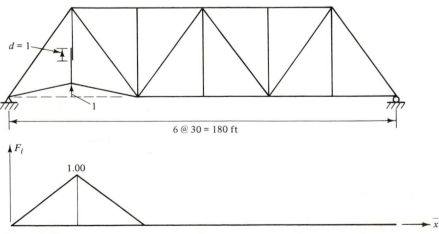

Figure 5.6b

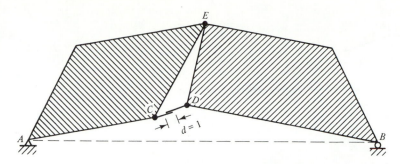

Figure 5.6c

2. IL for F_2 (Fig. 5.6c). With member 2 cut (to destroy the capability to resist axial force) in the implementation of the Muller-Breslau principle, truss portions ACE and BDE can be treated as rigid bodies (all of the induced displacement occurs in member 2). Point B must roll toward point A and the rigid bodies must rotate about points A and B; the hinged interconnection point E of the rigid bodies must rise when point D is displaced a unit amount in the direction of a tensile force acting on the cut section. Line segments AC, CD, and DB are the qualitative IL for F_2. The easiest procedure for obtaining the IL ordinates at C and D is by the equations of statics; that is, place the unit load at C (Fig. 5.6d) and find F_2; then place the unit load at D (Fig. 5.6e) and find F_2. Since only one member force is needed, the Method of Sections for truss analysis is employed in the equations of statics (Fig. 5.6d, and e, respectively).

$$\sum M_Z = 0 \text{ at } E: \qquad 90 * \tfrac{1}{3} - 40 * F_2 = 0; \qquad F_2 = \tfrac{30}{40} = 0.75 \text{ (IL ordinate at } C)$$
$$\sum M_Z = 0 \text{ at } E: \qquad 40 * F_2 - 90 * \tfrac{1}{2} = 0; \qquad F_2 = \tfrac{45}{40} = 1.125 \text{ (IL ordinate at } D)$$

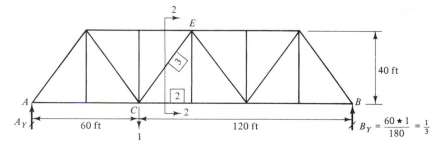

Figure 5.6d

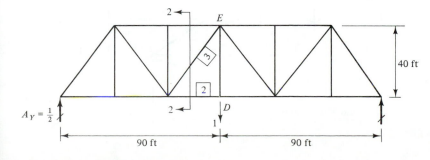

Figure 5.6e

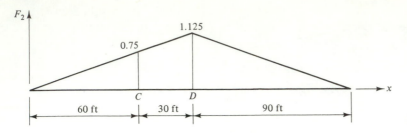

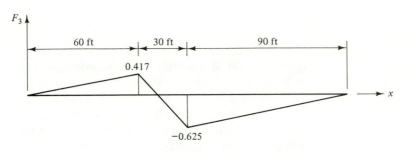

Figure 5.6f

Figure 5.6g Note: minus sign denotes compression.

3. IL for F_3. The FBD used to obtain F_2 are also used to obtain F_3. FBD section 2–2 right:

$$\sum F_Y = 0: \qquad \tfrac{1}{3} - \tfrac{4}{5} * F_3 = 0; \qquad F_3 = \tfrac{5}{12} = 0.417 \text{ (IL ordinate at } C)$$

FBD section 2–2 left:

$$\sum F_Y = 0: \qquad \tfrac{1}{2} + \tfrac{4}{5} * F_3 = 0; \qquad F_3 = -\tfrac{5}{8} = -0.625 \text{ (IL ordinate at } D)$$

Note that F_3 changes from tension to compression when the unit load travels on the platform from point C to point D (Fig. 5.6g).

EXAMPLE 5.7 ——

For the truss shown in Figure 5.7a, obtain the influence lines for the axial force in the members labeled 1, 2, and 3. The platform on which the unit load rolls sits on the panel points of the bottom chord of the truss.

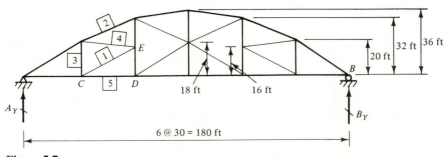

Figure 5.7a

Solution.   The Method of Sections and equations of statics approach are used to find the IL ordinates at C and D. Note that joint E is a K joint.

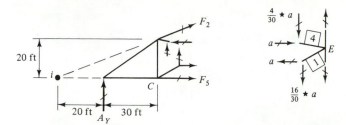

Figure 5.7b

1. IL for F_1. Cut a Y-direction section in panel CD. Locate the intersection of the lines of action of F_2 and F_5; name this point of intersection point i (Fig. 5.7b). When the unit load is placed at C in Figure 5.7a,

$$A_Y = \frac{50*1}{60} = \frac{5}{6}$$

Also, the unit load is located at C in Figure 5.7b:

$\sum M_Z = 0$ at i: $20*A_Y - 50*1 + 50*(\frac{16}{30}a) + 50*(\frac{4}{30}a) + 20*a = 0$

$A_Y = \frac{5}{6}$

$a = 1.247$

$F_1 = \frac{34}{30}*a = 1.413$ (IL ordinate at C)

When the unit load is placed at D in Figure 5.7a,

$$A_Y = \frac{40*1}{60} = \frac{2}{3}$$

In this case the unit load is not on Figure 5.7b:

$\sum M_Z = 0$ at i: $20*A_Y + 50*(\frac{16}{30} + \frac{4}{30})*a + 20*a = 0$

$A_Y = \frac{2}{3}$

$a = -0.499$

$F_1 = \frac{34}{30}*a = -0.565$ (IL ordinate at D)

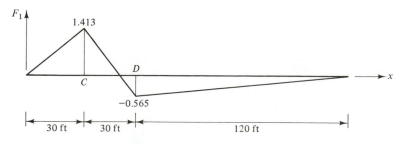

Figure 5.7c

2. IL for F_2 (Fig. 5.7d). The FBD (Fig. 5.7b) used to find F_1 can also be used to find F_2.

$$\sum M_Z = 0 \text{ at } C \text{ when the unit load is at } C$$
$$-(30*A_Y + 20*\tfrac{30}{34}*F_2) = 0$$
$$A_Y = \tfrac{5}{6}$$
$$F_2 = -1.417$$
$$\sum M_Z = 0 \text{ at } C \text{ when the unit load is at } D$$
$$-(30*A_Y + 20*\tfrac{30}{34}*F_2) = 0$$
$$A_Y = \tfrac{2}{3}$$
$$F_2 = -1.133$$

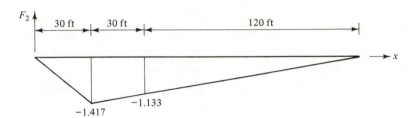

Figure 5.7d

3. IL for F_3 (Fig. 5.7e). FBD is the joint at point C. When the unit load is at point C,

$$\sum F_Y = 0: \qquad F_3 + \tfrac{16}{34}*1.413 - 1 = 0$$
$$F_3 = 0.335$$

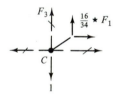

When the unit load is at point D,

$$\sum F_Y = 0: \qquad F_3 + \tfrac{16}{34}*(-0.565) = 0$$
$$F_3 = 0.266$$

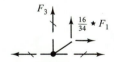

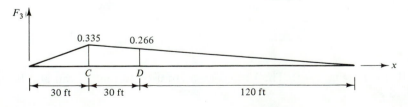

Figure 5.7e

5.6 USES OF INFLUENCE LINES

Consider Figure 5.8 which shows some influence lines for a structure.

If a single concentrated load of 60^k is placed at point D on the structure of Figure 5.8:

1. $A_Y = 60*(-0.20) = -12.0^k$ ($A_Y = 12.0^k$ acting down at A).
2. $C_Y = 60*1.20 = 72.0^k$ ($C_Y = 72.0^k$ acting upward at C).
3. $V_{C(\text{left})} = 60*(-0.20) = -12.0^k$. The 60^k load at D causes a negative shear to the left of point C.
4. $M_C = 60^k*(-10.0 \text{ ft}) = -600.0 \text{ ftk}$. The 60^k load at D causes a negative moment at point C.

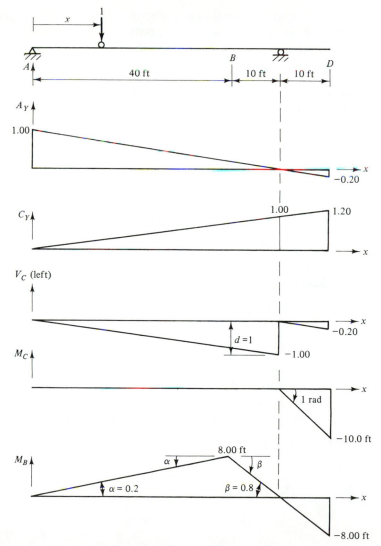

Figure 5.8 Influence lines.

5. $M_B = 60*(-8.00) = -480.0$ ftk. The 60^k load at D causes a negative moment at point C.

If the following set of concentrated loads (Fig. 5.9) is placed on the structure of Figure 5.8:

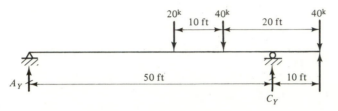

Figure 5.9

1. $C_Y = 40*1.20 + 40*(\frac{40}{50}*1.00) + 20*(\frac{30}{50}*1.00) = 92.0^k$.
2. $V_{C(\text{left})} = 40*(-0.20) + 40*(-1.00) + 20*(\frac{30}{50}*(-1.00)) = -60.0^k$.
3. $M_B = 40*(-8.00) + 40*(8.00) + 20*(\frac{30}{40}*8.00) = 120$ ftk.

That is, the value of a structural response function for a set of concentrated loads is obtained by superposition (summation of the individual load effects).

If a uniformly distributed load of 2 k/ft (Fig. 5.10) is placed on the structure of Figure 5.8:

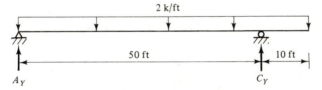

Figure 5.10

Every ordinate of each IL in Figure 5.8 is multiplied by 2 k/ft. Thus, the area under an influence line in Figure 5.8 multiplied by 2 k/ft is the value of the structural response function due to a distributed load of 2 k/ft. For example:

1. $A_Y = [\frac{1}{2}*50*1.00 + \frac{1}{2}*10*(-0.20)]*2$ k/ft $= 24$ ft$*2$ k/ft $= 48.0^k$.
2. $C_Y = [\frac{1}{2}*60*1.20]*2$ k/ft $= 36*2 = 72.0^k$.
3. $M_C = [\frac{1}{2}*10$ ft$*(-10.0$ ft$)]*2$ k/ft $= -50.0$ ft^{2*2} k/ft $= -100.0$ ftk.

Where should a uniformly distributed load of 2 k/ft be placed on the structure of Figure 5.8 to cause the maximum positive value of each structural response function? Examine each IL in Figure 5.8 to determine the region(s) where the ordinates are positive; place the 2 k/ft on the structure in that region (or regions). For example:

1. Maximum value for an upward A_Y is:

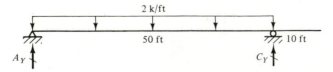

$A_Y = [\text{area from } A \text{ to } C \text{ under IL for } A_Y]*(2\,\text{k/ft}) = [\frac{1}{2}*50\,\text{ft}*1.00]*(2\,\text{k/ft}) = 50^k$

2. Maximum positive M_B due to 2 k/ft is:

$$M_B = [\text{area from } A \text{ to } C \text{ under IL for } M_B] * 2 \text{ k/ft}$$
$$M_B = [\tfrac{1}{2} * 50 \text{ ft} * 8.00 \text{ ft}] * 2 \text{ k/ft} = 400 \text{ ftk}$$

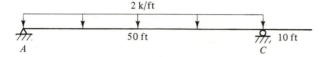

Where should the 2 k/ft be placed on the structure of Figure 5.8 to produce the minimum nonzero value of each structural response function? Place the 2 k/ft in the region(s) where the IL ordinates are negative. For example:

1. The minimum M_B due to 2 k/ft is:

$$M_B = [\text{area from } C \text{ to } D \text{ under IL for } M_B] * 2 \text{ k/ft}$$
$$M_B = [\tfrac{1}{2} * 10 * (-8.00)] * 2 \text{ k/ft} = -80.0 \text{ ftk}$$

2. The minimum $V_{C(\text{left})}$ due to 2 k/ft is:

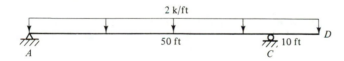

$$V_{C(\text{left})} = [\text{area from } A \text{ to } D \text{ under IL for } V_{C(\text{left})}] * 2 \text{ k/ft}$$
$$V_{C(\text{left})} = [\tfrac{1}{2} * 50 * (-1.00) + \tfrac{1}{2} * 10 * (-0.20)] * 2 \text{ k/ft} = -52.0^k$$

If there are two or more concentrated live loads, it is not always possible to tell by inspection which concentrated load should be placed at the maximum positive ordinate of the influence line in order to produce the maximum positive value of a structural response function. In general, the procedure is to try a position for the set of live loads and calculate the value of the structural response function. Then, the set of live loads is moved forward or/and backward to determine by trial the position of the loading configuration that produces the maximum value of a particular structural response function.

PROBLEMS

For Problems 5.1 through 5.14, draw qualitative ILs for the listed response functions. Obtain qualitative ordinates at the ends of each straight line segment on the IL. Answers not given for problems preceded by an asterisk.

5.1. IL for the reactions, shear at points 2, 3 (left), 3 (right), and the moment at points 2 and 3.

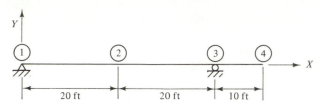

                                                                                                    Figure P5.1

* 5.2. IL for the reactions, shear at 2 (left), shear at 2 (right), and the moment at points 2 and 3.

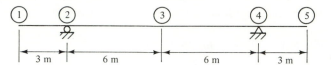

                                                                                                    *Figure P5.2

* 5.3. IL for the reactions (Y_1, Y_4, M_4), M_2, and the shear at 3 (left) and 3 (right).

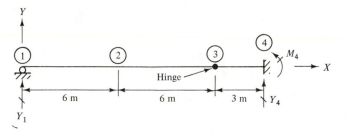

                                                                                                    *Figure P5.3

5.4. IL for the reactions, moment at points 3 and 4, and the shear at 3 (left) and 3 (right).

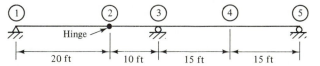

                                                                                                    Figure P5.4

* 5.5. IL for the reactions (M_1, Y_1, Y_3), M_3, and shear at 3 (left) and 2 (right).

                                                                                                    *Figure P5.5

* 5.6. IL for Y_1, Y_2, M_2, shear at 2 (left) and shear at 2 (right).

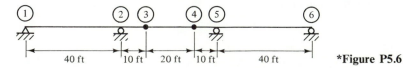

                                                                                                    *Figure P5.6

5.7. IL for the reactions, shear at 2 (right), and moment at 3. The platform is from point 1 to point 3 only.

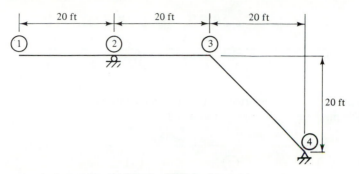

Figure P5.7

5.8. Platform is from point 1 to point 5. Find IL for A_Y, M_C, and shear in panel BC.

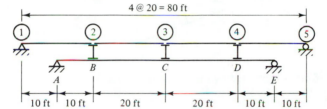

Figure P5.8

5.9. Platform is from point 1 to point 4. Find IL for A_Y, shear in panel AB, Y_4 and M_3.

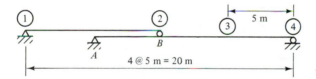

*Figure P5.9

5.10. Platform is from point 1 to point 5. Find IL for A_Y, M_B, and shear in panel AB.

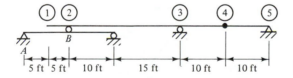

Figure P5.10

5.11. Platform is from point 1 to point 7; a gap in the platform occurs at point 3. Find the IL for the reactions (A_Y, B_Y).

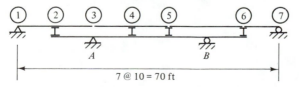

Figure P5.11

5.12. Platform is from point 1 to point 9; a gap in the platform occurs at point 7. Find the IL for A_Y, M_A, and the shear in panel FG.

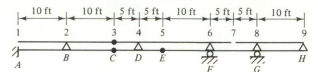

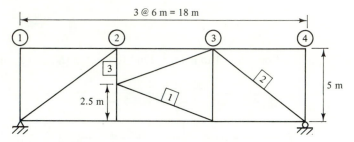

Figure P5.12

* 5.13. Platform is supported by the top panel points on the truss. Find the IL for the axial forces in the truss members 1 through 3.

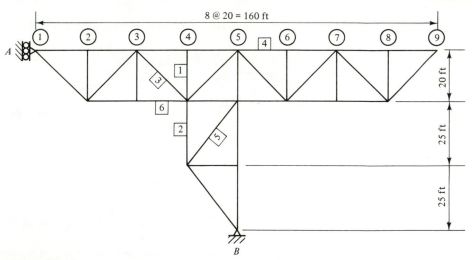

*Figure P5.13

* 5.14. Platform is supported by the top panel points on the truss. Find the IL for the reactions and the axial forces in members 1 through 6.

*Figure P5.14

Chapter 6

Displacements of Determinate Structures

6.1 INTRODUCTION

Displacements is a generalized term which includes all of the possible independent movement directions of a particular point named i in a structure. Since the purpose of this text is to progress gradually toward using computerized solutions of structures, as explained below it is necessary to choose the generalized term of displacements for the movements at the joints on the structure. Also, it is important to adopt a notation and sign convention for the joint displacements (and the member-end forces—defined in Chapter 8) which are applicable to all types of structures. Consequently, we must start our discussion by considering a three-dimensional structure which is the highest order type of structure. Planar structures are either components of or a degenerative case of a three-dimensional structure.

In this text, the highest order type of structure considered is a space frame (see Fig. 6.1) which is a three-dimensional structure composed of members and joints. A *member* is a portion of the structure for which the member-end force-deformation relations can be derived. In this chapter, for conceptual ease, a member can be conceived of as being a straight lined segment in a structure. A *space frame member* is capable of resisting an axially directed tensile or compressive force; shears in each of the two principal axis directions of the member's cross section; a torque; and bending moments about each of the two principal axes of the member's cross section. An *interior joint* is an interior point on the structure at which two or more members interconnect. An *exterior joint* is a point where one or more members cease to exist on the boundary of a structure. Since each interior joint must be examined as a free body diagram to ensure that the joint is in statical equilibrium, an interior joint is a point of intersecting members plus an infinitesimal portion of the intersecting member ends.

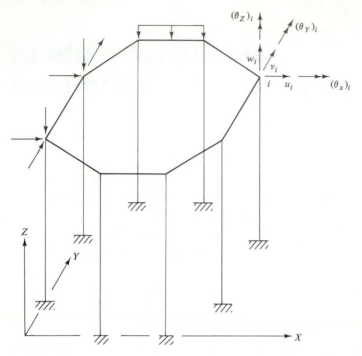

Figure 6.1 Space frame.

Prior to any application of loads (including the members' self weight), an interior joint of a space frame is located by means of X, Y, Z coordinates from the chosen origin of a set of right-handed, orthogonal, reference axes named X, Y, and Z (see Fig. 6.1). In this chapter, an interior joint of a space frame is considered to be infinitesimal and rigid (nondeformable), but capable of translating parallel to and rotating about each of the X, Y, Z reference axes. Therefore, at any interior joint i in a space frame there are six possible independent movement directions (kinematic degrees of freedom), namely, three translational displacements $(u, v, w)_i$ parallel to X, Y, Z, respectively, and three rotational displacements $(\theta_X, \theta_Y, \theta_Z)_i$. The preceding notation is convenient for a person to use whenever the person does not have to communicate with a computer. In a computerized solution of a space frame, all joint displacements at every interior joint must be stored in an array and the array can have only one name which the author chooses as D (for displacements). For a space frame having a total of n joint displacements, the computer array named D contains n entries (D_1 through D_n).

Consider a plane frame (see Fig. 6.2) lying in the X, Y plane. In this chapter, an interior joint of a plane frame is considered to be infinitesimal and rigid (nondeformable), but capable of translating parallel to X and Y and rotating about Z. Therefore, a typical interior joint i in a plane frame has three independent displacements $(u, v, \theta)_i$—two translations and one rotation. Since there is no possibility of ambiguity, for notational convenience the subscript Z is omitted from the rotational displacement symbol θ. A *plane frame member* is capable of resisting an axially directed tensile or compressive force; a shear in the X, Y plane; and, a bending moment about the principal axis of the member's cross section that is parallel to Z.

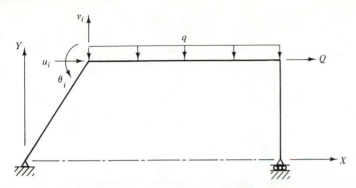

Figure 6.2 Plane frame lying in XY plane.

Trusses are structures having all pinned ended members. A truss member is capable of resisting only an axially directed tensile or compressive force. Therefore, an interior joint i in a space truss (see Fig. 6.3) is a pin that does not resist any rotation about the X, Y, Z axes and only has three independent displacements $(u, v, w)_i$. For a plane truss lying in the X, Y plane (see Fig. 6.4), an interior joint i is a pin that does not resist any rotation about the Z axis and has only two independent displacements $(u, v)_i$.

Displacements must be considered in the design of almost every structure. In the interest of minimizing the dead weight of highrise structures, structural designers

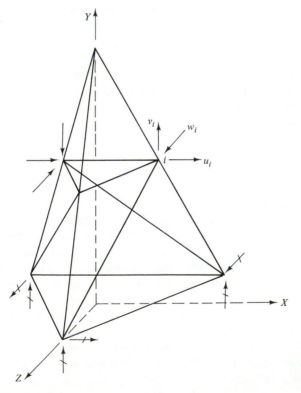

Figure 6.3 Space truss. Note: Vectors with a slash on them are reactions.

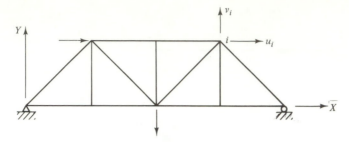

Figure 6.4 Plane truss lying in XY plane.

are using higher and higher grades of concrete and steel for the structural members wherever it is economically and structurally possible. Consider, for example, a prismatic steel member which is 20 ft long. The member weight is a function of only one variable, the cross-sectional area. As the member weight decreases, the cross-sectional area and, usually, the moments of inertia decrease, and the member becomes more flexible and permits larger displacements to occur. Consequently, controlling displacements becomes more of a problem when the dead weight is minimized.

It is common practice for a structural designer to put absolute limits on the displacements of a structure and on the displacements of an individual member. If a highrise structure sways too much or too rapidly, the occupants become nauseated or frightened at best although no structural damage may occur. Similarly, the public becomes alarmed if a floor system of a building or a bridge is too flexible and noticeably sags more than a tolerable amount. Also there are situations, such as a beam over a plate glass window or a water pipe, where excessive displacements can cause considerable damage if they are not controlled by the structural designer. If the beams in a flat roof sag too much, water ponds on the roof, causing additional sagging and more ponding which can rupture the roof and cause extensive water damage to the contents of the structure. Consequently, it is not unusual for displacements to be the controlling factor in the design of a structure or a member in the structure.

Construction can only be done within tolerable limits—for example, columns cannot be perfectly plumbed and foundations cannot be placed perfectly in plan view or in elevation view. During the construction of a structure displacements must be controlled to ensure the safety of the structure, the construction workers, and the public. Total collapses of partially constructed structures have occurred because the structures were not adequately braced to limit displacements during construction.

The St. Louis arch memorial was constructed by cantilevering independently from the two foundations toward the crown of the arch. The last prefabricated segment of the arch was inserted between the two independently erected cantilevered parts of the arch. Since the arch lies in the north–south plane, the top of the southern half of the arch was directly exposed to the sun at high noon, whereas the top of the northern half of the arch was shaded (due to the curvature of the arch). At high noon the differential displacements at the tips of the two independently erected cantilevers were approaching their maximum amounts. The contractor preferred to insert the last segment at the crown shortly after daybreak when the temperatures of the two cantilevers would be nearly the same. However, for maximum public relations purposes, public officials decided to have the last segment of the arch inserted at high noon and arranged for the St. Louis Fire Department to hose down the southern half of the arch with water to minimize the differential displacements caused at the cantilevered tips by the two different sun exposures.

Unless handbook solutions are available, displacements must be computed in the analysis of every indeterminate structure since the equations of statics alone do not provide sufficient conditions to render a solution for the member-end forces needed to check the design of a structure. If a structural designer can adequately predict the deformed behavior of an indeterminate structure to be designed, points of inflection on the deflected shape can be approximately located and other displacement conditions can be used to develop approximate methods of analysis needed in the preliminary design stage. Reasonably good estimates of the member sizes are needed as input to a computerized solution of an indeterminate structure, otherwise the cost of an iterated computerized solution may be too expensive.

The deflected shape of a structure due to service conditions can sometimes be economically controlled by precambering the structure. Precambering is achieved by erecting the structure with built-in deformations such that the structure deflects to or slightly below its theoretical no-load shape when the worst design loads occur on the structure. For example, the bottom chords of a simply supported plane truss can be deliberately fabricated too short. When the truss is assembled, the interior truss joints displace upwardly. When the service loads are applied, the interior truss joints displace downwardly to or slightly below the no-load positions.

Current structural design practice is to design the structural members to have adequate strength to resist axial force, shear, and bending moment. Then the structure is checked for displacements to ensure serviceability. Some of the common service-ability problems are [6.1]:

1. Local damage of nonstructural elements (for example, windows, ceilings, partitions, walls) occurs due to displacements caused by loads, temperature changes, moisture, shrinkage, and creep.
2. Equipment (for example, elevators) does not function normally due to excessive displacements.
3. Displacements are so noticeable that occupants become alarmed.
4. Extensive nonstructural damage occurs due to a tornado or a hurricane.
5. Structural deterioration occurs due to age and usage (for example, deterioration of bridges and parking decks due to deicing salt).
6. Motion sickness of the occupants occurs due to excessive floor vibrations caused by routine occupant activities or lateral vibrations due to the effects of wind or an earthquake.

These serviceability problems can be categorized as a function of either the gravity direction displacement or the lateral displacement. Let L = span length of a floor or roof member, h = story height, and H = building height. The units of the displacement are identical to the units of L, h, and H.

Displacement index	Typical serviceability behavior
$h/1000$	Not visible cracking of brickwork
$h/500$	Not visible cracking of partition walls
$h/300$ or $L/300$	Visible: Architectural damage Cracks in reinforced walls Ceiling and floor damage Leaks in the structural facade
$L/200$ to $L/300$ ⎫ $h/200$ to $h/300$ ⎭	Visible: Cracks are visually annoying Damage to partitions and large, plate glass windows
$L/100$ to $L/200$ ⎫ $h/100$ to $h/200$ ⎭	Visible: Damage to structural finishes Doors, windows, sliding partitions, and elevators do not function properly

It is customary steel design practice to limit the Deflection Index to:

1. $L/360$ due to live load on a floor or snow load on a roof when the beam supports a plastered ceiling.
2. $L/240$ due to live load or snow load if the ceiling is not plastered.
3. $h/667$ to $h/200$ for each story due to the effects of wind or earthquakes—only a range of limiting values can be given for many reasons (type of facade, activity of the occupants, routine design, innovative design, structural designer's judgment and experience).
4. $H/715$ to $H/250$ for entire building height H due to the effects of wind or earthquakes—comment in item 3 applies here too.

It is important to note that the Deflection Index limits for lateral displacements are about the same as the accuracy that can be achieved in the erection of the structure. Also, it is important to note that the largest tolerable translational displacement due to live load is 0.5% of the member length. Consequently, the displacements are grossly exaggerated for clarity on the sketches of the deflected structure in this text.

Numerous techniques have been developed for calculating displacements. The author chooses to restrict the discussion in this text to small displacement theory and linearly elastic structural behavior. Also, the author chooses to emphasize those techniques that he considers the most useful: (1) Conjugate Beam Method for beams; (2) Virtual Work Method for beams, frames, and trusses. In order to adequately explain the Conjugate Beam Method, the author chooses first to briefly discuss the Integration Method which is also useful to obtain displacements for curved beams and arches. In order to present the Integration Method in the best manner, the Moment–Curvature Relation which is needed in the Integration Method is derived as a separate topic.

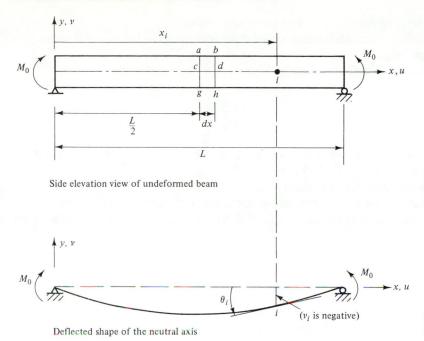

Side elevation view of undeformed beam

Deflected shape of the neutral axis

$(v_i$ is negative)

Figure 6.5 Prismatic beam subjected to pure bending.

6.2 MOMENT–CURVATURE RELATION

Since live load displacements are the ones of primary concern to the designer, the dead weight of the member is ignored in the following derivation. The beam shown in Figure 6.5 is subjected to pure bending about the z axis (not shown). Axes y and z are the principal axes of the member's cross section and the x axis is the neutral axis of the member. Prior to the application of the member-end moments, any point i on the neutral axis is located by a known value of x. After the member-end moments are applied, the neutral axis is an arc of a circle and the displacements of point i are $(u,v,\theta)_i$ which are positive in the x,y,z directions, respectively. For a beam, the u displacement is negligible.

The following definitions and sign conventions are used in deriving the Moment–Curvature Relation:

1. A positive moment causes compression in the positive y region of a differential element:

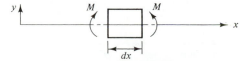

2. σ and ϵ, respectively, denote stress and strain.
3. Tensile stresses and strains are positive (compressive stresses and strains are negative).
4. Slope of the deflection curve at i is

$$\theta_i = \left(\frac{dv}{dx}\right)_i$$

(a counterclockwise θ is positive). The slope is so small that $\theta = \sin \theta = \tan \theta$; that is, $\theta \leq 0.0963$ rad.
5. E is the modulus of elasticity.
6. I is the moment of inertia about the z axis of the cross section.
7. ρ is the radius of curvature of a differential element and is derived in most elementary calculus texts:

$$\rho = \frac{[1 + (dv/dx)^2]^{3/2}}{d^2v/dx^2}$$

This fundamental differential equation of bending is commonly known as the Bernoulli-Euler beam theory equation [6.2]. For the pure bending case shown in Figure 6.5, ρ is the same for all differential elements along the beam since the neutral axis bends into an arc of a circle.
8. Curvature, ϕ, is the reciprocal of the radius of curvature. Since dv/dx must be less than 0.0963 for small slope theory to be applicable, $(dv/dx)^2$ is negligible and $\phi \cong d^2v/dx^2$.

Since all differential elements in Figure 6.5 have the same radius of curvature, consider the differential element that begins at midspan of the beam. An enlarged sketch of this differential element in the deformed state is shown in Figure 6.6.

$\bar{y}$ is an absolute value of y.

y values are negative below line cd.

Elongations and tensile stresses have a positive sign.

$$ff' = \bar{y} * d\theta$$

$$\text{strain} = \epsilon = \frac{ff'}{cd} = \frac{\bar{y} * d\theta}{\rho * d\theta} = \frac{-y}{\rho}$$

$$\epsilon = \frac{-y}{\rho} = \frac{\sigma}{E}$$

$$\frac{1}{\rho} = \frac{-\sigma}{Ey} \quad \text{or} \quad \sigma = \frac{-Ey}{\rho}$$

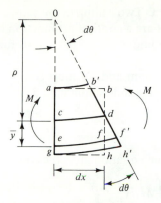

(a) Deformed differential element

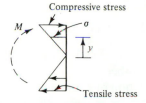

(b) Bending stress distribution   **Figure 6.6**

M is positive in the direction shown. σ is negative in the region where y is positive. σ is positive in the region where y is negative. Consequently, the product of y and σ is always negative.

$$M = \int_A -y * \sigma * dA = \frac{E}{\rho} \int_A y^2 \, dA = \frac{EI}{\rho}$$

$$\frac{1}{\rho} = \frac{M}{EI} = \phi = \frac{d^2v}{dx^2} = \frac{d\theta}{dx}$$

$$d\theta = \frac{M * dx}{EI}$$

$$\boxed{\text{The Moment–Curvature Relation is } \phi = \frac{d^2v}{dx^2} = \frac{M}{EI}}$$

6.3 INTEGRATION METHOD

The Moment–Curvature Relation, $\phi = d^2v/dx^2 = M/EI$, can be used to obtain the equation of the y-direction deflection, v, due to flexural deformations only by integrating the M/EI equation which is a function of x. Two examples are shown to illustrate the method: (1) The loading and the moment are a continuous function of x; and (2) the loading is a single concentrated load (loading is not a continuous function of x and two moment expressions must be used).

For a prismatic member, EI is a constant and the most convenient form of the Moment–Curvature Relation is the following differential equation:

$$EI\frac{d^2v}{dx^2} = M$$

The first integration of this differential equation gives the equation for the slope:

$$EI\frac{dv}{dx} = \int M\,dx + C_1$$

and the second integration gives the deflection

$$EIv = \int\left(\int M\,dx\right)dx + C_1 x + C_2$$

where C_1 and C_2 are constants of integration and they must be determined from the boundary conditions of deflection for the member.

If the loading is not one continuous function of x, more than one moment equation must be obtained and integrated to obtain the equations for the deflection curve.

EXAMPLE 6.1 ──

Use the Integration Method to obtain the equation of the deflection curve for the prismatic beam shown in Figure 6.7a.

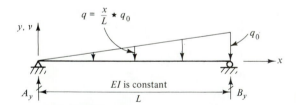

Figure 6.7a

Solution. First, we need to find A_y in order to obtain the moment equation by statics at an arbitrary value of x (Fig. 6.7b).

$$Q = \frac{1}{2}*x*q = \frac{q_0 x^2}{2L} \quad \text{and} \quad A_y = \frac{q_0 L}{6}$$

$$M = x*A_y - \frac{x}{3}*Q = \frac{q_0}{6L}(xL^2 - x^3)$$

$$EI\frac{d^2v}{dx^2} = \left[M = \frac{q_0}{6L}(xL^2 - x^3)\right]$$

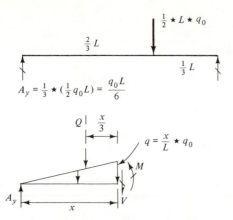

Figure 6.7b

$$EI\frac{dv}{dx} = \frac{q_0}{6L}\int (xL^2 - x^3)\,dx + C_1 = \frac{q_0}{6L}\left(\frac{x^2L^2}{2} - \frac{x^4}{4}\right) + C_1$$

$$EIv = \frac{q_0}{6L}\int \left(\frac{x^2L^2}{2} - \frac{x^4}{4}\right)dx + C_1\int dx + C_2$$

$$EIv = \frac{q_0}{6L}\left(\frac{x^3L^2}{6} - \frac{x^5}{20}\right) + C_1 x + C_2$$

The deflection boundary conditions are:

1. $v = 0$ at $x = 0$.
2. $v = 0$ at $x = L$.

Satisfaction of the first boundary condition requires

$$0 = 0 + 0 + C_2 \quad \text{which gives} \quad C_2 = 0$$

Satisfaction of the second boundary condition requires

$$0 = \frac{q_0}{6L}\left(\frac{L^5}{6} - \frac{L^5}{20}\right) + C_1 L \quad \text{which gives} \quad C_1 = -\frac{7}{360}q_0 L^3$$

Therefore, the deflection and slope equations are

$$v = \frac{-q_0}{360EIL}(3x^5 - 10L^2x^3 + 7L^4x)$$

$$\theta = \frac{dv}{dx} = \frac{-q_0}{360EIL}(15x^4 - 30L^2x^2 + 7L^4)$$

$$\theta_A = \frac{-7q_0L^3}{360EI} \qquad \theta_B = \frac{q_0L^3}{45EI}$$

The deflection having the maximum magnitude occurs where the slope is zero (Fig. 6.7c):

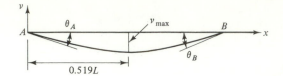

Figure 6.7c

$$15x^4 - 30L^2x^2 + 7L^4 = 0 \quad \text{at} \quad x = 0.519L \text{ (by trial and error)}$$

$$v_{\text{max}} = \frac{-0.00652q_0L^4}{EI}$$

EXAMPLE 6.2 _____

Use the Integration Method to obtain the equation of the deflection curve for the prismatic beam shown in Figure 6.8.

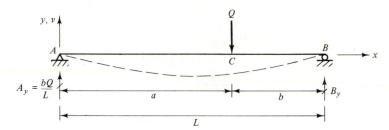

Figure 6.8 *EI* is constant.

Solution. For beam segment AC, $0 \le x \le a$:

$$M = x \star \frac{Qb}{L} = \frac{Qbx}{L}$$

$$EI\frac{dv}{dx} = \frac{Qb}{L}\int x\,dx + C_1 = \frac{Qbx^2}{2L} + C_1 \tag{6.1}$$

$$EIv = \frac{Qb}{2L}\int x^2\,dx + C_1\int dx + C_2 = \frac{Qbx^3}{6L} + C_1x + C_2 \tag{6.2}$$

For beam segment CB, $a \le x \le L$:

$$M = \frac{Qbx}{L} - Q(x - a)$$

$$EI\frac{dv}{dx} = \frac{Qb}{L}\int x\,dx - Q\int (x-a)\,dx + C_3 = \frac{Qbx^2}{2L} - Q\left(\frac{x^2}{2} - ax\right) + C_3 \tag{6.3}$$

$$EIv = \frac{Qb}{2L}\int x^2\,dx - Q\int \left(\frac{x^2}{2} - ax\right)dx + C_3\int dx + C_4$$

$$EIv = \frac{Qb}{2L}\left(\frac{x^3}{3}\right) - Q\left(\frac{x^3}{6} - \frac{ax^2}{2}\right) + C_3 x + C_4 \tag{6.4}$$

The deflection boundary conditions are:

1. $v = 0$ at $x = 0$. Substitution of zero for v and x in Eq. 6.2 gives $0 = 0 + 0 + C_2$. Therefore,

$$C_2 = 0 \tag{6.5}$$

2. $v = 0$ at $x = L$. Substitution of zero for v and L for x in Eq. 6.4 gives

$$0 = \frac{QbL^2}{6} - Q\left(\frac{L^3}{6} - \frac{aL^2}{2}\right) + C_3 L + C_4 \tag{6.6}$$

The compatibility conditions of the beam segments at point C are:

1. $v_{CA} = v_{CB}$ at $x = a$ (deflection of segments CA and CB is the same at point C). Substitution of $x = a$ into Eq. 6.2 and 6.4 and equating the resulting expressions gives

$$\frac{Qba^3}{6L} + C_1 a = \frac{Qba^3}{6L} - Q\left(\frac{a^3}{6} - \frac{a^3}{2}\right) + C_3 a + C_4 \tag{6.7}$$

2. $\theta_{CA} = \theta_{CB}$ at $x = a$ (slope of segments CA and CB is the same at point C). Satisfaction of this condition requires

$$\frac{Qba^2}{2L} + C_1 = \frac{Qba^2}{2L} - Q\left(\frac{a^2}{2} - a^2\right) + C_3$$

which reduces to

$$C_1 - C_3 = \frac{Qa^2}{2} \tag{6.8}$$

Substitution of $Qa^2/2$ for $C_1 - C_3$ in Eq. 6.7 gives

$$C_4 = \frac{Qa^3}{6} \tag{6.9}$$

Substitution of $Qa^3/6$ for C_4 in Eq. 6.6 gives

$$C_3 = \frac{Q}{6L}[L^4 - (3a + b)L^3 - a^3] \tag{6.10}$$

Substitution of C_3 into Eq. 6.8 gives

$$C_1 = \frac{Q}{6L}[L^4 - (3a + b)L^3 + 3a^2 - a^3] \tag{6.11}$$

Since $L = a + b$ and $a = L - b$ can be substituted into Eq. 6.9 through 6.11, the equations for deflection and slope are:

$0 \le x \le a$:

$$v = \frac{-Qbx}{6EIL}(L^2 - b^2 - x^2)$$

$$\theta = \frac{dv}{dx} = \frac{-Qb}{6EIL}(L^2 - b^2 - 3x^2)$$

$a \le x \le L$:

$$v = \frac{-Qa(L-x)}{6EIL}[2bL - b^2 - (L-x)^2]$$

$$\theta = \frac{dv}{dx} = \frac{-Qa}{6EIL}[b^2 - 2bL + 3(L-x)^2]$$

The end slopes are

$$\theta_A = \frac{-Qb}{6EIL}(L^2 - b^2)$$

$$\theta_B = \frac{Qab(L+a)}{6EIL}$$

and the worst deflection is at the point of zero slope:

$$x = \sqrt{\frac{a(a+2b)}{3}} \quad \text{when} \quad a > b$$

$$v_{max} = \frac{Qab(L+b)\sqrt{3a(L+b)}}{27EIL} \quad \text{when} \quad a > b$$

and the deflection at the concentrated load point is

$$v_C = \frac{Qa^2b^2}{3EIL}$$

6.4 MOMENT DIAGRAM BY PARTS

In Sections 6.5 and 6.6, the reader will be shown how to obtain the y-direction displacement at any point i on a beam by the Conjugate Beam Method and the Virtual Work Method. In each of these methods, the area under the M/EI diagram between two points is needed, EI is constant between the two points, and the centroid of the area must be determined.

Since the shear diagram is not needed, the approach used in Chapter 4 to obtain the Moment Diagram is not the best approach to use in obtaining only the Moment Diagram for purposes of the Conjugate Beam Method and Virtual Work Method. The Moment Diagram can be obtained directly by superposition using whichever one of the two following approaches is most convenient in any given problem:

1. Superposition of Moment Diagrams for simply supported beams.
2. Superposition of Moment Diagrams for cantilevered beams.

In both approaches, the Moment Diagrams to be superimposed have one of the following shapes: (1) rectangles; (2) triangles; and (3) areas that can be related to the general equation $y = ax^n$ where n is an integer and $n \geq 2$.

1. For a rectangle:

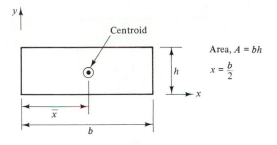

Area, $A = bh$

$x = \dfrac{b}{2}$

2. For a triangle:

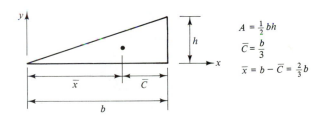

$A = \frac{1}{2}bh$

$\bar{C} = \dfrac{b}{3}$

$\bar{x} = b - \bar{C} = \frac{2}{3}b$

3. For $y = ax^n$, where n is an integer and $n \geq 2$, it is easily shown by calculus (see Appendix A) that:

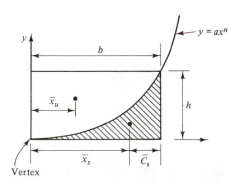

For the shaded area:

$$A_s = \frac{bh}{n+1}$$

$$\bar{C}_s = \frac{b}{n+2}$$

$$\bar{x}_s = b - \bar{C}_s = \frac{n+1}{n+2} * b$$

For the unshaded area:

$$A_u = bh - A_s = \frac{n}{n+1} * bh$$

$$\bar{x}_u = \frac{1}{2}\bar{x}_s = \frac{n+1}{n+2} * \frac{b}{2}$$

Let A_i denote an area under the moment curve and $\bar{x}_i$ denote the location of the centroid for A_i. Examples of the Moment Diagram, its area(s), and centroid(s) for individual loads are:

1. For a simply supported beam:

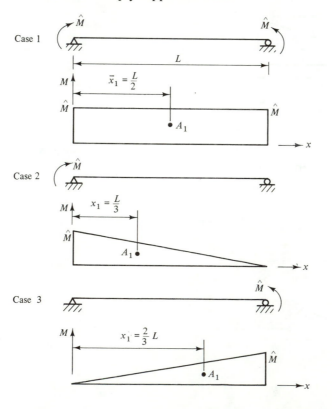

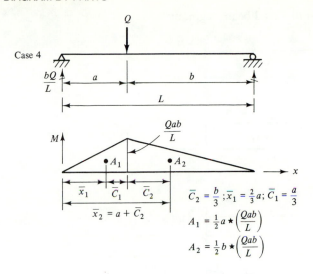

Case 4

$$\bar{C}_2 = \frac{b}{3} \;;\bar{x}_1 = \frac{2}{3}a; \bar{C}_1 = \frac{a}{3}$$

$$A_1 = \frac{1}{2}a \star \left(\frac{Qab}{L}\right)$$

$$A_2 = \frac{1}{2}b \star \left(\frac{Qab}{L}\right)$$

$$\bar{x}_2 = a + \bar{C}_2$$

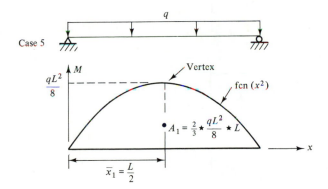

Case 5

Sometimes only half of the area and the centroid for it are needed.

$$\bar{C}_1 = \bar{C}_2 = \frac{3}{8} \star \frac{L}{2}; \qquad \bar{x}_1 = \frac{5}{8} \star \frac{L}{2}; \qquad A_1 = A_2 = \frac{2}{3} \star \frac{qL^2}{8} \star \frac{L}{2}$$

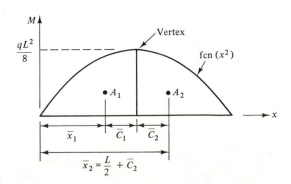

2. For a cantilevered beam segment:

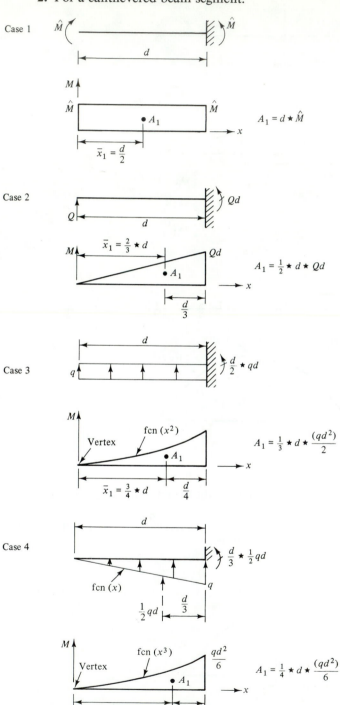

Case 1

$A_1 = d \star \hat{M}$

$\bar{x}_1 = \dfrac{d}{2}$

Case 2

$A_1 = \dfrac{1}{2} \star d \star Qd$

$\bar{x}_1 = \dfrac{2}{3} \star d$

Case 3

$A_1 = \dfrac{1}{3} \star d \star \dfrac{(qd^2)}{2}$

$\bar{x}_1 = \dfrac{3}{4} \star d$

Case 4

$A_1 = \dfrac{1}{4} \star d \star \dfrac{(qd^2)}{6}$

$\bar{x}_1 = \dfrac{4}{5} \star d$

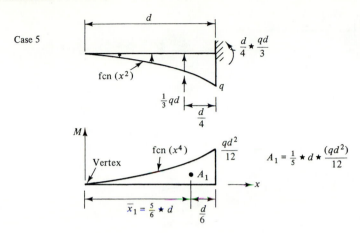

Case 5

$$\frac{d}{4} \star \frac{qd}{3}$$

fcn (x^2)

$$\frac{1}{3}qd \qquad \frac{d}{4}$$

M

Vertex

fcn (x^4)

$$\frac{qd^2}{12} \qquad A_1 = \frac{1}{5} \star d \star \frac{(qd^2)}{12}$$

$\bullet A_1$

$\longrightarrow x$

$$\bar{x}_1 = \frac{5}{6} \star d \qquad \frac{d}{6}$$

Case 6

$a \qquad b$

$$\left(b + \frac{a}{2}\right) \star (qa)$$

q

q

$$\frac{(a+b)}{2} \star [q(a+b)] - \frac{b}{2} \star (qb)$$

q

M

fcn (x^2)

$$\frac{(a+b)}{4} \qquad \frac{q(a+b)^2}{2}$$

$\bullet A_1$

$\bullet A_2$

$\longrightarrow x$

fcn (x^2)

$$\frac{-qb^2}{2}$$

$$A_1 = \frac{1}{3} \star (a+b) \star \frac{q(a+b)^2}{2} \qquad \frac{b}{4}$$

$$A_2 = -\frac{1}{3} \star b \star \frac{qb^2}{2}$$

For case 7
$$
\begin{cases}
M_R = \dfrac{(a+b)^3 q}{6a} - \dfrac{qb^2}{2} - \dfrac{qb^3}{6a} = \dfrac{qa(a+3b)}{6} \\[3mm]
A_1 = \dfrac{1}{4}(a+b) * \dfrac{(a+b)^3 q}{6a} = \dfrac{(a+b)^4 q}{24a} \\[3mm]
A_2 = -\dfrac{1}{3} b * \dfrac{(b^2 q)}{2} = -\dfrac{qb^3}{6} \\[3mm]
A_3 = -\dfrac{1}{4} b * \left(\dfrac{b^3 q}{6a}\right) = -\dfrac{qb^4}{24a}
\end{cases}
$$

Case 7

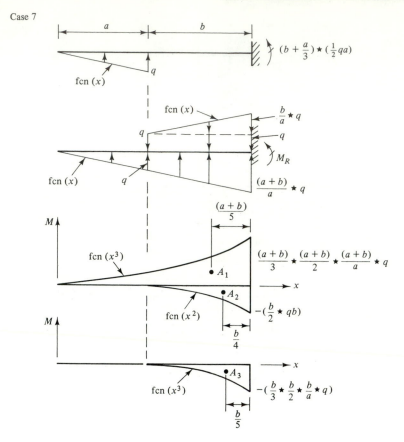

$(b + \frac{a}{3}) \star (\frac{1}{2}qa)$

fcn (x)

fcn (x)

$\frac{b}{a} \star q$

fcn (x)

M_R

fcn (x)

$\frac{(a+b)}{a} \star q$

$\frac{(a+b)}{5}$

fcn (x^3)

$\frac{(a+b)}{3} \star \frac{(a+b)}{2} \star \frac{(a+b)}{a} \star q$

A_1

fcn (x^2)

A_2

$-(\frac{b}{2} \star qb)$

$\frac{b}{4}$

fcn (x^3)

A_3

$-(\frac{b}{3} \star \frac{b}{2} \star \frac{b}{a} \star q)$

$\frac{b}{5}$

EXAMPLE 6.3 ───

For the structure shown in Figure 6.9a, obtain the Moment Diagram by parts using superposition of moment diagrams for a simply supported beam. From an indeterminate analysis for EI = constant:

$$M_A = 1201 \text{ ftk}; \qquad A_Y = 110.02^k \text{ (say } 110^k); \qquad B_Y = 49.98^k \text{ (say } 50^k)$$

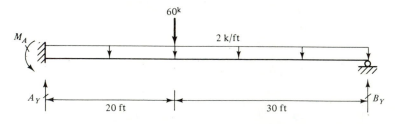

Figure 6.9a

Solution

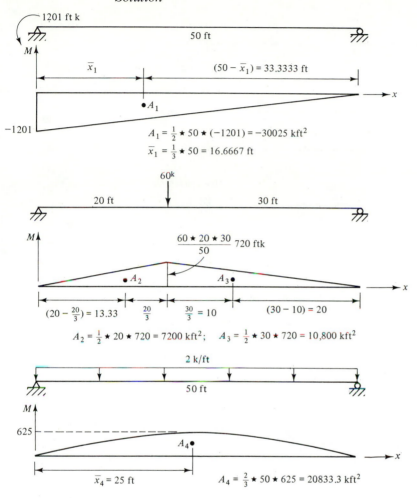

$$A_1 = \tfrac{1}{2} \star 50 \star (-1201) = -30025 \text{ kft}^2$$
$$\bar{x}_1 = \tfrac{1}{3} \star 50 = 16.6667 \text{ ft}$$

$$A_2 = \tfrac{1}{2} \star 20 \star 720 = 7200 \text{ kft}^2; \quad A_3 = \tfrac{1}{2} \star 30 \star 720 = 10{,}800 \text{ kft}^2$$

$$A_4 = \tfrac{2}{3} \star 50 \star 625 = 20833.3 \text{ kft}^2$$

Figure 6.9b

EXAMPLE 6.4

For the structure shown in Figure 6.10a, obtain the Moment Diagram by parts using superposition of moment diagrams for a cantilevered beam with the cantilever support being at point *B*.

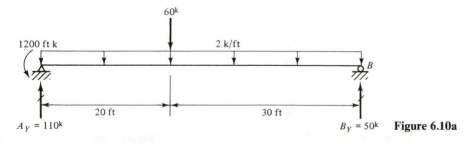

Figure 6.10a

Solution

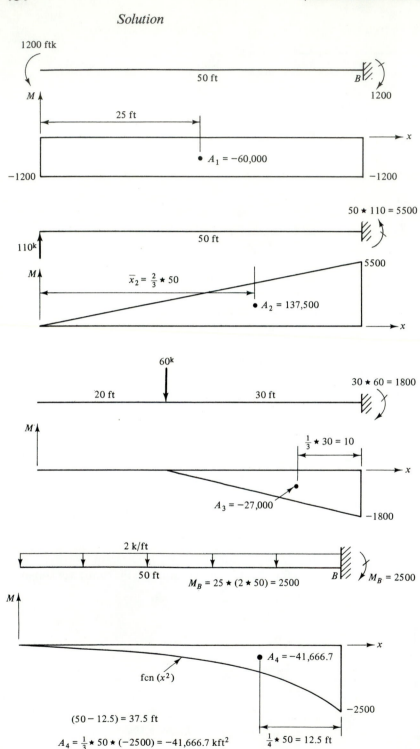

$(50 - 12.5) = 37.5$ ft

$A_4 = \frac{1}{3} \star 50 \star (-2500) = -41,666.7$ kft^2

Figure 6.10b

Check the arithmetic—total $M_B = 0$, if the arithmetic is correct:

$$M_B = -1200 + 5500 - 1800 - 2500 = 0 \text{ (arithmetic checks)}$$

Suppose the point at $x = 20$ ft is named point C and we wish to find M_C.

$$M_C = -1200 + \tfrac{20}{50} * 5500 + 0 + (\tfrac{20}{50})^2 * (-1800) = 600 \text{ ftk}$$

The total Moment Diagram ordinate at any other point can be found in a similar manner (by superposition).

EXAMPLE 6.5

For the structure shown in Figure 6.11, obtain the Moment Diagram by parts using superposition of moment diagrams for a cantilevered beam with the cantilever support being at point C.

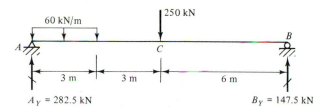

Figure 6.11a

Solution

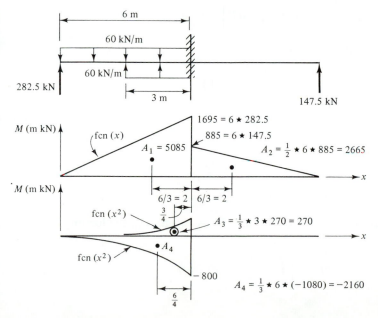

Figure 6.11b

6.5 CONJUGATE BEAM METHOD

In Chapter 4, the following relations were found to exist between the transverse loading, shear, and bending moment in a region where the transverse loading was a continuous function of x (Fig. 6.12):

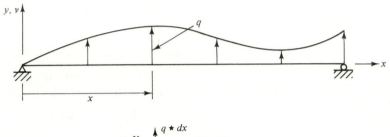

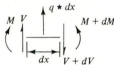

Figure 6.12

$$\frac{dM}{dx} = V \tag{6.12}$$

$$\left[(M_2 - M_1) = \int_{M_1}^{M_2} dM \right] = \int_{x_1}^{x_2} V\, dx \tag{6.13}$$

$$\frac{dV}{dx} = q \tag{6.14}$$

$$\left[(V_2 - V_1) = \int_{V_1}^{V_2} dV \right] = \int_{x_1}^{x_2} q\, dx \tag{6.15}$$

Equation 6.12 can be differentiated to obtain

$$\frac{d}{dx}\left(\frac{dM}{dx} \right) = \frac{d}{dx}(V)$$

$$\frac{d^2 M}{dx^2} = \frac{dV}{dx} \tag{6.16}$$

Therefore, from Eqs. 6.14 and 6.16

$$\frac{d^2 M}{dx^2} = q \tag{6.17}$$

Consequently, M can be obtained from Eq. 6.17 by integrating the loading function twice:

$$M = \int \left[\int q\, dx \right] dx + C_1 x + C_2 \tag{6.18}$$

where C_1 and C_2 are constants of integration which must be evaluated from the boundary conditions on moment. This approach is comparable to the Integration Method in Section 6.3 used to obtain solutions of the type

$$v = \int \left[\int \frac{M}{EI} dx \right] dx + C_1 x + C_2 \qquad (6.19)$$

However, in Chapter 4 it was found to be much easier to obtain the Moment Diagram as follows:

1. $(V_2 - V_1) =$ [area under the load curve from point 1 to point 2]. Starting at the beam boundary where V_1 was known; $V_2 = V_1 +$ [area under the load curve from point 1 to point 2].
2. $(M_2 - M_1) =$ [area under the shear curve from point 1 to point 2]. Starting at the beam boundary where M_1 was known; $M_2 = M_1 +$ [area under the shear curve from point 1 to point 2].

Thus, if we properly define a *conjugate beam* (see footnote),

$$(v_2 - v_1) = \int \left[\int \frac{M}{EI} dx \right] dx$$

can be dealt with in the same fashion that

$$(M_2 - M_1) = \int \left[\int q\, dx \right] dx$$

was dealt with in Chapter 4 which includes the FBD and equilibrium equation approach.

The Conjugate* Beam Method was published first in 1868 by Otto Mohr, a German professor and engineer. Definitions needed in the Conjugate Beam Method are:

1. The *conjugate beam* is a fictitious beam for which the length is identical to the length of the real beam.
2. The boundary conditions of the conjugate beam are the slope and deflection at the boundary points on the deflection curve of the real beam.
3. The "loading" on the conjugate beam is either the M/EI diagram for real loads or the curvature diagram for temperature changes and prestrain problems (precambering, initially crooked members, misfits due to fabrication or construction errors). Examples pertaining to temperature changes and prestrain are shown in Chapter 14. In those examples, the conjugate beam is loaded with either the curvature diagram or concentrated angle changes. In

---

* Webster dictionary definitions of *Conjugate* are: (1) having the same derivation and, therefore, usually some likeness in meaning; (2) having features in common but opposite or inverse in some particular(s); (3) joined together in pairs (coupled).

this text, all displacements obtained by the Conjugate Beam Method are due to flexural deformations only.

4. Conjugate beam "shear" at point i is θ_i which is the slope at point i on the deflection curve of the real beam.
5. Conjugate Beam "Bending Moment" at point i is v_i which is the deflection at point i on the deflection curve of the real beam.
6. The positive sign conventions for the conjugate beam "loading," "shear," and "bending moment" are the same as those defined in Chapter 4 for the loading, shear, and bending moment of a real beam:

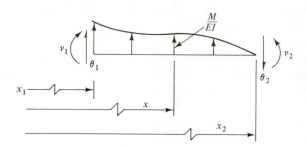

7. The boundary conditions and any special conditions at interior points on the conjugate beam are obtained by examining the slopes and deflections at the corresponding points on a sketch of the deflection curve of the real beam subjected to the real loading. This procedure is illustrated in the following four cases in which the deflection curves were deliberately chosen such that the slopes and deflections at the boundary points and special interior points are all positive. This was done to emphasize the positive sign convention on the conjugate beam; all unknowns on the conjugate beams have been shown as acting in their positive definition directions in each of the four cases shown in Figure 6.13.

EXAMPLE 6.6 ——

For the structure shown in Figure 6.14a, use the Conjugate Beam Method to find θ_B and v_B. EI is constant from A to B.

Solution. The Moment Diagram is shown in Figure 6.14b. The Boundary Conditions (BC) for the conjugate beam are the slopes and deflections at points A and B on the deflection curve in Figure 6.14a:

1. At point A, the slope and deflection are zero ($\theta_A = 0$; $v_A = 0$).
2. At point B, the slope and deflection are not zero ($\theta_B \neq 0$; $v_B \neq 0$). Point B translates downward (v_B is negative) and rotates in the clockwise direction (θ_B is negative). However, the author recommends that the unknowns θ_B and v_B be shown as acting in their positive "shear" and "bending moment" directions on the conjugate beam at B. Then, the computed values of θ_B and v_B will turn out to be negative (see Fig. 6.14c).

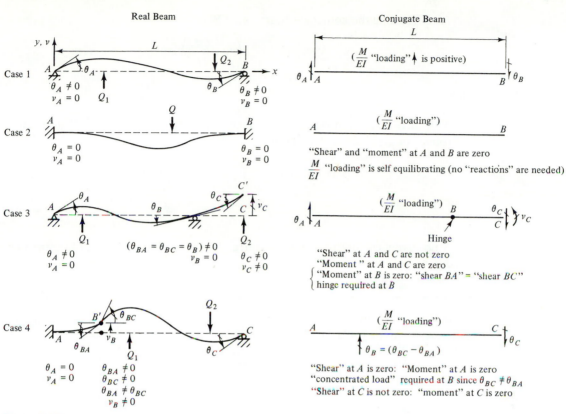

Real Beam / Conjugate Beam

Case 1

$\theta_A \neq 0$
$v_A = 0$

$\theta_B \neq 0$
$v_B = 0$

$(\frac{M}{EI}$ "loading" ↑ is positive$)$

Case 2

$\theta_A = 0$
$v_A = 0$

$\theta_B = 0$
$v_B = 0$

$(\frac{M}{EI}$ "loading"$)$

"Shear" and "moment" at A and B are zero
$\frac{M}{EI}$ "loading" is self equilibrating (no "reactions" are needed)

Case 3

$\theta_A \neq 0$
$v_A = 0$

$(\theta_{BA} = \theta_{BC} = \theta_B) \neq 0$
$v_B = 0$

$\theta_C \neq 0$
$v_C \neq 0$

$(\frac{M}{EI}$ "loading"$)$

Hinge

"Shear" at A and C are not zero
"Moment" at A and C are zero
$\begin{cases}$ "Moment" at B is zero: "shear BA" = "shear BC" \\ hinge required at $B \end{cases}$

Case 4

$\theta_A = 0$
$v_A = 0$

$\theta_{BA} \neq 0$
$\theta_{BC} \neq 0$
$\theta_{BA} \neq \theta_{BC}$
$v_B \neq 0$

$(\frac{M}{EI}$ "loading"$)$

$\theta_B = (\theta_{BC} - \theta_{BA})$

"Shear" at A is zero: "Moment" at A is zero
"concentrated load" required at B since $\theta_{BC} \neq \theta_{BA}$
"Shear" at C is not zero: "moment" at C is zero

Figure 6.13

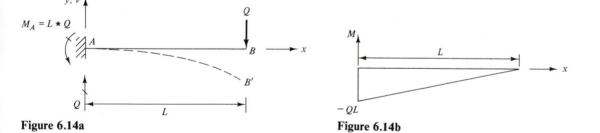

$M_A = L \star Q$

Figure 6.14a

Figure 6.14b

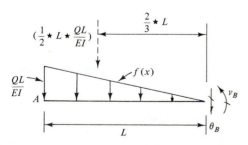

$(\frac{1}{2} \star L \star \frac{QL}{EI})$

$\frac{2}{3} \star L$

$\frac{QL}{EI}$

$f(x)$

v_B

θ_B

Figure 6.14c Conjugate beam for Figure 6.14a "loaded" with the M/EI diagram.

Treating the conjugate beam as a FBD:

$$\sum F_y = 0: \qquad -\left(\theta_B + \frac{QL^2}{2EI}\right) = 0$$

$$\theta_B = -\frac{QL^2}{2EI}$$

A positive θ_B is counterclockwise; since θ_B is negative, point B in Figure 6.14a rotates in the clockwise direction.

$$\sum M_z = 0 \text{ at } B: \qquad v_B + \frac{2L}{3}*\left(\frac{QL^2}{2EI}\right) = 0$$

$$v_B = -\frac{QL^3}{3EI}$$

A positive v_B is upward; since v_B is negative, point B in Figure 6.14a translates downward.

EXAMPLE 6.7 ——

For the structure shown in Figure 6.15a, EI is constant from A to C. Find θ_A, θ_B, v_B, and the equation for v at x for $0 \le x \le a$ and for $a \le x \le L$. Use the Conjugate Beam Method.

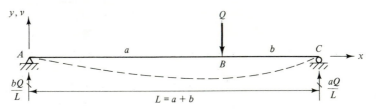

Figure 6.15a

Solution

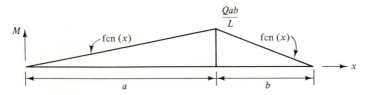

Figure 6.15b

The slopes and deflections at points A and C in Figure 6.15a are the BC for the conjugate beam (Fig. 6.15c):

1. $\theta_A \ne 0$; conjugate "shear at A" $\ne 0$. $v_A = 0$; conjugate "bending moment at A" $= 0$.

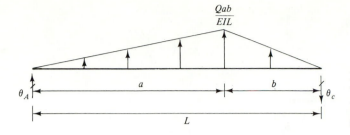

Figure 6.15c Conjugate beam for Figure 6.15a "loaded" with M/EI diagram.

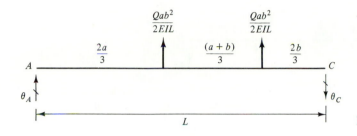

Figure 6.15d

2. $\theta_C \neq 0$; conjugate "shear at C" $\neq 0$. $v_C = 0$; conjugate "bending moment at C" $\neq 0$.

For purposes of finding the "reactions" of the conjugate beam (Fig. 6.15d):

$$\sum M_z = 0 \text{ at } A: \qquad \frac{2a}{3} * \frac{Qa^2b}{2EIL} + \left(L - \frac{2b}{3}\right) * \frac{Qab^2}{2EIL} - L * \theta_C = 0$$

$$\theta_C = \frac{Qab(L+a)}{6EIL}$$

$$\sum F_y = 0: \qquad \theta_A + \frac{Qa^2b}{2EIL} + \frac{Qab^2}{2EIL} - \theta_C = 0$$

$$\theta_A = -\frac{Qb(L^2 - b^2)}{6EIL}$$

The above solution for the slopes at A and C were obtained using only symbols to provide an honest comparison to the Integration Method solution (see Example 6.2). As an independent check on the algebra involved in the above solution, repeat the derivation using the following numerical values: $Q = 60^k$; $a = 24$ ft; $b = 12$ ft; $L = 36$ ft (Fig. 6.15e).

$$\sum M_z = 0 \text{ at } A: \qquad 36 * \theta_C = 16 * \frac{5760}{EI} + 28 * \frac{2880}{EI}; \qquad \theta_C = \frac{4800}{EI}$$

$$\sum F_y = 0: \qquad \theta_A = \theta_C - \left(\frac{5760}{EI} + \frac{2880}{EI}\right) = \frac{-3840}{EI}$$

Direct substitution of $Q = 60^k$, $a = 24$ ft, $b = 12$ ft, and $L = 36$ ft into the derived formulas gives

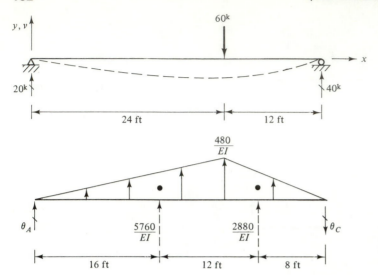

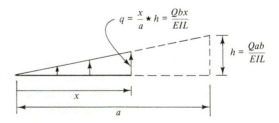

Figure 6.15e Conjugate beam "loaded" with M/EI diagram.

$$\theta_C = \frac{60*24*12*(36+24)}{6EI*36} = \frac{4800}{EI} \text{ (checks)}$$

$$\theta_A = \frac{-60*12*[(36)^2-(12)^2]}{6EI*36} = \frac{-3840}{EI} \text{ (checks)}$$

To find formulas for v_B and v at any value of x, we return to the M/EI diagram expressed in terms of Q, a, b, and L (Fig. 6.15f).

$$\sum M_z = 0 \text{ at } B: \qquad v_B = \frac{a}{3}*\frac{Qa^2b}{2EIL}+a*\theta_A = \frac{Qa^2b^2}{3EIL}$$

For $0 \le x \le a$ (Fig. 6.15g):

$$\sum M_z = 0 \text{ at } x: \qquad v = \frac{x}{3}*\frac{qx}{2}+x*\theta_A$$

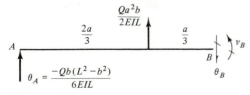

Figure 6.15f

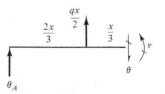

Figure 6.15g

Substitution of $q = Qbx/EIL$ and $\theta_A = -Qb(L^2 - b^2)/6EIL$ gives

$$v = \frac{-Qbx}{6EIL}(L^2 - b^2 - x^2)$$

For $a \le x \le L$ (Fig. 6.15h):

$$\sum M_z = 0 \text{ at } x: \qquad v = \frac{(b-x)}{3} * \frac{q(b-x)}{2} + (b-x) * \theta_C$$

Substitution for q, θ_C, and $(b - x) = (L - a - x)$ gives

$$v = \frac{-Qa(L-x)}{6EIL}[2bL - b^2 - (L-x)^2]$$

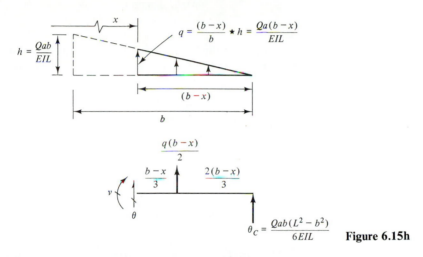

Figure 6.15h

EXAMPLE 6.8 ───

For the structure shown in Figure 6.16, use the Conjugate Beam Method to find the midspan deflection. EI is constant from A to B.

Solution. The Moment Diagram is obtained by parts with the loads placed on a cantilever supported at point B (Fig. 6.16b). From the deflected shape in Figure 6.16a, the observed BC of the conjugate beam are conjugate "moment" at A and B are zero since v_A and v_B are zero in Figure 6.16a; conjugate "shear" at A and B are not zero ($\theta_A \ne 0$ and $\theta_B \ne 0$). The M diagram by parts is divided by EI and placed on the conjugate beam (Fig. 6.16c).

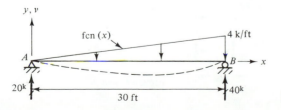

Figure 6.16a

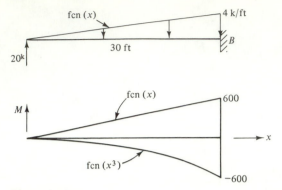

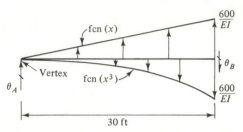

Figure 6.16b

Figure 6.16c Conjugate beam "loaded" with M/EI diagram.

For purposes of computing the "reactions" of the conjugate beam, the areas of the M/EI are computed and located at the centroids of the areas (Fig. 6.16d).

$$A_1 = \frac{1}{2} * 30 * \frac{600}{EI} = \frac{9000}{EI}$$

$$\bar{C}_1 = \frac{1}{3} * 30 = 10 \text{ ft}$$

$$\bar{C}_2 = \frac{1}{5} * 30 = 6 \text{ ft}$$

$$A_2 = \frac{1}{4} * 30 * \frac{600}{EI} = \frac{4500}{EI}$$

$$\sum M_z = 0 \text{ at } C: \qquad 6 * \frac{4500}{EI} - 10 * \frac{9000}{EI} - 30 * \theta_A = 0$$

$$\theta_A = \frac{-2100}{EI}$$

$$\sum F_y = 0: \qquad \frac{9000}{EI} - \frac{2100}{EI} - \frac{4500}{EI} - \theta_B = 0$$

$$\theta_B = \frac{2400}{EI}$$

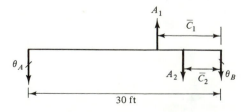

Figure 6.16d

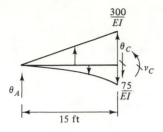

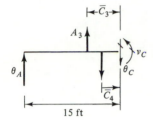

Figure 6.16e

Name the point at midspan of Figure 6.16a point C to compute v_C (Fig. 6.16e).

$$\bar{C}_3 = \frac{1}{3} * 15 = 5 \, \text{ft}$$

$$A_3 = \frac{1}{2} * 15 * \frac{300}{EI} = \frac{2250}{EI}$$

$$\bar{C}_4 = \frac{1}{5} * 15 = 3 \, \text{ft}$$

$$A_4 = \frac{1}{4} * 15 * \frac{75}{EI} = \frac{1125}{4EI}$$

$$\sum M_z = 0 \, \text{at} \, C: \quad v_C = 5 * \frac{2250}{EI} + 15 * \left(\frac{-2100}{EI} \right) - 3 * \frac{1125}{4EI}$$

$$v_C = \frac{-84375}{4EI}$$

This value of v_C agrees with the solution at $x = 15$ ft of Example 6.1 for $q_0 = 4$ k/ft and $L = 30$ ft.

EXAMPLE 6.9

For the structure shown in Figure 6.17a, use the Conjugate Beam Method to find θ_{BA}, θ_{BC}, v_B, θ_C, v_C, θ_D, θ_E, and v_E. Note that the beam segments from B to E have the property of EI being constant and segment AB has the property of $2EI$.

Solution. Due to the variation in EI, the Moment Diagram is obtained by cantilevered parts with the cantilever support at B and divided by the appropriate

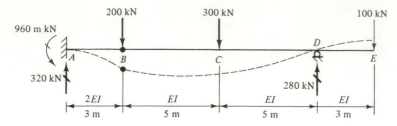

Figure 6.17a

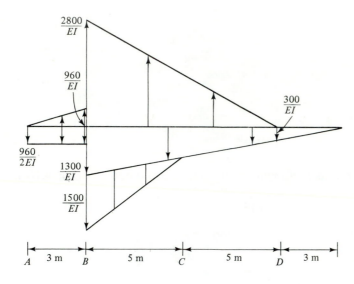

Figure 6.17b

property to obtain the M/EI "loading." See Figure 6.17b. The BC of the conjugate beam are (Fig. 6.17c):

1. $\theta_A = 0$ and $v_A = 0$.
2. There is an unknown concentrated angle change at B: $\hat{\theta}_B = \theta_{BC} - \theta_{BA}$.
3. A hinge is needed at D since $v_D = 0$ and $\theta_D \neq 0$.
4. $\theta_E \neq 0$ and $v_E \neq 0$.

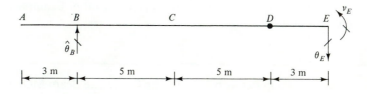

Figure 6.17c Conjugate beam boundary conditions.

To find the "reactions" of the conjugate beam, consider a FBD of segment AD "loaded" with the M/EI diagram areas located at their centroids (Fig. 6.17d):

$$A_1 = \frac{1}{2} * 3 * \frac{960}{2EI} = \frac{720}{EI}; \qquad A_2 = \frac{1}{2} * 10 * \frac{2800}{EI} = \frac{14000}{EI}; \qquad A_3 = 3 * \frac{960}{2EI} = \frac{1440}{EI}$$

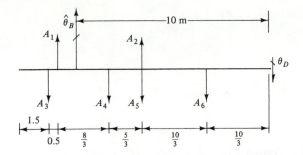

Figure 6.17d

$$A_4 = \frac{1}{2} * 5 * \frac{1500}{EI} = \frac{3750}{EI}; \qquad A_5 = \frac{1}{2} * 10 * \frac{1300}{EI} = \frac{6500}{EI};$$

$$A_6 = \frac{1}{2} * 10 * \frac{300}{EI} = \frac{1500}{EI}$$

$\sum M_z = 0$ at D:

$$11.5 * A_3 + \frac{25}{3} * A_4 + \frac{20}{3} * A_5 + \frac{10}{3} * A_6 - 11 * A_1 - 10 * \hat{\theta}_B - \frac{20}{3} * A_2 = 0$$

$$\hat{\theta}_B = \frac{-511}{EI}$$

$$\sum F_y = 0: \qquad \theta_D = A_1 + A_2 + \theta_B - (A_3 + A_4 + A_5 + A_6) = \frac{1019}{EI}$$

Now, consider segment *DE* of the conjugate beam "loaded" with the *M/EI* area at its centroid as a FBD to find the "reactions" at point *E*:

$\theta_D = \dfrac{1019}{EI}$ v_E A_6 $\theta_{E'}$ 1 m 2 m

$$A_6 = \frac{1}{2} * 3 * \frac{300}{EI} = \frac{450}{EI}$$

$$\sum F_y = 0: \qquad \theta_E = \theta_D - A_6 = \frac{569}{EI}$$

$$\sum M_z = 0 \text{ at } E: \qquad v_E = 3 * \theta_D - 2 * A_6 = \frac{2157}{EI}$$

From the FBD of segment *AB*:

A_1 A_3 1.5 1

$$v_B = 1 \star A_1 - 1.5 \star A_3 = \frac{-1440}{EI}$$

$$\theta_{BA} = A_1 - A_3 = \frac{-720}{EI}$$

From the FBD of segment $AB + dx$:

$$\theta_{BC} = A_1 + \hat{\theta}_B - A_3 = \frac{-1231}{EI}$$

To find v_C and θ_C, either FBD of segment AC or segment CE can be used and the simpler FBD is segment CE. See Figure 6.17e.

$$A_7 = \frac{1}{2} * 5 * \frac{1400}{EI} = \frac{3500}{EI}$$

$$A_8 = \frac{1}{2} * 8 * \frac{800}{EI} = \frac{3200}{EI}$$

$$\sum F_y = 0: \qquad \theta_C = A_8 + \theta_E - A_7 = \frac{269}{EI}$$

$$\sum M_z = 0 \text{ at } C: \qquad v_C = \frac{5}{3} * A_7 + v_E - \frac{8}{3} * A_8 - 8 * \theta_E = \frac{-5095}{EI}$$

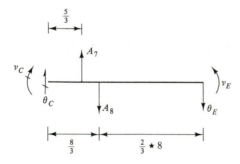

Figure 6.17e

6.6 VIRTUAL WORK METHOD FOR BEAMS AND FRAMES

In the derivation of this method, also known as the Dummy Unit Load Method, it is assumed that small displacement theory is applicable and the structural behavior is linearly elastic. Consequently, superposition is valid. Consider the structure shown in Figure 6.18. If all of the 40^k in Figure 6.18 were applied at the same instant of time, Q_1 would be a shock (or impact) load which would cause the beam to vibrate. While vibrations are occurring, the equations of statics are not applicable. Therefore, for statical analysis purposes, Q_1 must be considered as being slowly applied from zero to its value of 40^k. For the beam in Figure 6.18, the load-displacement curve shown in Figure 6.19 must be a straight line since the beam in Figure 6.18 is assumed to behave in a linearly elastic manner and small displacement theory is assumed to be applicable.

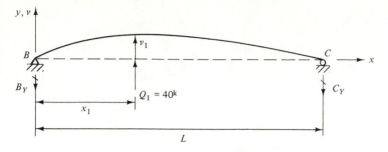

Figure 6.18

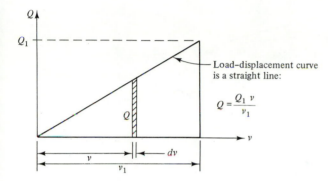

Figure 6.19

From similar triangles, $Q/v = Q_1/v_1$; therefore, the equation of the straight line is $Q = Q_1 * v/v_1$. The definition of the *external work done by Q_1* during its application is the area under the load-displacement curve in Figure 6.19:

$$(W_{ext} \text{ due to } Q_1) = \int_0^{v_1} Q\,dv = \frac{Q_1}{v_1}\int_0^{v_1} v\,dv = \frac{Q_1}{v_1}\left[\frac{v^2}{2}\right]_0^{v_1} = \frac{1}{2}Q_1 v_1$$

Similarly, if only a concentrated moment, $\hat{M}_1$, were slowly applied, the result would be as shown in Figure 6.20. Now, suppose we slowly apply $\bar{Q}_1$ as shown in Figure 6.21

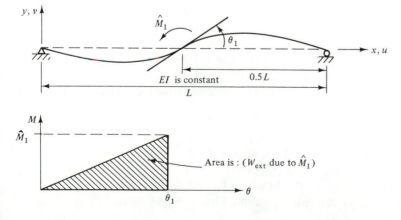

Figure 6.20

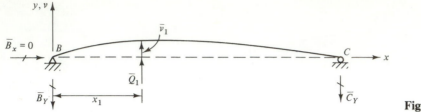

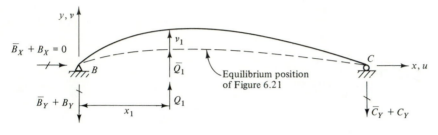

Figure 6.21

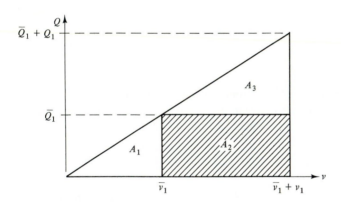

Figure 6.22

Equilibrium position
of Figure 6.21

Figure 6.23

$A_1 = \frac{1}{2}\bar{Q}_1\bar{v}_1 \quad (W_{est}$ done by $\bar{Q}_1$ in Fig. 6.21)

$\left.\begin{array}{l} A_2 = \bar{Q}_1 v_1 \\ A_3 = \frac{1}{2}Q_1 v_1 \end{array}\right\} (W_{est}$ done during application of Q_1 in Fig. 6.22)

and subsequently slowly apply Q_1 as shown in Figure 6.22; Figure 6.23 shows the external work done during the application of $\bar{Q}_1$ and Q_1.

The following definitions are needed in order to derive the Virtual Work Method:

1. $\bar{Q}_1$ in Figures 6.21 and 6.22 is a *virtual load.*
2. *Virtual* means not necessarily real.
3. $\bar{v}_1$ in Figures 6.21 and 6.22 is a *virtual displacement.*
4. *Virtual work* is either of the following:
   a. The product of a virtual load and a real displacement.
   b. The product of a real load and a virtual displacement.

The external virtual work, $\bar{W}_{\text{ext}}$, done in Figure 6.22 during the application of Q_1 (since $u_B = v_B = v_C = 0$ in Fig. 6.22) is

$$\bar{W}_{\text{ext}} = \bar{Q}_1 * v_1 + \bar{B}_Y * v_B + \bar{C}_Y * v_C + \bar{B}_X * u_B = \bar{Q}_1 * v_1$$

where $\bar{Q}_1$ is a virtual load and v_1 is a real displacement caused by the real load Q_1 in Figure 6.22. Note that a *superscript bar symbol is placed on each virtual variable.* All variables without the superscript bar symbol are real variables.

Consider only the bending deformation $d\theta$ of a differential element taken from Figure 6.22; an enlargement of this element is shown in Figure 6.24 and from Figure 6.6, $d\theta = (M/EI)dx$. The internal virtual work done in Figure 6.24 by $\bar{M}$ during the occurrence of $d\theta$ due only to M is $\bar{M} * d\theta$. Consequently, the total internal virtual work done by $\bar{Q}_1$ in Figure 6.22 during the application of Q_1 is

$$\bar{W}_{\text{int}} = \int_0^L \bar{M}\, d\theta = \int_0^L \bar{M}\left(\frac{M\, dx}{EI}\right)$$

Since Figure 6.22 is in statical equilibrium,

$$\bar{W}_{\text{ext}} = \bar{W}_{\text{int}}$$

$$\bar{Q}_1 * v_1 = \int_0^L \bar{M}\left(\frac{M}{EI}dx\right) \tag{6.20}$$

Equation 6.20 is the Virtual Work Method due to bending deformations only and it was developed by Johann Bernoulli in 1717. Since superposition is valid, terms can be added to Eq. 6.20 to account for axial deformations (see Section 6.7), shear deformations (see Section 14.6), and torsional deformations.

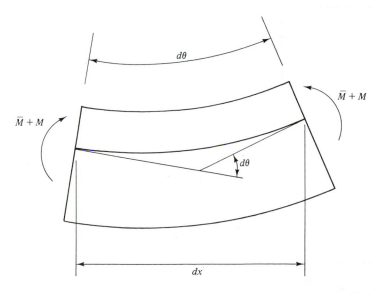

Figure 6.24 Note: $\bar{M}$ fully existed before M slowly occurred; M caused $d\theta$ to occur.

EXAMPLE 6.10

For the structure shown in Figure 6.25a, use the Virtual Work Method to find v_A due to bending deformations only. EI is constant from A to B.

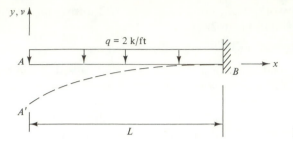

Figure 6.25a

Solution. Before the 2 k/ft is slowly applied in Figure 6.25a, a virtual load must be fully applied at the point where the displacement is desired due to the 2 k/ft. Since a positive v displacement is upward and v_A is desired, apply $\bar{Q}$ upward at A and obtain the Virtual Moment Diagram shown in Figure 6.25b. Now that $\bar{Q}$ has been placed at A, slowly add the 2 k/ft and find the Moment Diagram due only to the 2 k/ft. See Figure 6.25c.

$$\bar{W}_{\text{ext}} = \bar{Q} * v_A + \bar{B}_Y * (0) + \bar{M}_B * (0) = \bar{Q} * v_A$$

$$\bar{W}_{\text{ext}} = \bar{W}_{\text{int}}$$

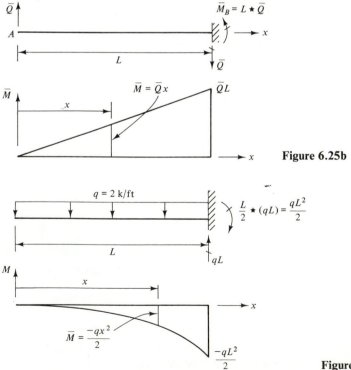

Figure 6.25b

Figure 6.25c

$$\bar{Q} * v_A = \int_0^L \bar{M}\left(\frac{M}{EI}dx\right) = \int_0^L (\bar{Q}x) * \left(\frac{-qx^2}{EI}dx\right)$$

$$\bar{Q} * v_A = \frac{-\bar{Q}q}{EI}\int_0^L x^3\,dx = \frac{-\bar{Q}q}{EI}\left[\frac{x^4}{4}\right]_0^L = \frac{-\bar{Q}qL^4}{4EI}$$

Since $\bar{Q} \neq 0$, divide both sides of the preceding equation by $\bar{Q}$ to obtain

$$v_A = \frac{-qL^4}{4EI}$$

It should be noted that v_A was assumed to be positive which is upward and $\bar{Q}$ was applied in the direction of a positive v_A. Therefore, the external virtual work was the product of two positive terms. However, the internal virtual work turned out to be negative. This requires that v_A be negative as shown in the last equation since $\bar{Q}$ was divided out on both sides of the preceding equation.

EXAMPLE 6.11 ———————————————————————————————————

For the structure shown in Figure 6.26a, use the Virtual Work Method to find v_C due to bending deformations only. EI is constant from A to B.

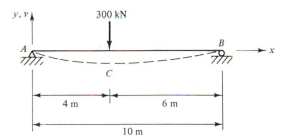

300 kN

Figure 6.26a

Solution. Before the 300 kN is slowly applied in Figure 6.26a, a virtual load $\bar{Q}$ must be fully applied at point C. Note that the last step in the solution of Example 6.10 was to divide both sides of the virtual work equation by $\bar{Q}$. Therefore, it is convenient to choose to use $\bar{Q} = 1$ instead of an unknown $\bar{Q}$ in the Virtual Work Method. The author did not forget to show the units of the unit force chosen for $\bar{Q}$. If $\bar{Q}$ were kept as an unknown variable and divided out on both sides of the virtual work equation, no units would be chosen for $\bar{Q}$. Consequently, the value of unity chosen for $\bar{Q}$ has no units for convenience (any chosen units would cancel eventually anyway). See Figure 6.26b.

For $0 \le x \le 4$, $\bar{M} = -0.6x$. For $4 \le x \le 10$, $\bar{M} = -(10 - x)*0.4$. Now, find the Moment Diagram due to the 300 kN load only. See Figure 6.26c. For $0 \le x \le 4$, $M = 180x$. For $4 \le x \le 10$, $M = 120*(10 - x)$.

$$\bar{W}_{ext} = \bar{W}_{int}$$

$$\bar{1} * v_C = \int_0^4 (-0.6x) * \left(\frac{180}{EI}dx\right) + \int_4^{10}[-0.4*(10-x)] * \left[\frac{120*(10-x)}{EI}dx\right]$$

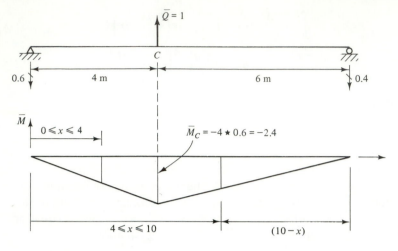

$$\bar{M}_C = -4 \star 0.6 = -2.4$$

Figure 6.26b

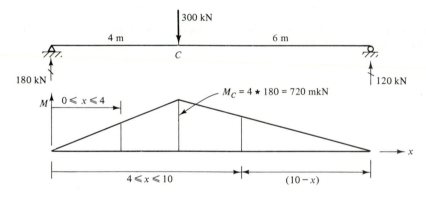

$$M_C = 4 \star 180 = 720 \text{ mkN}$$

Figure 6.26c

$$\bar{1} * v_C = \frac{-108}{EI}\left[\frac{x^3}{3}\right]_0^4 + \frac{48}{EI}\left[\frac{(10-x)^3}{3}\right]_4^{10}$$

$$\bar{1} * v_C = \frac{-2304}{EI} + \frac{16}{EI}[0 - (6)^3] = \frac{-5760}{EI}$$

The Virtual Work Method is used extensively in Chapter 7 to compute displacements due only to bending deformations for beams and frames. Having to obtain equations for $\bar{M}$ and M/EI and the integral of their product by calculus is tedious, purely mathematical, and highly susceptible to errors. Fortunately, it is possible to obtain the internal Virtual Work by Visual Integration as shown below [6.3].

In Section 6.4 it was shown that the Moment Diagram can be obtained by parts for which the areas and their centroids are easily computed. The virtual moment curve is always either a straight line or a series of straight lines. For any straight line segment from point j to point k on the $\bar{M}$ diagram, $\bar{M} = ax + b$; therefore, the internal virtual work is

$$\int_{x_j}^{x_k} \bar{M}\left(\frac{M}{EI}dx\right) = a \int_{x_j}^{x_k} x\left(\frac{M}{EI}dx\right) + b \int_{x_j}^{x_k} \frac{M}{EI}dx \qquad (6.21)$$

The second integral on the right-hand side of Eq. 6.21 is the area under the M/EI diagram between points j and k and can be denoted as

$$A_i = \int_{x_j}^{x_k} \frac{M}{EI}dx \qquad (6.22)$$

The first integral on the right-hand side of Eq. 6.21 is the first moment about point j of the area under the M/EI diagram between points j and k and can be denoted as

$$x_i A_i = \int_{x_j}^{x_k} x\left(\frac{M}{EI}dx\right) \qquad (6.23)$$

where x_i is the distance from point j to the centroid of A_i. Thus, Eq. 6.21 can be written as follows:

$$\int_{x_j}^{x_k} \bar{M}\left(\frac{M}{EI}dx\right) = a*(x_i A_i) + b*A_i = (ax_i + b)*A_i \qquad (6.24)$$

If we define $\bar{M}_i = ax_i + b$, then Eq. 6.24 can be written as

$$\int_{x_j}^{x_k} \bar{M}\left(\frac{M}{EI}dx\right) = \bar{M}_i * A_i \qquad (6.25)$$

where $\bar{M}_i$ is the ordinate on the $\bar{M}$ curve at x_i and x_i is the distance from point j to the centroid of A_i (Fig. 6.26d). Since the Moment Diagram by parts may consist of n areas between points j and k,

$$\int_{x_j}^{x_k} \bar{M}\left(\frac{M}{EI}dx\right) = \sum_{i=1}^{n} (\bar{M}_i * A_i) \qquad (6.26)$$

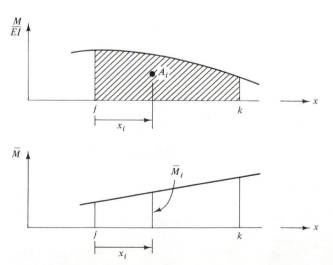

Figure 6.26d

Note: It must be emphasized that there can only be one straight line segment of $\bar{M}$ in each region $x_j < x < x_k$. Also, there are situations where the areas between the M/EI curve and the x axis in each region must be chosen such that all ordinates contained in an area must have the same sign. Consequently, the author recommends that all areas always be chosen such that all ordinates contained in each area have the same sign. The author prefers to obtain the M/EI diagram by parts such that the centroids of the chosen areas are readily known.

EXAMPLE 6.12 ───

For the structure shown in Figure 6.27a, use the Virtual Work Method and visual integration to find v_C and θ_B due to bending deformations only. EI is constant from A to B.

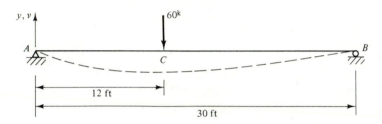

Figure 6.27a

Solution. Since the M/EI diagram is the same in the solution for v_C and θ_B, the areas on the M/EI diagram are chosen to be applicable in both solutions. First, the $\bar{M}$ diagrams are sketched and examined to determine the total number of straight line segments from A to B. Then the areas on the M/EI diagram are chosen such that each area does not span across a region that contains more than one straight line segment on each $\bar{M}$ diagram.

To find v_C, apply $\bar{Q} = 1$ upward at point C. See Figure 6.27b. To find θ_B, apply $\bar{Q} = 1$ counterclockwise at point B. See Figure 6.27c. Since $\bar{M}$ for $\bar{Q}_C$ has two straight line segments in the region from A to B, the areas must be chosen to span from A to C and from C to B which are the regions of the two straight line segments on $\bar{M}$ for $\bar{Q}_C$.

The areas of the M/EI diagram and the centroids of the areas are obtained in Figure 6.27d.

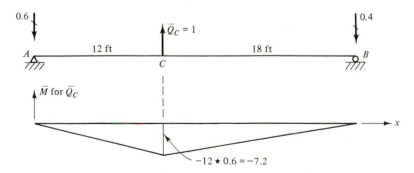

Figure 6.27b

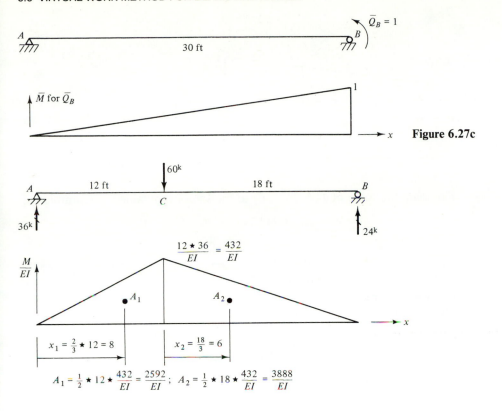

Figure 6.27c

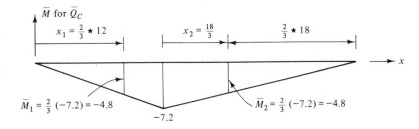

Figure 6.27d

$$\bar{W}_{\text{ext}} = \bar{W}_{\text{int}} \text{ (to compute } v_C\text{)}$$

$$\bar{Q}^1_C * v_C = \sum_{i=1}^{2} \bar{M}_i * A_i = (-4.8) * \frac{2592}{EI} + (-4.8) * \frac{3888}{EI} = \frac{-31104}{EI}$$

which is identically the same solution as obtained in Example 6.11.
To compute θ_B, the $\bar{M}_i$ are needed as shown in Figure 6.27e.

$$\bar{W}_{\text{ext}} = \bar{W}_{\text{int}} \text{ (to compute } \theta_B\text{)}$$

$$\bar{Q}^1_B * \theta_B = \sum_{i=1}^{2} \bar{M}_i * A_i = \left(\frac{4}{15}\right) * \frac{2592}{EI} + (0.6) * \frac{3888}{EI} = \frac{3024}{EI}$$

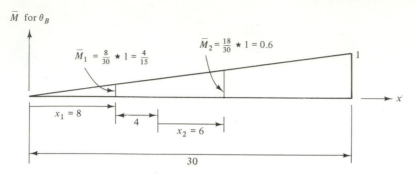

$\bar{M}$ for θ_B

$\bar{M}_1 = \frac{8}{30} \star 1 = \frac{4}{15}$

$\bar{M}_2 = \frac{18}{30} \star 1 = 0.6$

$x_1 = 8$

4

$x_2 = 6$

30

Figure 6.27e

Compare the value of $\theta_B = 3024/EI$ to the solution obtained in Example 6.2:

$$\theta_B = \frac{Qab(L+a)}{6EIL} = \frac{60 \star 12 \star 18 \star (30+12)}{6EI \star 30} = \frac{3024}{EI} \text{ (checks)}$$

To emphasize that an A_i of the M/EI diagram must not span across more than one straight line segment on the $\bar{M}$ diagrams, consider Figure 6.27f. This choice of A_i is permissible to use in finding θ_B since there is only one straight line segment of $\bar{M}$ for $\bar{Q}_B$ in the region from A to B:

$$\bar{1} \star \theta_B = \left[\frac{14}{30} \star 1\right] \star \frac{6480}{EI} = \frac{3024}{EI}$$

However, this choice of A_i is not permissible to use in finding v_C since there are two straight line segments of $\bar{M}$ for $\bar{Q}_C$ in the region from A to C. If this choice of A_i were used to compute v_C, the following erroneous value for v_C would be obtained:

$$\bar{1} \star v_C = \left[\frac{16}{18} \star (-7.2)\right] \star \frac{6480}{EI} = \frac{-41472}{EI}$$

whereas the correct value of $v_C = -31104/EI$.

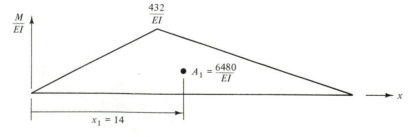

$\frac{M}{EI}$

$\frac{432}{EI}$

$A_1 = \frac{6480}{EI}$

$x_1 = 14$

Figure 6.27f

EXAMPLE 6.13

For the structure shown in Figure 6.28a, use the Virtual Work Method and visual integration to find the midspan deflection v_C due to bending deformations only. From B to D the beam's cross-sectional property is $2EI$; from A to B and D to E, the beam's cross-sectional property is EI.

y, v
3 k/ft
x
EI
6 ft
2 EI
12 ft
2 EI
12 ft
EI
6 ft
A
B
C
D
E

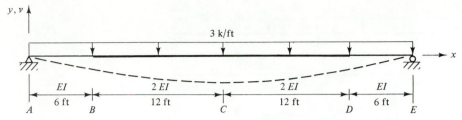

Figure 6.28a

Solution. Since $\bar{M}$ for $\bar{Q}_C$ is discontinuous at C, A_i for M/EI must span from A to C and from C to E (Fig. 6.28b). Since Figure 6.28a is symmetric with respect to point C, we can integrate from A to C and double the obtained result to account for integrating from C to E (Fig. 6.28c). Since EI is discontinuous at B, it is convenient to obtain the areas of the M/EI diagram by parts cantilevered at point B (Fig. 6.28d). Note that point A is a vertex for the fcn(x^2) from A to B and point C is a vertex for the fcn(x^2) from C to B.

$$A_1 = \frac{1}{2} * 6 * \frac{324}{EI} = \frac{972}{EI}$$

$$A_2 = 12 * \frac{243}{EI} = \frac{2916}{EI}$$

$$A_3 = \frac{1}{3} * 6 * \left(\frac{-54}{EI}\right) = \frac{-108}{EI}$$

$$A_4 = \frac{1}{3} * 12 * \left(\frac{-108}{EI}\right) = \frac{-432}{EI}$$

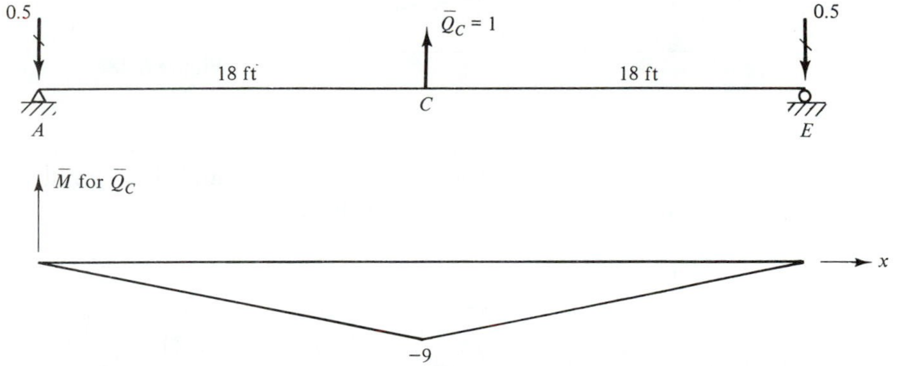

0.5
$\bar{Q}_C = 1$
0.5
18 ft
18 ft
A
C
E
$\bar{M}$ for $\bar{Q}_C$
x
−9

Figure 6.28b

3 k/ft
$M_C = \frac{3 \star (36)^2}{8} = 486$ ftk
A
EI
2 EI
54k
6 ft
12 ft
$V_C = 0$

Figure 6.28c

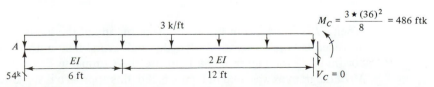

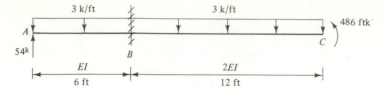

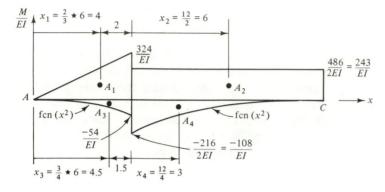

Figure 6.28d

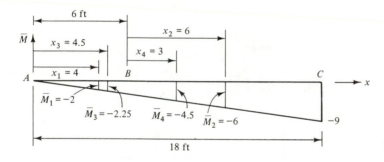

Figure 6.28e

Since we are only integrating from A to C and doubling the result, only the portion of $\bar{M}$ for $\bar{Q}_C$ from A to C is needed.

$$\bar{W}_{\text{ext}} = \bar{W}_{\text{int}}$$

$$\bar{1} * v_C = 2 * \left[(-2) * \frac{972}{EI} + (-6) * \frac{2916}{EI} + (-2.25) * \left(\frac{-108}{EI} \right) + (-4.5) * \left(\frac{-432}{EI} \right) \right]$$

$$v_C = \frac{-17253}{EI}$$

When EI is a variable in the M/EI diagram, any of the methods for computing displacements are necessarily lengthy endeavors. The author's preference for obtaining v_C for Figure 6.28a is to use the Conjugate Beam Method and the M/EI diagram as was used in Figure 6.28d; this approach is suggested as a possible problem in the problem set at the end of this chapter.

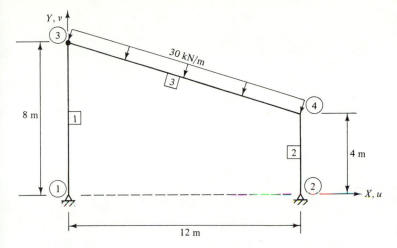

Figure 6.29a

EXAMPLE 6.14

For the structure shown in Figure 6.29a, use the Virtual Work Method and visual integration due to bending deformations only to find:

1. u_3, the horizontal displacement at joint 3.
2. $\hat{\theta}_3$, the total change in the interior angle between members 1 and 3 at joint 3.

The members are prismatic and $(EI)_3 = (EI)_2 = 2*(EI)_1$. Let $(EI)_1 = EI$. Note that there is an internal hinge at joint 3; members 1 and 3 are hinged ended at joint 3.

Solution. The length of member 3 is needed; $L_3 = \sqrt{(12)^2 + (4)^2} = \sqrt{160} = 12.649$ m. Since both $\bar{M}$ diagrams must be examined for discontinuities in order to obtain the areas of the M/EI diagram between these discontinuities, the $\bar{M}$ diagrams are obtained first. Then the A_i on the M/EI diagram are obtained, the

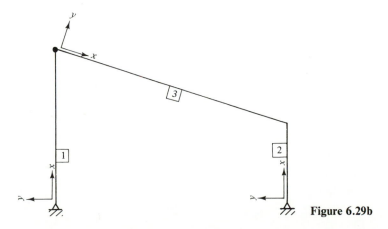

Figure 6.29b

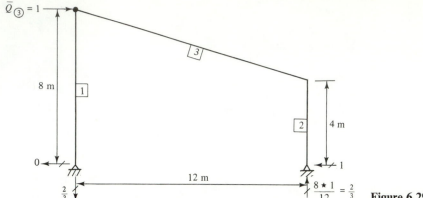

Figure 6.29c

centroids of A_i are computed, and finally the $\bar{M}_i$ are computed on each $\bar{M}$ diagram. However, to keep from having to draw the $\bar{M}$ diagrams twice (as was done in the previous examples for emphasis), the $\bar{M}_i$ values are shown on the $\bar{M}$ diagrams although they were computed after the A_i and x_i were computed on the M/EI diagram.

  The member reference axes are chosen as shown in Figures 6.29b and 6.29c. Positive M causes compression on the positive y-direction side of the member axis. $\bar{M}$ for $\bar{Q}_3$ for each member are shown in Figures 6.29d through 6.29f. Since the only $\bar{M}$ discontinuities occur at the member ends, the areas of the M/EI diagram must span between the member ends. The M/EI diagrams are obtained by superposition of simple beam parts (Figs. 6.29g and 6.29h).

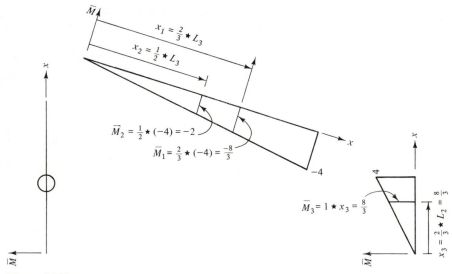

Figure 6.29d

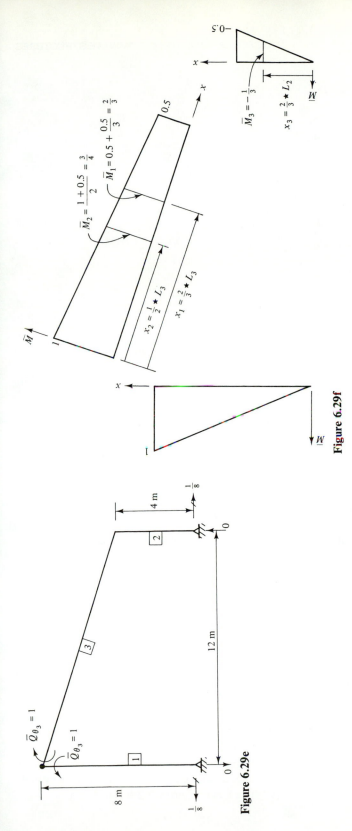

$$\bar{M}_2 = \frac{1 + 0.5}{2} = \frac{3}{4}$$

$$\bar{M}_1 = 0.5 + \frac{0.5}{3} = \frac{2}{3}$$

$$\bar{M}_3 = -\frac{1}{3}$$

$$x_3 = \frac{2}{3} \star L_2$$

$$x_2 = \frac{1}{2} \star L_3$$

$$x_1 = \frac{2}{3} \star L_3$$

Figure 6.29f

$$\bar{Q}_{\theta_3} = 1$$
$$\bar{Q}_{\theta_3} = 1$$

Figure 6.29e

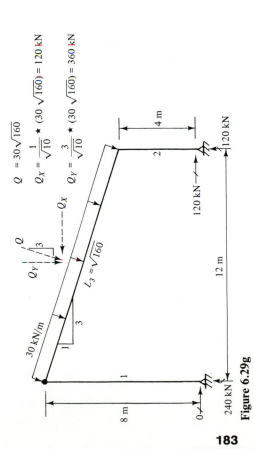

$$Q = 30\sqrt{160}$$

$$Q_X = \frac{1}{\sqrt{10}} \star (30\sqrt{160}) = 120 \text{ kN}$$

$$Q_Y = \frac{3}{\sqrt{10}} \star (30\sqrt{160}) = 360 \text{ kN}$$

$$L_3 = \sqrt{160}$$

$30\,kN/m$

120 kN

120 kN

120 kN

240 kN

Figure 6.29g

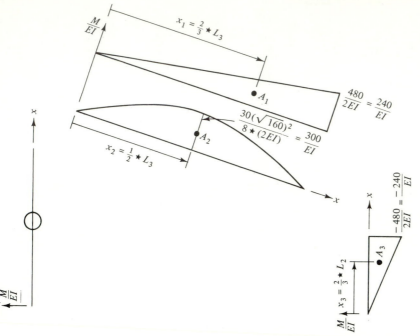

Figure 6.29h

$$A_1 = \frac{1}{2} * \sqrt{160} * \frac{240}{EI} = \frac{1517.9}{EI}$$

$$A_2 = \frac{2}{3} * \sqrt{160} * \frac{300}{EI} = \frac{2529.8}{EI}$$

$$A_3 = \frac{1}{2} * 4 * \left(\frac{-240}{EI}\right) = \frac{-480}{EI}$$

To find the horizontal displacement at joint 3, $\bar{M}_i$ for $\bar{Q}_{3u}$ are used:

$$\bar{1} * u_3 = \sum_{i=1}^{3} \bar{M}_i * A_i$$

$$u_3 = \left(\frac{-8}{3}\right) * \frac{1517.9}{EI} + (-2) * \frac{2529.8}{EI} + \left(\frac{8}{3}\right) * \left(\frac{-480}{EI}\right) = \frac{-10387.3}{EI}$$

To find the change in the internal angle at joint 3, $\bar{M}_i$ for $\bar{Q}_{3\theta}$ are used:

$$\bar{1} * \hat{\theta}_3 = \sum_{i=1}^{3} \bar{M}_i * A_i$$

$$\hat{\theta}_3 = \left(\frac{2}{3}\right) * \frac{1517.9}{EI} + \left(\frac{3}{4}\right) * \frac{2529.8}{EI} + \left(-\frac{1}{3}\right) * \left(\frac{-480}{EI}\right) = \frac{3069.3}{EI}$$

The angle between members 1 and 3 in Figure 6.29a before any loads are applied is $90° - \tan^{-1}\left(\frac{12}{36}\right) = 71.57°$; after the loads in Figure 6.29a have been applied, the internal angle is $71.57° - (\hat{\theta}_3 * 180/\pi)$. See Figures 6.29i and 6.29j.

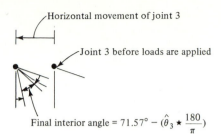

Horizontal movement of joint 3

Joint 3 before loads are applied

Final interior angle $= 71.57° - (\hat{\theta}_3 \star \frac{180}{\pi})$ **Figure 6.29i**

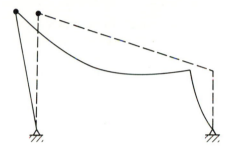

Figure 6.29j Deflected structure.

6.7 VIRTUAL WORK METHOD FOR TRUSSES

The truss subjected to the loads shown in Figure 6.30 is to be used in the discussion of the concepts that are involved in the computation of joint displacements of a truss. Suppose we are only interested in computing the vertical component of the displacement at joint 2. This displacement component is denoted as v_2 and, as shown in Figure 6.30, upward v displacements are positive.

In order for virtual work to be done during the application of the real loads shown in Figure 6.30, first we must apply a virtual load in the v direction at joint 2

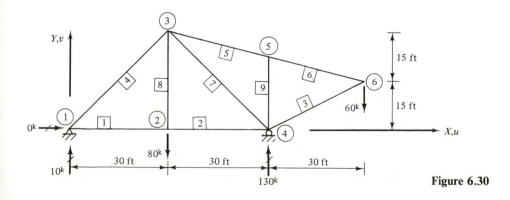

Figure 6.30

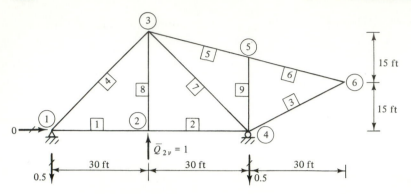

Figure 6.31

since v_2 is to be computed. For convenience we choose a unit value for the virtual load as shown in Figure 6.31; the virtual forces in all members due to this loading must be computed. The virtual force in member i is denoted as $\bar{F}_i$. Now, we slowly add the real loads shown in Figure 6.30 to the deformed structure caused by the virtual load in Figure 6.31. The real loads cause real structural deformations (either elongations or shortenings) to occur. Since the virtual load fully existed before the real loads were applied, the total external virtual work done during the application of the real loads is

$$\bar{W}_{\text{ext}} = \bar{Q}_{2v} * v_2$$

The total internal virtual work done during the application of the real loads is found by algebraically summing the internal virtual work done by each member. Consequently, we need to define the internal virtual work done by a typical member i. Consider Figure 6.32:

1. $\bar{F}_i$ is the virtual force in member i caused by the unit virtual load in Figure 6.31.
2. F_i is the real force in member i caused by the real loads in Figure 6.30.
3. $(dL)_i$ is the real deformation of member i which occurred during the application of the real loads.
4. $\bar{F}_i*(dL)_i$ is the internal virtual work done only by member i during the application of the real loads.

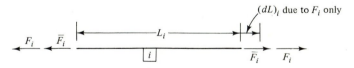

Figure 6.32 Internal virtual work $= \bar{W}_{\text{int}}$ $= \bar{F}_i * (dL)_i$, since the $\bar{F}_i$ fully existed before the F_i were slowly caused to occur by the real, applied external loads on the joints of the truss.

Therefore, the total internal virtual work done is

$$\bar{W}_{\text{int}} = \sum_{i=1}^{n} \bar{F}_i*(dL)_i$$

where n = total number of truss members

Since $(dL)_i$ is due to F_i, we must define $(dL)_i$ as a function of F_i. The positive sign convention definitions for a truss member are:

1. Tensile member-end forces are positive.
2. Elongation of a member is a positive member deformation.

The total elongation of member i is the integral of the elongation of a differential element taken from member i:

$$(dL)_i = \int_0^L \epsilon \, dx$$

where ϵ = the strain in member i caused by F_i (due to real loads)

$\epsilon = \dfrac{\sigma}{E}$ for linearly elastic structural behavior

$\sigma = \left(\dfrac{F}{A}\right)_i$

E = the modulus of elasticity

Thus,

$$(dL)_i = \int_0^L \epsilon \, dx = \left(\frac{F}{EA}\right)_i \int_0^L dx = \left(\frac{FL}{EA}\right)_i$$

Therefore, the total internal virtual work done is

$$\bar{W}_{int} = \sum_{i=1}^n \bar{F}_i * (dL)_i$$

and when the change in length of the members is due to real loads,

$$(dL)_i = \left(\frac{FL}{EA}\right)_i$$

For equilibrium, $\bar{W}_{ext} = \bar{W}_{int}$.

EXAMPLE 6.15 ───

For the truss shown in Figure 6.33a, use the Method of Virtual Work to find only the following joint displacement components and compare them to the computer solution shown in Example 3.2:

1. v_2
2. u_3

In the computer solution shown in Example 3.2 all members had the same area of 10 in.^{2} and the same E of 29,000 ksi.

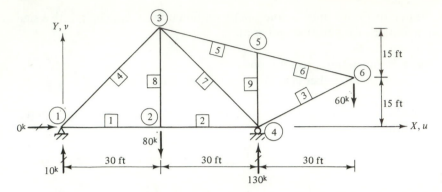

Figure 6.33a

Solution. $\bar{F}_i$ due to $\bar{Q}_{2v}$, $\bar{F}_i$ due to $\bar{Q}_{3u}$, and F_i due to the real loads are shown in Figures 6.33b through 6.33d. The virtual member forces, $\bar{F}_i$, due to each unit virtual load must exist in each member before the real loads are applied in order to have internal virtual work being done. F_i, the member forces due to the real loads, are needed to compute $(dL)_i$, the change in member length due to the real loads: $(dL)_i = (FL/EA)_i$.

$$\bar{W}_{ext} = \bar{W}_{int}$$

$$\bar{Q}_{2v} * v_2 = \sum_{i=1}^{9} (\bar{F}_i \text{ due to } \bar{Q}_{2v}) * (dL)_i$$

$$\bar{Q}_{3u} * u_3 = \sum_{i=1}^{9} (\bar{F}_i \text{ due to } \bar{Q}_{3u}) * (dL)_i$$

Since the $(dL)_i$ are the same in both virtual work equations, it is convenient to prepare the solution in tabular form. An example of the change in member length is:

$$(dL)_8 = \left(\frac{FL}{EA}\right)_8 = \frac{(80^k) * (360 \text{ in.})}{(29,000 \text{ ksi}) * (10.0 \text{ in.}^2)} = 0.0993103 \text{ in. (elongation)}$$

$$(dL)_4 = \frac{(-10\sqrt{2}) * (360\sqrt{2})}{29,000 * 10} = -0.0248276 \text{ in. (shortening)}$$

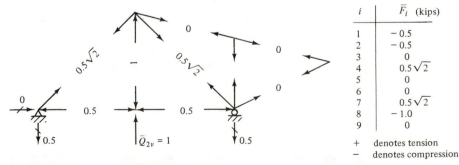

i	$\bar{F}_i$ (kips)
1	-0.5
2	-0.5
3	0
4	$0.5\sqrt{2}$
5	0
6	0
7	$0.5\sqrt{2}$
8	-1.0
9	0
$+$	denotes tension
$-$	denotes compression

Figure 6.33b $\bar{F}_i$ due to $\bar{Q}_{2v} = 1$.

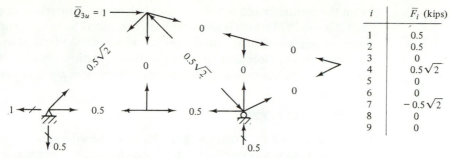

i	$\bar{F}_i$ (kips)
1	0.5
2	0.5
3	0
4	$0.5\sqrt{2}$
5	0
6	0
7	$-0.5\sqrt{2}$
8	0
9	0

Figure 6.33c $\bar{F}_i$ due to $\bar{Q}_{3u} = 1$.

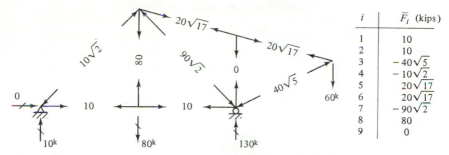

i	$\bar{F}_i$ (kips)
1	10
2	10
3	$-40\sqrt{5}$
4	$-10\sqrt{2}$
5	$20\sqrt{17}$
6	$20\sqrt{17}$
7	$-90\sqrt{2}$
8	80
9	0

Figure 6.33d F_i (due to real loads as obtained in Example 3.2).

The tabular form for the $\bar{W}_{\text{int}}$ summation is shown in Table 6.1; $E_i = 29,000$ ksi for all members; $(dL)_i = (FL/EA)_i$.

$$\bar{W}_{\text{ext}} = \bar{W}_{\text{int}}$$

$$\bar{1} * v_2 = \sum_{i=1}^{9} (\bar{F}_{i(2v)}) * (dL)_i = -0.2873 \text{ in.}$$

$$\bar{1} * u_3 = \sum_{i=1}^{9} (\bar{F}_{i(3u)}) * (dL)_i = 0.1529 \text{ in.}$$

TABLE 6.1

Member number i	A_i (in.^{2})	L_i (in.)	F_i (kips)	$(dL)_i$ (in.)	$\bar{F}_i$ due to $\bar{Q}_{2v} = 1$	$\bar{F}_i$ due to $\bar{Q}_{3u} = 1$	$\bar{F}_{i(2v)} * (dL)_i$ (in.)	$\bar{F}_{i(3u)} * (dL)_i$ (in.)
1	10	360	10	0.01241	-0.5	0.5	-0.006205	0.006205
2	10	360	10	0.01241	-0.5	0.5	-0.006205	0.006205
3	10	$180\sqrt{2}$	$-40\sqrt{5}$	-0.12414	0	0	0	0
4	10	$360\sqrt{2}$	$-10\sqrt{2}$	-0.02483	$0.5\sqrt{2}$	$0.5\sqrt{2}$	-0.01756	-0.01756
5	10	$90\sqrt{17}$	$20\sqrt{17}$	0.01055	0	0	0	0
6	10	$90\sqrt{17}$	$20\sqrt{17}$	0.01055	0	0	0	0
7	10	$360\sqrt{2}$	$-90\sqrt{2}$	-0.22345	$0.5\sqrt{2}$	$-0.5\sqrt{2}$	-0.15800	0.15800
8	10	360	80	0.09931	-1.0	0	-0.09931	0
9	10	270	0	0	0	0	0	0
							$\Sigma = -0.28728$	$\Sigma = 0.15285$

From the computerized solution shown below, the Y-direction displacement at joint 2 is -0.2873 which is identical to v_2 computed by virtual work and the X-direction displacement at joint 3 is 0.1529 which is also identical to u_3 computed by virtual work. Note that the computerized solution obtains all system displacements, whereas only one displacement component at a time can be obtained by virtual work.

SYSTEM DISPLACEMENTS (in.)

NOTE: ALL SUPPORT DISPLACEMENTS are shown as zero. If any PRE-
SCRIBED SUPPORT DISPLACEMENT was not zero for any load-
ing, the user will have to pencil in the non-zero PRE-
SCRIBED SUPPORT DISPLACEMENTS.

JOINT	LOAD NO.	X	Y	Z
1	1	.0000	.0000	.0000
2	1	.0124	−.2873	.0000
3	1	.1529	−.1880	.0000
4	1	.0248	.0000	.0000
5	1	.3086	.0000	.0000
6	1	.2403	−.7085	.0000

The heading entitled LOAD NO. is the abbreviation of Loading Number. Since there was only one loading (see Figure 6.33a), all of the Loading Numbers are 1. However, the solutions for the problems defined in Figures 6.33a, 6.34a, and 6.35a could have been solved in the same computer run. Then, there would have been loading numbers 1, 2, and 3.

The NOTE in the computer output is not applicable for this problem since the prescribed support displacements were all zero, namely, $u_1 = v_1 = v_4 = 0$. However, for example, if the analyst had prescribed $u_1 = v_4 = 0$ and $v_1 = 2$ in., the NOTE would be applicable. In that case, the computerized solution would give the correct SYSTEM DISPLACEMENTS except for the Y-direction displacement at joint 1 which should be penciled in as 2.0000 in. The reason for this peculiarity is that the computer program requires the user to input the data for a nonzero prescribed support displacement as member fixed ended forces. For $v_1 = 2$ in. $= 0.167$ ft, the user would have to calculate and input the member fixed ended forces shown in Figure 6.33e.

$$v_1 = 0.167 \text{ ft}; \qquad d_4 = v_1 * \sin 45° = 0.11785 \text{ ft}$$

Axial forces in members 1 and 4 due to v_1 being induced when all of the system DOF are zero are:

member 4: $\quad \left(\dfrac{EA}{L}\right)_4 * d_4 = \dfrac{(29{,}000 \text{ ksi}) * (10 \text{ in.}^2)}{(30\sqrt{2} \text{ ft}) * (0.11785 \text{ ft})} = 805.56^k$ (compression)

$\qquad\qquad F_1^f = 805.56$ and $F_4^f = -805.56$ are the fixed ended forces

member 1: $\quad$ the fixed ended forces are zero since v_1 is perpendicular to member 1 axis before v_1 is induced

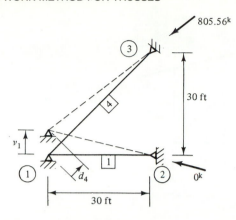

Figure 6.33e

EXAMPLE 6.16

During the assemblage of the truss shown in Figure 6.34a, it was found that member 7 had been fabricated 0.5 in. too short. Suppose the truss fabricator installs this member which is incorrectly too short. Assemblage of the truss is performed with the truss lying on a flat surface. Consequently, during truss assemblage the truss dead weight is perpendicular to the X, Y plane in Figure 6.34a and does not cause any member forces. Using the Virtual Work Method, find the u, v displacement components at joint 6 due only to $(dL)_7 = -0.5$ in.

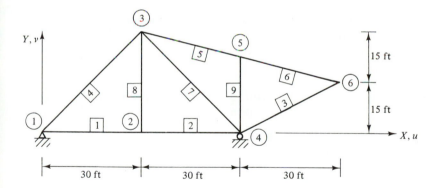

Figure 6.34a Member 7 is fabricated 0.5 in. too short and installed. All other members were correctly fabricated.

Solution. Since all $(dL)_i = 0$ except for $(dL)_7 = -0.5$ in., only the virtual force in member 7 is needed for each of the virtual loads shown in Figures 6.34b and 6.34c. The Method of Sections can be used to find the virtual force in member 7 for each of the virtual load cases; a section is passed vertically between joints 2 and 4 as well as between joints 3 and 5. Consider the FBD on the left side of the section and sum moments = 0 at the point where the extension of members 5 and 2 intersect.

$$\bar{W}_{\text{ext}} = \bar{W}_{\text{int}}$$

$$\bar{Q}_{u6} * u_6 = \sum_{i=1}^{9} \bar{F}_{i(u6)} * (dL)_i = (-0.589256) * (-0.5) = 0.294628$$

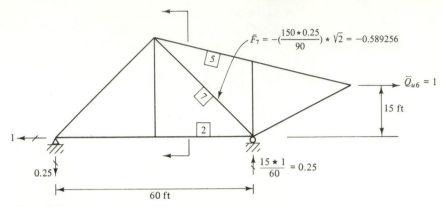

Figure 6.34b

$$u_6 = 0.2946 \text{ in.}$$

$$\bar{Q}_{v6} * v_6 = \sum_{i=1}^{9} \bar{F}_{i(v6)} * (dL)_i = (1.17851) * (-0.5) = -0.589256$$

$$v_6 = -0.5893 \text{ in.}$$

Six-digit precision was used in the preceding calculations in order to compare perfectly with the computerized solution which is shown below.

SYSTEM DISPLACEMENTS (in.)

NOTE: All SUPPORT DISPLACEMENTS are shown as zero. If any PRE-
       SCRIBED SUPPORT DISPLACEMENT was not zero for any load-
       ing, the user will have to pencil in the non-zero PRE-
       SCRIBED SUPPORT DISPLACEMENTS.

JOINT	LOAD NO.	X	Y	Z
1	1	.0000	.0000	.0000
2	1	.0000	−.3536	.0000
3	1	.3536	−.3536	.0000
4	1	.0000	.0000	.0000
5	1	.4419	.0000	.0000
6	1	.2946	−.5893	.0000

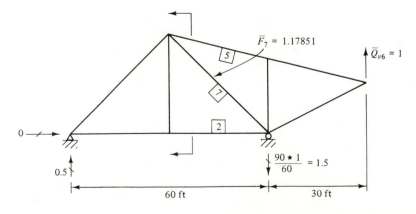

Figure 6.34c

All member forces were zero in the computer output and are not shown. This is always true for determinate trusses due to the installation of members whose lengths were incorrectly fabricated by a small amount. However, for indeterminate trusses an installed member whose member is incorrect does cause member forces to occur and they can be troublesome.

EXAMPLE 6.17

Use the Virtual Work Method to find the u,v displacement components at joint 6 in the truss of Figure 6.35a due only to an increase in temperature of 100°F in only members 4 through 6. For all members $E = 29000$ ksi, and the thermal coefficient of expansion is $\alpha = 0.0000065$ which has units of strain per degree Fahrenheit.

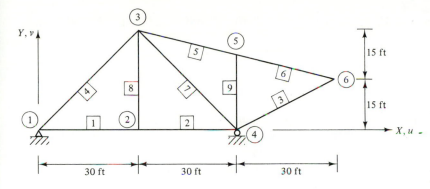

Figure 6.35a Only members 4 through 6 have an increase in temperature of 100°F.

Solution

$(dL)_i = \alpha * (dT)_i * L_i$ is the change in length due to temperature
$(dL)_4 = 0.0000065 * 100 * 360\sqrt{2} = 0.330926$ in.
$(dL)_5 = 0.0000065 * 100 * 90\sqrt{17} = 0.2412017$ in.
$(dL)_6 = 0.0000065 * 100 * 90\sqrt{17} = 0.2412017$ in.

All of the other $(dL)_i = 0$. Since all $(dL)_i$ except for $i = 4$ through 6 are zero, only the virtual force in members 4 through 6 are needed in the $\bar{W}_{int}$ summation terms due to each unit virtual load (Figs. 6.35b and 6.35c).

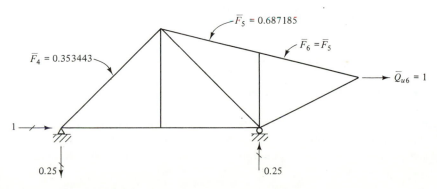

Figure 6.35b $\bar{F}_i$ due to $\bar{Q}_{u6}$—values shown are the only ones needed in $\bar{W}_{int}$ summation.

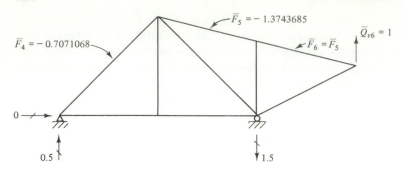

Figure 6.35c $\bar{F}_i$ due to $\bar{Q}_{v6}$— values shown are the only ones needed in $\bar{W}_{int}$ summation.

$$\bar{Q}_{u6} * u_6 = \sum_{i=1}^{9} \bar{F}_{i(u6)} * (dL)_i$$
$$u_6 = (0.353553) * (0.330926) + 2 * (0.687185) * (0.2412017) = 0.44850 \text{ in.}$$
$$\bar{Q}_{v6} * v_6 = \sum_{i=1}^{9} \bar{F}_{i(v6)} * (dL)_i$$
$$v_6 = (-0.707107) * (0.330926) + 2 * (-0.5153875) * (0.2412017)$$
$$= -0.89700 \text{ in.}$$

The computerized solution output results are shown below for the system displacements. All member forces were zero and are not shown. The member forces will always be zero for determinate trusses due to temperature changes. However, for indeterminate trusses the member forces due to temperature changes may be a significant part of the governing design loading.

SYSTEM DISPLACEMENTS (in.)

NOTE: All SUPPORT DISPLACEMENTS are shown as zero. If any PRE-
SCRIBED SUPPORT DISPLACEMENT was not zero for any load-
ing, the user will have to pencil in the non-zero PRE-
SCRIBED SUPPORT DISPLACEMENTS.

JOINT	LOAD NO.	X	Y	Z
1	1	.0000	.0000	.0000
2	1	.0000	.2340	.0000
3	1	.2340	.2340	.0000
4	1	.0000	.0000	.0000
5	1	.4241	.0000	.0000
6	1	.4485	−.8970	.0000

6.8 MAXWELL-BETTI RECIPROCAL THEOREM

The following proof of the Maxwell–Betti Reciprocal Theorem (hereafter abbreviated as MBRT wherever it is not confusing to do so) involves superposition and external work. Any structure for which superposition is valid can be chosen to demonstrate the proof. Also, any set of conditions causing deformations can be chosen, but for convenience the author chooses to use only concentrated loads. If distributed loads were used, integrals would have to be used in addition to summations in order to account for the total work done by all of the loads. Also, deformations due to temperature changes or prescribed support movements could be involved in the proof. However, the author chooses to involve only concentrated loads on a simple beam to keep the proof as short and as simple as possible.

For the case of a structure subjected to only concentrated loads, the Maxwell–Betti reciprocal theorem is:

> The external work done by a set of m forces, $\bar{F}$, during the deformation caused by a set of n forces, P, is equal to the external work done by the P forces during the deformation caused by the $\bar{F}$ forces.

MBRT stated in mathematical terms for concentrated loads is

$$\sum_{i=1*}^{m} \bar{F}_i D_i^P = \sum_{k=1}^{n} P_k D_k^{\bar{F}}$$

where D_i^P = displacement at and in the direction of $\bar{F}_i$ but due to P forces

   $D_k^{\bar{F}}$ = displacement at and in the direction of P_k but due to $\bar{F}$ forces

and the * attached to the subscript numbers for i will become helpful in the sketches (i and k are different subscripts, so their values must be uniquely different).

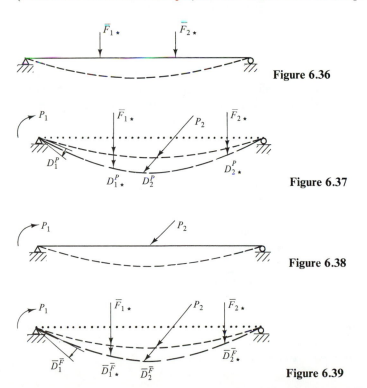

Figure 6.36

Figure 6.37

Figure 6.38

Figure 6.39

Proof of MBRT for Concentrated Loads

Since it is necessary to distinguish between real work, forces, and displacements and virtual work, forces, and displacements, a superscript bar is placed on all symbols representing a virtual quantity. *Virtual means not necessarily real;* sometimes it may be convenient for an analyst to think of a virtual force or displacement as being imaginary and at other times to think of a virtual force or displacement as being real. Symbols representing a real quantity have no superscript bar on them since they are always real.

See Figure 6.36. Apply the system of m virtual forces, $\bar{F}$, to the structure and allow equilibrium to be reached in the deflected position.

See Figure 6.37. To the equilibrium state of Figure 6.36, slowly apply the system of n forces, P, and allow equilibrium to be reached in another deflected position. The total external work done in moving from the equilibrium position in Figure 6.36 to the equilibrium position in Figure 6.37 is

$$W_{1-2} = \sum_{i=1^*}^{m} \bar{F}_i * D_i^P + \frac{1}{2} \sum_{k=1}^{n} P_k * D_k^P$$

and note that the first summation term is virtual work (all forces are virtual; all displacements are real; and their products contain only one Virtual quantity) whereas the second summation term is real work (all forces and displacements are real).

Since superposition is assumed to be valid (beam behavior is assumed to be linearly elastic and the same beam is used in each of Figures 6.36 through 6.39), we could have applied the P forces first and the $\bar{F}$ forces last as described below.

See Figure 6.38. Apply the system of n real forces, P, to the structure and allow equilibrium to be reached in the deflected position.

See Figure 6.39. To the equilibrium state of Figure 6.38, slowly apply the system of m virtual forces, $\bar{F}$, and allow equilibrium to be reached in another deflected position. The total external work done in moving from the equilibrium position in Figure 6.38 to the equilibrium position in Figure 6.39 is

$$W_{3-4} = \sum_{k=1}^{n} P_k * \bar{D}_k^{\bar{F}} + \frac{1}{2} \sum_{i=1^*}^{m} \bar{F}_i * \bar{D}_i^{\bar{F}}$$

and note that the first summation term is virtual work, whereas the second summation term is real work (this is obvious if we let the virtual forces and displacements be imaginary; the product of two imaginary math terms is real). Equating the virtual work terms of W_{1-2} and W_{3-4} gives

$$\sum_{i=1^*}^{m} \bar{F}_i * D_i^P = \sum_{k=1}^{n} P_k * \bar{D}_k^{\bar{F}}$$

Since the superscript bars were used only for convenience in keeping track of virtual items, the superscript bars can be removed now and we have proved the MBRT is also valid if all of the forces and displacements are real.

This form of the MBRT was first presented by E. Betti [6.4] in 1872 when he extended J. C. Maxwell's [6.5] 1864 presentation which involved only unit valued loads to derive the reciprocal deflection theorem. That is, Betti's reciprocal relation is a more generalized form of Maxwell's reciprocal deflection relation. The form presented by Betti is the one needed in Chapter 13 to prove that the system stiffness matrix and each member stiffness matrix are symmetric matrices.

Consider the following illustrative examples in which the MBRT is used. Example 6.18 is typical for the usual conditions for which MBRT is useful; that is, in the two involved systems of forces and displacements, only one force or displacement vector is unknown and all other vectors are known, but symbols instead of numerical values are shown on all known vectors whenever a formula derivation is desired. In order to identify which vectors are known and which vectors are unknown, a slash is placed on all unknown displacement and force vectors. Symbols on the vectors are scalars having no signs since each arrow denotes the correct vector direction (the arrow accounts for the sign of the vector). In the applications of the MBRT, actually the analyst needs only to denote whether the product of each force and displacement does positive or negative work.

EXAMPLE 6.18 ──

In each of Figures 6.40 and 6.41, the same simple beam of length L and constant EI is used. It should be noted that the only nonzero support displacements in those figures are the member-end rotations at the supports.

Use the MBRT to find the midspan deflection, v, in Figure 6.40 due to M. θ in Figure 6.41 can be easily found by using the Conjugate Beam Method, for example.

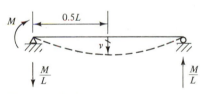

Figure 6.40 System 1.

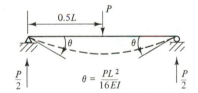

Figure 6.41 System 2.

Solution. According to the MBRT, the algebraic sum of the products of the forces in system 2 and their corresponding displacements in system 1 is equal to the algebraic sum of the products of the forces in system 1 and their corresponding displacements in system 2.

Apply MBRT:

$$P*v+\frac{P}{2}*0+\frac{P}{2}*0 = M*\theta+\frac{M}{L}*0+\frac{M}{L}*0 \text{ which is } P*v = M*\theta \text{ and } v=\frac{M}{P}*\theta$$

Substitute the expression for θ, and solve for v:

$$v=\frac{M}{P}*\left(\frac{PL^2}{16EI}\right)=\frac{ML^2}{16EI}$$

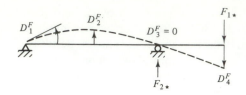

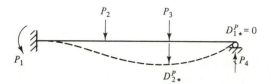

Figure 6.42a

EXAMPLE 6.19

In Figure 6.42a, the same beam of length L and known EI which may be variable along the length is used. Apply MBRT to the two systems.

Solution. Apply MBRT:

$$F_{2*}(-D_{2*}^P) + F_{1*}(0) = P_1(D_1^F) + P_2(-D_2^F) + P_3(0) + P_4(-D_4^F)$$

and, since P_1 and P_4 are unknown, we can only reduce this relation to the following form: $P_1 D_1^F - P_4 D_4^F = P_2 D_2^F - F_{2*}(D_{2*}^P)$. A wiser choice for system 1 is the following which enables one to compute P_4 in system 2 (Fig. 6.42b).

Apply MBRT:

$$-P_4 D_4^F + P_3 D_3^F + P_2 D_2^F = F_{1*}(0) \quad \text{which gives} \quad P_4 = \frac{P_2 D_2^F + P_3 D_3^F}{D_4^F}$$

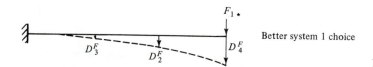

Better system 1 choice

Figure 6.42b

EXAMPLE 6.20

All of the previous discussion and examples of the MBRT involved only concentrated forces. The purpose of this example is to illustrate an *useful* application of the MBRT which involves distributed loads.

For the indeterminate structure shown in Figure 6.43a(i), assume that the member properties are known but the members are not necessarily prismatic. Also, assume that the member-end moments and rotations shown on Figures 6.43a(ii), 6.43b, and 6.43c are known.

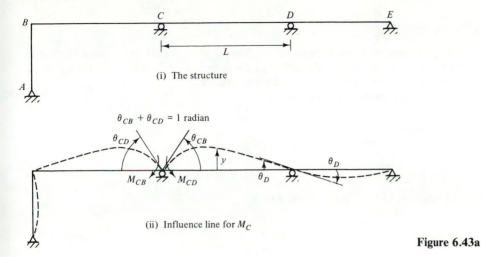

$\theta_{CB} + \theta_{CD} = 1$ radian

(i) The structure

(ii) Influence line for M_C

Figure 6.43a

The objective is to use the MBRT to find only the area under the influence line in span C to D of Figure 6.43a(ii) which is reproduced for convenience as Figure 6.43b.

Solution

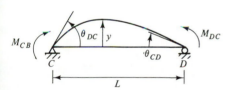

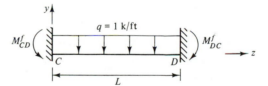

Figure 6.43b System 1 [span CD portion of the IL for M_C in Figure 6.43a(i)].

Figure 6.43c System 2 [span CD of the structure in Figure 6.43a(i)].

Apply the MBRT to Figures 6.43b and 6.43c:

$$\int_0^L (-1\,dz)y + M_{CD}^f * \theta_{CD} + M_{DC}^f * \theta_{DC} = 0$$

$$\int_0^L y\,dz = M_{CD}^f * \theta_{CD} + M_{DC}^f * \theta_{DC}$$

where the integral is the area under the deflected shape in Figure 6.43b which is the portion of the IL for M_C in span CD of Figure 6.43a(ii), and M_{CD}^f and M_{DC}^f are the member fixed ended moments due to $q = 1$ k/ft for the portion of the structure in Figure 6.43a(i).

PROBLEMS

Use either the Conjugate Beam Method or the Virtual Work Visual Integration Method as specified by the instructor and find the indicated displacements in Problems 6.1 through 6.15. Answers are not given for problems preceded by an asterisk.

6.1. Find v_A and θ_A. EI is constant from A to B.

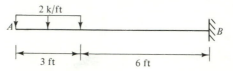

Figure P6.1

6.2. Find v_A and θ_A. EI is constant from A to B.

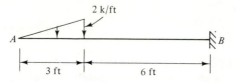

Figure P6.2

6.3. Find v_A and θ_A. EI is constant from A to B.

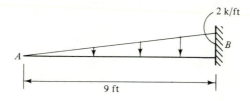

Figure P6.3

6.4. Find v_A and θ_A. EI is constant from A to B.

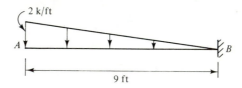

Figure P6.4

6.5. Find θ_A, θ_B, θ_C, and v_C. EI is constant from A to B. Sketch the deflection curve.

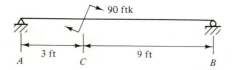

Figure P6.5

6.6. Find v_C. EI is constant from A to B.

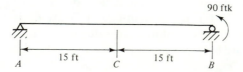

Figure P6.6

* 6.7. Find v_C. EI is constant from A to B.

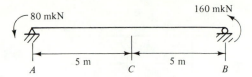

Figure P6.7

* 6.8. Find v_C. EI is constant from A to B.

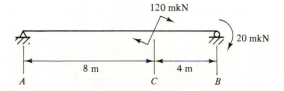

Figure P6.8

* 6.9. Find v_C. EI is constant from A to B.

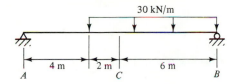

Figure P6.9

6.10. Find θ_C and v_C. $EI = 57{,}000$ kft²; $2EI = 114{,}000$ kft²

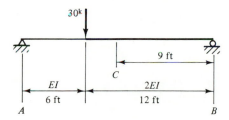

Figure P6.10

6.11. Find v_C. $EI = 312{,}152.78$ kft²; $2EI = 422{,}916.66$ kft². Hint: Find M_C (shear at $C = 0$). Find Moment Diagram by cantilevered parts at the points of discontinuity in EI; M diagram is symmetric with respect to point C.

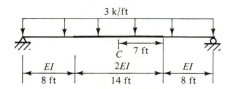

Figure P6.11

* 6.12. Find v_C. EI is constant from A to C.

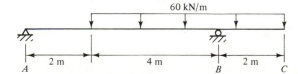

Figure P6.12

* 6.13. Find v_C. EI is constant from A to C.

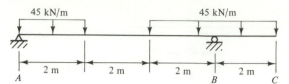

Figure P6.13

6.14. Find v_B, $\hat{\theta}_{BA}$, $\hat{\theta}_{BC}$, and v_E. $EI = 60{,}000$ kft^2; $2EI = 120{,}000$ kft^2.

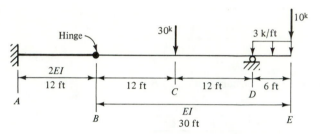

Figure P6.14

* 6.15. Solve Problem 6.14 for loading of 2 k/ft from A to E.

6.16. Use the calculus approach and the Virtual Work Method. Find the vertical displacement at the load point. $E = 29000$ ksi; $I = 405$ in.^{4}; $R = 25$ ft.

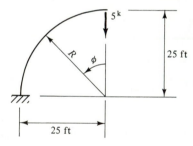

Figure P6.16

Use the Visual Integration approach and Virtual Work Method and find the indicated displacements in Problems 6.17 through 6.19.

6.17. Find θ_A, u_A, and v_A. $E = 29{,}000$ ksi; $I = 200$ in.^{4}

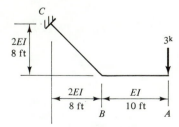

Figure P6.17

6.18. Find u_B, v_C, and u_E. $EI = 1 \times 10^9$ k in.^{2}

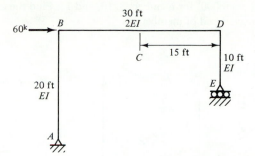

Figure P6.18

* **6.19.** Find v_C, $\hat{\theta}_D$, and u_D. EI is constant and the same for all members.

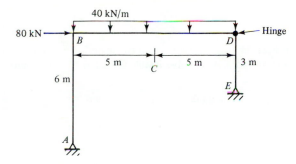

Figure P6.19

In Problems 6.20 through 6.31, find the indicated displacements by the Method of Virtual Work.

6.20. Find u_1, u_2, and v_3. $EA = 90,000^k$ for members 1 through 4. $EA = 60,000^k$ for member 5.

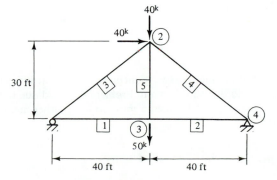

Figure P6.20

* **6.21.** Member 11 is disconnected at joint 2. $EA = 90,000^k$ for members 1 through 4, 6 through 9, 11, and 12. $EA = 60,000^k$ for members 5, 10, and 13. Find the distance between joint 2 and the disconnected end of member 11.

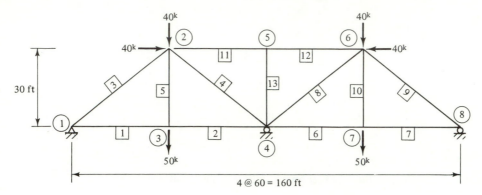

Figure P6.21

* **6.22.** In Problem 6.21, change support 1 to a roller and support 4 to a hinge. Use the results of Problem 6.20 and logic to solve Problem 6.21 with the revised boundary conditions.

* **6.23.** Find u_5 and v_5. $EA = 600,000$ kN for all members.

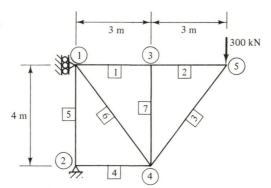

Figure P6.23

* **6.24.** In Problem 6.23, if the members were incorrectly fabricated by the following amounts, find u_5 and v_5 due to the load shown in Problem 6.23 and due to the fabrication errors. Members 1, 2, and 6 were fabricated 5 mm too long. Other members were fabricated 5 mm too short.

* **6.25.** Find v_3 and u_5. $EA = 29,000^k$ for all members.

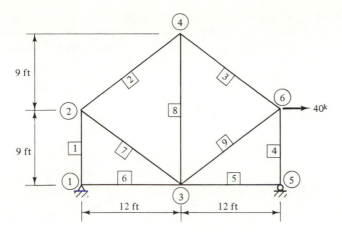

Figure P6.25

6.26. Find v_3 and u_6. For all members: $E = 10,000$ ksi and $L/A = 5$ in./in.²

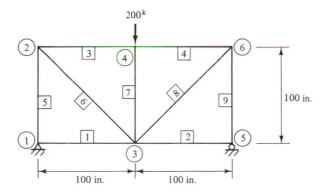

Figure P6.26

6.27. Find u_4. $E = 29,000$ ksi for all members. $A_1 = A_2 = 7.5$ in.²; $A_3 = A_4 = 4$ in.²; $A_5 = 3$ in.²

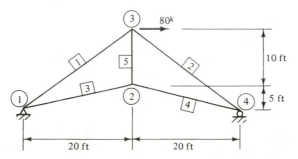

Figure P6.27

* **6.28.** Find u_2, v_2, and the displacement perpendicular to member 1 at 7.80 ft along member 1 from joint 1. Account for bending deformations and axial deformations. $E = 29,000$ ksi for both members. $A_1 = 11.8$ in.^{2}; $I_1 = 518$ in.^{4}; $A_2 = 3.83$ in.^{2}; $I_2 = 11.3$ in.^{4}

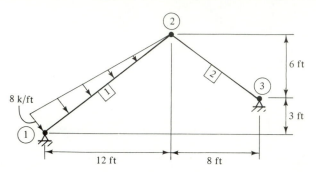

Figure P6.28

* **6.29.** Find θ_1. Account for bending deformations and axial deformations. $E_1 = E_2 = 20,000$ kN/cm^2; $A_1 = 150$ cm^2; $I_1 = 90,000$ cm^4; $A_2 = 30$ cm^2.

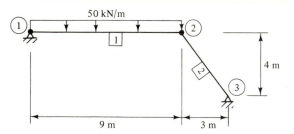

Figure P6.29

6.30. Find v_2 and u_4 accounting for bending deformations in members 1 and 2 and axial deformations in members 3 and 4. $E = 29,000$ ksi for all members. $I_1 = I_2 = 72.4$ in.^{4}; $A_3 = A_4 = 0.52$ in.^{2} Both ends of members 3 and 4 are pinned.

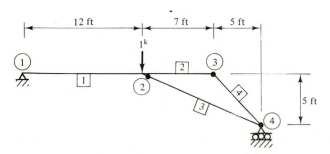

Figure P6.30

* **6.31.** Find v at the concentrated load point. $E = 20,000$ kN/cm^2 for all members. Cable: $A = 4$ cm^2. Beam: $A = 80$ cm^2; $I = 35,000$ cm^4.

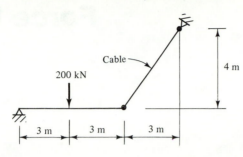

Cable

200 kN

4 m

3 m

3 m

3 m

Figure P6.31

Chapter 7

Force Method

7.1 INTRODUCTION

A statically indeterminate structure is a system of members for which the reactions (external forces) or/and the internal forces (stress resultants—axial, shear, moment, torsional) cannot be determined from the equations of static equilibrium alone. For example, an indeterminate plane truss may be:

1. Externally determinate, but internally indeterminate.
2. Externally indeterminate.
3. Externally and internally indeterminate.

In 1864 J. C. Maxwell [7.1] published the first unified method of analyzing statically indeterminate structures. His presentation included the reciprocal deflection theorem, was rather brief, contained no applications, and attracted little attention. Otto Mohr [7.2] described a similar method and showed some applications. In texts on classical methods of structural analysis, the method independently developed by these two men usually is referred to as the *Method of Redundants and Consistent Deformations,* but sometimes is referred to as the *Maxwell–Mohr Method.* In this chapter, that method will be dealt with using matrix notation and will be referred to as the *system approach of the Force Method* (some authors prefer the terminology Flexibility Method or Compatibility Method). Some matrix method of structural analysis textbooks [7.3, 7.4, 7.5, 7.6] also include a *member approach of the Force Method,* but this text will not present anything on that approach.

A statically determinate structure has the minimum number of reactions and internal forces required to produce a stable system of members that can be solved by the equations of static equilibrium alone. By adding extra reactions or internal forces to a determinate structure, an indeterminate structure is created. The number of extra reactions and internal forces is the degree of statical indeterminacy; these extra forces (a reaction is an external force) are referred to as *redundants.* In the Force Method, the chosen redundants are the unknowns in a set of simultaneous linear algebraic equations.

Analysis by the Force Method involves:

1. Choosing the necessary number, locations, and directions of the redundants.
2. Releasing the redundants to produce a stable and determinate structure referred to as the *released structure* (or primary structure).
3. Calculating deformations on the released structure at and in the directions of the redundants due to:
   a. Loads.
   b. Unit values of each redundant, one at a time (to compute the flexibility coefficients).
4. Using superposition to formulate a set of simultaneous linear algebraic equations; each equation restores the observed deformation compatibility of the original structure at a particular redundant location in the released structure.
5. Solving the set of deformation compatibility equations for the redundants.
6. Using the computed redundants, the original loads, and equations of static equilibrium to determine all reactions and internal forces needed to draw shear and moment diagrams, find axial forces, and so on.

A rigorous sign convention involving matrix notation will be established and used in all of the example problems solved by the Force Method. In his classroom presentations, the author introduces the sign convention, matrix notation, and the concepts of the Force Method in a series of carefully planned examples starting with very simple examples. None of these examples is one of the textbook examples. In the preparation of the remainder of this chapter, the author assumed that each teacher would:

1. Want a completely general conceptual example as an *overview for the teacher* and possibly as a summary review for the students.
2. Want a complete set of detailed procedural steps correlated to the conceptual example.
3. Use the textual material and examples in a supplementary manner; (make selective textual reading and example problem study assignments after devising and using other appropriate examples in the classroom presentations).
4. Be pleased to find some examples that are appropriate for inquisitive students who may be contemplating a structural engineering career.
5. Choose to present the material in the most logical sequence regardless of the author's textual sequence.

If the reader does not have a teacher to plan the textbook study assignments, the author recommends the following self-study plan for the reader:

1. Read Section 7.2 until the conceptual example appears. Then, skip immediately to the beginning of Section 7.4 and continue reading.

2. In Section 7.6, study Examples 7.1 through 7.4 involving only one redundant.
3. In the suggested problems at the end of Chapter 7, find and solve some similar problems.
4. Study Examples 7.5 through 7.7 involving two redundants.
5. See item 3.
6. Study Examples 7.8 through 7.11 involving three redundants.
7. See item 3.
8. Study the conceptual example and the subsequent material in Section 7.3.
9. Study the examples in Sections 7.7, 7.8, and 7.9 (follow the suggested study procedure described in items 2 through 7).

7.2 SIGN CONVENTION, NOTATION, AND DETAILED PROCEDURAL STEPS

When the number of redundants is three or more, the advantage of using a matrix formulation approach is apparent. Since the emphasis in this text is to formulate problems for hand solutions, most of the numerical examples will involve only one or two redundants. In this section, a conceptual example with three redundants will be used to define the sign convention, matrix notation, and detailed procedural steps for the Force Method. However, some discussion of releases for internal forces is needed prior to the conceptual example.

All chosen redundants must be removed from the statically indeterminate structure to create the released structure which must be stable and determinate. At each location of a redundant, the structure is released to permit motion freely only in the positive and negative directions of the redundant. Application of the preceding sentence for an external redundant is obvious, but some sketches showing releases for internal force redundants are needed. Consider a continuous member (Figure 7.1a) which can resist axial force, shear, and moment. Releases for each of these internal forces at point C are shown in Figure 7.1. Each release allows only one of the deformations for the deflected shape of the originally continuous member to be discontinuous. An axial release at point C removes only the capability of the member to resist axial force at C and allows the member either to separate or to overlap along its length direction, but capability to resist shear and moment is not destroyed by an axial release. Likewise, an angular release removes only the moment capacity and a transverse release removes only the shear capacity. Note that a release is not the same as a cut through the member. A cut removes the capability of the member to resist all of the internal forces at a point—continuity of all deformations at a point is destroyed by a cut.

CONCEPTUAL EXAMPLE (See Figure 7.2) ————————————————————————

Given: EI is constant from point A to point D; EA is constant from D to E. Supports A and E settle an identical amount (settlement is a downward vector); B settles 0.500 in. relative to A and E; and C settles 0.375 in. relative to A and E. Note: Grossly exaggerated deflected shapes are shown to demonstrate the deformation vector directions via an arrow (symbol on each arrow represents a magnitude). A slash is placed on each unknown force vector.

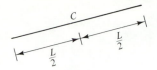

C

$\dfrac{L}{2}$ $\dfrac{L}{2}$

(a) Initially continuous member

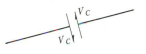

P_C P_C

Released force

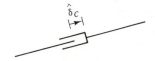

$\hat{\delta}_C$

Corresponding relative displacement

(b) Axial force release

V_C
V_C

Released force

(c) Shear force release

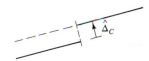

$\hat{\Delta}_C$

Corresponding relative displacement

M_C M_C

Released force

(d) Bending moment release

$\hat{\theta}_C$

Corresponding relative displacement

Figure 7.1 Definition of internal force releases for a plane frame member. Note: Superscript caret symbol denotes a relative displacement or a change in displacement corresponding to a force release.

For purposes of introducing the sign convention and matrix notation to be used in this chapter, consider the structure shown in Figure 7.2a. There are six unknown reaction forces (member DE is a two-force member) and only three equations of statics are available to find them. Therefore, the structure is statically indeterminate; $(6 - 3) = 3$ redundants need to be chosen and removed to produce a stable and determinate structure referred to as the released structure (Fig. 7.2b). The chosen redundants and the directions in which positive valued redundants are chosen to act are shown in Figure 7.2c. The redundants are the reaction at B (an external force); the internal moment at B; and the internal axial force in member DE. Note: In all subsequent examples, all of the following items will be shown on the same sketch: the released structure, redundant locations, and positive directions of the redundants. Deformations (axial separations or overlaps, relative changes in deflections, discontinuities in slope) of the released structure occur at the redundant locations as shown in Figures 7.2d–h. Each redundant's positive direction also defines the positive direction of the deformation at that location. For an internal redundant, a pair of positive

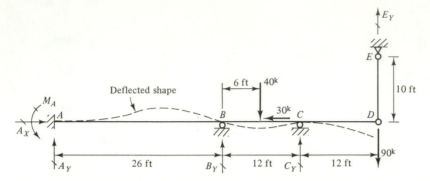

(a) Original structure (support settlements are not graphically shown)

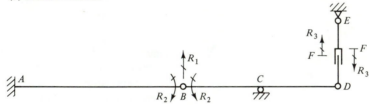

(b) Released structure

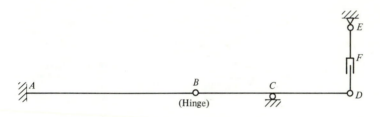

(c) Redundant locations and positive directions

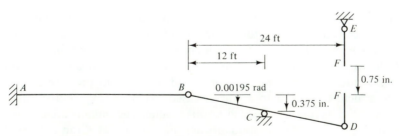

(d)   Deformations due to settlement of support C

(these deformations must be considered as a
part of the deformations due to loads; that is,
they must be algebraically added to those shown
in (e) below)

Figure 7.2 Conceptual example.

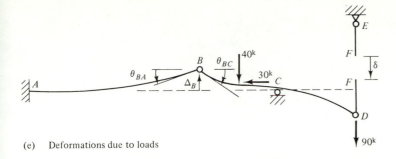

(e)   Deformations due to loads

The summation of the deformation vectors in (d) and (e) gives the deformation matrix due to loads:

$$\{D\} \quad \begin{Bmatrix} D_1 \\ D_2 \\ D_3 \end{Bmatrix} = \begin{Bmatrix} \Delta_B \\ -(0.00195 \text{ rad} + \theta_{BA} + \theta_{BC}) \\ -(0.75 \text{ in.} + \delta) \end{Bmatrix}$$

where $\Delta_B, \delta, \theta_{BA},$ and θ_{BC} are absolute valued symbols

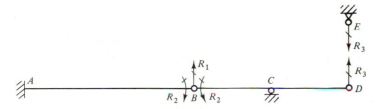

(c) Repeated – redundant locations and positive directions

Note: All values inside each dashlined box are multiplied by symbol on box's right

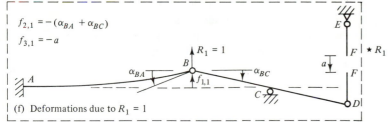

$f_{2,1} = -(\alpha_{BA} + \alpha_{BC})$

$f_{3,1} = -a$

(f) Deformations due to $R_1 = 1$

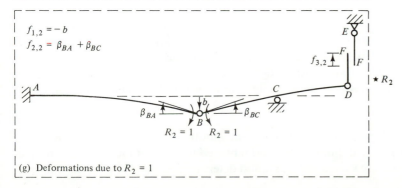

$f_{1,2} = -b$

$f_{2,2} = \beta_{BA} + \beta_{BC}$

(g) Deformations due to $R_2 = 1$

Figure 7.2 (*continued*)

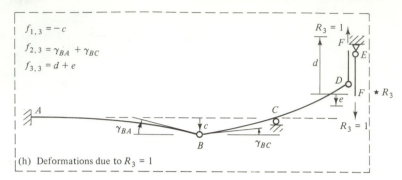

$f_{1,3} = -c$

$f_{2,3} = \gamma_{BA} + \gamma_{BC}$

$f_{3,3} = d + e$

(h) Deformations due to $R_3 = 1$

Figure 7.2 (*continued*)

directions exists at that location. Consider Figures 7.2c and 7.2e at point B: immediately to the left of point B, clockwise (CW) rotations are positive but the observed direction of vector θ_{BA} is counterclockwise (CCW); immediately to the right of point B, counterclockwise rotations are positive but the observed direction of vector θ_{BC} is CW; therefore, the angular deformation (discontinuity in slope) at point B in Figure 7.2e is $-(\theta_{BA} + \theta_{BC})$ since θ_{BA} and θ_{BC} are absolute values. Also consider member DE in Figures 7.2c and 7.2e: point E cannot move vertically, so point F on part EF does not move vertically in Figure 7.2e; on part DF, point F moves vertically downward and the positive direction of R_3 is upward at this location; therefore, the discontinuity in axial deformation at point F in Figure 7.2e is $-\delta$ since δ is an absolute value.

Since the soil conditions under each support are different and the reactions are not equal, unequal settlement of the supports occurs. Differential settlement of the supports causes internal forces which may need to be accounted for in addition to the internal forces due to loads. Suppose supports A and E settle 0.25 in. (vertical settlement directions are implicitly downward), support B settles 0.75 in., and support C settles 0.625 in. Structurally, this is the same as zero settlement of supports A and E, 0.50 in. settlement of support B, and 0.375 in. settlement of support C. Note that settlement of support B occurs at and in the direction of R_1; this settlement is a specified deformation, $D_1^s = 0.500$ in., in the Force Method. However, settlement of support C does not coincide with a redundant vector and must be treated as contributing to the deformations due to loads as shown in Figure 7.2d.

By superposition (see Fig. 7.2),

figure (a) = figure (d) + figure (e) + R_1 * figure (f) + R_2 * figure (g) + R_3 * figure (h)

where R_1, R_2, and R_3 are unknown values of the redundants which, when determined, will adjust (or scale) the indicated released structure deformations such that the algebraic summation of released structure deformations at each released location exactly agrees with the deformations, if any exist, in figure (a).

The detailed procedural steps involved in applying the system approach of the Force Method are:

Step 1. See Figures 7.2a–c. A sufficient number of redundants (R_1, R_2, . . . , R_n), are carefully chosen and removed from the statically indeterminate structure such that a stable and determinate structure remains. For convenience of discussion purposes, this structure is referred to as the *released structure.* Vector directions are chosen for each redundant—a pair of vector directions must be chosen for each internal force selected as a redundant.

Step 2. See Figures 7.2d and e. The given load case is applied to the released structure and the deformations $(D_1, D_2, \ldots, D_n)$, at and in the directions of the redundants are calculated. (Deformations are relative changes in deflections and discontinuities in slopes.) As a group these released structure deformations form a deformation matrix $\{D\}$. These deformations of the released structure are not compatible with the behavior of the actual structure and in step 4 they must be removed.

Step 3. See Figures 10.2f–h. Note: Differential support settlement vectors which do not coincide with a redundant vector position and line of action must be treated as contributing to the deformations due to loads (see Fig. 7.2d). A unit value for each of the redundants is applied *one at a time* to the released structure. At and in the direction of all redundants, deformations are computed due to each unit redundant. These deformations are referred to as *flexibility coefficients,* ($f_{i,j}$ where j ranges from 1 to n for each i value); as a group they form a flexibility matrix $[F]$. Note that Figure 7.2f is the deflected shape of the released structure due to $R_1 = 1$; the deformations in this figure define all of the elements of column 1 of matrix $[F]$. Similarly, Figures 7.2g and h, respectively define all of the elements of columns 2 and 3, respectively, of matrix $[F]$. That is, matrix $[F]$ is computed one column at a time. An arbitrary element in $[F]$ is denoted by $f_{i,j}$ which has the following meanings:

a. Structurally—i is the location of the deformation due to a unit redundant applied at j.

b. Mathematically—i is the matrix row number and j is the matrix column number of the position where $f_{i,j}$ is located in matrix $[F]$.

Step 4. Deformations computed in steps 2 and 3 do not exist in the original structure and they must be removed, that is, compatibility of deformation at each redundant location in the original structure must be restored. A compatibility of deformation equation is written at and in the direction of each redundant ensuring that the deflected shape of the original structure is not violated. Each compatibility equation is a linear combination of deformations at a redundant location. For generality, consider the ith equation which linearly combines deformations at the ith redundant location as follows: The ith deformation from step 2 is algebraically added to each of the ith deformations computed in step 3 times an unknown redundant value for each redundant and the resulting algebraic sum must match the specified deformation behavior, D_i^s, at the ith redundant location in the original structure.

For the conceptual example:

$D_1^s = -0.50$ in. (differential settlement of support B)

$D_2^s = 0$ (no discontinuity in slope exists at point B in Fig. 7.2a)

$D_3^s = 0$ (no discontinuity in axial deformation exists at point F in Fig. 7.2a)

and the compatibility equations of consistent deformations are

$$D_1 + f_{1,1}R_1 + f_{1,2}R_2 + f_{1,3}R_3 = D_1^s$$
$$D_2 + f_{2,1}R_1 + f_{2,2}R_2 + f_{2,3}R_3 = D_2^s$$
$$D_3 + f_{3,1}R_1 + f_{3,2}R_2 + f_{3,3}R_3 = D_3^s$$

Using matrix notation and the definition of matrix multiplication, these simultaneous equations can be written as follows:

$$\begin{Bmatrix} D_1 \\ D_2 \\ D_3 \end{Bmatrix} + \begin{bmatrix} f_{1,1} & f_{1,2} & f_{1,3} \\ f_{2,1} & f_{2,2} & f_{2,3} \\ f_{3,1} & f_{3,2} & f_{3,3} \end{bmatrix} \begin{Bmatrix} R_1 \\ R_2 \\ R_3 \end{Bmatrix} = \begin{Bmatrix} D_1^s \\ D_2^s \\ D_3^s \end{Bmatrix}$$

and, for matrix algebra purposes, the matrix equation of system compatibility is $\{D\} + [F]\{R\} = \{D^s\}$.

Step 5. The simultaneous equations of system compatibility formulated in step 4 are solved for the redundants.

Step 6. The redundants calculated in step 5 are treated as a part of the given loading for the original structure. Using as many free body diagrams as necessary, equations of static equilibrium are written and solved to determine enough of the other external or/and internal forces to enable one to draw the shear and moment diagrams, find member axial forces, and so on.

7.3 CHOOSING THE REDUNDANTS

Alternative sets of redundant choices for the conceptual example (Fig. 7.2) are shown in Figure 7.3. For the conceptual example, the author chose the redundants shown in Figure 7.3b to involve all of the concepts and notation needed in the Force Method. Figure 7.3c is the choice of redundants that is most easily visualized by the beginning student in regard to writing the system compatibility equations. Elongation of member DE occurs only due to R_3 and enters only in the calculation of $f_{3,3}$. If segment AB and segment BD had different EI values, Figure 7.3c would not be a good choice of redundants—cantilever beam A to D would involve calculations due to variable EI. However, the choice shown in Figure 7.3d would keep EI constant for beam segments between the redundants. Also, $f_{3,1} = 0$ and $f_{1,3} = 0$ for Figure 7.3d, but all flexibility coefficients for Figure 7.3c are nonzero. Figure 7.3e would be the poorest choice shown since elongation or shortening of member DE would occur for all three of those redundants.

Usually, by choosing internal moments at the joints in a frame or in a continuous beam and internal bar forces in a truss, less time is required for computing the flexibility coefficients than if one chooses the reactions as the redundants.

Consider the frame shown in Figure 7.4. Design requirements and structural economy may lead the designer to choose members such that $EI_c \neq EI_b$. In this case,

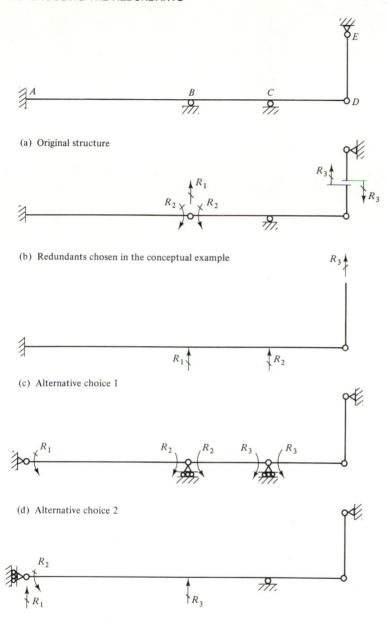

(a) Original structure

(b) Redundants chosen in the conceptual example

(c) Alternative choice 1

(d) Alternative choice 2

(e) Alternative choice 3

Figure 7.3 Alternative redundant choices for conceptual example.

the time required to compute the flexibility matrix, $[F]$, for the choice of redundants shown in Figure 7.4c is much less than for the choice shown in Figure 7.4b. In this chapter, the Method of Virtual Work and visual integration is used to compute $[F]$ for frame problems. Consider this approach of computing column 1 of $[F]$ for Figure 7.4b when $R_1 = 1$, $R_2 = 0$, $R_3 = 0$; all members will have bending moments due to

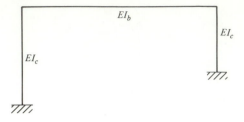

(a) Original structure

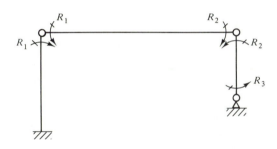

(b) Choice 1 of redundants

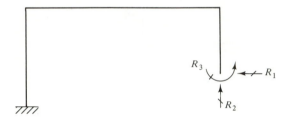

(c) Choice 2 of redundants

Figure 7.4 Redundant choices for a frame.

$R_1 = 1$. However, for $R_1 = 1$, $R_2 = 0$, $R_3 = 0$ in Figure 7.4c, the right-hand side member will not have any bending moment in it. Thus, it is readily seen that less time is required to solve for $[F]$ in Figure 7.4c than in Figure 7.4b. After the redundants are known, the final member-end forces, shear diagrams, and moment diagrams are more rapidly obtainable for the choice of redundants shown in Figure 7.4c than for Figure 7.4b. Therefore, students should resist the temptation to choose the redundants of Figure 7.4b simply because that choice is easier to visualize in regard to writing the system compatibility equations. More thinking is required but less computational time is required if one chooses the redundants shown in Figure 7.4c instead of those shown in Figure 7.4b. For the choice of redundants in Figure 7.5b: (1) none of the flexibility coefficients will be zero; and (2) near the center of the beam, the flexibility coefficients will be nearly equal which may give an ill-conditioned set of simultaneous equations. However, the choice of redundants shown in Figure 7.5c produces: (1) some zero valued flexibility coefficients (note: for a continuous beam of nine spans and eight redundants, 42 of the 64 flexibility coefficients would be zero); and (2) a

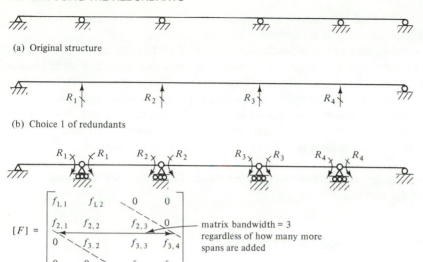

(a) Original structure

(b) Choice 1 of redundants

(c) Choice 2 of redundants and banded flexibility matrix

$$[F] = \begin{bmatrix} f_{1,1} & f_{1,2} & 0 & 0 \\ f_{2,1} & f_{2,2} & f_{2,3} & 0 \\ 0 & f_{3,2} & f_{3,3} & f_{3,4} \\ 0 & 0 & f_{4,3} & f_{4,4} \end{bmatrix}$$

matrix bandwidth = 3 regardless of how many more spans are added

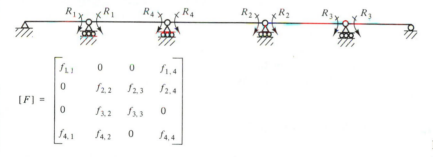

(d) Poor subscript sequence for choice 2 of redundants

$$[F] = \begin{bmatrix} f_{1,1} & 0 & 0 & f_{1,4} \\ 0 & f_{2,2} & f_{2,3} & f_{2,4} \\ 0 & f_{3,2} & f_{3,3} & 0 \\ f_{4,1} & f_{4,2} & 0 & f_{4,4} \end{bmatrix}$$

Figure 7.5 Choice of redundants for a continuous beam.

banded flexibility matrix occurs with the values on the main diagonal being much larger than the other values in any matrix row. In an efficiently organized banded matrix, the nonzero valued coefficients are tightly knitted in a band of constant width about the main diagonal of the matrix and external to the band all values are zero. A banded matrix is mathematically better conditioned, requires less computer memory, and is cheaper to solve than a full matrix. Therefore, the choice of redundants is an important issue. Also the numerical sequencing choice of the redundant subscripts is important as shown in Figure 7.5d. This choice gives a full flexibility matrix with some scattered zero values; that is, this poor numerical sequencing choice of the redundant subscripts does not give a banded flexibility matrix.

Guidelines for choosing the redundants are:

1. If the original structure is symmetric, choose the redundants such that the released structure is symmetric.

2. If possible, choose the redundants such that $[F]$ is banded with the minimum bandwidth and the maximum number of zero elements. For each chosen redundant, sketch the deflected shape of the released structure. Wherever it is possible, no deformations for each redundant should occur beyond the immediately adjacent redundant locations. In general, the best choice of redundants confines the deflected shape of the released structure for each redundant to the least neighborhood of redundant locations.

In the learning stage, the author specifies the choice of redundants to emphasize the principles of the method.

7.4 FORMULAS FOR SLOPES AND DEFLECTIONS

Information of the type shown in Figure 7.6 can be found in handbooks for a member of constant EI. These formulas for slopes and deflections will be used to determine the deformations due to loads $\{D\}$ and the flexibility coefficients $[F]$ in solving the example problems for continuous beams by the Force Method. Usage of these formulas allows the student to concentrate more fully on learning the principles of the method by minimizing the time needed for computations. Also, usage of these formulas causes the student to think about how the formulas can be used; learning how to think properly is the most important aspect of an engineering student's education. The last two sketches in Figure 7.6 are given to indicate that the released structure need not be statically determinate. Segments of the released structure may be statically indeterminate provided the needed deformations can be readily computed.

7.5 PURPOSE OF INDETERMINATE ANALYSIS

Structural analysis is performed for structural design purposes. In the design process, members must be chosen such that design specifications for deflection, shear, bending moment, and axial force are not violated. Design specifications are written in such a manner that separate analyses are needed for dead loads (permanent loads), live loads (position and/or magnitude vary with time), snow loads, and effects due to wind and earthquakes. Influence lines may be needed for positioning live loads to cause their worst effect. In addition, the structural designer may need to consider the effects due to fabrication and construction tolerances being exceeded; temperature changes; and differential settlement of supports. Numerical values of E and I are needed to perform continuous beam analyses of the latter type, but only relative values of EI are needed to perform analyses due to loads.

In the early stages of design, analyses for loads are performed using relative member properties (all members are related to a chosen base member) and trial member sizes are selected with known cross-sectional properties. Then, the analyses that require numerical values of E and I are performed and appropriately combined with the previous analyses to determine if the trial member sizes are okay to use. If the trial

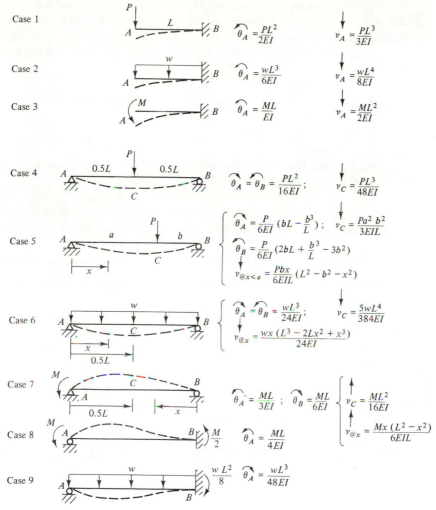

Figure 7.6 Formulas for slopes and deflections. In each sketch shown above, the member length is L and EI is constant. Arrows are used to denote the correct directions of the rotations, θ, and the vertical deflections, v.

sizes are not okay, new trial member sizes are selected and used in repeating all analyses. The member selection, analysis, and investigation process continues, if necessary, until the structural designer finds an acceptable set of member sizes.

In the examples, the author chooses to demonstrate separate analyses of the type that are performed in the first cycle of the iterative design procedure.

7.6 BEAM EXAMPLES

In the continuous beam examples, the author chooses internal moments at the supports as the redundants. The students should also solve some of these example problems by choosing reactions as the redundants. Hopefully, a full understanding of the method will thereby be obtained and the deduction will be made that the author's choice of redundants is computationally the better choice.

The example problems demonstrate analyses for the following conditions:

1. Gravity-type loads.
2. Connection misfits discovered during construction.
3. Temperature effects.
4. Settlement of supports.
5. Influence lines.
6. Member fixed ended moments due to loads and stiffness coefficients (needed for the Displacement Method in Chapters 8, 9, 13).

The formulas shown in Figure 7.6 are used to compute the flexibility matrix and, to the extent possible, the system deformations due to the above conditions.

EXAMPLE 7.1   ANALYSIS FOR A GRAVITY-TYPE LOADING ⎯⎯⎯⎯⎯⎯⎯⎯⎯⎯

EI is constant from A to C. Solve for the chosen redundant in Figure 7.7a.

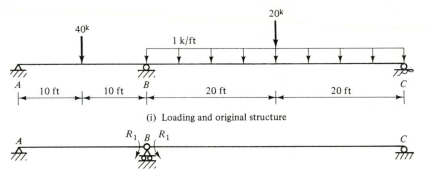

(i) Loading and original structure

(ii) Released structure and chosen redundant   **Figure 7.7a**

Solution

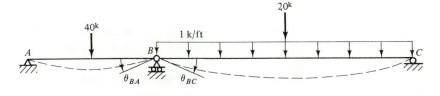

Figure 7.7b Deformation due to given loading is $D_1 = -(\theta_{BA} + \theta_{BC})$.

Figure 7.7c Flexibility coefficient due to $R_1 = 1$ is $f_{1,1} = \alpha_{BA} + \alpha_{BC}$.

Deformation due to given loading is $D_1 = -(\theta_{BA} + \theta_{BC})$. See Figure 7.7b. Flexibility coefficient due to $R_1 = 1$ is $f_{1,1} = \alpha_{BA} + \alpha_{BC}$. See Figure 7.7c. In Figure 7.7a (i) there is no discontinuity in slope at B; therefore, the specified deformation at the R_1 location is $D_1^s = 0$. The compatibility equation of deformation at B is

$$D_1 + f_{1,1}R_1 = D_1^s$$

Cases 4, 6, and 7 of Figure 7.6 are used to calculate D_1 and $f_{1,1}$. Using Cases 4 and 6 of Figure 7.6,

$$\theta_{BA} = \frac{(40^k)(20\,\text{ft})^2}{16EI}; \qquad \theta_{BC} = \frac{(20^k)(40\,\text{ft})^2}{16EI} + \frac{(1\,\text{k/ft})(40\,\text{ft})^3}{24EI}$$

$$D_1 = -(\theta_{BA} + \theta_{BC}) = -\left(\frac{1000}{EI} + \frac{4666.67}{EI}\right) = \frac{-5666.67\,\text{kft}^2}{EI}$$

Using Case 7 of Figure 7.6,

$$\alpha_{BA} = \frac{(1)(20\,\text{ft})}{3EI}; \qquad \alpha_{BC} = \frac{(1)(40\,\text{ft})}{3EI}$$

$$f_{1,1} = \alpha_{BA} + \alpha_{BC} = \frac{20\,\text{ft}}{EI}$$

The compatibility equation is:

$$\frac{-5666.67\,\text{kft}^2}{EI} + \frac{20\,\text{ft}}{EI}R_1 = 0$$

which gives

$$R_1 = 283.33\,\text{ftk}$$

Final member-end forces for Figure 7.7a (i) are obtained by statics and using the fact that $M_B = 283.33$ ftk. The reaction at B is the sum of the member-end shears which is $71.248^k\uparrow$. See Figure 7.7d. Note that a slash was placed on each unknown member-end force vector. Also the computed values are shown with more digits of precision than the designer will use in subsequent calculations. This was done deliberately to make a point—students should use $\geq$6-digit precision in electronic calculations *as a computerized solution would do*, but should never record less than 3-digit precision.

The moment diagram ordinates at 10-ft intervals are shown in Figure 7.7e.

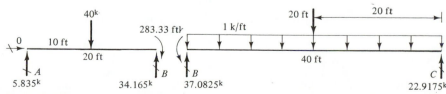

Figure 7.7d

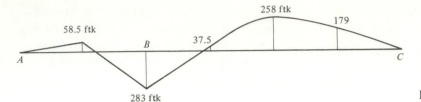

58.5 ftk

258 ftk

179

37.5

B

A

C

283 ftk

Figure 7.7e

EXAMPLE 7.2   ANALYSIS FOR MISFIT OF A MEMBER CONNECTION SPLICE ⎯⎯⎯⎯

See Figure 7.7a for the structure and choice of redundant, but a different loading is to be defined herein. Structural steel members are prepared by a steel fabricator in conformance with the structural engineering drawings and shipped to the construction site. Shipping constraints may not permit fabrication of one 60 ft long piece for the structure shown in Figure 7.7a. In that case, the fabricator cuts the 60 ft long piece into shippable lengths and prepares the pieces to be spliced on the construction site. All steel members arrive at the fabrication shop with some crookedness in them, but within acceptable tolerances. Fabrication is not perfect and is done within another set of tolerances. Consequently, the splice details on pieces being joined do not match perfectly. If the misfit exceeds normal tolerances, the structural engineer will have to decide if the as shipped pieces are acceptable. Suppose the steel erection crew butts two pieces together for splicing and finds that when the bottom of the pieces touch a 0.25-in. gap exists at the top of the pieces as shown in Figure 7.8a. Bolted splices are commonly used and in an end-plate connection splice the top bolts could be torqued to close the gap. This would create a discontinuity in slope of 0.0118315 rad at the connection splice point. If the splice point D is chosen to occur at point B in Figure 7.7a, $D_1^s = 0.0118315$ rad for a Force Method analysis since there is a redundant location and direction coincident with the angular mismatch *being removed* during

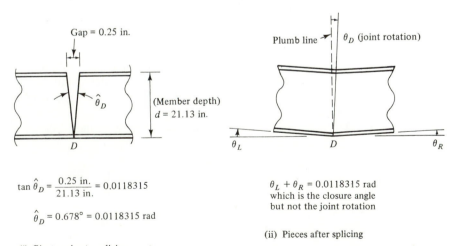

Gap = 0.25 in.

$\hat{\theta}_D$

(Member depth)
$d = 21.13$ in.

D

Plumb line

θ_D (joint rotation)

θ_L

D

θ_R

$$\tan \hat{\theta}_D = \frac{0.25 \text{ in.}}{21.13 \text{ in.}} = 0.0118315$$

$$\hat{\theta}_D = 0.678° = 0.0118315 \text{ rad}$$

$\theta_L + \theta_R = 0.0118315$ rad
which is the closure angle
but not the joint rotation

(ii) Pieces after splicing

(i) Pieces prior to splicing

Figure 7.8a

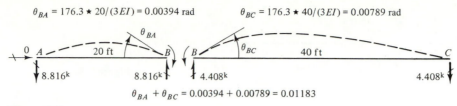

$\theta_{BA} = 176.3 \star 20/(3EI) = 0.00394$ rad $\qquad$ $\theta_{BC} = 176.3 \star 40/(3EI) = 0.00789$ rad

$\theta_{BA} + \theta_{BC} = 0.00394 + 0.00789 = 0.01183$

Figure 7.8b

splicing. Since vertical deflections are prevented at A, B, and C in Figure 7.7a, removal of the angular mismatch induces a moment at the splice point and causes bending of the pieces on each side of the splice point. The analysis by the Force Method for this removal of the splice gap occurring at point B in Figure 7.7a proceeds as follows. Since the original structure and the chosen redundant are the same as in Example 7.1, $f_{1,1} = 20$ ft/EI. The angular removal occurs at R_1; therefore, $D_1^s = 0.0118315$ rad. $D_1 = 0$ since there is no external loading. The compatibility equation, $D_1 + f_{1,1} R_1 = D_1^s$, is

$$0 + \frac{20\,\text{ft}}{EI} R_1 = 0.0118315\,\text{rad}$$

and $R_1 = 0.0118315 EI/20$ ft. In order to find R_1 in units of ftk, the numerical values of E and I must be known. Suppose $E = 29{,}000$ ksi for steel and $I = 1480$ in.[4] are used; $EI = 298{,}056$ kft². Then, $R_1 = 176.3$ ftk and the member-end forces caused during removal of the splice gap at B are as shown in Figure 7.8b. If the top bolts are not torqued more than enough to remove the angular gap, when the loading of Example 7.1 occurs $M_B = 283$ ftk and a partial reappearance of the top gap would occur. This is undesirable since the solution of Example 7.1 assumed there was no discontinuity in slope at B. If possible, splicing at a point of maximum moment is avoided. The more likely point for the splice is at 10 ft to the right of point B where the moment is small for the loading of Example 7.1 and gives two equal lengths of 30 ft to be joined. In that case, the angular discontinuity at point D does not occur at R_1 and the Force Method solution procedure is as shown in Figure 7.8c. D_1 can be obtained by placing the concentrated angle change of 0.0188315 rad on a conjugate beam and finding the left end slope, θ_{BD} (see Fig. 7.8d). The compatibility equation, $D_1 + f_{1,1} R_1 = D_1^s$, is

$$-0.0088736 + \frac{20\,\text{ft}}{EI} R_1 = 0$$

and $R_1 = 132.2$ ftk when $EI = 298056$ kft².

Final member-end forces due to removal of an angular misfit of 0.0188315 rad at 10 ft to the right of point B in the structure of Figure 7.7a are shown in Figure 7.8e.

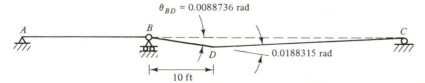

$\theta_{BD} = 0.0088736$ rad

Figure 7.8c

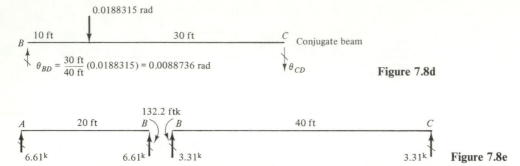

Figure 7.8d

Figure 7.8e

EXAMPLE 7.3   ANALYSIS FOR EFFECTS DUE TO TEMPERATURE CHANGE

See Figure 7.7a for the structure and choice of redundant, but a different loading is to be defined herein. Suppose the top surface of the member from A to C in Figure 7.7a is heated by 60°F more than the bottom surface is heated and the temperature variation from the top surface to the bottom surface is linear. Consider a differential element at any arbitrary location between A and B or between B and C in the released structure. For the described temperature condition, the differential element deforms as shown in Figure 7.9a where a steel member of depth, $d = 21.13$ in. is assumed. Member-end rotations for the released structure subjected to the described temperature change can be found by placing the curvature on conjugate beams as shown in Figure 7.9b; note that the curvature is constant for the described temperature condition.

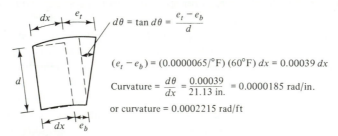

$$d\theta = \tan d\theta = \frac{e_t - e_b}{d}$$

$$(e_t - e_b) = (0.0000065/°F)(60°F)\,dx = 0.00039\,dx$$

$$\text{Curvature} = \frac{d\theta}{dx} = \frac{0.00039}{21.13\text{ in.}} = 0.0000185\text{ rad/in.}$$

or curvature = 0.0002215 rad/ft

Figure 7.9a

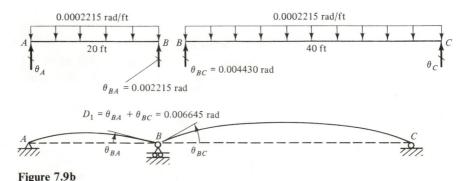

Figure 7.9b

Figure 7.9c

Since the original structure and the chosen redundant are the same as in Example 7.1, $f_{1,1} = 20$ ft/EI and $D_1^s = 0$ since no discontinuity in slope occurs in the original structure at B (location of R_1) when the temperature change takes place. The compatibility equation, $D_1 + f_{1,1}R_1 = D_1^s$ becomes

$$0.006645 \text{ rad} + \frac{20 \text{ ft}}{EI}R_1 = 0$$

and $R_1 = -99.0$ ftk when $EI = 298{,}056$ kft^2. Final member-end forces due to the described temperature change are shown in Figure 7.9c.

EXAMPLE 7.4   ANALYSIS FOR SETTLEMENT OF SUPPORTS

See Figure 7.7a for the structure and choice of redundant, but a different loading is to be defined herein. Suppose that settlement of the supports occurs as shown in Figure 7.10a for the original structure of Figure 7.7a; total settlement of point B is 1.90 in. Relative settlement of the supports, 0.70 in. at point B, causes the released structure to deform as shown in Figure 7.10b. Since the original structure and the chosen redundant are the same as in Example 7.1, $f_{1,1} = 20$ ft/EI and $D_1^s = 0$ since no discontinuity in slope occurs in the original structure at B when the differential settlement happens. The compatibility equation, $D_1 + f_{1,1}R_1 = D_1^s$, becomes

$$0.004375 \text{ rad} + \frac{20 \text{ ft}}{EI}R_1 = 0$$

and $R_1 = -65.2$ ftk when $EI = 298{,}056$ kft^2.

Final member-end forces due to the relative settlement are shown in Figure 7.10c.

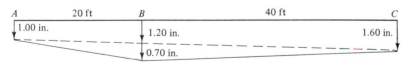

Straight line from A to C **Figure 7.10a**

Figure 7.10b

Figure 7.10c

EXAMPLE 7.5 ———

Find the fixed ended forces for the structure shown in Figure 7.11a.

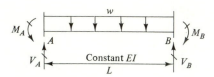

(i) Original structure

(ii) Released structure and chosen redundants

Figure 7.11a

Solution

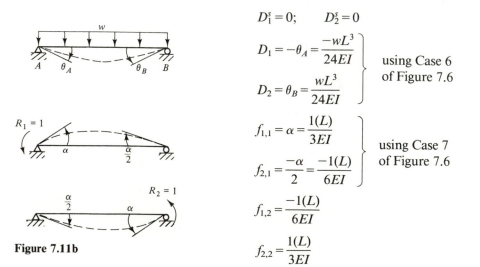

Figure 7.11b

$$D_1^s = 0; \qquad D_2^s = 0$$

$$\left.\begin{array}{l} D_1 = -\theta_A = \dfrac{-wL^3}{24EI} \\[2ex] D_2 = \theta_B = \dfrac{wL^3}{24EI} \end{array}\right\} \begin{array}{l}\text{using Case 6} \\ \text{of Figure 7.6}\end{array}$$

$$\left.\begin{array}{l} f_{1,1} = \alpha = \dfrac{1(L)}{3EI} \\[2ex] f_{2,1} = \dfrac{-\alpha}{2} = \dfrac{-1(L)}{6EI} \end{array}\right\} \begin{array}{l}\text{using Case 7} \\ \text{of Figure 7.6}\end{array}$$

$$f_{1,2} = \dfrac{-1(L)}{6EI}$$

$$f_{2,2} = \dfrac{1(L)}{3EI}$$

The matrix equation of system compatibility, $\{D\} + [F]\{R\} = \{D^s\}$ is

$$\frac{wL^3}{24EI}\begin{Bmatrix} -1 \\ 1 \end{Bmatrix} + \frac{L}{6EI}\begin{bmatrix} 2 & -1 \\ -1 & 2 \end{bmatrix}\begin{Bmatrix} R_1 \\ R_2 \end{Bmatrix} = \begin{Bmatrix} 0 \\ 0 \end{Bmatrix}$$

which gives

$$R_1 = \frac{wL^2}{12} \quad \text{and} \quad R_2 = \frac{-wL^2}{12}$$

and the final member fixed ended forces are

$$M_A = \frac{wL^2}{12} \;\big\rangle; \qquad M_B = \frac{wL^2}{12} \;\big\rangle; \qquad V_A = V_B = \frac{wL}{2} \uparrow$$

EXAMPLE 7.6 _____

Find the member-end moments required to induce a known value of θ_A and $\theta_B = 0$.
See Figure 7.12.

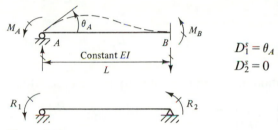

$$D_1^s = \theta_A$$
$$D_2^s = 0$$

Figure 7.12

Solution. $D_1 = 0$ and $D_2 = 0$ since θ_A is coincident with R_1 and there are no
loads between points A and B. Also, the flexibility matrix, $[F]$, is the same as in
Example 7.5 since the member properties and chosen redundants are the same
as in Figure 7.11a. The matrix equation of system compatibility is

$$\begin{Bmatrix} 0 \\ 0 \end{Bmatrix} + \frac{L}{6EI} \begin{bmatrix} 2 & -1 \\ -1 & 2 \end{bmatrix} \begin{Bmatrix} R_1 \\ R_2 \end{Bmatrix} = \begin{Bmatrix} \theta_A \\ 0 \end{Bmatrix}$$

which gives

$$M_A = R_1 = \frac{4EI}{L}\theta_A \;\big\rangle \quad \text{and} \quad M_B = R_2 = \frac{2EI}{L}\theta_A \;\big\rangle$$

Note that $M_B = \frac{1}{2}M_A$. From statics,

$$V_A = \frac{M_A + M_B}{L} = \frac{6EI}{L^2}\theta_A \uparrow \quad \text{and} \quad V_B = \frac{6EI}{L^2}\theta_A \downarrow$$

EXAMPLE 7.7 _____

Find the fixed ended forces required to induce a relative settlement of Δ at the member
ends. See Example 7.5 for $[F]$ and Figure 7.13a.

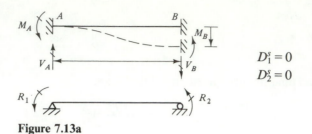

$$D_1^s = 0$$
$$D_2^s = 0$$

Figure 7.13a

Solution

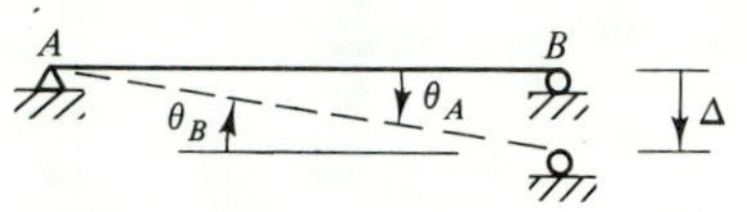

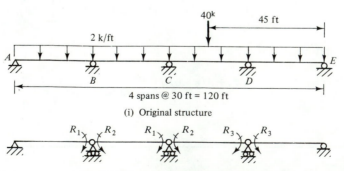

$$\theta_A = \theta_B = \frac{\Delta}{L}$$

$$D_1 = -\theta_A$$

Figure 7.13b

$$D_2 = -\theta_B$$

The matrix equation of system compatibility is (see Figs. 7.13a and 7.13b)

$$\frac{-\Delta}{L}\begin{Bmatrix}1\\1\end{Bmatrix} + \frac{L}{6EI}\begin{bmatrix}2 & -1\\-1 & 2\end{bmatrix}\begin{Bmatrix}R_1\\R_2\end{Bmatrix} = \begin{Bmatrix}0\\0\end{Bmatrix} \quad \text{and} \quad \begin{Bmatrix}R_1\\R_2\end{Bmatrix} = \frac{6EI\Delta}{L^2}\begin{Bmatrix}1\\1\end{Bmatrix}$$

The fixed ended forces are

$$M_A = \frac{6EI\Delta}{L^2} \;\rotatebox{45}{\(\nwarrow\)}; \qquad M_B = \frac{6EI\Delta}{L^2} \;\rotatebox{45}{\(\nwarrow\)};$$

$$V_A = \frac{M_A + M_B}{L} = \frac{6EI\Delta}{L^3}\;\uparrow; \qquad V_B = \frac{6EI\Delta}{L^3}\;\downarrow$$

EXAMPLE 7.8 ——

Solve for the chosen redundants. *EI* is constant and same for all spans (Fig. 7.14a).

Figure 7.14a

Solution

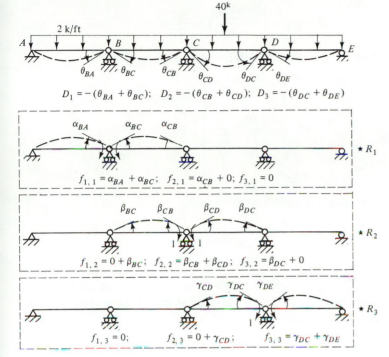

Figure 7.14b

Since all spans have the same L and same EI, using Cases 4, 6, 7 of Figure 7.6:

1. The member-end rotations due to the uniform load only $= \dfrac{2*(30)^3}{24EI} = \dfrac{2250}{EI}$.

2. The member-end rotations due to the concentrated load only $= \dfrac{40*(30)^2}{16EI} = \dfrac{2250}{EI}$.

3. $f_{1,1} = f_{2,2} = f_{3,3} = \dfrac{1*30}{3EI} + \dfrac{1*30}{3EI} = \dfrac{20}{EI}$.

4. All member-end rotations on the end opposite of the applied unit moment are equal to $\dfrac{1*(30)}{6EI} = \dfrac{5}{EI}$.

Therefore,

$$D_1 = -\left(\frac{2250}{EI} + \frac{2250}{EI}\right) = \frac{-4500}{EI}$$

$$D_2 = -\left(\frac{2250}{EI} + \frac{4500}{EI}\right) = \frac{-6750}{EI}$$

$$D_3 = -\left(\frac{4500}{EI} + \frac{2250}{EI}\right) = \frac{-6750}{EI}$$

The matrix equation of system compatibility, $\{D\} + [F]\{R\} = \{D^s\}$, is

$$\begin{Bmatrix} \dfrac{-4500}{EI} \\[2mm] \dfrac{-6750}{EI} \\[2mm] \dfrac{-6750}{EI} \end{Bmatrix} + \begin{bmatrix} \dfrac{20}{EI} & \dfrac{5}{EI} & 0 \\[2mm] \dfrac{5}{EI} & \dfrac{20}{EI} & \dfrac{5}{EI} \\[2mm] 0 & \dfrac{5}{EI} & \dfrac{20}{EI} \end{bmatrix} \begin{Bmatrix} R_1 \\ R_2 \\ R_3 \end{Bmatrix} = \begin{Bmatrix} 0 \\ 0 \\ 0 \end{Bmatrix}$$

Solution of the simultaneous equations gives

$$R_1 = 168.75 \text{ ftk}$$
$$R_2 = 225.00 \text{ ftk}$$
$$R_3 = 281.25 \text{ ftk}$$

The reactions are found by considering FBD of each span to find the member-end shears—the algebraic sum of the member-end shears at the internal support locations gives the reactions. See Figure 7.14c.

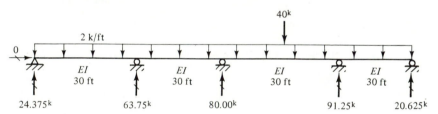

Figure 7.14c

EXAMPLE 7.9

See Figure 7.14a for the structure and chosen redundants. Solve for the redundants due to the support settlements shown in Figure 7.15a. $EI = 368,542 \text{ kft}^2$.

Solution. Since the relative support settlement vectors do not coincide with the

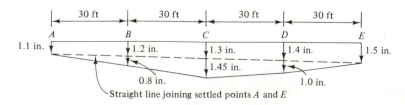

Figure 7.15a

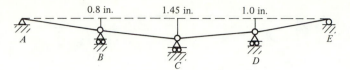

Figure 7.15b

Figure 7.15c

chosen redundant vectors, the relative settlements must be treated as a loading contribution and the specified deformations are zero. See Figure 7.15b.

$$D_1 = \theta_{BA} + \theta_{BC} = \frac{0.80\,\text{in.}}{360\,\text{in.}} + \frac{0.80 - 1.45}{360} = 0.000416667 \text{ rad}$$

$$D_2 = \theta_{CB} + \theta_{CD} = \frac{1.45 - 0.8}{360} + \frac{1.45 - 1.00}{360} = 0.00305556 \text{ rad}$$

$$D_3 = \theta_{DC} + \theta_{DE} = \frac{1.00 - 1.45}{360} + \frac{1.00}{360} = 0.00152778 \text{ rad}$$

The flexibility matrix is the same as in Example 7.8 with $EI = 368{,}542 \text{ kft}^2$. Therefore, the matrix equation of system compatibility is

$$
\begin{Bmatrix} 0.000416667 \\ 0.00305556 \\ 0.00152778 \end{Bmatrix} +
\begin{bmatrix} 0.0000542679 & 0.0000135670 & 0 \\ 0.0000135670 & 0.0000542679 & 0.0000135670 \\ 0 & 0.0000135670 & 0.0000542679 \end{bmatrix}
\begin{Bmatrix} R_1 \\ R_2 \\ R_3 \end{Bmatrix} =
\begin{Bmatrix} 0 \\ 0 \\ 0 \end{Bmatrix}
$$

and $R_1 = 5.85$ ftk; $R_2 = -54.1$ ftk; $R_3 = -14.6$ ftk. The reactions are as shown in Figure 7.15c.

EXAMPLE 7.10

See Figure 7.14a for the structure and chosen redundants. The influence line ordinates for M_B are desired at 5-ft intervals from A to E. $EI = 368{,}542 \text{ kft}^2$.

Solution. According to the Muller-Breslau principle, the influence line for M_B is the deflection curve for a specified unit deformation at B; that is, $D_1^s = 1$ rad is induced at point B, but $D_2^s = 0$ and $D_3^s = 0$.

The matrix equation of system compatibility is

$$
\begin{Bmatrix} 0 \\ 0 \\ 0 \end{Bmatrix} +
\begin{bmatrix} 0.0000542679 & 0.0000135670 & 0 \\ 0.0000135670 & 0.0000542679 & 0.0000135670 \\ 0 & 0.0000135670 & 0.0000542679 \end{bmatrix}
\begin{Bmatrix} R_1 \\ R_2 \\ R_3 \end{Bmatrix} =
\begin{Bmatrix} 1 \\ 0 \\ 0 \end{Bmatrix}
$$

which gives $M_B = R_1 = 19{,}743$ ftk; $M_C = R_2 = -5265$ ftk; $M_D = R_3 = 1316$ ftk. Using these member-end moments for each span and case 7 of Figure 7.6, the influence line ordinates for M_B are computed and shown in Figure 7.16. Note that for spans BC and CD, case 7 is used for each end of the span and the algebraic sum of the deflections for the member-end moments is obtained.

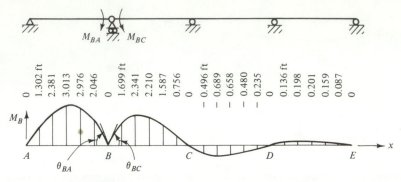

Influence line for M_B: $\theta_{BA} + \theta_{BC} = 1$ rad; $M_{BA} = M_{BC} = 19743$ ftk   **Figure 7.16**

EXAMPLE 7.11

See Figure 7.14a for the structure and chosen redundants. The influence line for M_F is desired and point F is 10 ft to the right of point B. $EI = 368,542$ kft².

Solution. The loading is a pair of unknown moments at F to induce a discontinuity in slope of 1 rad at F. Since the desired unity discontinuity does not coincide with a redundant, deformations at the redundant locations must be computed as shown in Figure 7.17a.

The matrix equation of system compatibility is

$$\left\{ \begin{array}{c} 0.666667 \\ 0.333333 \\ 0 \end{array} \right\} + \left[\begin{array}{ccc} 0.0000542679 & 0.0000135670 & 0 \\ 0.0000135670 & 0.0000542679 & 0.0000135670 \\ 0 & 0.0000135670 & 0.0000542679 \end{array} \right] \left\{ \begin{array}{c} R_1 \\ R_2 \\ R_3 \end{array} \right\} = \left\{ \begin{array}{c} 0 \\ 0 \\ 0 \end{array} \right\}$$

which gives $M_B = R_1 = -11,407$ ftk; $M_C = R_2 = -3510$ ftk; $M_D = R_3 = 877.5$ ftk; $M_F = 8775$ ftk; $\Delta_F = 4.52$ ft↑. Using these member-end moments for each span and case 7 of Figure 7.6, the influence line ordinates for M_F are computed and shown in Figure 7.17b.

7.7 FRAME EXAMPLES

Examples are shown for frames having only bending members which are assumed to be axially rigid. If the released structure's deflected shape involves joint translation,

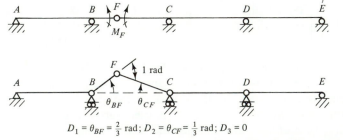

$D_1 = \theta_{BF} = \frac{2}{3}$ rad; $D_2 = \theta_{CF} = \frac{1}{3}$ rad; $D_3 = 0$ **Figure 7.17a**

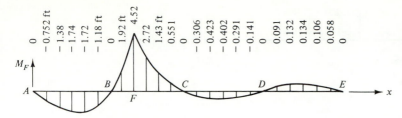

Figure 7.17b

usage of the formulas shown in Figure 7.6 may be too complicated or not even possible. In either event, the author recommends using the Virtual Work Method and visual integration to compute the system deformations and flexibility matrix. This provides additional opportunities for the student to master statics and drawing moment diagrams; also this approach directly establishes the correct sign of each computed deformation.

EXAMPLE 7.12 _____

EI is constant and the same for both members. Solve for the chosen redundant using the formulas of Figure 7.6, find the reactions, and draw the moment diagram. Ignore axial deformations of the members. Sketch the qualitative deflected shape. See Figure 7.18a.

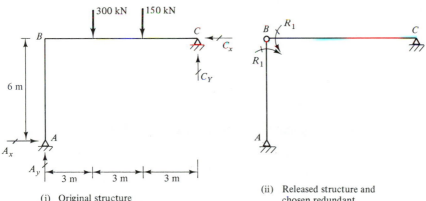

(i) Original structure

(ii) Released structure and chosen redundant

Figure 7.18a

Solution

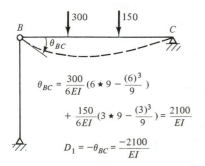

$$\theta_{BC} = \frac{300}{6EI}(6 \star 9 - \frac{(6)^3}{9})$$

$$+ \frac{150}{6EI}(3 \star 9 - \frac{(3)^3}{9}) = \frac{2100}{EI}$$

$$D_1 = -\theta_{BC} = \frac{-2100}{EI}$$

$$\theta_{BC} = \frac{1 \star 9}{3EI} = \frac{3}{EI}$$

$$\theta_{BA} = \frac{1 \star 6}{3EI} = \frac{2}{EI}$$

$$f_{1,1} = \theta_{BA} + \theta_{BC} = \frac{5}{EI}$$

Figure 7.18b

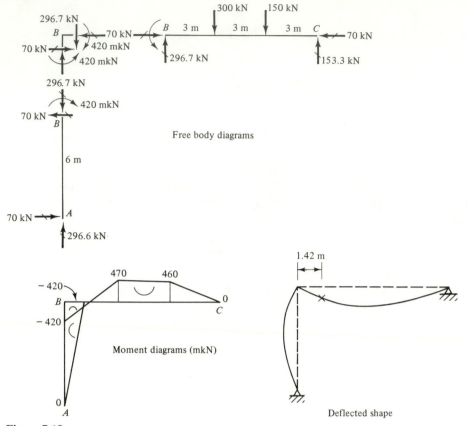

Free body diagrams

Moment diagrams (mkN)

Deflected shape

Figure 7.18c

See Figure 7.18b. The compatibility equation, $D_1 + f_{1,1} R_1 = D_1^s$, is

$$\frac{-2100}{EI} + \frac{5}{EI} R_1 = 0$$

which gives $R_1 = 420$ mkN.

Member-end shears are readily obtained by considering free body diagrams of the individual members and using the equations of statics. Axial forces are found by considering a free body diagram of joint B; alternatively, in this structure, the entire structure can be the free body diagram used to find the reactions after the member-end shear A_x is obtained. The moment diagram is drawn on the compression side of the neutral axis for each member. See Figure 7.18c.

EXAMPLE 7.13 ————————————————————————————————

EI is constant and the same for all members. Solve for the chosen redundant, find the reactions, and draw the moment diagram. Use Method of Virtual Work and visual integration to compute all needed deformations. See Figure 7.19a.

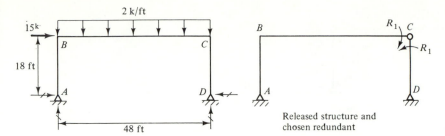

Released structure and
chosen redundant

Figure 7.19a

Solution. The released structure is subjected to the given loading to find D_1.

Areas of the $\dfrac{M}{EI}$ diagrams shown in Figure 7.19b are:

$$A_1 = \frac{1}{2} * 18 * \frac{270}{EI} = \frac{2430}{EI}$$

$$A_2 = \frac{1}{2} * 48 * \frac{2574}{EI} = \frac{61{,}776}{EI}$$

$$A_3 = \frac{1}{3} * 48 * \left(\frac{-2304}{EI}\right) = \frac{-36{,}864}{EI}$$

Since the Virtual Work Method is being used to find D_1 (the deformation at the location of R_1) and since R_1 is a pair of moment vectors, the released structure

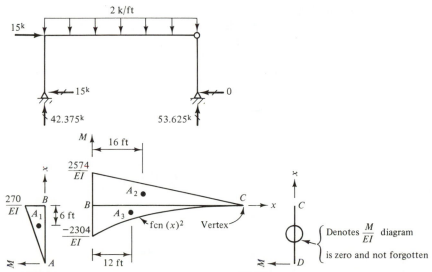

Figure 7.19b M/EI diagrams drawn by parts for each member—note the $+M$ and x direction choices for each member.

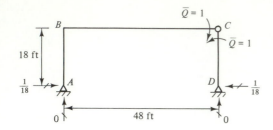

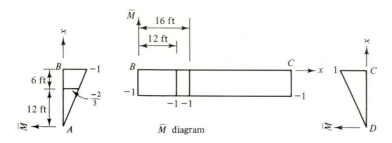

$$\bar{Q} \star D_1 = \sum_{i=1}^{3} \bar{M}_i \star A_i = \frac{-2}{3} \star \frac{2430}{EI} + (-1) \star \frac{61,776}{EI} + (-1) \star \frac{-36,864}{EI} = \frac{-26,532}{EI}$$

Since $\bar{Q} = 1$, $D_1 = \dfrac{-26,532}{EI}$

Figure 7.19c

must be subjected to a pair of virtual moment vectors $\bar{Q}$ ($\bar{Q} = 1$ is chosen for computational convenience) coincident with the R_1 vectors to find the virtual moment diagrams. Ordinates of the virtual moment diagrams are needed at the locations of the centroids for the areas of the M/EI diagrams due to the real loading. Signs of the $\bar{M}$ ordinates must be consistent with the choice of signs made for the M/EI diagram. See Figure 7.19c.

The areas and their centroidal locations of the M/EI diagram due to $R_1 = 1$ are needed to find $f_{1,1}$.

Areas of $\dfrac{M}{EI}$ diagrams in Figure 7.19d are:

$$A_4 = \frac{1}{2} * 18 * \frac{-1}{EI} = \frac{-9}{EI}$$

$$A_5 = 48 * \frac{-1}{EI} = \frac{-48}{EI}$$

$$A_6 = \frac{1}{2} * 18 * \frac{1}{EI} = \frac{9}{EI}$$

Virtual moment diagram ordinates at the centroidal locations of A_4 through A_6 are needed for $\bar{Q} = R_1 = 1$. See Figure 7.19e.

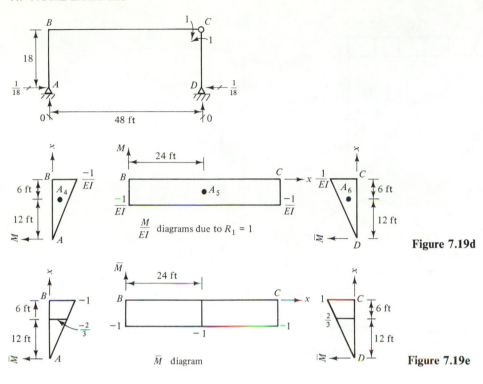

$$\bar{Q}*f_{1,1} = \sum_{i=4}^{6} \bar{M}_i * \Lambda_i = \frac{-2}{3}*\frac{-9}{EI} + (-1)*\frac{-48}{EI} + \frac{2}{3}*\frac{9}{EI} = \frac{60}{EI}$$

and $f_{1,1} = 60/EI$. The compatibility equation, $D_1 + f_{1,1}R_1 = D_1^s$, is

$$\frac{-26{,}532}{EI} + \frac{60}{EI}R_1 = 0 \quad \text{and} \quad R_1 = 442.2\,\text{ftk}$$

Using the fact that $M_c = R_1 = 442.2$ ftk and a free body diagram of member CD, the member-end shear at D is obtained. Considering the entire structure as a free body diagram, the other reactions are obtained. The reactions and Moment Diagram due to the given loading on the original structure are shown in Figure 7.19f.

EXAMPLE 7.14 ──

EI is constant but different for each member. Solve for the chosen redundants and the reactions; draw the moment diagram. See Figure 7.20a.

Solution. The total gravity direction load is 2.025 k/ft $*$ 33.941 ft = 68.73[k] and needs to be resolved into components perpendicular and parallel to member BC. See Figure 7.20b.

$$\theta = \sin^{-1}\left(\frac{12\,\text{ft}}{36\,\text{ft}}\right) = 19.47°$$

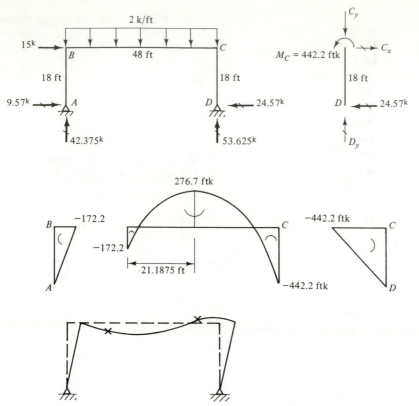

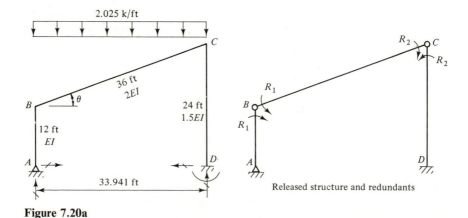

Figure 7.19f

Placement of the given loading on the released structure gives the following M/EI diagrams for each member. Note the choice of $+M$ sign convention for each member. See Figure 7.20c.

To find the deformations due to loads and the flexibility matrix, the following virtual moment diagram ordinates (Figs. 7.20d and 7.20e) are needed if

Figure 7.20a

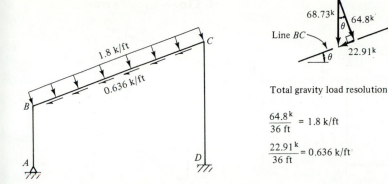

Total gravity load resolution

$$\frac{64.8^k}{36 \text{ ft}} = 1.8 \text{ k/ft}$$

$$\frac{22.91^k}{36 \text{ ft}} = 0.636 \text{ k/ft}$$

Figure 7.20b

the Virtual Work Method is used to compute the deformations. Only the $\bar{M}_1$ ordinates are needed for the deformations due to loads; the $\bar{M}_2$ through $\bar{M}_6$ ordinates are needed to compute the flexibility matrix.

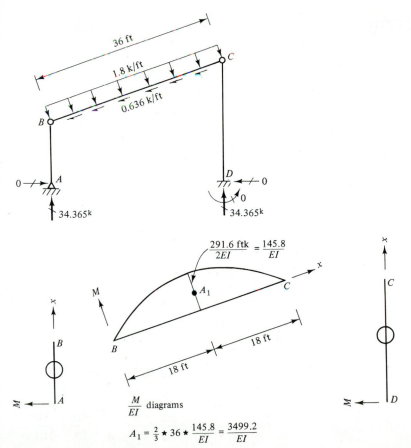

$\dfrac{M}{EI}$ diagrams

$$A_1 = \tfrac{2}{3} \star 36 \star \frac{145.8}{EI} = \frac{3499.2}{EI}$$

Figure 7.20c

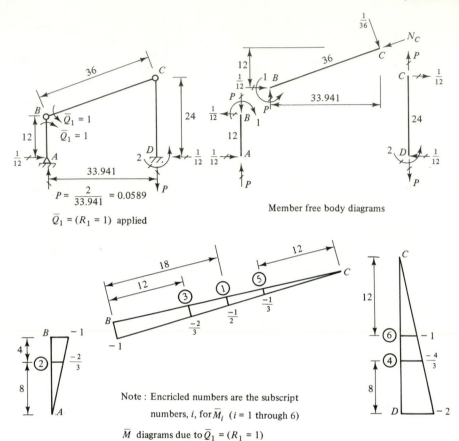

Member free body diagrams

$\bar{Q}_1 = (R_1 = 1)$ applied

Note : Encricled numbers are the subscript

numbers, i, for $\bar{M}_i$ ($i = 1$ through 6)

$\bar{M}$ diagrams due to $\bar{Q}_1 = (R_1 = 1)$

Figure 7.20d

$$\bar{Q}_1 * D_1 = \sum_{i=1}^{1} (\bar{M}_i \text{ due to } \bar{Q}_1) * A_i = \frac{-1}{2} * \frac{3499.2}{EI} = \frac{-1749.6}{EI} \quad \text{and} \quad \bar{Q}_1 = 1;$$

$$D_1 = \frac{-1749.6}{EI}$$

$$\bar{Q}_2 * D_2 = \sum_{i=1}^{1} (\bar{M}_i \text{ due to } \bar{Q}_2) * A_i = \frac{-1}{2} * \frac{3499.2}{EI} = \frac{-1749.6}{EI} \quad \text{and} \quad \bar{Q}_2 = 1;$$

$$D_2 = \frac{-1749.6}{EI}$$

To find the flexibility matrix, the areas of the M/EI diagrams due to $R_1 = 1$ and $R_2 = 1$ are needed. See Figure 7.20f.

$$\bar{Q}_1 * f_{1,1} = \sum_{i=2}^{4} (\bar{M}_i \text{ due to } \bar{Q}_1) * A_i = \frac{-2}{3} * \frac{-6}{EI} + \frac{-2}{3} * \frac{-9}{EI} + \frac{-4}{3} * \frac{-16}{EI} = \frac{94}{3EI}$$

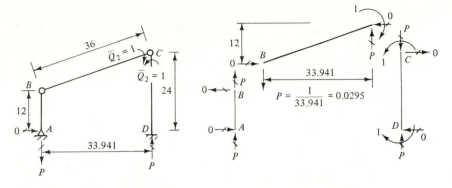

$\bar{Q}_2 = (R_2 = 1)$ applied

Member free body diagrams

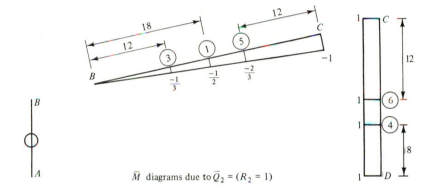

$\bar{M}$ diagrams due to $\bar{Q}_2 = (R_2 = 1)$

Figure 7.20e

$$\bar{Q}_2 * f_{2,1} = \sum_{i=2}^{4} (\bar{M}_i \text{ due to } \bar{Q}_2) * A_i = 0 + \frac{-1}{3} * \frac{-9}{EI} + 1 * \frac{-16}{EI} = \frac{-39}{3EI}$$

$$\bar{Q}_1 * f_{1,2} = \sum_{i=5}^{6} (\bar{M}_i \text{ due to } \bar{Q}_1) * A_i = \frac{-1}{3} * \frac{-9}{EI} + (-1) * \frac{16}{EI} = \frac{-39}{3EI}$$

$$\bar{Q}_2 * f_{2,2} = \sum_{i=5}^{6} (\bar{M}_i \text{ due to } \bar{Q}_2) * A_i = \frac{-2}{3} * \frac{-9}{EI} + 1 * \frac{16}{EI} = \frac{66}{3EI}$$

The matrix equation of system compatibility is

$$\frac{-1749.6}{EI} \begin{Bmatrix} 1 \\ 1 \end{Bmatrix} + \frac{1}{3EI} \begin{bmatrix} 94 & -39 \\ -39 & 66 \end{bmatrix} \begin{Bmatrix} R_1 \\ R_2 \end{Bmatrix} = \begin{Bmatrix} 0 \\ 0 \end{Bmatrix}$$

which gives

$$\begin{Bmatrix} R_1 \\ R_2 \end{Bmatrix} = \begin{Bmatrix} 117.7 \text{ ftk} \\ 149.1 \text{ ftk} \end{Bmatrix}$$

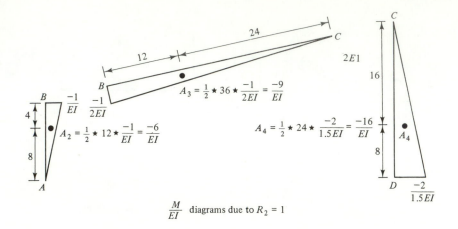

$$\frac{M}{EI} \quad \text{diagrams due to } R_2 = 1$$

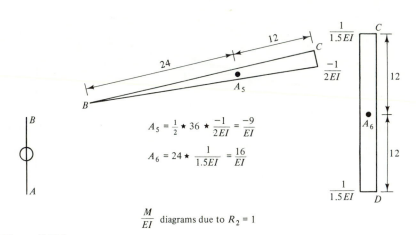

$$\frac{M}{EI} \quad \text{diagrams due to } R_2 = 1$$

Figure 7.20f

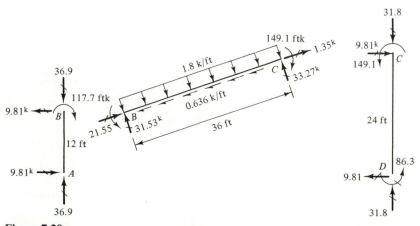

Figure 7.20g

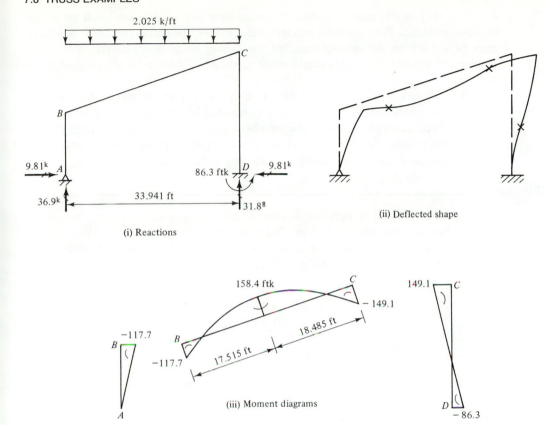

Figure 7.20h

Only the member-end shears and M_D were found by using the free body diagrams in Figure 7.20g; the axial forces at A and D were found by solving for the vertical reactions in Figure 7.20h(i). To find the axial force at B in member BC, for example, the shear and axial force at B in member BA are reversed and resolved into components along the direction BC.

7.8 TRUSS EXAMPLES

Trusses may be statically indeterminate for the following reasons:

1. *Externally indeterminate* because there are more than enough reaction components for static equilibrium.
2. *Internally indeterminate* because there are more than enough members for static equilibrium.
3. *Externally and internally indeterminate* which is a combination of items 1 and 2.

For internally indeterminate trusses, the extra member axial forces must be chosen as the redundants. For externally indeterminate trusses, the extra reaction components may be chosen as the redundants, but the author finds it computationally more convenient to choose an appropriate set of member axial forces as the redundants in this case.

The Virtual Work Method is used to compute deformations due to loads and due to unit redundants. Axial forces are computed by the Method of Joints using correctly directed vectors to depict the member-end actions on the joints. In the virtual work equation, tensile axial forces and member elongation displacements are positive; compressive axial forces and member shortening displacements are negative.

EXAMPLE 7.15 _____

Choose the axial force in member 5 (assume it is in compression) as the redundant for the externally indeterminate truss. Solve for the redundant, the reactions, and all bar forces. All members have the same $E = 29,000$ ksi; areas are A_1 through $A_4 = 10.0$ in.^{2} and $A_5 = 5.00$ in.^{2} See Figure 7.21a.

$$L_1 = L_2 = 28.844 \text{ ft} = 346 \text{ in.}$$
$$L_3 = L_4 = 20.0 \text{ ft} = 240 \text{ in.}$$
$$L_5 = 12 \text{ ft} = 144 \text{ in.}$$

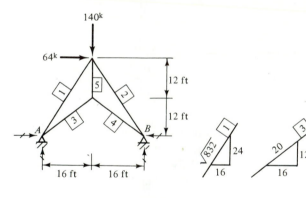

Figure 7.21a

Solution

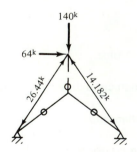

Released structure F_i due to loads

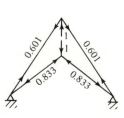

Released structure $\bar{F}_i$ due to $\bar{Q} = (R_1 = 1)$ **Figure 7.21b**

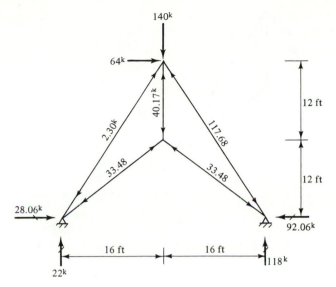

Figure 7.21c

$$\bar{Q}*D_1 = \sum_{i=1}^{5} \bar{F}_i*(\Delta L)_i = \sum_{i=1}^{5} \bar{F}_i*\left(\frac{FL}{EA}\right)_i$$

$$= 0.601*\frac{346}{10E}(-26.44 - 141.82) + 0 + 0 + 0 = -0.12065 \text{ in.}$$

$$\bar{Q}*f_{1,1} = \sum_{i=1}^{5} \bar{F}_i*\left(\frac{\bar{F}L}{EA}\right)_i = \frac{2}{10E}[(0.601)^2*346 + (0.833)^2*240] + \frac{(1)^2*144}{5E}$$

$$= 0.00300351 \text{ in./k}$$

Since $\bar{Q} = 1$, $D_1 = -0.12065$ in. and $f_{1,1} = 0.00300351$ in/k. The compatibility equation, $D_1 + f_{1,1}R_1 = D_1^s$, restoring continuity of released member 5 is

$$-0.12065 + 0.00300351R_1 = 0$$

which gives $R_1 = 40.17^k$ (final bar force in member 5 is 40.17^k compression). Using the Method of Joints and the computed value $F_5 = 40.17^k$ compression, the other bar forces and the reactions are determined as shown in Figure 7.21c.

EXAMPLE 7.16 ──

The structure and the choice of redundant is the same as for Example 7.15, but a different loading is to be defined. Suppose the members are fabricated for bolted connections and the truss is to be assembled at the construction site. Assume that member 5 is fabricated 0.250 in. longer than it should have been, but all other members are correctly fabricated. Assuming that the steel erection crew is able to force the joints apart enough to insert member 5, find the bar forces due to the fabrication error.

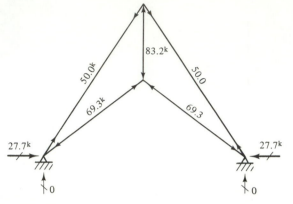

Figure 7.22

Solution. $(\Delta L)_5 = 0.250$ in. and for members 1 through 4, $\Delta L = 0$.

$$\bar{Q} * D_1 = \sum_{i=1}^{5} \bar{F}_i * (\Delta L)_i = 0 + 0 + 0 + 0 + (-1) * 0.250 \text{ in.} = -0.250 \text{ in.}$$

and $D_1 = -0.250$ in.

$$f_{1,1} = 0.00300351 + \frac{(1)^2 * 0.25}{5E} = 0.00300351 + 0.0000017 = 0.00300521$$

Note that the added term accounts for the extra length of member 5 for $f_{1,1}$. There is no appreciable change in $f_{1,1}$ and the added term could be ignored. Since $D_1^s = 0$, the compatibility equation is

$$-0.250 + 0.00300521 R_1 = 0$$

which gives $R_1 = 83.2^k$. See Figure 7.22.

EXAMPLE 7.17 ───

See Example 7.15 for the structure and choice of redundant. Assume that at any time, all members have the same temperature and the temperature is constant along each member length. Find the axial forces due to a temperature rise of 60°F; thermal coefficient of expansion = 0.0000065 in./in./°F.

Solution

$$(\Delta L)_1 = (\Delta L)_2 = (0.0000065 \text{ in./in./°F}) * (60°F) * (346 \text{ in.}) = 0.135 \text{ in.}$$
$$(\Delta L)_3 = (\Delta L)_4 = 0.0000065 * 60 * 240 = 0.0936 \text{ in.}$$
$$(\Delta L)_5 = 0.0000065 * 60 * 144 = 0.0562 \text{ in.}$$
$$\bar{Q} * D_1 = \sum_{i=1}^{5} \bar{F}_i * (\Delta L)_i$$
$$= 2 * [0.601 * 0.135 + (-0.833) * 0.0936] + (-1) * 0.0562 = -0.0499 \text{ in.}$$

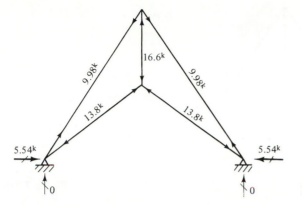

Figure 7.23

Since $D_1^s = 0$ and $f_{1,1} = 0.0030051$ (same as for Example 7.15), the compatibility equation is

$$-0.0499 + 0.0030051 R_1 = 0$$

which gives $R_1 = 16.6^k$. See Figure 7.23.

EXAMPLE 7.18 ——

See Example 7.15 for the structure and choice of redundant. Suppose the site location crew improperly locates the foundation for the right support 0.500 in. farther from the left support than it should have been located. Find the axial forces due to this construction error. Note that this is the same as supposing the right support shifts 0.500 in. more to the right than the left support shifts due to some loading.

Solution. Note that points B and C in the released structure will displace 0.250 in. to the right when point D shifts 0.500 in. to the right (Fig. 7.24a). The vertical movements of points B and C will not be the same, however; the difference in vertical movements of points B and C is D_1 which must be computed. The rigid body rotation of line AC is $\alpha = 0.25$ in./288 in. $= 0.000868$ rad which gives a downward displacement of point C of 192 in. $*0.000868$ rad $= 0.167$ in. The rigid body rotation of line AB is $\beta = 0.25$ in./144 in. $= 0.0017361$ rad which

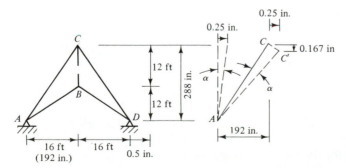

Figure 7.24a

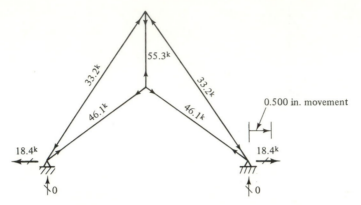

Figure 7.24b

gives a downward displacement of point B of 192 in. $* 0.0017361$ rad $= 0.333$ in. Therefore, a gap of $D_1 = 0.333$ in. $- 0.167$ in. $= 0.166$ in. occurs in member 5 at the released point. The compatibility equation is $0.166 + 0.00300351R_1 = 0$ which gives $R_1 = -55.3^k$. See Figure 7.24b.

EXAMPLE 7.19 _____

All of the truss members are identical: $L = 20$ ft; $A = 8.00$ in.^{2}; $E = 29,000$ ksi. The truss was designed as a simple span truss of 80 ft subjected to concentrated loads of 20^k at each top chord joint. During construction the truss is shored at the interior bottom chord joints and loaded as shown in Figure 7.25a. Choose bar forces 1 through 3 as the redundants and a positive redundant as a tension force. Solve for the bar forces and the reactions.

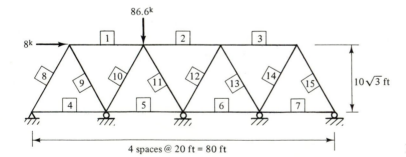

4 spaces @ 20 ft = 80 ft

Figure 7.25a

Solution. See Figure 7.25b. The released structure axial member deformations due to the given loading are computed, $(\Delta L)_i = (FL/EA)_i$ for $i = 1$ through 15, and stored in Table 7.1 for ease of later computations. The bar forces in Figure 7.25c due to unit values for each redundant are stored in Table 7.1 and are needed for the following computations:

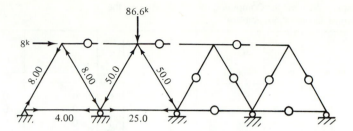

Released structure bar forces, F_i, due to given loading

Figure 7.25b

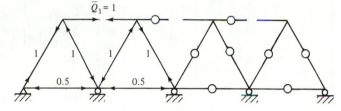

Released structure bar forces, $\bar{F}_2$, due to $\bar{Q}_2 = (R_2 = 1)$

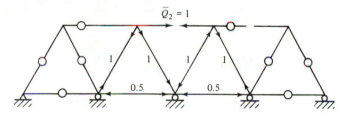

Released structure bar forces, $\bar{F}_1$, due to $\bar{Q}_1 = (R_1 = 1)$

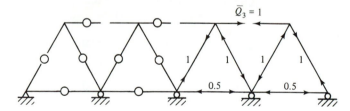

Released structure bar forces, $\bar{F}_3$, due to $\bar{Q}_3 = (R_3 = 1)$

Figure 7.25c

$$\bar{Q}_1 * D_1 = \sum_{i=1}^{15} (\bar{F}_1)_i * (\Delta L)_i = 0.001553$$

$$\bar{Q}_2 * D_2 = \sum_{i=1}^{15} (\bar{F}_2)_i * (\Delta L)_i = -0.01293$$

$$\bar{Q}_3 * D_3 = \sum_{i=1}^{15} (\bar{F}_3)_i * (\Delta L)_i = 0$$

TABLE 7.1 Convenient Tabular Scheme for Computing the Coefficients of the Matrix Equation of System Compatibility

Member number i	L (in.)	A (in.²)	$\frac{L}{EA}$ (in./k)	F (kips)	$\Delta L = \frac{FL}{EA}$ (in.)	$\bar{F}_1$	$\bar{F}_2$ (no units)	$\bar{F}_3$	$\bar{F}_1*\Delta L$ ($\leftarrow$	$\bar{F}_2*\Delta L$ inches	$\bar{F}_3*\Delta L$ $\rightarrow$)
1	240	8.00	0.0010345	0	0	1	0	0	0	0	0
2				0	0	0	1	0	0	0	0
3				0	0	0	0	1	0	0	0
4				4	0.004138	−0.5	0	0	−0.002069	0	0
5				25	0.02586	−0.5	−0.5	0	−0.01293	−0.01293	0
6				0	0	0	−0.5	−0.5	0	0	0
7				0	0	0	0	−0.5	0	0	0
8				8	0.008276	1	0	0	0.008276	0	0
9				−8	−0.008276	−1	0	0	0.008276	0	0
10				−50	−0.05172	−1	1	0	0.05172	−0.05172	0
11				−50	−0.05172	1	−1	0	−0.05172	0.05172	0
12				0	0	0	−1	1	0	0	0
13				0	0	0	1	−1	0	0	0
14				0	0	0	0	−1	0	0	0
15	240	8.00	0.0010345	0	0	0	0	1	0	0	0
Σ	NA	NA	NA	NA	NA	NA	NA	NA	0.001553	−0.01293	0

$$\bar{Q}_1*f_{1,1} = \sum_{i=1}^{15} (\bar{F}_1)_i * \left(\frac{\bar{F}_1 L}{EA}\right)_i = 0.0056897$$

$$\bar{Q}_2*f_{2,1} = \sum_{i=1}^{15} (\bar{F}_2)_i * \left(\frac{\bar{F}_1 L}{EA}\right)_i = -0.0018104$$

$$\bar{Q}_3*f_{3,1} = \sum_{i=1}^{15} (\bar{F}_3)_i * \left(\frac{\bar{F}_1 L}{EA}\right)_i = 0$$

$$\bar{Q}_2*f_{2,2} = \sum_{i=1}^{15} (\bar{F}_2)_i * \frac{\bar{F}_2 L}{EA_i} = 0.0056897$$

$$\bar{Q}_3*f_{3,2} = \sum_{i=1}^{15} (\bar{F}_3)_i * \frac{\bar{F}_2 L}{EA_i} = -0.0018104$$

$$\bar{Q}_3*f_{3,3} = \sum_{i=1}^{15} (\bar{F}_3)_i * \frac{\bar{F}_3 L}{EA_i} = 0.0056897$$

Since the flexibility matrix is always symmetric for linearly elastic structural behavior, the following relations are obtained from matrix symmetry: $f_{1,2} = f_{2,1}$; $f_{1,3} = f_{3,1}$; $f_{2,3} = f_{3,2}$. Also it should be noted that *for this particular released structure,* columns 2 and 3 of the flexibility matrix could have been deduced from the coefficients for column 1 since the geometry, member properties, and bar forces are identical for each unit redundant solution. However, the author chose to show the general approach which is needed when *L, A,* and geometry vary.

$\dfrac{\bar{F}_1^2 L}{EA}$	$\dfrac{\bar{F}_2 \bar{F}_1 L}{EA}$	$\dfrac{\bar{F}_3 \bar{F}_1 L}{EA}$	$\dfrac{\bar{F}_2^2 L}{EA}$	$\dfrac{\bar{F}_3 \bar{F}_2 L}{EA}$	$\dfrac{\bar{F}_3^2 L}{EA}$
($\longleftarrow$			in./k		$\longrightarrow$)
0.0010345	0	0	0	0	0
0	0	0	0.0010345	0	0
0	0	0	0	0	0.0010345
0.0002586	0	0	0	0	0
0.0002586	0.0002586	0	0.0002586	0	0
0	0	0	0.0002586	0.0002586	0.0002586
0	0	0	0	0	0.0002586
0.0010345	0	0	0	0	0
0.0010345	0	0	0	0	0
0.0010345	−0.0010345	0	0.0010345	0	0
0.0010345	−0.0010345	0	0.0010345	0	0
0	0	0	0.0010345	−0.0010345	0.0010345
0	0	0	0.0010345	−0.0010345	0.0010345
0	0	0	0	0	0.0010345
0	0	0	0	0	0.0010345
0.0056897	−0.0018104	0	0.0056897	−0.0018104	0.0056897

The matrix equation of system compatibility is

$$
\begin{Bmatrix} 0.001553 \\ -0.01293 \\ 0 \end{Bmatrix} +
\begin{bmatrix} 0.0056897 & -0.0018104 & 0 \\ -0.0018104 & 0.0056897 & -0.0018104 \\ 0 & -0.0018104 & 0.0056897 \end{bmatrix}
\begin{Bmatrix} R_1 \\ R_2 \\ R_3 \end{Bmatrix} =
\begin{Bmatrix} 0 \\ 0 \\ 0 \end{Bmatrix}
$$

which gives

$$
\begin{Bmatrix} R_1 \\ R_2 \\ R_3 \end{Bmatrix} =
\begin{Bmatrix} 0.599^k \\ 2.741^k \\ 0.872^k \end{Bmatrix}
$$

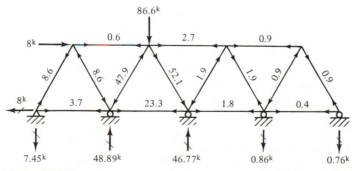

Figure 7.25d

EXAMPLE 7.20 _____

The first interior support from the left end of the truss in Example 7.19 was missing a shim of 0.0625 in. when the loading was applied. This means that the support deflected 0.0625 in. before making contact with the shoring. Determine the bar forces due to a prescribed settlement of 0.0625 in. at the location of the missing shim; these bar forces would have to be added to those of Example 7.19 to obtain the effect of the missing shim when the loading was applied.

Solution. The flexibility matrix is the same as in Example 7.19; since the settlement of the support does not coincide with any of the redundant vector directions, the settlement must be treated as shown in Figure 7.26a. Note that the settlement is grossly exaggerated for clarity in the sketch.

$$\theta_L = \frac{0.0625 \text{ in.}}{240 \text{ in.}} = 0.0002604 \text{ rad}; \qquad a = 17.32 * 12 * \theta_L = 0.054127 \text{ in.}$$

$\theta_R = \theta_L; \qquad b = c = a = 0.054127 \text{ in.}$

$D_1 = a + b = 0.10825 \text{ in.}$ which is the overlap occurring at R_1 release point

$D_2 = -c = -0.054127 \text{ in.}$ which is the gap occurring at R_2 release point

$D_3 = 0$ since no movement occurs at R_3 release point

The matrix equation of system compatibility is

$$\begin{Bmatrix} 0.10825 \\ -0.054127 \\ 0 \end{Bmatrix} + \begin{bmatrix} 0.0056897 & -0.0018104 & 0 \\ -0.0018104 & 0.0056897 & -0.0018104 \\ 0 & -0.0018104 & 0.0056897 \end{bmatrix} \begin{Bmatrix} R_1 \\ R_2 \\ R_3 \end{Bmatrix} = \begin{Bmatrix} 0 \\ 0 \\ 0 \end{Bmatrix}$$

which gives

$$\begin{Bmatrix} R_1 \\ R_2 \\ R_3 \end{Bmatrix} = \begin{Bmatrix} -17.65^k \\ 8.68^k \\ 2.76^k \end{Bmatrix}$$

and the bar forces due only to the settlement are shown in Figure 7.26b. Note that at the support where settlement occurred:

1. Reaction due to settlement is 34.32^k downward.
2. Reaction due to loading in Example 7.19 without settlement is 48.89^k upward.
3. The combined reaction is 14.57^k upward; therefore, the truss makes contact with the shoring after deflecting 0.0625 in. downward.

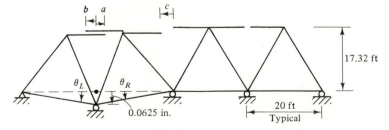

Figure 7.26a

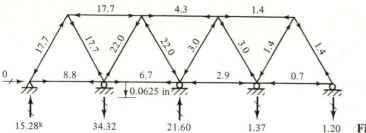

17.7   4.3   1.4

17.7   17.7   22.0   22.0   3.0   3.0   1.4   1.4

0   8.8   6.7   2.9   0.7

↓0.0625 in

15.28k   34.32   21.60   1.37   1.20   **Figure 7.26b**

EXAMPLE 7.21

Choose the bar forces (assume tension) in members 12 and 15 as the redundants to analyze the truss. Members 1 through 8 have an area of 4 in.^{2}; members 9 through 15 have an area of 3 in.^{2}; $E = 29{,}000$ ksi for all members. See Figure 7.27a.

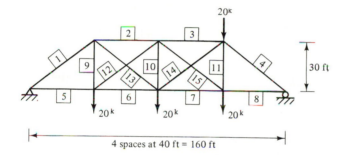

20$^\mathrm{K}$

30 ft

20$^\mathrm{k}$ 20$^\mathrm{k}$ 20$^\mathrm{k}$

4 spaces at 40 ft = 160 ft   **Figure 7.27a**

Solution. Analyses by the Method of Joints are performed for the released structure to find the bar forces due to the following loading conditions:

1. The given loading shown in Figure 7.27a.
2. $R_1 = 1$ (member 12 has a unit tension force in it).
3. $R_2 = 1$ (member 15 has a unit tension force in it).

The bar forces for each of the three load cases are shown in Figure 7.27b and are used as follows:

1. Bar forces for case 1 are needed to compute the elongation or shortening of the members; the change in length of the members are multiplied by case 2 forces and case 3 forces to find D_1 and D_2, respectively, by the Virtual Work Method.
2. Bar forces for cases 2 and 3 are needed to compute the elongation or shortening of the members due to unit values of the redundants; also they are needed to find the flexibility matrix coefficients by the Virtual Work Method.

$$\bar{Q}_1 * D_1 = \sum_{i=1}^{15} (\bar{F}_1)_i * (\Delta L)_i = 0.188967 \quad \text{(see Table 7.2 for the summation)}$$

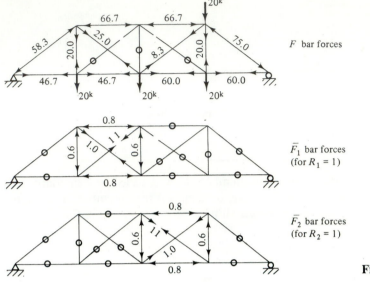

F bar forces

$\bar{F}_1$ bar forces
(for $R_1 = 1$)

$\bar{F}_2$ bar forces
(for $R_2 = 1$)

Figure 7.27b

$$\bar{Q}_2 * D_2 = \sum_{i=1}^{15} (\bar{F}_2)_i * (\Delta L)_i = 0.029768 \quad \text{(see Table 7.2 for the summation)}$$

$$\bar{Q}_1 * f_{1,1} = \sum_{i=1}^{15} (\bar{F}_1)_i * \left(\frac{\bar{F}_1 L}{EA}\right)_i = 0.022069 \quad \text{(see Table 7.2 for the summation)}$$

$$\bar{Q}_2 * f_{2,1} = \sum_{i=1}^{15} (\bar{F}_2)_i * \left(\frac{\bar{F}_1 L}{EA}\right)_i = 0.0014896 \quad \text{(see Table 7.2 for the summation)}$$

$$\bar{Q}_2 * f_{2,2} = \sum_{i=1}^{15} (\bar{F}_2)_i * \left(\frac{\bar{F}_2 L}{EA}\right)_i = 0.022069 \quad \text{(see Table 7.2 for the summation)}$$

The matrix equation of system compatibility is

$$\begin{Bmatrix} 0.188967 \\ 0.029768 \end{Bmatrix} + \begin{bmatrix} 0.022069 & 0.0014896 \\ 0.0014896 & 0.022069 \end{bmatrix} \begin{Bmatrix} R_1 \\ R_2 \end{Bmatrix} = \begin{Bmatrix} 0 \\ 0 \end{Bmatrix}$$

which gives

$$\begin{Bmatrix} R_1 \\ R_2 \end{Bmatrix} = \begin{Bmatrix} -8.51^k \\ -0.77^k \end{Bmatrix}$$

The final bar forces are shown in Figure 7.27c.

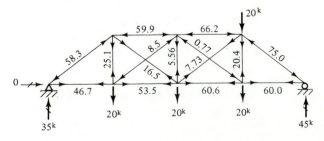

Figure 7.27c

TABLE 7.2 Tabular Scheme for Computing the Matrix Equation Coefficients for Example 7.21

Member number i	L (in.)	A (in.^{2})	$\dfrac{L}{EA}$ (in./k)	F (kips)	$\Delta L = \dfrac{FL}{EA}$ (in.)	$\bar{F}_1$ (no units)	$\bar{F}_2$	$\bar{F}_1 * \Delta L$ (in.)	$\bar{F}_2 * \Delta L$ (in.)	$\dfrac{\bar{F}_1^2 L}{EA}$ (in./k)	$\dfrac{\bar{F}_2 \bar{F}_1 L}{EA}$ (in./k)	$\dfrac{\bar{F}_2^2 L}{EA}$ (in./k)
1	600	4	0.0051724	−58.3	−0.30155	0	0	0	0	0	0	0
2	480	4	0.0041379	−66.7	−0.27600	−0.8	0	0.220800	0	0.0026483	0	0
3	480	4	0.0041379	−66.7	−0.27600	0	−0.8	0	0.220800	0	0	0.0026483
4	600	4	0.0051724	−75.0	−0.38793	0	0	0	0	0	0	0
5	480	4	0.0041379	46.7	0.19324	0	0	0	0	0	0	0
6	480	4	0.0041379	46.7	0.19324	0.8	0	−0.154593	0	0.0026483	0	0
7	480	4	0.0041379	60.0	0.24827	0	0.8	0	−0.198619	0	0	0.0026483
8	480	4	0.0041379	60.0	0.24827	0	0	0	0	0	0	0
9	360	3	0.0041379	20.0	0.08276	−0.6	0	−0.049655	0	0.0014896	0	0
10	360	3	0.0041379	0	0	−0.6	−0.6	0	0	0.0014896	0.0014896	0.0014896
11	360	3	0.0041379	20.0	0.08276	0	−0.6	0	−0.049655	0	0	0
12	600	3	0.0068966	0	0	1.0	0	0.172415	0	0.0068966	0	0
13	600	3	0.0068966	25.0	0.17242	1.0	0	0	0	0.0068966	0	0
14	600	3	0.0068966	8.3	0.05724	0	1.0	0	0.057242	0	0	0.0068966
15	600	3	0.0068966	0	0	0	1	0	0	0	0	0.0068966
Σ	NA	NA	NA	NA	NA	NA	NA	0.188967	0.029768	0.0220690	0.0014896	0.0220690

7.9 EXAMPLES INVOLVING CABLES AS BEAM OR FRAME SUPPORTS

A prospective building owner may desire an outside canopy or an inside balcony with no obstructions beneath the canopy or balcony. In some buildings, this may be accomplished by extending a portion of a floor slab or a roof slab to create a cantilevered overhang. The span or the load or combinations of the span and load may cause the free end of the cantilevered overhang to deflect excessively for an economically feasible slab depth. Cables or sag rods may be used to suspend the free end of the cantilevered overhang from the main structure. Cables or sag rods are capable of developing only a tension force; their shear and moment capacities are nearly zero. Elongation of the cables or sag rods usually needs to be determined to decide whether or not the deflected structure is properly designed. If columns (axial compression members) are used to prop the free end of the cantilevered overhang, shortening of the columns is usually negligible. Columns must be prevented from buckling due to compressive load; this circumstance dictates that the area of a column is much larger than a cable or sag rod would need to be for the same axial force and member length. Elongation or shortening of axially loaded members is inversely proportional to the area of the member. For cable-supported structures, the final force in any cable must be a tension force, otherwise the cable is inactive in regard to providing any support. However, in the *mathematical phase* of structural analysis cables must be allowed to resist compression; the choice of redundants, the type of loading, and geometry may cause them to be in compression at some particular phase of the mathematical solution. Only after the final force is obtained should the structural analyst ask "Is the cable in tension or is it inactive?"

EXAMPLE 7.22 ——

Member BC is a high-strength, stranded wire, steel cable for which $EA = 7337^k$ and the maximum recommended design load is 39^k; member AB is a steel beam for which $EI = 422{,}917$ kft^2. Choose the tension in the cable as the redundant and perform indeterminate analyses for:

1. Account for axial elongation of the cable.
2. Ignore axial elongation of the cable.

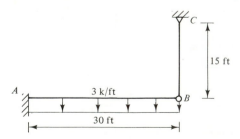

Figure 7.28a

Solution

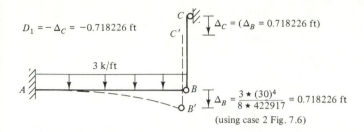

$D_1 = -\Delta_C = -0.718226$ ft

$\Delta_C = (\Delta_B = 0.718226$ ft$)$

3 k/ft

$\Delta_B = \dfrac{3 * (30)^4}{8 * 422917} = 0.718226$ ft

(using case 2 Fig. 7.6)

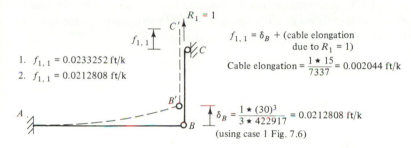

$R_1 = 1$

$f_{1,1}$

1. $f_{1,1} = 0.0233252$ ft/k
2. $f_{1,1} = 0.0212808$ ft/k

$f_{1,1} = \delta_B +$ (cable elongation due to $R_1 = 1$)

Cable elongation $= \dfrac{1 * 15}{7337} = 0.002044$ ft/k

$\delta_B = \dfrac{1 * (30)^3}{3 * 422917} = 0.0212808$ ft/k

(using case 1 Fig. 7.6)

Figure 7.28b

The equation of deformation compatibility, $D_1 + f_{1,1} R_1 = D_1^s$, is:

1. Accounting for axial elongation of the cable $-0.718226 + 0.0233252 R_1 = 0$
   which gives $R_1 = 30.79^k$. The cable elongation is $\dfrac{30.79 * (15 * 12)}{7337} = 0.828$ in.;
   therefore, point B in the original structure deflects downward by 0.828 in.
   and $M_A = 426.3$ ftk.
2. Ignoring axial elongation of the cable $-0.718226 + 0.0212808 R_1 = 0$ which
   gives $R_1 = 33.75^k$ and $M_A = 337.5$ ftk.

Note that $33.75^k/30.79^k = 1.096$ and 337.5 ftk/426.3 ftk $= 0.792$; that is, the
solution that ignores the cable elongation gives a cable tension 9.6% larger than
the correct value and a M_A 20.8% less than the correct value (Fig. 7.28b). In the
structural design of member AB, the designer should use $M_A = 426.3$ ftk in
checking the bending strength requirements to avoid overstressing member AB
by 26.3%. If the designer chooses to use a 1.5-in.-diameter threaded sag rod of
A36 steel, $EA = 51243^k$, instead of the high-strength cable, the sag rod
tension $= 33.29^k$; the sag rod elongation $= 0.117$ in.; and $M_A = 351.3$ ftk. If the
elongation of the sag rod were ignored, the correct value of M_A would be underestimated by only 4.1% (this would be slightly larger if the weight of the sag
rod, clevis, and turnbuckle weight of about 140 lb were accounted for in the
elongation of the sag rod).

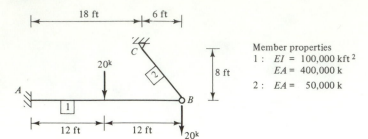

Member properties
1 : EI = 100,000 kft^2
       EA = 400,000 k
2 : EA = 50,000 k

Figure 7.29a

EXAMPLE 7.23

Choose M_A (assume CCW direction) as the redundant. Use Virtual Work Method to compute the needed deformations and do not ignore axial elongations or shortenings of the members. Member 1 spans from A to B. See Figure 7.29a.

Solution

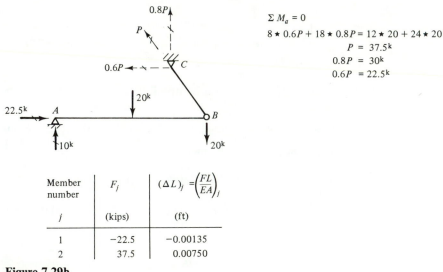

$$\Sigma M_a = 0$$
$$8 \star 0.6P + 18 \star 0.8P = 12 \star 20 + 24 \star 20$$
$$P = 37.5^k$$
$$0.8P = 30^k$$
$$0.6P = 22.5^k$$

Member number	F_j	$(\Delta L)_j = \left(\dfrac{FL}{EA}\right)_j$
j	(kips)	(ft)
1	−22.5	−0.00135
2	37.5	0.00750

Figure 7.29b

$$\bar{Q}*D_1 = \overset{1}{\underset{i=1}{\sum}} \bar{M}_i A_i + \overset{2}{\underset{j=1}{\sum}} \bar{F}_j(\Delta L)_j$$
$$1*D_1 = [(-0.5)*0.01440] + [0.03124*(-0.00135)+(-0.05208)*(0.00750)]$$
$$D_1 = -0.0076328 \text{ rad}$$

To compute the flexibility coefficient, $f_{1,1}$, Figures 7.29b and 7.29c and Figure 7.29d are needed:

$$\bar{Q}*f_{1,1} = \overset{2}{\underset{i=2}{\sum}} \bar{M}_i A_i + \overset{2}{\underset{j=1}{\sum}} \bar{F}_j(\delta L)_j$$

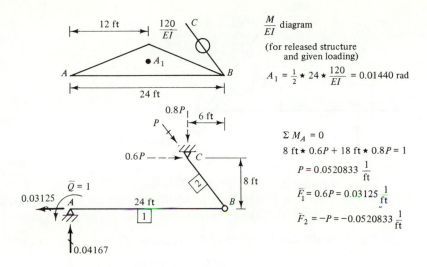

$\dfrac{M}{EI}$ diagram

(for released structure and given loading)

$$A_1 = \tfrac{1}{2} \star 24 \star \dfrac{120}{EI} = 0.01440 \text{ rad}$$

$$\Sigma M_A = 0$$
$$8 \text{ ft} \star 0.6P + 18 \text{ ft} \star 0.8P = 1$$
$$P = 0.0520833 \ \dfrac{1}{\text{ft}}$$
$$\bar{F_1} = 0.6P = 0.03125 \ \dfrac{1}{\text{ft}}$$
$$\bar{F_2} = -P = -0.0520833 \ \dfrac{1}{\text{ft}}$$

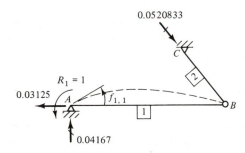

$\bar{M}$ diagram
(due to $\bar{Q} = (R_1 = 1)$)

Figure 7.29c

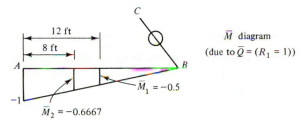

Member number	F_j	$(\delta L)_j = \left(\dfrac{FL}{EA}\right)_j$
j	(1/ft)	(ft/k)
1	0.03125	0.00000188
2	−0.05208	−0.00001042

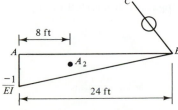

$\dfrac{M}{EI}$ diagram

(for $R_1 = 1$)

$$A_2 = \tfrac{1}{2} \star 24 \star \dfrac{-1}{EI} = -0.00012 \text{ rad/ftk}$$

Figure 7.29d

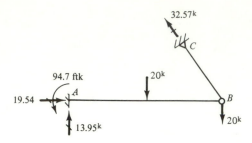

Figure 7.29e

$$1 * f_{1,1} = [(-0.6667)*(-0.00012)]$$
$$+ [0.03125*0.00000188 + (-0.05208)*(-0.00001042)]$$
$$f_{1,1} = 0.00008061 \text{ rad/ftk}$$

The compatibility equation of deformation, $D_1 + f_{1,1}R_1 = D_1^s$, is

$$-0.0076328 + 0.00008061R_1 = 0$$

which gives $R_1 = 94.7$ ftk and the final member-end forces are shown in Figure 7.29e.

EXAMPLE 7.24 ——

Choose the internal moments at B and D as the redundants. Account for elongation or shortening of the members. See Figure 7.30a. The loading is 2 k/ft acting downward on members 1 through 3 in Figure 7.30a.

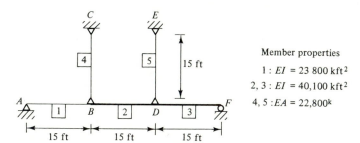

Member properties

1 : EI = 23 800 kft²

2, 3 : EI = 40,100 kft²

4, 5 : EA = 22,800ᵏ

Figure 7.30a

Solution. To the extent possible, Figure 7.6 formulas will be used to compute the needed deformations. By this approach, the formulas of Figure 7.6 give the deformations for no elongation or shortening of members 4 and 5; subsequently, the deformations due to elongation or shortening of members 4 and 5 must be

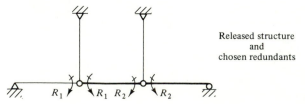

Released structure
and
chosen redundants

Figure 7.30b

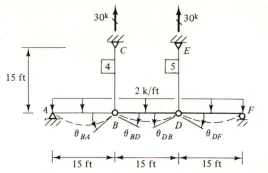

Figure 7.30c

found and added to obtain the total deformations. For no elongation of members 4 and 5 due to the 2 k/ft, the deformations at the released points are shown in Figure 7.30c.

$$\theta_{BA} = \frac{2*(15)^3}{24*23,800} = 0.0118\,\text{rad}$$

$$\theta_{BD} = \frac{2*(15)^3}{24*40,100} = 0.00701\,\text{rad}$$

$$\theta_{DB} = \theta_{DF} = \theta_{BD}$$

Due to the 2 k/ft loading, members 4 and 5 elongate by the amount of $\Delta = 30*15/22,800 = 0.01974$ ft; therefore, points B and D deflect downward the amount Δ, causing the additional deformations shown in Figure 7.30d.

The total deformations due to 2 k/ft are

$$D_1 = -(\theta_{BA} + \theta_{BD}) + \phi_{BA} = -0.0188 + 0.001316 = -0.0175\,\text{rad}$$
$$D_2 = -(\theta_{DB} + \theta_{DF}) + \phi_{DF} = -0.0140 + 0.001316 = -0.0127\,\text{rad}$$

Deformations needed to obtain the flexibility matrix coefficients for no elongation or shortening of members 4 and 5 are shown in Figures 7.30e and 7.30f.

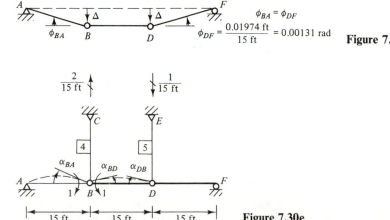

$$\phi_{BA} = \phi_{DF}$$
$$\phi_{DF} = \frac{0.01974\,\text{ft}}{15\,\text{ft}} = 0.00131\,\text{rad}$$

Figure 7.30d

Figure 7.30e

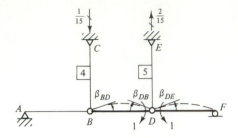

Figure 7.30f

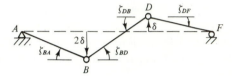

Figure 7.30g

For $R_1 = 1$ (see Figure 7.30e),

$$\alpha_{BA} = \frac{1*15}{3*23,800} = 0.000210084 \text{ rad/ftk}$$

$$\alpha_{BD} = \frac{1*15}{3*40,100} = 0.000124688 \text{ rad/ftk}$$

$$\alpha_{DB} = \tfrac{1}{2}\alpha_{BD} = 0.0000623441 \text{ rad/ftk}$$

For $R_2 = 1$ (see Figure 7.30f),

$$\beta_{DB} = \frac{1*15}{3*40,100} = 0.000124688$$

$$\beta_{BD} = \tfrac{1}{2}\beta_{DB} = 0.0000623441$$

$$\beta_{DF} = \beta_{DB}$$

Deformations to be added to account for nonaxially rigid members 4 and 5 are shown in Figure 7.30g where

$$\delta = \frac{(1/15\text{ ft})*15\text{ ft}}{22,800^k} = 0.00004386 0 \,1/k$$

$$\zeta_{BA} = \frac{2\delta}{15\text{ ft}} = 0.000005848 \,1/\text{ftk}$$

$$\zeta_{BD} = \frac{3\delta}{15\text{ ft}} = 0.000008772 \,1/\text{ftk}$$

$$\zeta_{DF} = \frac{\delta}{15\text{ ft}} = 0.000002924 \,1/\text{ftk}$$

$$\zeta_{DB} = \zeta_{BD}$$

The flexibility matrix coefficients obtained by superposition are

$$f_{1,1} = \alpha_{BA} + \alpha_{BD} + \zeta_{BA} + \zeta_{BD} = 0.000349392 \text{ rad/ftk}$$
$$f_{2,1} = \alpha_{DB} - \zeta_{DB} - \zeta_{DF} = 0.000050648 \text{ rad/ftk}$$

$$f_{2,2} = \beta_{DB} + \beta_{DF} + \zeta_{BA} + \zeta_{BD} = 0.000263996 \text{ rad/ftk}$$
$$f_{1,2} = f_{2,1} = 0.000136384 \text{ rad/ftk}$$

The matrix equation of system compatibility is:

$$\begin{Bmatrix} -0.0175 \\ -0.0127 \end{Bmatrix} + \begin{bmatrix} 0.000349392 & 0.000050648 \\ 0.000050648 & 0.000263996 \end{bmatrix} \begin{Bmatrix} R_1 \\ R_2 \end{Bmatrix} = \begin{Bmatrix} 0 \\ 0 \end{Bmatrix}$$

and

$$\begin{Bmatrix} R_1 \\ R_2 \end{Bmatrix} = \begin{Bmatrix} 44.35 \text{ ftk} \\ 39.60 \text{ ftk} \end{Bmatrix}$$

The final reactions are shown in Figure 7.30h. Member 4 elongates by the amount $33.28 * 15 * 12/22800 = 0.263$ in.

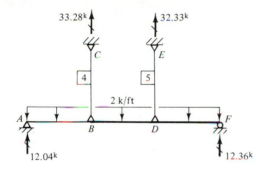

Figure 7.30h Member 4 elongates by the amount $\dfrac{33.28 * 15 * 12}{22800} = 0.263$ in.

PROBLEMS

Solve each problem by the Force Method using the indicated choice of redundant(s). Wherever they are applicable, use the formulas given in Figure 7.6 to compute the deformations due to loads and the flexibility matrix coefficients. Otherwise, use the Conjugate Beam Method for the beam problems and the Virtual Work Method for the frame problems to compute the deformations and the flexibility matrix coefficients. If the member properties are not shown, assume that all members are prismatic and have the same EI or EA. Also calculate the reactions. For the beam and frame problems, except for the problems involving example problems, draw the final shear and Moment Diagrams. Answers are not given for problems preceded by an asterisk.

7.1. Solve Example 7.1 choosing the reaction at B as the redundant.

7.2. Solve Example 7.2 choosing the reaction at B as the redundant.

7.3. Solve Example 7.3 choosing the reaction at B as the redundant.

7.4. Solve Example 7.4 choosing the reaction at B as the redundant.

7.5. Choose Y_2 as the redundant.

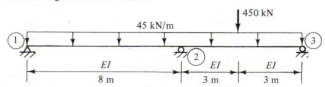

Figure P7.5

7.6. Choose M_2 as the redundant in Problem 7.5.

* 7.7. Choose M_2 as the redundant.

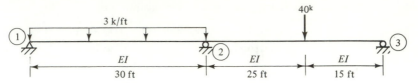

Figure P7.7

7.8. Choose M_2 as the redundant.

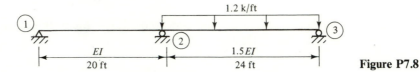

Figure P7.8

* 7.9. Choose M_2 as the redundant.

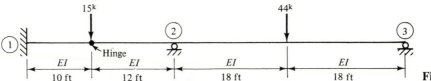

Figure P7.9

7.10. Choose Y_4 as the redundant.

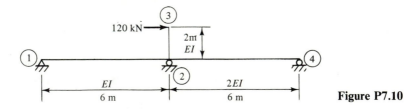

Figure P7.10

* 7.11. Choose M_2 as the redundant.

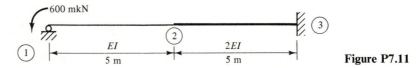

Figure P7.11

* 7.12. Choose M_2 as the redundant.

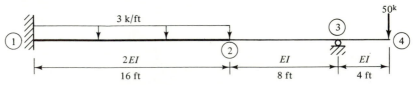

Figure P7.12

* 7.13. Choose M_1 and M_2 as the redundants.

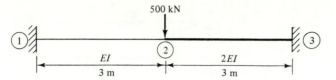

Figure P7.13

* 7.14. Choose M_2 and M_3 as the redundants.

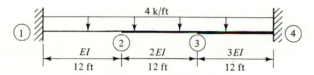

Figure P7.14

7.15. Choose M_2 and M_3 as the redundants.

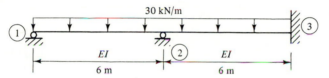

Figure P7.15

* 7.16. Choose M_2 and M_3 as the redundants.

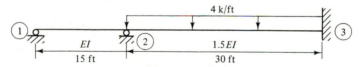

Figure P7.16

7.17. Choose M_1 and M just to the right of point 2 as the redundants.

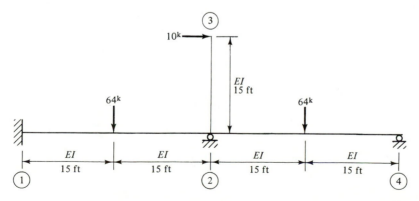

Figure P7.17

7.18. Choose M_2 and M_3 as the redundants.

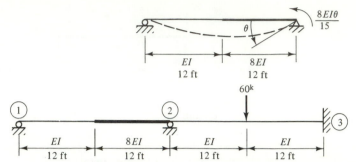

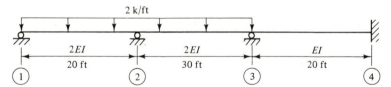

Figure P7.18

7.19. Solve Example 7.8 choosing the reactions at B, C, and D as the redundants.

* 7.20. Choose M_2, M_3, M_4 as the redundants. Positive redundants cause tension in the top of the beam. Only formulate the matrix equation of system compatibility.

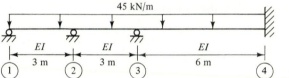

Figure P7.20

7.21. Choose M_2, M_3, and M_4 as the redundants.

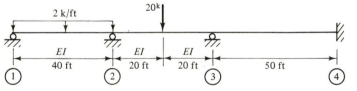

Figure P7.21

* 7.22. Choose M_2, M_3, and M_4 as the redundants.

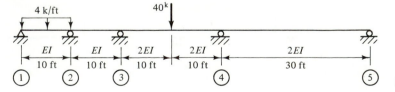

Figure P7.22

* 7.23. Choose M_2, M_3, and M_4 as the redundants.

Figure P7.23

* **7.24.** Choose Y_2 as the redundant. $E = 29,000$ ksi and $I = 612$ in.^{4} Support 2 settles 1 in. downward. Zero settlement occurs at Supports 1 and 3.

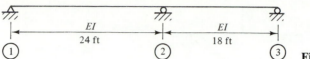

Figure P7.24

* **7.25.** Solve Problem 7.24 choosing M_2 as the redundant.

7.26. Choose M_2 and M_3 as the redundants. $EI = 160,000$ kN/m^2. Support 2 settles 15 mm downward. Zero settlement occurs at supports 1 and 3.

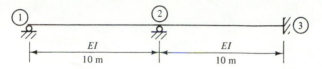

Figure P7.26

7.27. Choose Y_2 and M_3 as the redundants. $EI = 400,000$ kft^2. Support 2 settles 0.06 ft downward. Support 3 settles 0.0524 rad counterclockwise. No other support settlements occur. M_3 produces tension in top of beam. Only formulate the matrix equation of system compatibility due to the loading shown below plus the prescribed support movements stated above.

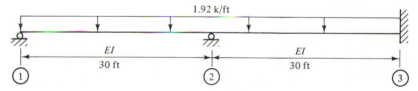

Figure P7.27

7.28. Solve Example 7.9 choosing the reactions at B, C, and D as the redundants.

7.29. Solve Example 7.12 choosing C_X as the redundant.

7.30. Solve Example 7.13 choosing D_X as the redundant.

7.31. Choose M_1 as the redundant. $EA = $ infinity for all members.

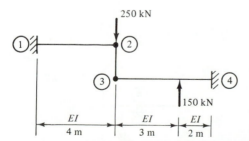

Figure P7.31

In Problems 7.32 through 7.51, ignore axial deformations of all members.

* **7.32.** Choose M_2 as the redundant.

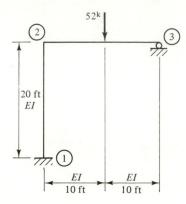

Figure P7.32

7.33. Choose M_2 as the redundant.

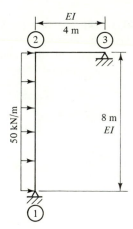

Figure P7.33

* **7.34.** Choose M_2 as the redundant.

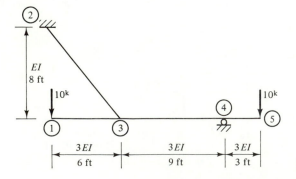

Figure P7.34

7.35. Choose M_2 as the redundant.

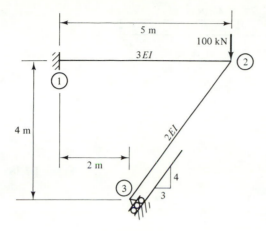

Figure P7.35

7.36. Choose M_2 as the redundant.

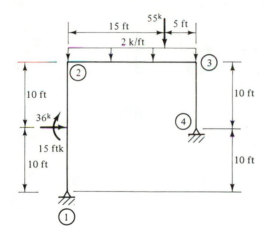

Figure P7.36

* **7.37.** Choose M_2 as the redundant.

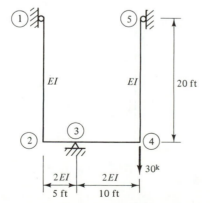

Figure P7.37

7.38. Choose M_2 as the redundant.

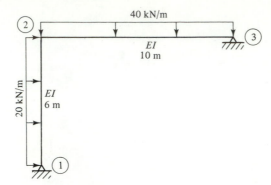

Figure P7.38

7.39. Choose M_1 as the redundant.

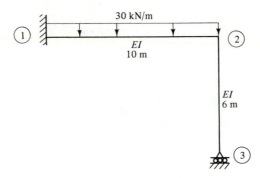

Figure P7.39

7.40. Choose M_3 as the redundant.

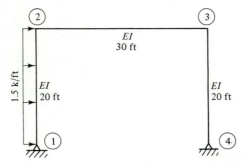

Figure P7.40

7.41. Choose M_2 as the redundant.

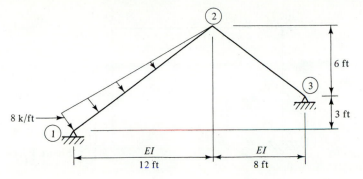

Figure P7.41

7.42. Choose M_2 as the redundant.

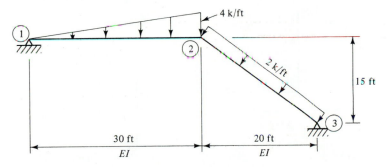

Figure P7.42

7.43. Choose M_2 as the redundant.

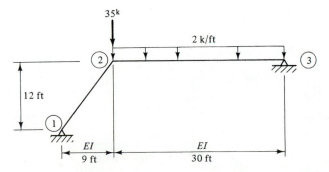

Figure P7.43

7.44. Choose M_2 and M_3 as the redundants.

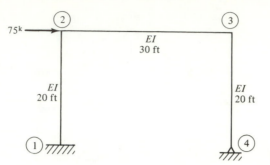

Figure P7.44

* 7.45. Choose M_2 and M_3 as the redundants.

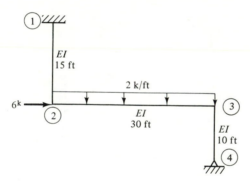

Figure P7.45

* 7.46. Choose M_2 and M_3 as the redundants.

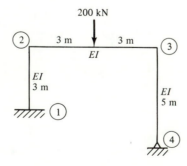

Figure P7.46

* **7.47.** Choose M_3 and M_4 as the redundants.

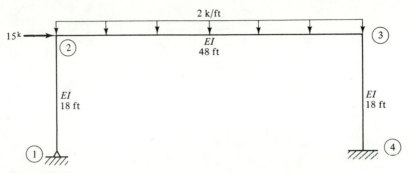

Figure P7.47

7.48. Choose M_2 and M_3 as the redundants.

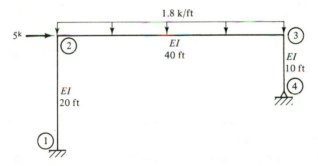

Figure P7.48

7.49. Choose M_2 and M_3 as the redundants.

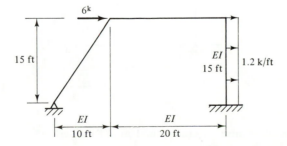

Figure P7.49

7.50. Choose $M_{2(\text{right})}$ and M_4 as the redundants.

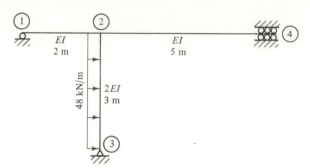

Figure P7.50

* **7.51.** Choose M_2, M_3, and M_4 as the redundants.

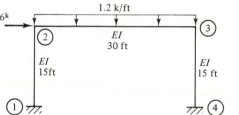

Figure P7.51

7.52. Choose B_X as the redundant in Example 7.15.

7.53. Choose B_X as the redundant in Example 7.16.

7.54. Choose B_X as the redundant in Example 7.17.

* **7.55.** Choose Y_2 as the redundant. EA is same for all members.

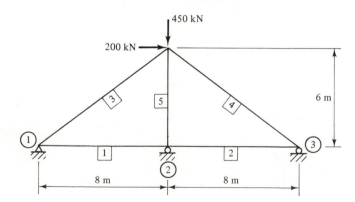

Figure P7.55

* **7.56.** Choose force in member 4 as the redundant in Problem 7.55.

7.57. Choose Y_2 as the redundant. $L/EA = 0.0005$ in./kip for all members.

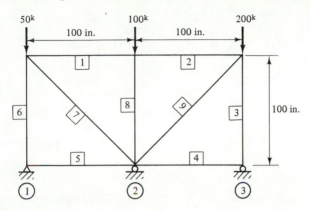

Figure P7.57

7.58. Solve Problem 7.57 choosing force in member 9 as the redundant.

* **7.59.** Choose the reaction at joint 2 as the redundant. EA is the same for all members.

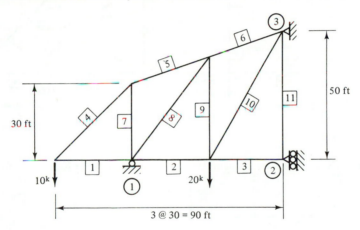

Figure P7.59

7.60. Choose force in member 5 as the redundant. All members have the same EA.

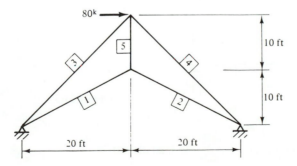

Figure P7.60

7.61. Choose Y_2 as the redundant. EA is the same for all members.

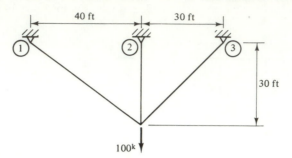

40 ft · 30 ft

30 ft

100^k

Figure P7.61

7.62. Choose Y_3 as the redundant. $E = 29,000$ ksi for all members. $A = 6$ in.^{2} for members 1 through 8. $A = 3$ in.^{2} for members 9 through 16.

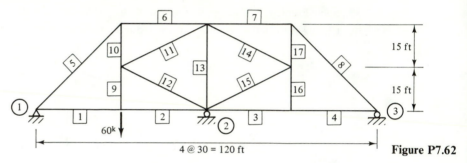

15 ft

15 ft

60^k

4 @ 30 = 120 ft

Figure P7.62

7.63. Choose the force in member 2 as the redundant. EA is the same for all members.

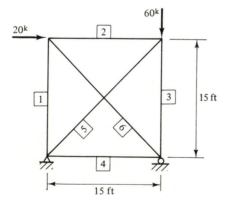

60^k

20^k

15 ft

15 ft

Figure P7.63

7.64. Solve Problem 7.63 with the 60^k load deleted.

7.65. Choose the force in members 9 and 11 as the redundants. $E = 29,000$ ksi for all members. $A = 6$ in.^{2} for members 1 through 13. $A = 4$ in.^{2} for members 14 through 23.

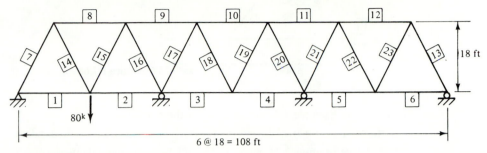

Figure P7.65

7.66. Choose the force in members 7 and 10 as the redundants. $E = 2 \times 10^8$ kN/m^2 for all members. $A = 40$ cm^2 for members 1 through 4. $A = 30$ cm^2 for members 5 and 6. $A = 20$ cm^2 for members 7 through 10.

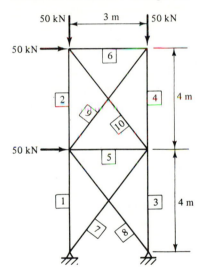

Figure P7.66

7.67. Choose the force in members 12 and 15 as the redundants. $E = 29,000$ ksi for all members. $A = 8$ in.^{2} for members 1 through 8. $A = 4$ in.^{2} for members 9 through 11. $A = 6$ in.^{2} for members 12 through 15.

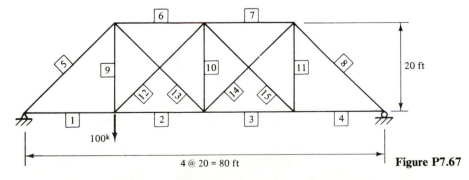

Figure P7.67

7.68. Choose X_2 as the redundant. EI and EA are constant from 1 to 2.

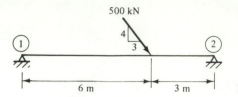

Figure P7.68

7.69. Solve Example 7.22 choosing M_A as the redundant.

7.70. Solve Example 7.23 choosing the force in member 2 as the redundant.

7.71. Solve Example 7.24 choosing M_B and Y_E as the redundants.

7.72. Choose the moment to the right of point 3 as the redundant. $EI = 25,000$ kft² for members 1 and 2. $EA = 4000^k$ and $EI = 0$ for member 3.

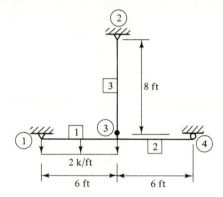

Figure P7.72

7.73. Choose M_1 as the redundant. $E = 2 \times 10^8$ kN/m² for all members. $I_1 = I_2 = 12,000$ cm⁴. $A_3 = 0.2$ cm²; $I_3 = 0$.

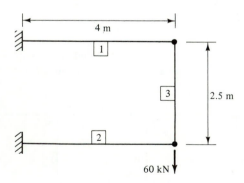

Figure P7.73

7.74. Choose the force in member 3 as the redundant in Problem 7.73.

7.75. Choose Y_3 as the redundant. $EI = 1296 \times 10^4$ ksi for members 1 and 2. $EA = 15,000^k$ for member 3.

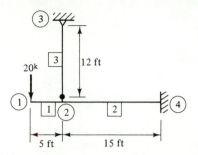

Figure P7.75

7.76. Find the prefabricated length of member 3 such that the length of member 3 is precisely 144 in. when the load shown in Problem 7.75 is applied.

* 7.77. Choose M_3 as the redundant. $EI = 276,458$ kft² and $EA = $ infinity for members 1 through 3. $EA = 60,000^k$ and $EI = 0$ for member 4.

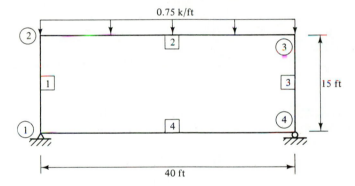

Figure P7.77

* 7.78. Choose M_1 as the redundant. $EI = 150,000$ kft² and $EI = $ infinity for member 1. $EA = 12,800^k$ and $EI = 0$ for member 2.

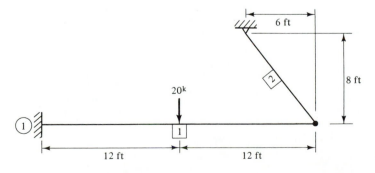

Figure P7.78

7.79. Choose the force in member 5 as the redundant. Member 5 is a pinned ended compression member. Members 3 and 4 are a cable. $E = 29{,}000$ ksi for all members. $A = 9.26$ in.^{2} and $I = 215.8$ in.^{4} for members 1 and 2. $A = 2$ in.^{2} for members 3 and 4. $A = 3.82$ in.^{2} for member 5.

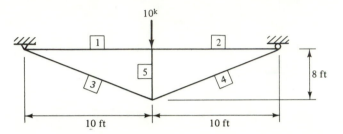

Figure P7.79

7.80. Choose M_1 and Y_4 as the redundants. $EI = 80{,}000$ kN m^2 for members 1 through 3. $EA = 2000$ kN for member 4.

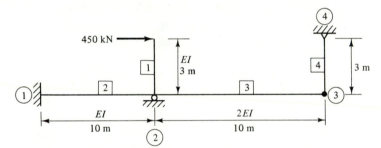

Figure P7.80

* 7.81. Choose M_2 and Y_5 as the redundants. $E = 2 \times 10^8$ kN/m^2. $I_1 = 25{,}000$ cm^4. $I_2 = I_3 = 42{,}000$ cm^4. $A_4 = A_5 = 6$ cm^2; $I_4 = I_5 = 0$.

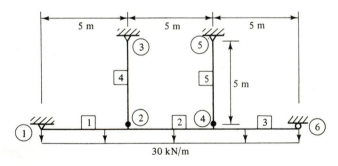

Figure P7.81

Chapter 8

Displacement Method— Noncomputerized Approach

8.1 INTRODUCTION

The Force Method of analyzing statically indeterminate structures was presented in Chapter 7. In that method the unknowns to be solved for in the simultaneous equations are forces, the redundants. Another approach for solving indeterminate structures is the *Displacement Method* and it is presented in this chapter. In the Displacement Method the unknowns to be solved for in the simultaneous equations are *joint displacements* (rotations and translations). Some analysts choose to refer to this method as the *Stiffness Method,* since the coefficients of the unknowns are stiffness coefficients; however, the author chooses to refer to it as the Displacement Method. Although the Displacement Method is generally used to solve indeterminate structural problems, it can be applied to statically determinate structures. This is possible since a structural analyst can choose artificial or unnecessary joints at any number of points along each member that can develop a bending moment when the structure is loaded. Each joint, whether real or artificial, has a certain number of degrees of freedom in regard to how it can move when loads are applied to the structural system.

In 1880 Heinrich Manderla [8.1] used joint displacements as the unknowns to solve an indeterminate structural problem. His formulation involved the effect of the axial force on the bending behavior of each member—today this approach is referred to as a *Second Order Analysis by the Displacement Method.* Due to the complexity of the method and lack of a computer, it was not appropriate for widespread usage at that time. In 1892 Otto Mohr [8.2] presented a good approximate method involving joint displacements. The true forerunner of today's *First Order Analysis by the Displacement Method* was called the *Slope-Deflection Method* when it was first presented in Germany by Alex Bendixen [8.3] in 1914 and in the United States of America by George Maney [8.4] in 1915. Later, in 1930, Hardy Cross [8.5] published the Moment

Distribution Method which is an iterative method based on the Displacement Method. Moment Distribution is discussed in Chapter 10.

This chapter focuses on the first order analysis of structures by the Displacement Method. In a first order analysis, the effect of the axial force on the bending behavior of each member is assumed to be negligible. A set of linear simultaneous equations of system equilibrium are formulated in terms of the unknown joint displacements and solved for the joint displacements. Subsequently, the member-end forces and reactions are obtained. With the availability of a computer to perform the mathematical operations, solving simultaneous equations that can be routinely formulated by a computerized method of analysis is not a major problem unless the memory of the computer is exceeded. The efficient means of mathematical organization for a computerized solution is to use matrix notation in the formulation. Although this chapter does not fully present the computerized formulation of the Displacement Method, the basic fundamentals are dealt with in a manner such that in an advanced course the student can learn how to put the pieces together and write a computer program for the Displacement Method of Analysis.

8.2 SOME BASIC CONCEPTS

Consider the deflected shape of the continuous beam supported by a cable at point C and loaded as shown in Figure 8.1a. Rotations (θ_B , θ_C , θ_D) occur at points B, C, and D; a translation (Δ_C) occurs at point C.

Span BC of the continuous beam is considered in the following discussion. As shown in Example 7.5, the solution for no rotations and no translations at ends B and C can be obtained for the fixed ended moments in Figure 8.1b. Also, as shown in Figures 8.1c through 8.1e, there are end moments at B and C due to each of the following: θ_B , θ_C , and Δ_C; these end moments can be obtained as shown in Examples 7.6 and 7.7 provided numerical values of θ_B, θ_C, Δ_C and the structural properties for span BC are known.

In (b) through (e) of Figure 8.1, the correct directions of the end moments are shown. These correct directions are easily determined from the deflected shapes of the deformed structure as described in Figure 8.2. For graphical convenience, only the neutral axis line of the structure was shown in Figure 8.1. If the depth of the member had been shown in (b) and (c) of Figure 8.1, for example, these figures would look as shown in Figure 8.2.

If one draws the end moment arrow direction externally on the member ends, the tail of the arrow is on the tension side of the neutral axis and the head of the arrow is on the compression side of the neutral axis.

Suppose counterclockwise end moments are chosen to be positive member-end moments. The moment at end B of span BC in Figure 8.1a can be obtained by algebraically adding the moments at end B in Figures 8.1b through 8.1e. Similarly, the moment at end C of span BC in Figure 8.1a can be obtained by algebraically adding the moments at end C in Figures 8.1b through 8.1e.

In this text, the portion of the structure spanning from B to C is referred to as a *member*. The algebraic summation process described in the previous paragraph

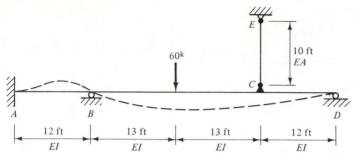

$$\theta_B\rangle, \theta_C\rangle, \Delta_C\downarrow, \theta_D\rangle$$

Points A through E are joints: joint movement vectors are: $\theta_B\rangle, \theta_C\rangle, \Delta_C\downarrow, \theta_D\rangle$

(a) Original structure – dashed line shows deflected shape

60^k

195 ftk

13 ft   13 ft

B   C

$$\frac{60^k \star 26 \text{ ft}}{8} = 195 \text{ ftk}$$

(b) Fixed ended moments for span BC

θ_B

$\frac{4EI}{26 \text{ ft}} \star \theta_B$ θ_B $\frac{2EI}{26 \text{ ft}} \star \theta_B$

(c) Member end moments due to θ_B for span BC

θ_C

$\frac{2EI}{26 \text{ ft}} \star \theta_C$ θ_C $\frac{4EI}{26 \text{ ft}} \star \theta_C$

(d) Member end moments due to θ_C for span BC

$\frac{6EI}{(26 \text{ ft})^2} \star \Delta_C$ $\frac{6EI}{(26 \text{ ft})^2} \star \Delta_C$

Δ_C

(e) Member end moments due to Δ_C for span BC **Figure 8.1** Span BC end moments.

defines the *member-end force-deformation relation* for the member spanning from point B to point C. Similarly, member-end force-deformation relations can be obtained for the other continuous beam members in Figure 8.1a. The cable is a tension member that elongates allowing point C to deflect downward; the member-end force-deformation relation for this type of member behavior will be derived later.

A *member* is defined as a portion of a structure for which the force-deformation relation can be derived at the ends of the structural portion. It should be noted that prismatic structural portions that are curved in an elevation view such as an arc of a circle can be defined as members since member-end force-deformation relations can

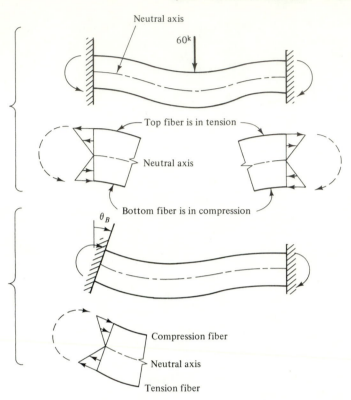

Figure 8.2 If one draws the end moment arrow direction externally on the structure ends, the tail of the arrow is on the tension side of the neutral axis and the head of the arrow is on the compression side of the neutral axis.

be derived for such structural portions. However, in this text all examples deal with only members that are straight lines lying in a common plane and the structural loads also lie in that common plane.

An *internal joint* is defined as a junction in a structure where two or more member ends coincide (points *B* and *C* in Figure 8.1a are internal joints). An *external joint* or *boundary joint* is defined as a location on a structure's boundary where the structure ceases to exist (points *A, D,* and *E* in Figure 8.1a are external joints or boundary joints). In this text, all joints are considered to be *infinitesimal in size and nondeformable.* That is, the joints are treated as *infinitesimal rigid bodies* and *rigid body mechanics* can be invoked at each joint to write equilibrium equations for each joint. The joint equilibrium equations are functions of the member-end forces *reversed to act on the joints.* Since member-end force deformations can be derived for each member, the set of simultaneous equations for all joints that either rotate or/and translate due to known loads can be expressed in terms of the unknown joint displacements (rotations and translations).

Solution of this set of linear algebraic equations gives the unknown joint displacements. Then the process described for Figure 8.1 can be conducted for each member in the structure to compute the final or total member-end moments. Member-end shears can then be found for each member by using equations of statics.

In this chapter, the structures dealt with are customarily referred to by one of the following names:

1. A *plane truss* when all members only elongate or shorten due to applied joint loads.
2. A *plane frame* when some or all of the members bend due to either applied member loads or/and applied joint loads.

8.3 NOTATION AND SIGN CONVENTION

A rigorous sign convention involving matrix notation will be established and used later to describe a detailed set of procedural steps for the Displacement Method. Although the emphasis in Chapters 8 and 9 is on noncomputerized formulations and solutions of plane frames, the adopted sign convention and notation will be of the type needed for a computerized solution to enable students to gradually progress toward using computerized solutions. In a computerized solution, all joints need to be sequentially numbered; likewise all members need to be sequentially numbered. Consequently, all joint numbers are encircled and each member number is enclosed within a rectangular box attached to the member.

In this text, all plane frames or plane trusses will be shown as lying in the XY plane of a right-handed XYZ coordinate system. This restriction is chosen for simplicity in the textual discussions.

At a typical interior joint in a plane frame (see Figure 8.3), the joint is a rigid body confined to move in the XY plane. These rigid body motions are a translation parallel to the X axis; a translation parallel to the Y axis; and a rotation about an axis parallel to the Z axis. The author chooses to adopt the most commonly used sign convention in computerized solutions for joint movements at a typical internal joint of a plane frame. That sign convention shown in Figure 8.4 for a typical internal joint j is as follows:

1. At joint j, construct X', Y', Z' axes parallel to and in the direction of the X, Y, Z axes.
2. Positive directions of joint j translational movements are in the directions of X' and Y'.
3. The positive direction of the joint j rotation is a right-handed screw rule vector coincident with Z'.

These movements at all of the internal joints are collectively referred to as *system displacements* or *system degrees of freedom* (system DOF). A computerized solution allows for only one unique symbol, an array name in which all computed joint movements are stored, to be used to describe the system displacements. The author chooses D as the array name and $\{D\}$ as the matrix symbol for the system displacements; in expanded form

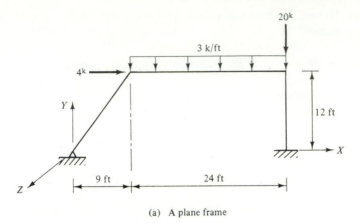

(a)   A plane frame

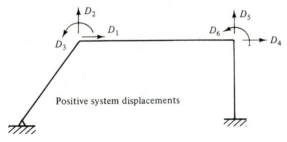

(b)   Frame joint and member numbers

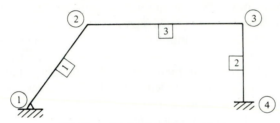

Positive system displacements

(c) Positive directions and locations of system displacements

Figure 8.3 Example showing system notation definitions.

$$\{D\} = \begin{Bmatrix} D_1 \\ D_2 \\ D_3 \\ \vdots \\ D_n \end{Bmatrix}$$

where n = NSD (*N*umber of *S*ystem *D*isplacements).

For a noncomputerized solution, the subscript numbers for $\{D\}$ begin at the lowest internal joint number as shown in Figure 8.3c; the X-direction translation is numbered first, then the Y-direction translation is numbered next, and the Z-direction rotation is numbered last. Then one moves repeatedly to the next internal joint number

Axes directions   Positive directions of joint movements   **Figure 8.4**

and continues to increase the subscript numbers for $\{D\}$ in X,Y,Z direction order at each internal joint. (In a computerized solution, a rotational DOF would be chosen at joint 1, a boundary joint, in Figures 8.3b and 8.3c; in this case, there would be 7 DOF—that is, NSD = 7—and the subscript numbers for $\{D\}$ would begin with the rotational DOF at joint 1.)

Applied joint loads, $\{P^a\}$, at and in the direction of each positive system displacement are also positive. For the example shown in Figure 8.3a, the applied joint loads are

$$\{P^a\} = \left\{ \begin{array}{c} P^a_1 \\ P^a_2 \\ P^a_3 \\ P^a_4 \\ P^a_5 \\ P^a_6 \end{array} \right\} = \left\{ \begin{array}{c} 4^k \\ 0 \\ 0 \\ 0 \\ -20^k \\ 0 \end{array} \right\}$$

For a plane truss (see Figure 8.5), each joint is defined to be a pin that cannot resist any Z-direction rotation. Consequently, at each internal joint in the truss, only translations parallel to the X axis and Y axis can be chosen as DOF. At joint 1, a boundary joint, the X-direction translation is prevented by the roller support, but the joint can deflect in the vertical direction.

Some texts refer to the positive system displacements, $\{D\}$ in Figures 8.3c and Figure 8.5c, as either the *degrees of freedom (DOF)* or the *degrees of kinematical indeterminacy*. The author simply prefers to refer to them as the *positive system displacements* chosen to define the internal joint movements (for a truss, any boundary joint translations must be accounted for in numbering the positive system displacements). Also the author finds it convenient sometimes to use the terminology DOF for brevity instead of writing out the words "positive system displacements."

8.4 MEMBER-END FORCE-DEFORMATION RELATIONS FOR A PRISMATIC MEMBER

Consider a member (see Fig. 8.6) which spans between two internal joints i and j in a plane frame (not shown). Each end of the member is attached to a joint (a rigid body which can translate in the X' and Y' directions and can rotate about the Z' axis). Since the member is fully attached at each end to either joint i or joint j, the member-end movements are identically the same as the movements of joints i and j in the

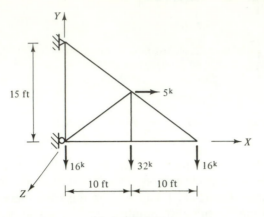

(a) A plane truss

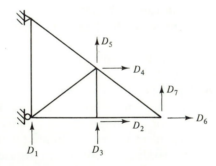

(b) Truss joint and member numbers

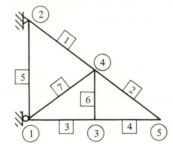

(c) Positive sytem displacements

Figure 8.5 Truss example showing system notation.

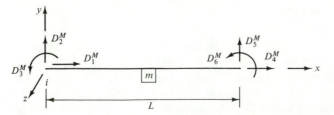

Figure 8.6 Member-end displacements—positive directions shown.

plane frame structure. For notational convenience, let $\{D^M\}_m$ denote the member-end displacements for member m.

Member m may have some load(s) on it between joints i and j. For example, the member weight is a uniformly distributed load acting in the negative y direction and there may be other loads on the member. Note that each member has its own coordinate system designated by small letters x, y, z; this is necessary for a computerized solution and is called the *member coordinate system* (or the *local coordinate system*).

As shown in Figure 8.1, the *final* (or *total*) member-end moments are the algebraic sum of the member-end moments due to:

1. The fixed ended moments caused by the member loads.
2. Each of the member-end rotations.
3. The relative movement of the member's ends in the y direction.

Fixed ended moments for frequently encountered member loadings are shown in Figure 8.7 and they can also be found in handbooks. Member-end moments and

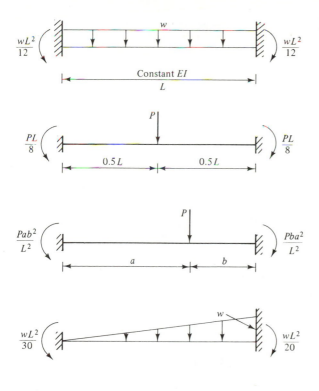

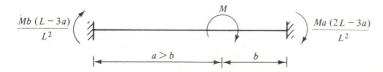

Figure 8.7 Fixed ended moments for a prismatic member.

shears *due to unit valued member-end rotations and y-direction translations* are shown in Figure 8.8; they are referred to as stiffness coefficients due to bending. Later, when the computed values of D_2^M, D_3^M, D_5^M, and D_6^M are available, the member-end moments and shears due to member-end movements can be found by multiplying the member-end forces in Figures 8.8a through 8.8d respectively by the computed values of D_3^M, D_6^M, D_2^M, and D_5^M.

It was shown in Chapter 6 that the change in length caused by an axial force in a truss member is $\Delta = PL/EA$; therefore, $P/\Delta = EA/L$. Figure 8.9 shows the member-end axial forces due to *unit valued* member-end x-direction translations for a perfectly straight, prismatic member.

The member-end forces due to *unit valued member-end displacements,* see Figures 8.8 and 8.9, are referred to as member-end *stiffness coefficients.*

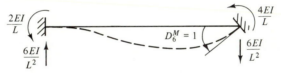

(a) Member-end forces due to $D_3^M = 1$

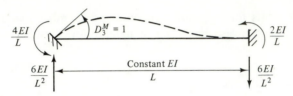

(b) Member-end forces due to $D_6^M = 1$

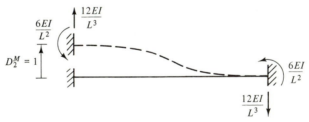

(c) Member-end forces due to $D_2^M = 1$

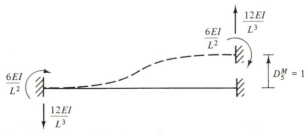

(d) Member-end forces due to $D_5^M = 1$

Figure 8.8 Member-end stiffness coefficients.

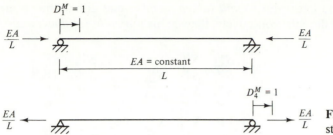

Figure 8.9 Member-end axial stiffness coefficients.

At this stage in the development of the method in his classroom presentations, the author does not prefer to use any symbols that can be avoided. However, since some teachers will not share this point of view, the author chooses to make the following definitions at this stage of the discussion. In a computerized solution, matrix symbols are needed to distinguish between each of the member-end force types—namely total or final; fixed ended; and those due to member-end displacements. This is necessary in order to perform mathematical manipulations on them. Consequently, the rigorous sign convention shown in Figure 8.10 is adopted to enable the definitions of the matrix symbols for the various member-end forces to be made. Note in Figure 8.10 that only the subscript numbers of the member-end forces are shown for the adopted locations and positive directions of these forces; each of the subsequently defined matrix symbols for member-end forces has these subscripts numbers attached as part of their definitions. Also note that the location numbers and positive directions in Figure 8.10 for the member-end forces are identically the same as in the definition for member-end displacements (see Figure 8.6).

Member fixed ended forces due to member loads for member m are $\{F^f\}_m$, and in expanded form they are

$$\{F^f\}_m = \begin{Bmatrix} F^f_1 \\ F^f_2 \\ F^f_3 \\ F^f_4 \\ F^f_5 \\ F^f_6 \end{Bmatrix}_m$$

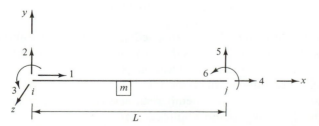

Figure 8.10 Positive member-end forces subscript numbers.

Member-end forces due to *unit valued* member-end displacements are referred to as stiffness coefficients; collectively, they define a *member stiffness matrix:*

$$[S^M]_m = \begin{bmatrix} S_{1,1}^M & S_{1,2}^M & S_{1,3}^M & S_{1,4}^M & S_{1,5}^M & S_{1,6}^M \\ S_{2,1}^M & S_{2,2}^M & S_{2,3}^M & S_{2,4}^M & S_{2,5}^M & S_{2,6}^M \\ S_{3,1}^M & S_{3,2}^M & S_{3,3}^M & S_{3,4}^M & S_{3,5}^M & S_{3,6}^M \\ S_{4,1}^M & S_{4,2}^M & S_{4,3}^M & S_{4,4}^M & S_{4,5}^M & S_{4,6}^M \\ S_{5,1}^M & S_{5,2}^M & S_{5,3}^M & S_{5,4}^M & S_{5,5}^M & S_{5,6}^M \\ S_{6,1}^M & S_{6,2}^M & S_{6,3}^M & S_{6,4}^M & S_{6,5}^M & S_{6,6}^M \end{bmatrix}_m$$

The final or total member-end forces, $\{F\}$, are the algebraic sum of the member fixed ended forces and the member-end forces due to member-end displacements:

$$\{F\} = \{F^f\}_m + [S^M]_m\{D^M\}_m$$

This matrix equation is referred to as the *member-end force-deformation relation;* the expanded form of this relation is

$$\begin{Bmatrix} F_1 \\ F_2 \\ F_3 \\ F_4 \\ F_5 \\ F_6 \end{Bmatrix}_m = \begin{Bmatrix} F_1^f \\ F_2^f \\ F_3^f \\ F_4^f \\ F_5^f \\ F_6^f \end{Bmatrix}_m + \begin{bmatrix} S_{1,1}^M & S_{1,2}^M & S_{1,3}^M & S_{1,4}^M & S_{1,5}^M & S_{1,6}^M \\ S_{2,1}^M & S_{2,2}^M & S_{2,3}^M & S_{2,4}^M & S_{2,5}^M & S_{2,6}^M \\ S_{3,1}^M & S_{3,2}^M & S_{3,3}^M & S_{3,4}^M & S_{3,5}^M & S_{3,6}^M \\ S_{4,1}^M & S_{4,2}^M & S_{4,3}^M & S_{4,4}^M & S_{4,5}^M & S_{4,6}^M \\ S_{5,1}^M & S_{5,2}^M & S_{5,3}^M & S_{5,4}^M & S_{5,5}^M & S_{5,6}^M \\ S_{6,1}^M & S_{6,2}^M & S_{6,3}^M & S_{6,4}^M & S_{6,5}^M & S_{6,6}^M \end{bmatrix}_m \begin{Bmatrix} D_1^M \\ D_2^M \\ D_3^M \\ D_4^M \\ D_5^M \\ D_6^M \end{Bmatrix}_m$$

In the member force-deformation relation, note that the member-end forces due to member-end displacements are $[S^M]_m \{D^M\}_m$. Since no mathematical operator is shown between the two matrices, $[S^M]_m$ and $\{D^M\}_m$, it is implicitly understood that matrix multiplication of the two matrices is to be performed. That is, the individual entries in each row of matrix $[S^M]_m$ are to be multiplied times the corresponding entries in column matrix $\{D^M\}_m$ and the scalar products are algebraically summed giving a scalar member-end force value which is to be added algebraically to the corresponding member fixed ended force value to obtain the corresponding final or total member-end force.

The customary matrix algebra notations for the subscripts of matrix $[S]$ apply; namely, the first subscript number is the matrix row number and the second subscript number is the matrix column number. An arbitrary entry in $[S^M]_m$ is denoted as $S_{i,j}^M$ for member m, where i is the matrix row number and j is the matrix column number.

Also a structural connotation is attached to the subscript numbers of matrix $[S^M]_m$. The first subscript number, i, is the location of a member-end force, $S_{i,j}^M$, and the second subscript number j, is the location of the unit valued member-end displacement which caused the member-end force, $S_{i,j}^M$ for member m.

For a prismatic member, the member stiffness coefficients given in Figures 8.8 and 8.9 define the following member stiffness matrix:

$$[S^M]_m = \begin{bmatrix} \dfrac{EA}{L} & 0 & 0 & \dfrac{-EA}{L} & 0 & 0 \\[2.2ex] 0 & \dfrac{12EI}{L^3} & \dfrac{6EI}{L^2} & 0 & \dfrac{-12EI}{L^3} & \dfrac{6EI}{L^2} \\[2.2ex] 0 & \dfrac{6EI}{L^2} & \dfrac{4EI}{L} & 0 & \dfrac{-6EI}{L^2} & \dfrac{2EI}{L} \\[2.2ex] \dfrac{-EA}{L} & 0 & 0 & \dfrac{EA}{L} & 0 & 0 \\[2.2ex] 0 & \dfrac{-12EI}{L^3} & \dfrac{-6EI}{L^2} & 0 & \dfrac{12EI}{L^3} & \dfrac{-6EI}{L^2} \\[2.2ex] 0 & \dfrac{6EI}{L^2} & \dfrac{2EI}{L} & 0 & \dfrac{-6EI}{L^2} & \dfrac{4EI}{L} \end{bmatrix}_m$$

Note that all of the entries in column 3 of $[S^M]_m$ were obtained from Figure 8.8a which gives the member-end forces due to $D_3^M = 1$ for member m. Also, Figure 8.8b gives the member-end forces due to $D_6^M = 1$ for member m and these are the entries for column 6 of $[S^M]_m$. Similarly, the entries for each of the other columns of $[S^M]_m$ are obtained from Figures 8.8 and 8.9 by using the definitions shown in Figures 8.6 and 8.10 and the expanded mathematical form of the member stiffness matrix,

---

For noncomputerized solutions, member-end force-deformation relations are also needed when one end of the member is hinged. These relations are obtained as shown in Figure 8.11 by superposition of relations shown in Figure 8.8; similarly, the fixed ended moment solutions for frequently encountered loadings shown in Figure 8.12 can be found for a member with one end hinged. After the fixed ended moment has been *numerically evaluated,* the member-end shears can be computed by using equations of statics. Usage of these particular relations enables the analyst to involve fewer DOF and fewer simultaneous equations in a noncomputerized solution; this saves the analyst formulation and solution time in the Displacement Method of analysis. However, in a computerized approach it is cheaper not to use these special relations for one end of the member being hinged—by avoiding special definitions and using only one definition, it is unnecessary to program the computer to choose which definitions to use; the computer can solve a few more simultaneous equations quicker than it can make decisions to select the proper definition from a set of definitions.

8.5 CONCEPTUAL EXAMPLE AND DETAILED PROCEDURAL STEPS

The author recommends that the reader treat this section in much the same manner as one does in learning how to play a brand new game. Suppose the rules of the new game are verbally explained and visually demonstrated by an experienced player; in this case, the beginner probably would be told to do the following:

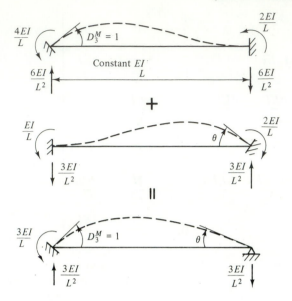

$+$

$=$

Similarly, it can be shown that

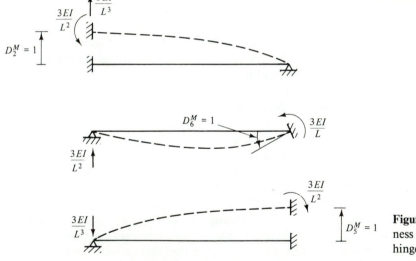

Figure 8.11 Member-end stiffness coefficients for member hinged on one end.

1. Pay attention to the basic definitions, terminology, and general drift of the game; in the beginning do not worry about the finer points and details of the game—they will be more easily learned after some game experience.
2. Play a dry run game with the experienced player coaching the beginner.
3. Play a dry run game with the beginner asking for advice on what to do only when the beginner needs help with definitions, rules, and strategy.
4. Play the game with the beginner only allowed to ask for clarification of the rules (no help with game strategy would be allowed, however).

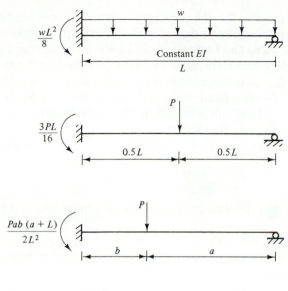

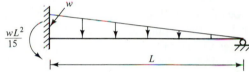

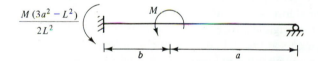

Figure 8.12 Fixed ended moment for member hinged on one end.

5. Ask about and discuss the finer points of the game; discuss and compare alternative game strategies.

6. Practice, practice, practice (play the game competitively against friendly but more experienced players, if possible).

Analogously speaking, the prior sections in this chapter have been presented to take care of item 1 in the "learning how to play a new game" description. The purposes of this section are:

1. To go through a "general drift" dry run using the matrix symbol definitions and the rigorous sign convention.

2. To give the needed set of detailed procedural steps for the noncomputerized Displacement Method analysis.

However, in the figures alongside the symbol definition implementations, the author shows detailed descriptions of how values for each of the symbol definitions can be calculated. This is done so that the beginner can return later as needed to this section for clarification purposes during the study of the example problems in subsequent sections of this chapter. The beginner should keep in mind that the following is a conceptual example with 3 DOF chosen in order to fully describe the needed matrix notation. In the *first pass* through the figures and discussion, the reader should ignore the information on the right-hand side of each equal mark. Simpler learning examples involving only 1 or 2 DOF are provided in Chapter 9 to illustrate how to calculate the information for each expanded matrix symbol.

The **detailed procedural steps** for a noncomputerized Displacement Method analysis are:

1. Eliminate any determinate part of the structure; (Fig. 8.13b illustrates this step).
2. Identify $\{D\}$, the positive directions and locations of the unknown system displacements which must be involved in the solution; rotations at boundary joints which are hinges need not be involved since the needed member-end stiffness coefficients and fixed ended moments for a member with one end hinged can be obtained; (Fig. 8.13c illustrates this step).
3. Write down $\{P^a\}$, the applied joint loads; they are coincident with $\{D\}$; (Fig. 8.13e illustrates this step).
4. Modify the original structure such that all system displacements are zero, $\{D\} = \{0\}$; place all member loads on the modified structure; find the member fixed ended forces and the fixed joint loads, $\{P^f\}$, which are the restraining joint forces required to make $\{D\} = \{0\}$. Note: $P_j^f =$ summation of all member fixed ended forces at and in the direction of D_j (Fig. 8.13f illustrates this step).
5. Modify the original structure such that $D_1 = 1$ and all other unknown system displacements are zero; find the inducing joint force at D_1 and the restraining joint forces at all other unknown system displacement locations; the inducing and restraining joint forces are system stiffness coefficients and they define column 1 of the system stiffness matrix, $[S]$; repeat this step for $D_j = 1$ and all other system displacements being zero to find column j of $[S]$. Note: j is to take on each of the values 2 through NSD one at a time; $[S]$ is a symmetric matrix (Figs. 8.13h, j, *l* illustrate this step).
6. Use superposition to perform the algebraic summation of joint forces at and in the direction of each DOF to satisfy statical equilibrium of the system— the force from step 4 at D_j plus the forces from step 5 at D_j times the corresponding unknown system displacement symbol D_j for each unit-induced displacement must add up to the applied joint load at D_j found in step 3; the result is a matrix equation of system equilibrium (a set of simultaneous equations in which the unknowns are $\{D\}$ and the coefficients of the unknowns are $[S]$); the matrix equation of system equilibrium is

$$\{P^f\} + [S]\{D\} = \{P^a\}$$

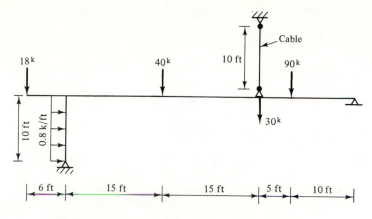

(a) Complete structure (see figure (d) for member properties)

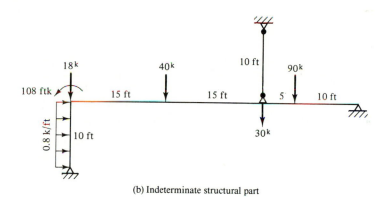

(b) Indeterminate structural part

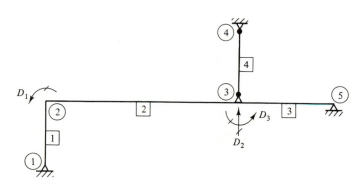

(c) Joint numbers, member numbers and positive system replacements

Figure 8.13 Conceptual example.

E = 29,000 ksi for all members

I = 1330 in.^{4} for members 1 through 3; EI = 38,570,000 k in.^{2} = 267,847 kft^2

A = 5 in.^{2} for member 4

$(EI/L)_1$ = 26,785 ftk

$(EI/L)_2$ = 8928 ftk

$(EI/L)_3$ = 17,856 ftk

$(EA/L)_4$ = 14,500 k/ft

Member 1: fixed ended moment $= \dfrac{wL^2}{8} = \dfrac{(0.8\,\text{k/ft}) * (10\,\text{ft})^2}{8} = 10$ ftk

Member 2: fixed ended moments $= \dfrac{PL}{8} = \dfrac{40^k * 30\,\text{ft}}{8} = 150$ ftk

Member 3: fixed ended moment $= \dfrac{Pab(a+L)}{2L^2} = \dfrac{90^k * 10\,\text{ft} * 5\,\text{ft} * (10\,\text{ft} + 15\,\text{ft})}{2 * (15\,\text{ft})^2} = 250$ ftk

(Formulas for fixed ended moments were given in Figures 8.7 and 8.11)

(d) Member properties and FEM

Applied joint loads at and in the direction of $\{D\}$ are referred to as $\{P^a\}$; P_1^a = 108 ftk; P_2^a = -30^k; P_3^a = 0.

(e) Applied joint loads

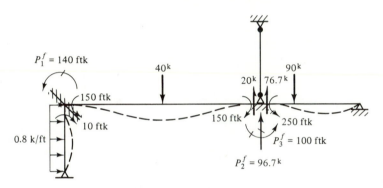

Fixed joint forces, $\{P^f\}$, are obtained from member fixed ended forces

(f) Solution for $D_1 = D_2 = D_3 = 0$

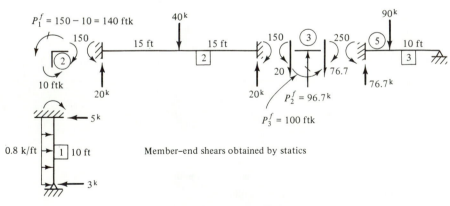

Member–end shears obtained by statics

(g) Expanded version of figure (f).

Figure 8.13 (*continued*)

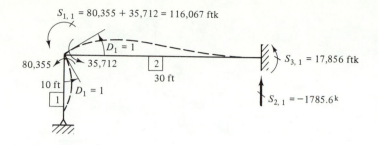

$S_{1,1} = 80,355 + 35,712 = 116,067$ ftk

80,355   35,712 $D_1 = 1$ $S_{3,1} = 17,856$ ftk

10 ft $D_1 = 1$ 30 ft   2

1     $S_{2,1} = -1785.6^k$

(h) Solution for $D_1 = 1$ and $D_2 = D_3 = 0$

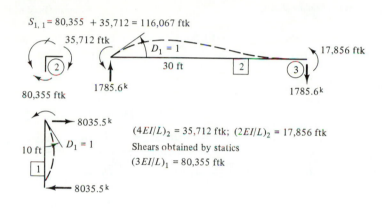

$S_{1,1} = 80,355 + 35,712 = 116,067$ ftk

35,712 ftk $D_1 = 1$ 17,856 ftk

2   30 ft   2   3

1785.6^k 1785.6^k

80,355 ftk

8035.5^k

10 ft $D_1 = 1$ $(4EI/L)_2 = 35,712$ ftk; $(2EI/L)_2 = 17,856$ ftk

1   Shears obtained by statics

$(3EI/L)_1 = 80,355$ ftk

8035.5^k

(i) Expanded version of figure (h)

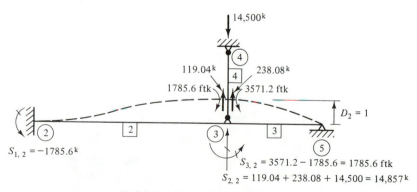

14,500^k

4

119.04^k 4   238.08^k

1785.6 ftk   3571.2 ftk

$D_2 = 1$

2   2   3   3

$S_{1,2} = -1785.6^k$ 5

$S_{3,2} = 3571.2 - 1785.6 = 1785.6$ ftk

$S_{2,2} = 119.04 + 238.08 + 14,500 = 14,857^k$

(j) Solution for $D_2 = 1$ and $D_1 = D_3 = 0$

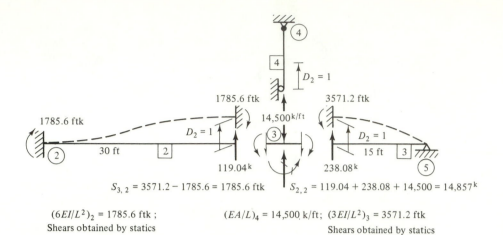

$S_{3,2} = 3571.2 - 1785.6 = 1785.6$ ftk $\quad S_{2,2} = 119.04 + 238.08 + 14,500 = 14,857^k$

$(6EI/L^2)_2 = 1785.6$ ftk ; $(EA/L)_4 = 14,500$ k/ft; $(3EI/L^2)_3 = 3571.2$ ftk
Shears obtained by statics   Shears obtained by statics

(k) Expanded version of figure (j)

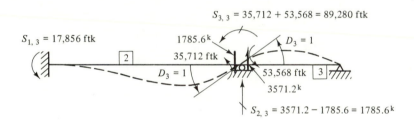

(l) Solution for $D_3 = 1$ and $D_1 = D_2 = 0$

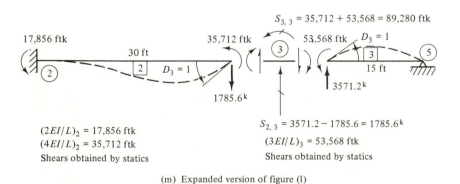

$(2EI/L)_2 = 17,856$ ftk
$(4EI/L)_2 = 35,712$ ftk
Shears obtained by statics

$(3EI/L)_3 = 53,568$ ftk
Shears obtained by statics

(m) Expanded version of figure (l)

Figure 8.13 (*continued*)

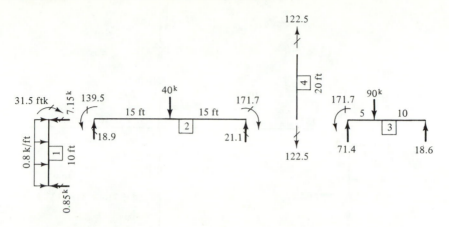

The member-end moments shown above, with counterclockwise moments being positive, are obtained from: Figures 8.13 $[f + D_1*h + D_2*j + D_3*L]$:

Member 1: moment at top end $= -10 + 80{,}355*(-0.00026766) + 0 + 0 = -31.5$ ftk

Member 2: moment at left end $= 150 + 35{,}712*(-0.00026766) + (-1785.6)*(-0.00845227)$
$$+ 17{,}856*(-0.00089749) = 139.5$$

moment at right end $= -150 + 17{,}856*(-0.00026766) + (-1785.6)*(-0.00845227)$
$$+ 35{,}712*(-0.00089749) = -171.7$$

Member 3: moment at left end $= 250 + 0 + 3571.2*(-0.00845227) + 53{,}568*(-0.00089749)$
$$= 171.7$$

The member-end shears shown above are obtained by:

1. summation of moments = 0 at one end to find other end shear
2. then summation of forces perpendicular to member = 0

Member 1: at bottom end: $(5*10*0.8 - 31.5)/10 = 0.85$
    at top end: $10*0.8 - 0.85 = 7.15$

Member 2: at left end: $(139.5 - 171.7 + 15*40)/30 = 18.9$
    at right end: $40 - 18.9 = 21.1$

Member 3: at left end: $(171.7 + 10*90)/15 = 71.4$
    at right end: $90 - 71.4 = 18.6$

(n) Final member-end shears and moments

Since the only unknowns in the preceding matrix equation are the system displacements, $\{D\}$, it is convenient to define $\{P^e\} = \{P^a\} - \{P^f\}$ and to write the matrix equation of system equilibrium as

$$[S]\{D\} = \{P^e\}$$

which is the form needed for solving the simultaneous equations (this step is performed after Figure 8.13m is obtained, but it is located in the text after Figure 8.13o.

7. Solve the matrix equation of system equilibrium for the unknown system displacements.

8. Multiply the unit displacement solutions in step 5 by the actual displacements found in step 7; algebraically add these results onto the member fixed ended moments found in step 4 to find the total (or final) member-end moments;

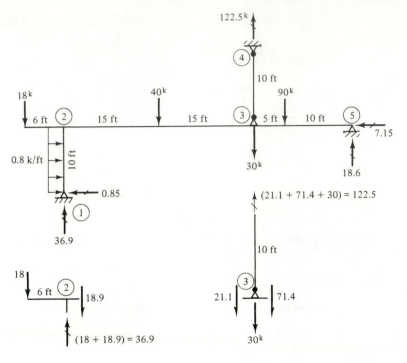

Joint 2 vertical equilibrium   Joint 3 vertical equilibrium

(o) Reactions

Figure 8.13 (*continued*)

find the member-end shears from statics for each member; find the member-end axial forces from joint equilibrium; find the reactions by statics; (Figs. 8.13n, o show the results for this step).

Step 6 Illustrated Summation of the external moments at joint 2 in the direction of D_1 from Figures 8.13 [f, $h*D_1$, $j*D_2$, $L*D_3$] must add up to the externally applied moment at the location of D_1 in Figure 8.13b. This gives system equilibrium equation number 1, namely

$$P_1^f + S_{1,1}D_1 + S_{1,2}D_2 + S_{1,3}D_3 = P_1^a$$

Likewise, summation of the external vertical forces at joint 3 in the direction of D_2 from Figures 8.13 [f, $h*D_1$, $j*D_2$, $L*D_3$] must add up to the externally applied vertical force at the location of D_2 in Figure 8.13b. This gives system equilibrium equation number 2, namely

$$P_2^f + S_{2,1}D_1 + S_{2,2}D_2 + S_{2,3}D_3 = P_2^a$$

Similarly, summation of the external moments at joint 3 in the direction of D_3 from Figures 8.13 [f, $h*D_1$, $j*D_2$, $L*D_3$] must add up to the externally applied moment at the location of D_3 in Figure 8.13b. This gives system equilibrium equation 3, namely

$$P_3^f + S_{3,1} D_1 + S_{3,2} D_2 + S_{3,3} D_3 = P_3^a$$

In these system equilibrium equations, only D_1, D_2, and D_3 are unknown. Therefore, it is convenient to define $\{P^e\}$, the effective joint loads, as follows:

$$\{P^e\} = \{P^a\} - \{P^f\}$$

and in expanded form this definition is

$$P_1^e = P_1^a - P_1^f$$
$$P_2^e = P_2^a - P_2^f$$
$$P_3^e = P_3^a - P_3^f$$

The simultaneous equations of system equilibrium written in expanded matrix form, for this example, are as follows:

$$\begin{bmatrix} S_{1,1} & S_{1,2} & S_{1,3} \\ S_{2,1} & S_{2,2} & S_{2,3} \\ S_{3,1} & S_{3,2} & S_{3,3} \end{bmatrix} \begin{Bmatrix} D_1 \\ D_2 \\ D_3 \end{Bmatrix} = \begin{Bmatrix} P_1^e \\ P_2^e \\ P_3^e \end{Bmatrix}$$

and in condensed matrix form are

$$[S]\{D\} = \{P^e\}$$

The last two equations are referred to as the matrix equation of system equilibrium in expanded matrix form and condensed matrix form, respectively.

For this conceptual example, the numerically formulated matrix equation of system equilibrium is

$$\begin{bmatrix} 116067 & -1785.6 & 17856 \\ -1785.6 & 14857 & 1785.6 \\ 17856 & 1785.6 & 89289 \end{bmatrix} \begin{Bmatrix} D_1 \\ D_2 \\ D_3 \end{Bmatrix} = \begin{Bmatrix} -32 \\ -126.7 \\ -100 \end{Bmatrix}$$

and the solution of these simultaneous equations is

$$\begin{Bmatrix} D_1 \\ D_2 \\ D_3 \end{Bmatrix} = \begin{Bmatrix} -0.00026766 \text{ rad} \\ -0.00845227 \text{ ft} \\ -0.00089749 \text{ rad} \end{Bmatrix}$$

By using these computed values of D_1, D_2, and D_3, the member-end moments for each member in Figures 8.13 [$h*D_1$, $j*D_2$, and $L*D_3$] are algebraically added to the member fixed ended moments for each member in Figure 8.13f to obtain the total (or final) member-end moments for each member in Figure 8.13b (these results are shown in Figure 8.13n).

After the final member-end moments have been computed for each member, free body diagrams of each member in Figure 8.13b can be solved by statics for the member-end shears (see Figure 8.13n). Finally, using statics and free body diagrams of the joints in Figure 8.13b, the axial forces in each member can be obtained and the reactions can be computed (see Figure 8.13o). Axial, shear, and moment diagrams can be drawn for each member now that the final member-end forces are available.

Chapter 9

Displacement Method— Noncomputerized Solutions

9.1 EXAMPLES WITHOUT ANY JOINT TRANSLATIONS

In the following examples, it is assumed that no translational movements can occur at any support (prescribed support movements are dealt with in Section 9.3). Some analysts would prefer that the author say these are *examples without any sidesway;* they define sidesway as joint translational displacements which allow some member ends to move perpendicular to the direction of the member length axis. For a beam continuous over several supports, if the supports cannot displace perpendicular to the length axis of the beam and if all member loads are perpendicular to the length axis of the beam, the only DOF are joint rotations. In single story, single or multibay frames the axial deformations of the members can be neglected (members are treated as axially rigid—axial stiffness = infinity) without causing any serious errors in the calculated behavior of the structure. An example of a frame is shown in which there are no joint translations since the members are treated as axially rigid and all translational support movements are prevented; in this case, the only DOF are joint rotations.

In the example problem solutions, the author shows the deflected shape of the structure in the figures for the solution. The deflected shapes enhance the understanding of the concepts; also they help immensely to determine the correct directions of the member-end moments and the signs that need to be used for them in the numerical evaluation of the fixed joint loads and the system stiffness matrix coefficients. The author is convinced that the deflected shapes can play as important a role in the understanding of the Displacement Method concepts and sign convention as the free body diagram does in writing the statical equilibrium equations for a structural part.

Also, since the objective is to begin to prepare students for the eventual usage of an available Displacement Method computer program in actual design situations, the author emphasizes the usage of numerical values for the stiffness coefficients. For noncomputerized solutions, it is possible to use relative values of *EI* for the members if axial deformations are ignored and if there are no prescribed support movements. However, in a computerized approach, the computer user must input absolute values for each parameter in almost all computer programs for the Displacement Method. Furthermore, the author firmly believes that usage of absolute values instead of relative values for the stiffness coefficients and the solution of the simultaneous equations for the system displacements enhances the learning process—the beginner understands absolute values better than relative values, particularly in regard to the system displacements. By obtaining absolute values (instead of values that contain *EI* as a symbol) for the system displacements, the beginner has information in the form that can best help an aspiring structural engineer to acquire "a feel for the structural behavior" and is in a better position to decide whether or not the system displacement solutions look reasonable.

Since there is a need to communicate structural properties in the problem statement of each example, the author chooses to identify the member numbers and joint numbers in the original structure instead of waiting to define them in a solution step. Except for this slight modification, the previously given detailed procedural steps are strictly followed and demonstrated in the example problems.

EXAMPLE 9.1

Solve by the Displacement Method. The only DOF to be involved in the system equilibrium equation is the rotation of joint 2. $(EI)_1 = (EI)_2 = 573,958$ kft^2. See Figure 9.1a.

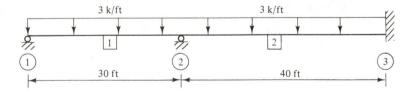

Figure 9.1a Structure with loading on it.

Solution

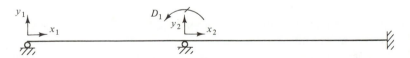

Figure 9.1b Positive system displacement and member axes.

Figure 9.1c Applied joint loads.

$P_1^a = 0$ (there is no externally applied moment at D_1 location in Figure a)

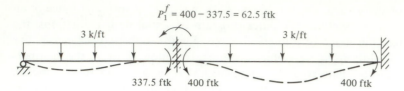

See Figures 8.7 and 8.12 for the fixed ended moment formulas used below.

For member 1, the fixed ended moment is $\dfrac{3 \text{ k/ft} \star (30 \text{ ft})^2}{8} = 337.5$ ftk

For member 2, the fixed ended moments are $\dfrac{3 \text{ k/ft} \star (40 \text{ ft})^2}{12} = 400$ ftk

Figure 9.1d Solution for $D_1 = 0$ (to find the fixed joint load, P_1^f).

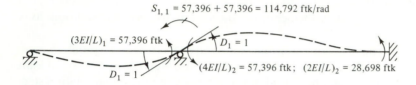

The structural analyst must choose the units that the analyst prefers for the system displacements — the author prefers rotational units of radians and translational units of feet. If rotational units of degrees are chosen, all of the rotational forces must be divided by 57.2958° and these calculations are too time consuming. However, it should be noted that the imposed unit rotation, $D_1 = 1$ rad, must be treated as a small angle in formulating the stiffness coefficients.

Figure 9.1e Solution for $D_1 = 1$.

The matrix equation of system equilibrium, $[S]\{D\} = \{P^e\}$ where $\{P^e\} = \{P^a\} - \{P^f\}$, in this problem is $S_{1,1}D_1 = P_1^e$ since there is only one DOF. From the above calculations, this equilibrium equation becomes

$$114{,}792 D_1 = (0 - 62.5 = -62.5)$$

and the solution of this equation gives

$$D_1 = -0.000544 \text{ rad} = -0.0312°$$

The final member-end moments are found for each member by algebraically summing those in Figure 9.1d and D_1 times those in Figure 9.1e. After the member-end moments are obtained, the member-end shears are found by using the following free body diagram equilibrium equations for each member:

1. At one member end, summation of moments equal to zero gives the shear at the other member end.
2. Summation of vertical forces equal to zero gives the other member-end shear.

Final member-end forces are shown in Figures 9.1f and 9.1g. Reactions, shear and moment diagrams, and the deflected shape are shown in Figure 9.1h.

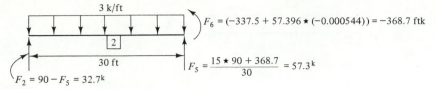

$$F_6 = (-337.5 + 57.396 \star (-0.000544)) = -368.7 \text{ ftk}$$

$$F_5 = \frac{15 \star 90 + 368.7}{30} = 57.3^k$$

$$F_2 = 90 - F_5 = 32.7^k$$

Figure 9.1f

$$F_3 = 400 + 57,396 \star (-0.000544) = 368.8 \text{ ftk}$$

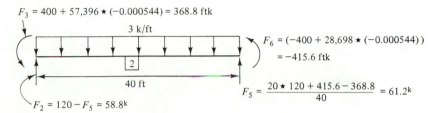

$$F_6 = (-400 + 28,698 \star (-0.000544))$$
$$= -415.6 \text{ ftk}$$

$$F_5 = \frac{20 \star 120 + 415.6 - 368.8}{40} = 61.2^k$$

$$F_2 = 120 - F_5 = 58.8^k$$

Figure 9.1g

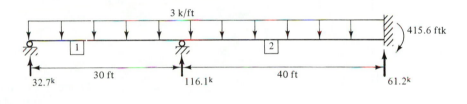

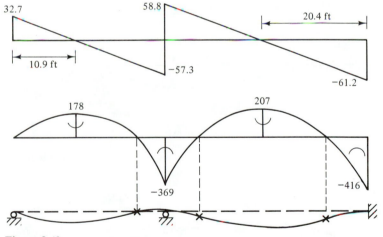

Figure 9.1h

EXAMPLE 9.2

Solve by the Displacement Method for the DOF shown in Figure 9.2b; $(EI)_1 = 166,180$ kNm^2; $(EI)_2 = 332,360$ kNm^2. Note that axial deformations of all members are to be ignored.

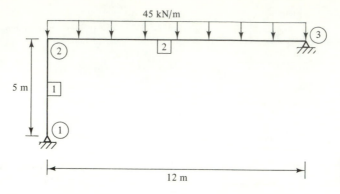

Figure 9.2a Original structure with loading on it.

Solution

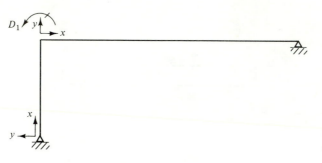

Figure 9.2b Positive system displacement and member axes.

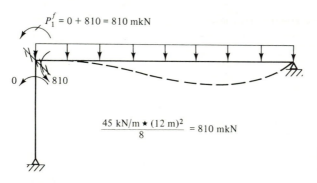

$P_1^f = 0 + 810 = 810$ mkN

$$\frac{45 \text{ kN/m} \star (12 \text{ m})^2}{8} = 810 \text{ mkN}$$

Figure 9.2c Solution for $\{D\} = \{0\}$.

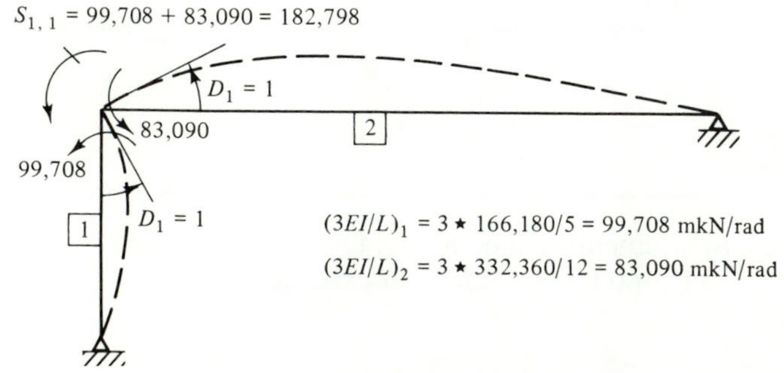

$S_{1,1} = 99{,}708 + 83{,}090 = 182{,}798$

$D_1 = 1$

$83{,}090$

$99{,}708$

$D_1 = 1$

$(3EI/L)_1 = 3 \star 166{,}180/5 = 99{,}708$ mkN/rad

$(3EI/L)_2 = 3 \star 332{,}360/12 = 83{,}090$ mkN/rad

Figure 9.2d Solution for $D_1 = 1$.

Since there is no applied external moment at D_1 in Figure 9.2a, $P_1^q = 0$ and $P_1^e = P_1^q - P_1^f = 0 - 810 = -810$ mkN. The matrix equation of system equilibrium, $[S]\{D\} = \{P^e\}$, is

$$182{,}798D_1 = -810$$

which gives $D_1 = -0.0044311$ rad. The final member-end moments for each member are the algebraic sum of those in Figure 9.2c and D_1 times those in Figure 9.2d. Final member-end shears are found for each member by using free body diagram equilibrium equations after the member-end moments are available. Axial forces in each member are found by joint equilibrium as shown in Figure 9.2e; that is, once the left end shear of member 2 is known, it can be transferred and reversed on joint 2; vertical equilibrium of joint 2 gives the axial force in member 1. Similarly, the top end shear of member 1 is reversed on joint 2; horizontal equilibrium of joint 2 gives the axial force in member 2. The reactions are shown in Figure 9.2f.

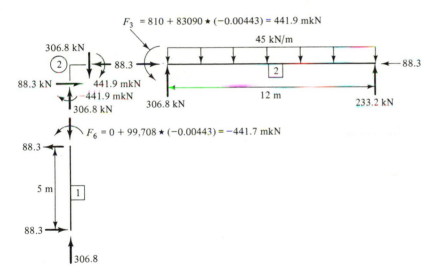

Figure 9.2e Final member-end forces.

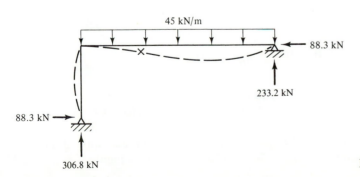

Figure 9.2f Reactions.

EXAMPLE 9.3

Solve by the Displacement Method involving only the rotations of joints 2 and 3 as the DOF. $EI = 312,153$ kft^2 for all members. See Figure 9.3a.

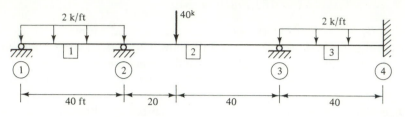

Figure 9.3a Original structure and loading.

Solution

Figure 9.3b Positive system displacements.

Figure 9.3c Member axes.

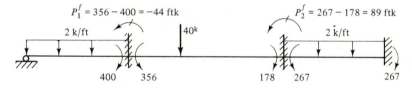

Figure 9.3d Solution for $\{D\} = \{0\}$ (to find fixed joint loads, $\{P^f\}$).

Fixed ended moments are:

member 1: $\dfrac{2\,\text{k/ft} * (40\,\text{ft})^2}{8} = 400\,\text{ftk}$

member 2:   at left end: $\dfrac{40^k * 20\,\text{ft} * (40\,\text{ft})^2}{(60\,\text{ft})^2} = 355.56\,\text{ftk} = 356\,\text{ftk}$

at right end: $\dfrac{40^k * 40\,\text{ft} * (20\,\text{ft})^2}{(60\,\text{ft})^2} = 177.78\,\text{ftk} = 178\,\text{ftk}$

member 3: $\dfrac{2\,\text{k/ft} * (40\,\text{ft})^2}{12} = 267\,\text{ftk}$

The matrix equation of system equilibrium, $[S]\{D\} = \{P^e\}$ where $\{P^e\} = \{P^a\} - \{P^f\}$, is numerically formulated one row at a time as follows:

1. At and in the direction of D_1, summation of the externally applied moments in Figure 9.3d plus D_1 times Figure 9.3e plus D_2 times Figure 9.3f must add

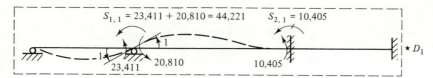

$(3EI/L)_1 = 3 \star 312,153 \text{ kft}^2/40 \text{ ft} = 23411 \text{ ftk}$

$(4EI/L)_2 = 4 \star 312,153/60 = 20810 \text{ ftk}; \quad (2EI/L)_2 = 10,405 \text{ ftk}$

Figure 9.3e Solution for $D_1 = 1$ (to find column 1 of $[S]$).

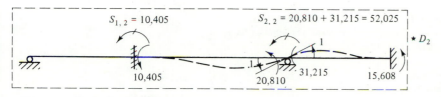

$(4EI/L)_3 = 4 \star 312,153/40 = 31,215 \text{ ftk}; \quad (2EI/L)_3 = 15,608 \text{ ftk}$

Figure 9.3f Solution for $D_2 = 1$ (to find column 2 of $[S]$).

up to the externally applied moment in the original structure; this gives $P_1^f + S_{1,1}D_1 + S_{1,2}D_2 = P_1^a$ and in convenient simultaneous equation format becomes $S_{1,1}D_1 + S_{1,2}D_2 = P_1^e$ and $P_1^e = P_1^a - P_1^f$.

2. At and in the direction of D_2, summation of the externally applied moments in Figure 9.3d plus D_1 times Figure 9.3e plus D_2 times Figure 9.3f must add up to the externally applied moment in the original structure; this gives $P_2^f + S_{2,1}D_1 + S_{2,2}D_2 = P_2^a$ which can be written as $S_{2,1}D_1 + S_{2,2}D_2 = P_2^e$.

Since

$$\begin{Bmatrix} P_1^e \\ P_2^e \end{Bmatrix} = \begin{Bmatrix} P_1^a \\ P_2^a \end{Bmatrix} - \begin{Bmatrix} P_1^f \\ P_2^f \end{Bmatrix} = \begin{Bmatrix} 0 \\ 0 \end{Bmatrix} - \begin{Bmatrix} -44 \\ 89 \end{Bmatrix} = \begin{Bmatrix} 44 \\ -89 \end{Bmatrix}$$

the matrix equation of system equilibrium is

$$\begin{bmatrix} 44221 & 10405 \\ 10405 & 52025 \end{bmatrix} \begin{Bmatrix} D_1 \\ D_2 \end{Bmatrix} = \begin{Bmatrix} 44 \\ -89 \end{Bmatrix}$$

Solution of the simultaneous equations gives

$$\begin{Bmatrix} D_1 \\ D_2 \end{Bmatrix} = \begin{Bmatrix} 0.00146654 \text{ rad} \\ -0.00200402 \text{ rad} \end{Bmatrix}$$

The final member-end forces are found by first finding the member-end moments for each member—at a member-end location, numerically obtain the algebraic sum of the member-end moment in Figure 9.3d plus D_1 times the corresponding one in Figure 9.3e plus D_2 times the corresponding one in Figure 9.3f. Then, consider each member as a free body diagram to find the member-end shears. These calculations are numerically illustrated in Figure 9.3g.

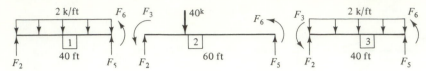

For member 1:

$F_6 = -400 + 23{,}411 * 0.00146654 = -365.67$ ftk
from statics: $F_2 = 30.86^k$ and $F_5 = 49.14^k$

For member 2:

$F_3 = 356 + 20{,}810 * 0.00146654 + 10{,}405 * (-0.00200402) = 365.67$ ftk
$F_6 = -178 + 10{,}405 * 0.00146654 + 20{,}810 * (-0.00200402) = -204.44$ ftk
from statics: $F_2 = 29.35^k$ and $F_5 = 10.65^k$

For member 3:

$F_3 = 267 + 31{,}215 * (-0.00200402) = 204.44$ ftk
$F_6 = -267 + 15{,}608 * (-0.00200402) = -298.28$ ftk
from statics: $F_2 = 27.65^k$ and $F_5 = 42.35^k$

Figure 9.3g Final member-end forces.

The reactions at joints 2 and 3 are found by algebraically summing the final member-end shears. These reactions as well as the shear and moment diagrams are shown in Figure 9.3h. Note that the conventional load-shear-moment sign convention is used in Figure 9.3h; that is, the Displacement Method sign convention is abandoned in drawing the shear and moment diagrams.

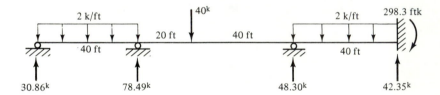

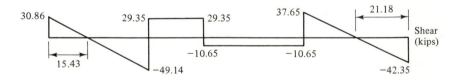

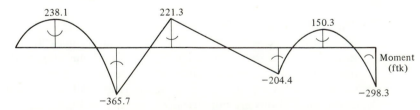

Figure 9.3h Reactions, shear and moment diagrams.

Figure 9.3i Deflected structure.

9.2 EXAMPLES WITH UNKNOWN JOINT TRANSLATIONS

In the following examples, it is assumed that the members are axially rigid and that all translational movements of the supports are prevented. These assumptions are made to simplify the problem concepts in this section; examples in subsequent sections will demonstrate how to account for prescribed support movements and axial deformations.

EXAMPLE 9.4

Solve for the reactions by the Displacement Method. $(EI)_1 = 300{,}000$ kft^2; $(EI)_2 = 600{,}000$ kft^2. See Figure 9.4a.

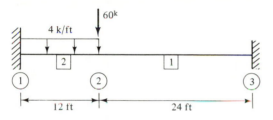

Figure 9.4a Original structure and loading.

Solution

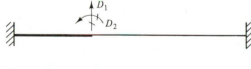

Figure 9.4b Positive system displacements.

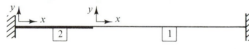

Figure 9.4c Member axes.

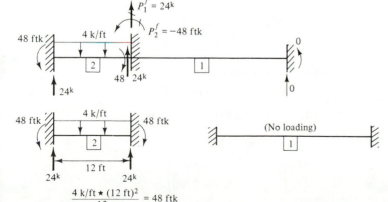

$$\frac{4 \text{ k/ft} \star (12 \text{ ft})^2}{12} = 48 \text{ ftk}$$

Figure 9.4d Solution for $\{D\} = \{0\}$ (to find fixed joint loads, $\{P^f\}$).

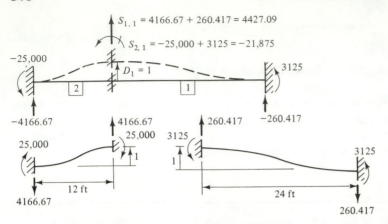

$(6EI/L^2)_2 = 6 \star 600{,}000/(12)^2 = 25{,}000; \quad 2 \star 25{,}000/12 \text{ ft} = 4166.67$

$(6EI/L^2)_1 = 6 \star 300{,}000/(24)^2 = 3125; \quad 2 \star 3125/24 \text{ ft} = 260.417$

Figure 9.4e Solution for $D_1 = 1$ (to find column 1 of $[S]$).

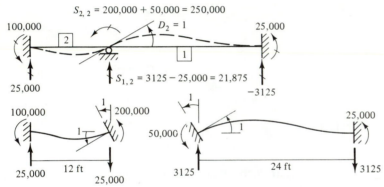

$(4EI/L)_2 = 4 \star 600{,}000/12 = 200{,}000; \quad (2EI/L)_2 = 100{,}000;$

$(4EI/L)_1 = 4 \star 300{,}000/24 = 50{,}000; \quad (2EI/L)_1 = 25{,}000;$

Figure 9.4f Solution for $D_2 = 1$ (to find column 2 of $[S]$).

Since $P_1^a = -60^k$ and $P_2^a = 0$, the right-hand side of the matrix equation of system equilibrium is

$$\begin{Bmatrix} P_1^e \\ P_2^e \end{Bmatrix} = \begin{Bmatrix} -60 \\ 0 \end{Bmatrix} - \begin{Bmatrix} 24 \\ -48 \end{Bmatrix} = \begin{Bmatrix} -84 \\ 48 \end{Bmatrix}$$

and the matrix equation of system equilibrium is

$$\begin{bmatrix} 4427.09 & -21875 \\ -21875 & 250000 \end{bmatrix} \begin{Bmatrix} D_1 \\ D_2 \end{Bmatrix} = \begin{Bmatrix} -84 \\ 48 \end{Bmatrix}$$

Solution of these simultaneous equations gives

$$\begin{Bmatrix} D_1 \\ D_2 \end{Bmatrix} = \begin{Bmatrix} -0.0317545 \text{ ft} \\ -0.0025865 \text{ rad} \end{Bmatrix}$$

The reactions are the algebraic sum of the reactions in Figure 9.4d plus D_1 times those in Figure 9.4e plus D_2 times those in Figure 9.4f.

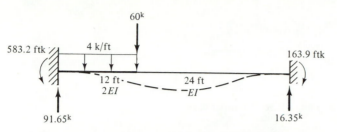

583.2 ftk   4 k/ft   60^k 163.9 ftk

12 ft – 2EI   24 ft – EI

91.65^k 16.35^k **Figure 9.4g** Reactions.

EXAMPLE 9.5

Solve for the reactions by the Displacement Method.

$$(EI/L)_1 = 1667\,\text{ftk}; \quad (EI/L)_2 = 1389\,\text{ftk}; \quad (EI/L)_3 = 1042\,\text{ftk}$$

Ignore axial deformations of all members. See Figure 9.5a.

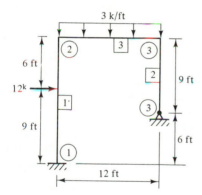

Figure 9.5a Structure and loading.

Solution

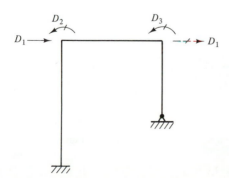

Figure 9.5b Positive system displacements.

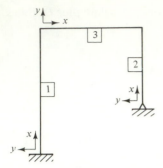

Figure 9.5c Member axes.

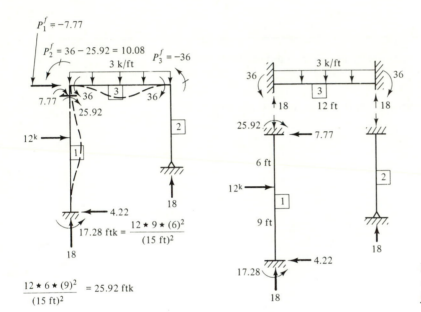

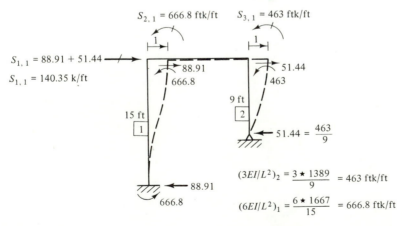

Figure 9.5d Solution for $\{D\} = \{0\}$ (to find the fixed joint loads, $\{P^f\}$).

$$\frac{12 \star 6 \star (9)^2}{(15 \text{ ft})^2} = 25.92 \text{ ftk}$$

$S_{2,1} = 666.8 \text{ ftk/ft}$ $\quad$ $S_{3,1} = 463 \text{ ftk/ft}$

$S_{1,1} = 88.91 + 51.44$

$S_{1,1} = 140.35 \text{ k/ft}$

$$(3EI/L^2)_2 = \frac{3 \star 1389}{9} = 463 \text{ ftk/ft}$$

$$(6EI/L^2)_1 = \frac{6 \star 1667}{15} = 666.8 \text{ ftk/ft}$$

Member–end shears are obtained by statics: $2 \star 666.8/15 = 88.91 \text{ k/ft}$

Figure 9.5e Solution for $D_1 = 1$ (to find column 1 of $[S]$).

318

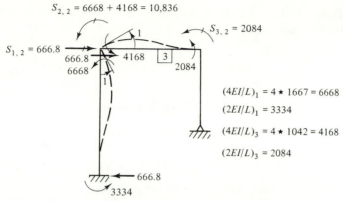

Figure 9.5f Solution for $D_2 = 1$ (to find column 2 of [S]).

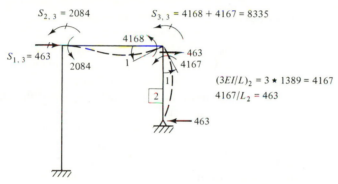

Figure 9.5g Solution for $D_3 = 1$ (to find column 3 of [S]).

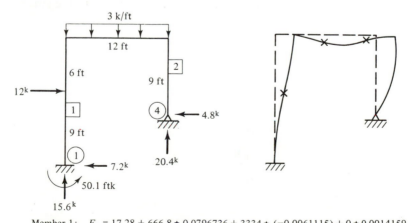

Member 1: $F_3 = 17.28 + 666.8 \star 0.0796736 + 3334 \star (-0.0061115) + 0 \star 0.0014159 = 50.1$ ftk
Member 2: $F_2 = 0 + 51.44 \star 0.0796736 + 0 \star (-0.0061115) + 463 \star 0.0014159 = 4.8^k$

Figure 9.5h Reactions.

From Figure 9.5a, $P_1^a = 0$, $P_2^a = 0$, and $P_3^a = 0$. The right-hand side of the matrix equation of system equilibrium is

$$\begin{Bmatrix} P_1^e \\ P_2^e \\ P_3^e \end{Bmatrix} = \begin{Bmatrix} 0 \\ 0 \\ 0 \end{Bmatrix} - \begin{Bmatrix} -7.77 \\ 10.08 \\ -36 \end{Bmatrix} = \begin{Bmatrix} 7.77 \\ -10.08 \\ 36 \end{Bmatrix}$$

and the matrix equation of system equilibrium is

$$\begin{bmatrix} 140.35 & 666.8 & 463 \\ 666.8 & 10836 & 2084 \\ 463 & 2084 & 8335 \end{bmatrix} \begin{Bmatrix} D_1 \\ D_2 \\ D_3 \end{Bmatrix} = \begin{Bmatrix} 7.77 \\ -10.08 \\ 36 \end{Bmatrix}$$

Solution of these simultaneous equations gives:

$$\begin{Bmatrix} D_1 \\ D_2 \\ D_3 \end{Bmatrix} = \begin{Bmatrix} 0.0796736 \text{ ft} \\ -0.0061115 \text{ rad} \\ 0.0014159 \text{ rad} \end{Bmatrix}$$

If an analyst only wishes to find the reactions now that the system displacements are known, the quickest procedure in this problem is to find the moment at joint 1 and the horizontal force at joint 4; then moment equilibrium about joint 1 of the entire system as a free body diagram gives the vertical reaction at joint 4 and vertical equilibrium gives the vertical reaction at joint 1. The moment at joint 1 and the horizontal force at joint 4 are found by algebraically summing the values in Figure 9.5d plus D_1 times the values in Figure 9.5e plus D_2 times the values in Figure 9.5f plus D_3 times the values in Figure 9.5g.

After all of the reactions are known, the structural system is determinate and the final member-end forces at the interior joint locations can be readily found by performing a determinate analysis of the structure.

EXAMPLE 9.6 ———

Only find columns 1 and 2 of the system stiffness matrix for the structure and DOF shown; axial deformations are ignored.

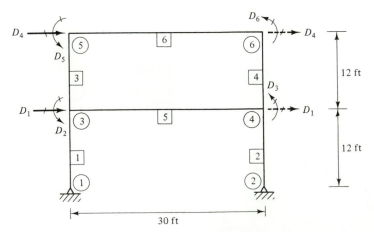

Figure 9.6a Positive system displacements and structural configuration.

$$\text{members 1 through 4:} \qquad \frac{EI}{L} = 13{,}500 \text{ ftk}$$

$$\text{members 5 through 6:} \qquad \frac{EI}{L} = 10{,}800 \text{ ftk}$$

Solution. Member-end stiffness values that are needed in the solution are (Figs. 9.6b and 9.6c):

$$\text{for members 1 and 2:} \qquad \frac{3EI}{L} = 40{,}500; \qquad \frac{3EI}{L^2} = 3375; \qquad \frac{3EI}{L^3} = 281.25$$

$$\text{for members 3 and 4:} \qquad \frac{4EI}{L} = 54{,}000; \qquad \frac{2EI}{L} = 27{,}000$$

$$\frac{6EI}{L^2} = 6750; \qquad \frac{12EI}{L^3} = 1125$$

$$\text{for members 5 and 6:} \qquad \frac{4EI}{L} = 43{,}200; \qquad \frac{2EI}{L} = 21{,}600$$

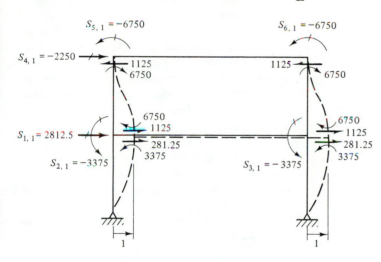

Figure 9.6b Solution for $D_1 = 1$ (to find column 1 of $[S]$).

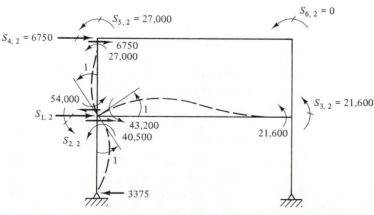

Figure 9.6c Solution for $D_2 = 1$ (to find column 2 of $[S]$).

$S_{1,2} = 3375 - 6750 = -3375; \qquad S_{2,2} = 40{,}500 + 54{,}000 + 43{,}200 = 137{,}700$

9.3 EXAMPLES OF PRESCRIBED SUPPORT MOVEMENTS

In an actual structural design situation, the structural designer may need to explore the consequences of certain magnitudes of assumed support movements at various locations. For example, the soil beneath a structural foundation may settle due to the continuously applied loads or heave due to frost action in arctic regions. In that case, the structural designer may choose to perform the structural analyses assuming that support movements are prevented. Separate analyses would also be performed to determine the response of the structure due only to certain specified support movements which might be reasonably expected to occur. The results of analyses performed due to specified support movements only would be superimposed onto the results of analyses which presumed no support movements occurred in order to obtain the final results. Such a procedure allows the structural designer to determine what the consequences of support movements only are for the structure. Later, the foundations engineer may find that the foundations cost will be too expensive unless the structural designer can tolerate more support movement than is customarily agreed on in a routine design situation. At that time, the structural analysis due to support movements only would be redone for the revised support movements that need to be tolerated. Another situation that can happen is that greater than expected support movements can occur sometime during the life of the structure. Actual support movement measurements would be used to perform an analysis due to support movements only and these results would be added to the analyses results which presumed no support movements occurred. If the structural properties have not deteriorated appreciably, these latter analyses would not need to be rerun since the analyses for the final design check would have been saved by the structural designer.

Sufficient reasons have been given to show that it is customary to treat foundation movement problems as a separate analysis. Even if there were no better reason for choosing to do as the author does in this section, it is conceptually easier to deal with prescribed support movements as a separate "loading condition" as the reader will see in the following examples.

If the soil beneath all foundations of a structure settles the same amount in the gravity direction, there are no increases or decreases in the member-end axial forces, shears, or bending moments for any member in the structure. However, if one foundation settles more than the neighboring ones do, there may be significant changes in the member-end axial forces, shears, or bending moments for some of the members in the structure. Relative foundation movements are what may cause significant increases or decreases in the member-end forces for members in the vicinity of the supports. In the following example problems, the author chooses to list only the relative support movements in order to emphasize that those are the ones that must be used in the prescribed support movement analysis.

EXAMPLE 9.7 ───

The structure shown in Figure 9.7a is the same one as in Example 9.1 and is shown here for convenience (to avoid having to find it in the previous location in the text). The objective of this example is to find the member-end forces due only to the following prescribed support movements:

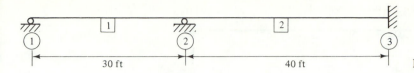

Figure 9.7a Structure.

1. 0.5 in. downward = 0.0417 ft downward at joint 2.

2. 0.001 rad counterclockwise at joint 3.

Solution. The solution procedure shown is the one that the author recommends for a noncomputerized solution. This procedure enables the analyst to use the previously computed system stiffness matrix; this saves computational time. Each of the prescribed support movements is treated separately in this example, but both of them could be treated as a single case, if desired. Each support movement is treated as a "loading" causing member fixed ended forces and fixed joint loads; then, the DOF must be allowed to occur; superposition of these two solutions gives the final member-end forces.

Solution for Part 1. The DOF involved in Example 9.1 is shown in Figure 9.7b for convenience. From Example 9.1, $S_{1,1} = 114{,}792$ ftk/rad. For a prescribed downward movement of 0.0417 ft at joint 2, the induced member fixed ended moments are as shown in Figure 9.7c. Since $P_1^a = 0$, $P_1^e = P_1^a - P_1^f = 0 - (-10.03) = 10.03$, the matrix equation of system equilibrium is

$$114{,}792 D_1 = 10.03$$

which gives $D_1 = 0.0000874$ rad. The final member-end forces due only to the prescribed 0.0417 ft downward movement at joint 2 are shown in Figure 9.7d.

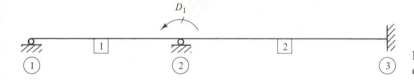

Figure 9.7b Positive system displacement.

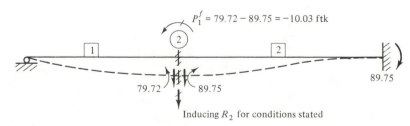

$P_1^f = 79.72 - 89.75 = -10.03$ ftk

79.72   89.75

89.75

Inducing R_2 for conditions stated

$(3EI/L^2)_1 \star 0.0417 \text{ ft} = 3 \star 573{,}958/(30)^2 \star 0.0417 = 79.72$ ftk

$(6EI/L^2)_2 \star 0.0417 \text{ ft} = 6 \star 573{,}958/(40)^2 \star 0.0417 = 89.75$ ftk

Figure 9.7c Solution for $D_1 = 0$ and prescribed 0.0417 ft downward at joint 2 (to find the value of P_1^f, the fixed joint load).

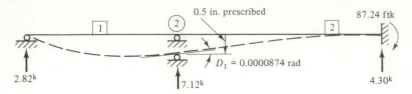

Member 1: right end moment = 79.72 + 57,396 ⋆ 0.0000874 = 84.73 ftk

Figure 9.7d Solution for part a of Example 9.7.

Solution for Part 2. For a prescribed counterclockwise rotation of 0.001 rad at joint 3, the induced member fixed ended moments are as shown in Figure 9.7e. $P_1^e = 0 - 28.7 = -28.7$ and the matrix equation of system equilibrium is

$$114,792D_1 = -28.7$$

which gives $D_1 = -0.00025$ rad. The final member-end forces due only to the prescribed 0.001 rad counterclockwise rotation at joint 3 are shown in Figure 9.7f.

If an analyst needed or desired the solution for the loading used in Example 9.1 to occur simultaneously with a prescribed support movement of 0.5 in. downward at joint 2, for example, the solution could be obtained by algebraically adding the solution of Example 9.1 to the part 1 solution of Example 9.7 since superposition is valid.

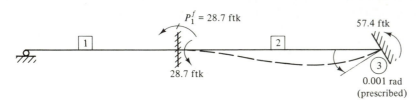

Member 2: $(4EI/L) \star 0.001$ rad $= 57.4$ ftk
$(2EI/L) \star 0.001$ rad $= 28.7$ ftk

Figure 9.7e Solution for $D_1 = 0$ and prescribed 0.001 rad at joint 3.

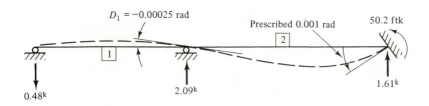

Member 1: right end moment $= 0 + 57,396 \star (-0.00025) = -14.35$ ftk
Member 2: right end moment $= 57.4 + 28,698 \star (-0.00025) = 50.2$ ftk

Figure 9.7f Solution for part b of Example 9.7.

EXAMPLE 9.8

The structure shown in Figure 9.8a is the same one as in Example 9.2 and is shown here for reference convenience. The objective of this example is to find the member-end forces due only to a prescribed downward movement of 0.038 m at joint 1.

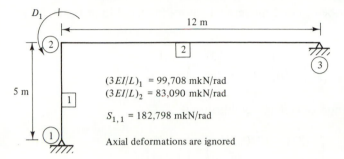

$(3EI/L)_1 = 99,708$ mkN/rad
$(3EI/L)_2 = 83,090$ mkN/rad

$S_{1,1} = 182,798$ mkN/rad

Axial deformations are ignored

Figure 9.8a Structure and positive system displacement.

Solution. Since axial deformations are being ignored (members are axially rigid), a prescribed downward movement of 0.038 m at joint 1 with $\{D\} = \{0\}$ causes the fixed joint load shown in Figure 9.8b. $P_1^e = 0 - (-263.1) = 263.1$ and the matrix equation of system equilibrium is

$$182,798D_1 = 263.1$$

which gives $D_1 = 0.00144$ rad. The final member-end forces due only to the prescribed support movement are shown in Figure 9.8c.

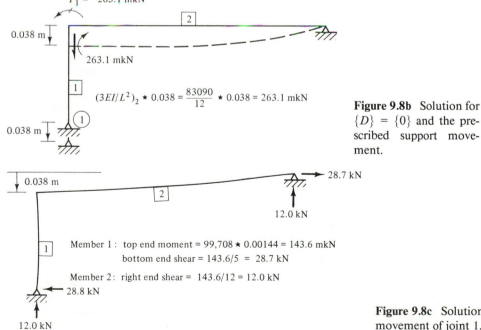

$P_1^f = -263.1$ mkN

0.038 m

263.1 mkN

$(3EI/L^2)_2 \star 0.038 = \dfrac{83090}{12} \star 0.038 = 263.1$ mkN

0.038 m

Figure 9.8b Solution for $\{D\} = \{0\}$ and the prescribed support movement.

0.038 m

28.7 kN

12.0 kN

Member 1: top end moment = 99,708 $\star$ 0.00144 = 143.6 mkN
bottom end shear = 143.6/5 = 28.7 kN

Member 2: right end shear = 143.6/12 = 12.0 kN

28.8 kN

12.0 kN

Figure 9.8c Solution for prescribed movement of joint 1.

EXAMPLE 9.9 _____

The structure shown below is the same one as in Example 9.3 and is shown here for reference convenience. The objective of this example is to find the member-end forces due only to a prescribed downward movement of 1.5 in. = 0.125 ft at joint 2.

Solution. The prescribed joint 2 movement causes the fixed joint loads shown in Figure 9.9b.

$$\begin{Bmatrix} P_1^e \\ P_2^e \end{Bmatrix} = \begin{Bmatrix} 0 \\ 0 \end{Bmatrix} - \begin{Bmatrix} 8.13 \\ -65.03 \end{Bmatrix} = \begin{Bmatrix} -8.13 \\ 65.03 \end{Bmatrix}$$

and the matrix equation of system equilibrium is

$$\begin{bmatrix} 44221 & 10405 \\ 10405 & 52025 \end{bmatrix} \begin{Bmatrix} D_1 \\ D_2 \end{Bmatrix} = \begin{Bmatrix} -8.13 \\ 65.03 \end{Bmatrix}$$

which gives

$$\begin{Bmatrix} D_1 \\ D_2 \end{Bmatrix} = \begin{Bmatrix} -0.0005016 \text{ rad} \\ 0.0013503 \text{ rad} \end{Bmatrix}$$

The reactions for the deflected shape due to a prescribed downward movement of 0.125 ft at joint 2 are shown in Figure 9.9c.

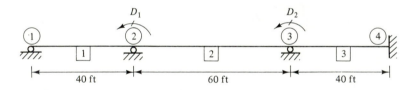

$(3EI/L)_1 = 23{,}411$ ftk; $(4EI/L)_2 = 20{,}810$ ftk; $(4EI/L)_3 = 31{,}215$ ftk

$$[S] = \begin{bmatrix} 44221 & 10405 \\ 10405 & 52025 \end{bmatrix}$$

Figure 9.9a   Structure and DOF.

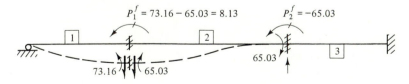

$(3EI/L^2)_1 \star 0.125$ ft $= 73.16$ ftk; $(6EI/L^2)_2 \star 0.125$ ft $= 65.03$ ftk

Figure 9.9b   Solution for $\{D\} = \{0\}$ and the prescribed support movement.

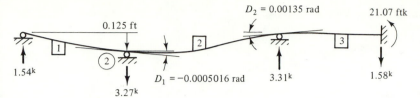

Member 1: right end moment = $73.16 + 23,411 \star (-0.0005016) = 61.42$ ftk
left end shear = $61.42/40 = 1.54^k$

Member 3: left end moment = $31,215 \star 0.00135 = 42.15$ ftk
right end moment = $42.15/2 = 21.07$ ftk
right end shear = $(42.15 + 21.07)/40 = 1.58^k$

Member 2: end shears = $(61.42 + 42.15)/60 = 1.73^k$

Figure 9.9c Solution for prescribed movement of joint 2.

9.4 EXAMPLES INVOLVING CABLES AS BEAM SUPPORTS

Cables have negligible bending stiffness, but the elongation of a cable should be accounted for in a Displacement Method analysis. In the previous numerical examples, axial deformations of the members were ignored. For tall, multistory buildings subjected to wind effects, axial deformations of the columns contribute significantly to the lateral displacements of the structure in the direction of wind. The objective of this section is to demonstrate how to account for axial deformations of members whenever axial deformations are not ignored.

EXAMPLE 9.10

A continuous beam, members 1 and 2, is supported at joint 2 by a cable, member 3 (Fig. 9.10a). Solve for the reactions by the Displacement Method; account for the axial deformation of the high strength cable. Sketch the deflected structure. $(EI/L)_1 = 10,573$ kft^2; $(EI/L)_2 = 14,097$ kft^2; $(EA/L)_3 = 935$ k/ft.

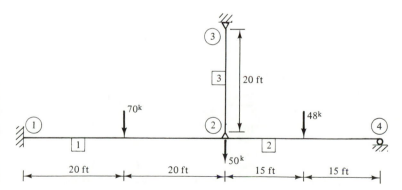

Figure 9.10a Structure for Example 9.10.

Solution

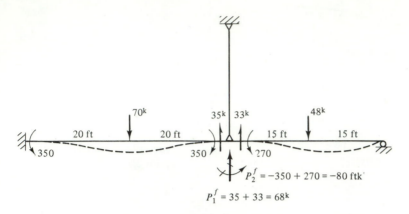

Member 1: fixed ended moments $= 70 \star 40/8 = 350$ ftk
shears $= 70/2 \quad = 35^k$

Member 2: fixed ended moment $= 3 \star 48 \star 30/16 = 270$ ftk
left end shear $= 48/2 + 270/30 = 33^k$

Figure 9.10b Solution for $\{D\} = \{0\}$ (to find $\{P^f\}$).

In the computations of the system stiffness matrix coefficients, a cable must be treated as having the same compressive stiffness as it has in tension. The reason for this is that the choice of DOF directions is arbitrary and in the stiffness matrix formulation of a complex problem it is not always possible to choose DOF directions such that all cables are in tension. However, after the final member-end forces are obtained, each cable must be in tension or else the cable is ineffective. In case a cable is ineffective, it must be removed and the analysis must be repeated to obtain the correct solution.

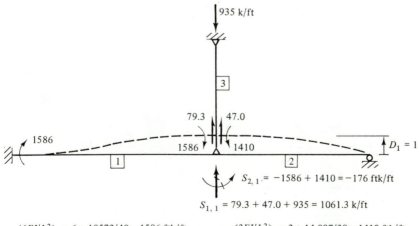

$(6EI/L^2)_1 = 6 \star 10573/40 = 1586$ ftk/ft

$2 \star 1586/40 = 79.3$ k/ft

$(3EI/L^2)_2 = 3 \star 14,097/30 = 1410$ ftk/ft

$1410/30 = 47.0$ k/ft

Figure 9.10c Solution for $D_1 = 1$ (to find column 1 of $[S]$).

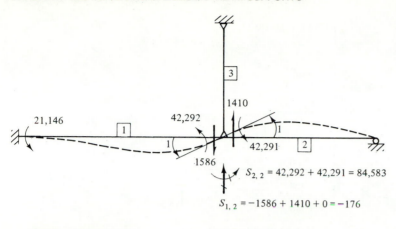

$$S_{2,2} = 42,292 + 42,291 = 84,583$$

$$S_{1,2} = -1586 + 1410 + 0 = -176$$

$(4EI/L)_1 = 42,292$ ftk/rad

$(2EI/L)_1 = 21,146$ ftk/rad

$(21,146 + 42,292)/40 = 1586$ k/rad

$(3EI/L)_2 = 42,291$ ftk/rad

$42,291/30 = 1410$ k/rad

Figure 9.10d Solution for $D_2 = 1$ (to find column 2 of $[S]$).

Observing that $P_1^a = -50^k$ and $P_2^a = 0$,

$$\begin{Bmatrix} P_1^e \\ P_2^e \end{Bmatrix} = \begin{Bmatrix} -50 \\ 0 \end{Bmatrix} - \begin{Bmatrix} 68 \\ -80 \end{Bmatrix} = \begin{Bmatrix} -118 \\ 80 \end{Bmatrix}$$

The matrix equation of system equilibrium is

$$\begin{bmatrix} 1061.3 & -176 \\ -176 & 84,583 \end{bmatrix} \begin{bmatrix} D_1 \\ D_2 \end{bmatrix} = \begin{Bmatrix} -118 \\ 80 \end{Bmatrix}$$

Solution of the simultaneous equations gives

$$\begin{Bmatrix} D_1 \\ D_2 \end{Bmatrix} = \begin{Bmatrix} -0.111 \text{ ft} \\ 0.000715 \text{ rad} \end{Bmatrix}$$

The reactions are shown in Figure 9.10e and it is noted that the cable is in tension. Therefore, the cable is effective and the solution is correct.

A final comment on the results of Example 9.10 is appropriate. Note that the cable elongated 0.111 ft = 1.33 in. This elongation contributed $1586*0.111 = 176$ ftk to the moment at joint 1 which is more than 30% of the final moment shown in Figure 9.10e. If the cable axial deformation had been ignored, the final cable tension = 118^k and the moment at joint 1 = 370 ftk. Since $370/541 = 0.684$, by ignoring the cable axial deformation one would underestimate the moment at joint 1 by 32%. The author deliberately chose a high strength cable to allow the axial deformation of the cable to be as large as possible. If a conventional steel threaded rod of 50 ksi yield strength were used instead of a cable, EA/L of the threaded rod would be 7118 k/ft as compared to the 935 k/ft for the high-strength cable. Usage of the threaded rod would allow joint 2 to deflect 0.01627 ft = 0.195 in. downward as compared to

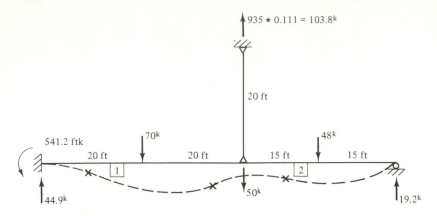

Member 1: left end moment = 350 + (−1586) ∗ (−0.111) + 21,146 ∗ 0.000715 = 541.2 ftk

Member 2: left end moment = 270 + 1410 ∗ (−0.111) + 42,291 ∗ 0.000715 = 143.7 ftk

All member-end shears obtained from statics of individual member free bodies

Figure 9.10e Reactions for Example 9.10.

the 1.33 in. allowed by the high strength cable. If the axial deformation of the threaded rod were ignored, the moment at joint 1 would be underestimated by 1586 ∗ 0.01627 = 25.8 ftk or 7% underestimation of the moment at joint 1.

EXAMPLE 9.11 ───

See Figure 8.1. The reader now should be fully capable of understanding the detailed calculations given in the conceptual example. See pages 299 through 305 for the solution of this example.

9.5 MULTIPLE LOAD CASES

A structure must be capable of resisting its own weight and the superimposed loads. Building codes categorize the weight of the structure as dead load and the superimposed loads as live loads; snow loads; wind loads; and earthquake loads. In addition the structure may need to be analyzed for one or more prescribed support movement conditions. Live loads are defined as those loads excluding the effects due to wind, earthquake, and snow which vary in magnitude and/or in position on the structure during the life of the structure. The structural designer has the responsibility to position the live loads on the structure to cause the worst moments, shears, and so on, within the structure. All of the building code loading conditions are independent conditions. Building codes also specify that certain combinations of the independent loadings are to be made; these loading combinations are referred to as dependent loadings.

The purpose of this section is to show an example with the live load positioned to cause the worst moment at each support and the worst moment within each span. Qualitative influence lines obtained by the Muller-Breslau principle are used to determine where to position the live load to cause the worst moment for each of the desired locations. These various live loading cases are combined with dead loads to create loading combinations after the solution of the simultaneous equations for the independent loadings. In a first order analysis, the system stiffness matrix is not a function of the loadings. Therefore, all of the independent loading solutions can be formulated as multiple loadings in the solution of the simultaneous equations of system equilibrium. After the final member-end forces for the independent loadings are obtained, solutions for the loading combinations are obtained by superposition of the appropriate solutions due to the independent loadings.

In Figures 9.11g through 9.11l, the influence lines are drawn such that in the span(s) of the induced unit angle change, the ordinates lie above the structural axis. Live load is to be placed on all spans that have ordinates above the structural axis in order to maximize the moment for which the influence line was drawn.

EXAMPLE 9.12 _____

$(EI/L)_1 = 41,285$ ftk; $(EI/L)_2 = 20,089$ ftk; $(EI/L)_3 = (EI/L)_4 = 15,066$ ftk. The structure shown in Figure 9.11a is subjected to the dead load shown in Figure 9.11b. Also the structure is to be subjected to live loads of 90^k at midspan of member 1 and 2 k/ft on the other members in such a manner that the live loads produce the following:

1. Maximum moment at support 1.
2. Maximum moment at support 2.
3. Maximum moment at support 3.
4. Maximum moment at support 4.
5. Maximum moment within the span of members 1 and 3.
6. Maximum moment within the span of members 2 and 4.

After the solution for the independent loadings due to dead load and live loads are obtained, the dead load solution is to be added to each of the live load solutions. Note that there are seven independent loadings and six dependent loadings.

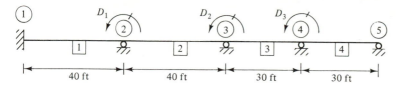

Figure 9.11a Structure and DOF.

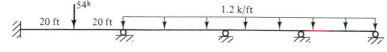

Figure 9.11b Loading 1—dead load.

Solution. Since $\{P^a\} = \{0\}$ for each independent loading in this example, the $\{P^f\}$ are as shown in Figures 9.11f through 9.11l, and $\{P^e\} = \{P^a\} - \{P^f\}$, the matrix coefficients for the matrix equation of system equilibrium, $[S]\{D\} = \{P^e\}$, are

$$[S] = \begin{bmatrix} 245496 & 40178 & 0 \\ 40178 & 140620 & 30132 \\ 0 & 30132 & 105462 \end{bmatrix}$$

$$
\{P^e\} = \begin{Bmatrix}
\begin{array}{ccccccc}
1 & 2 & 3 & 4 & 5 & 6 & 7 \\
110 & 450 & 183.3 & -266.7 & 450 & 450 & -266.7 \\
70 & -150 & 266.7 & 116.7 & -150 & -150 & 266.7 \\
45 & 150 & -225 & \underset{\textbf{I}}{\llcorner}150 & -75 & 150 & -225
\end{array}
\end{Bmatrix}
$$

$\lceil$ - - - - - - independent loading case numbers - - - - - - - $\rceil$

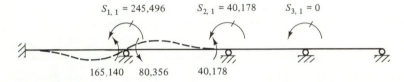

$S_{1,1} = 245{,}496 \qquad S_{2,1} = 40{,}178 \qquad S_{3,1} = 0$

$165{,}140 \quad 80{,}356 \qquad 40{,}178$

Figure 9.11c Solution for $D_1 = 1$ and $D_2 = D_3 = 0$ to find column 1 of $[S]$.

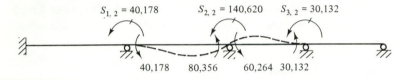

$S_{1,2} = 40{,}178 \qquad S_{2,2} = 140{,}620 \qquad S_{3,2} = 30{,}132$

$40{,}178 \quad 80{,}356 \qquad 60{,}264 \ 30{,}132$

Figure 9.11d Solution for $D_2 = 1$ and $D_1 = D_3 = 0$ to find column 2 of $[S]$.

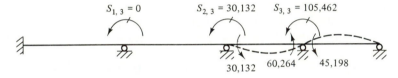

$S_{1,3} = 0 \qquad S_{2,3} = 30{,}132 \qquad S_{3,3} = 105{,}462$

$30{,}132 \quad 60{,}264 \qquad 45{,}198$

Figure 9.11e Solution for $D_3 = 1$ and $D_1 = D_2 = 0$ to find column 3 of $[S]$.

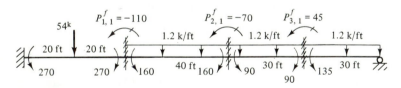

54^k

$P^f_{1,1} = -110 \qquad P^f_{2,1} = -70 \qquad P^f_{3,1} = 45$

20 ft $\quad$ 20 ft $\quad$ 1.2 k/ft $\quad$ 1.2 k/ft $\quad$ 1.2 k/ft

270 $\quad$ 270 $\quad$ 160 $\quad$ 40 ft $\ 160$ $\quad$ 90 $\quad$ 30 ft $\ 135$ $\quad$ 30 ft

90

Figure 9.11f Loading 1 (dead load) solution for $\{D\} = \{0\}$.

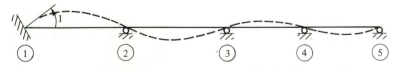

Influence line for moment at support 1

Figure 9.11g Loading 2 (live load) solution for $\{D\} = \{0\}$.

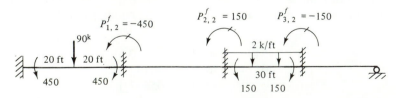

$P^f_{1,2} = -450 \qquad P^f_{2,2} = 150 \qquad P^f_{3,2} = -150$

90^k

20 ft $\quad$ 20 ft $\quad$ 2 k/ft

450 $\quad$ 450 $\quad$ 30 ft

150 $\quad$ 150

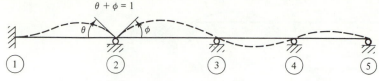

Influence line moment at support 2

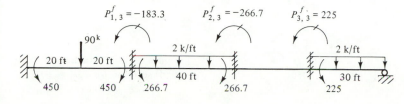

$P^f_{1,3} = -183.3$ $P^f_{2,3} = -266.7$ $P^f_{3,3} = 225$

Figure 9.11h Loading 3 (live load) solution for $\{D\} = \{0\}$.

Influence line for moment at support 3

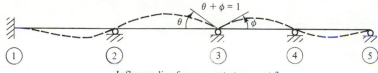

$P^f_{1,4} = 266.7$ $P^f_{2,4} = -116.7$ $P^f_{3,4} = 150$

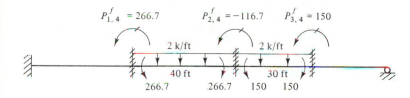

Figure 9.11i Loading 4 (live load) solution for $\{D\} = \{0\}$.

The dependent loadings are

$$\text{loading } 8 = \text{loading } 1 + \text{loading } 2$$
$$\text{loading } 9 = \text{loading } 1 + \text{loading } 3$$
$$\text{loading } 10 = \text{loading } 1 + \text{loading } 4$$
$$\text{loading } 11 = \text{loading } 1 + \text{loading } 5$$
$$\text{loading } 12 = \text{loading } 1 + \text{loading } 6$$
$$\text{loading } 13 = \text{loading } 1 + \text{loading } 7$$

If the solution for this example were pursued by using the noncomputerized approach employed in this text in all previous examples, the following steps would be performed:

1. The matrix equation of system equilibrium would be solved for the system displacements for each of the independent load cases.
2. The final member-end forces would be obtained for each of the independent load cases.
3. The solution for the system displacements and final member-end forces for the dependent load cases would be obtained by algebraically adding the results from the independent load cases; for example, the solution for loading 8 would be obtained by algebraically adding the results of loading 1 and loading 2.

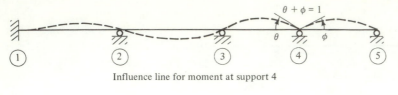

Influence line for moment at support 4

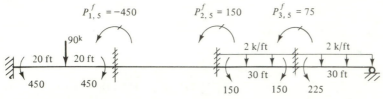

Figure 9.11j Loading 5 (live load) solution for $\{D\} = \{0\}$.

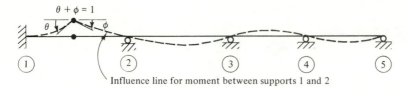

Influence line for moment between supports 1 and 2

Influence line for moment between supports 3 and 4

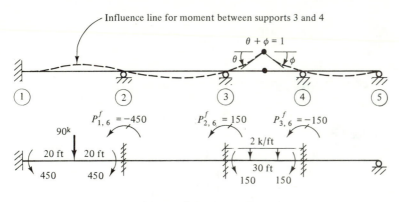

Figure 9.11k Loading 6 (live load) solution for $\{D\} = \{0\}$. Note that loading 6 is identical to loading 2 in Figure 9.11g.

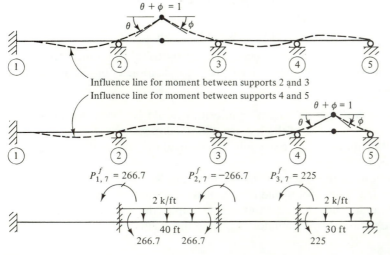

Influence line for moment between supports 2 and 3

Influence line for moment between supports 4 and 5

Figure 9.11l Loading 7 (live load) solution for $\{D\} = \{0\}$.

Obviously, the noncomputerized approach entails lots of calculations in this example. Therefore, the author chooses to show the results for this example as obtained from a computerized solution. The results are shown in Figures 9.11m through 9.11o; note that the computerized solution involved 4 DOF with D_4 being the rotational DOF at joint 5.

OUTPUT DATA

SYSTEM DISPLACEMENTS

JOINT	LOADING	X DISP	Y DISP	Z THETA
1	1	0.00E+00	0.00E+00	0.00E+00
	2	0.00E+00	0.00E+00	0.00E+00
	3	0.00E+00	0.00E+00	0.00E+00
	4	0.00E+00	0.00E+00	0.00E+00
	5	0.00E+00	0.00E+00	0.00E+00
	6	0.00E+00	0.00E+00	0.00E+00
	7	0.00E+00	0.00E+00	0.00E+00
	8	0.00E+00	0.00E+00	0.00E+00
	9	0.00E+00	0.00E+00	0.00E+00
	10	0.00E+00	0.00E+00	0.00E+00
	11	0.00E+00	0.00E+00	0.00E+00
	12	0.00E+00	0.00E+00	0.00E+00
	13	0.00E+00	0.00E+00	0.00E+00
2	1	0.00E+00	0.00E+00	3.63E−04
	2	0.00E+00	0.00E+00	2.18E−03
	3	0.00E+00	0.00E+00	3.54E−04
	4	0.00E+00	0.00E+00	−1.24E−03
	5	0.00E+00	0.00E+00	2.10E−03
	6	0.00E+00	0.00E+00	2.18E−03
	7	0.00E+00	0.00E+00	−1.58E−03
	8	0.00E+00	0.00E+00	2.54E−03
	9	0.00E+00	0.00E+00	7.17E−04
	10	0.00E+00	0.00E+00	−8.76E−04
	11	0.00E+00	0.00E+00	2.46E−03
	12	0.00E+00	0.00E+00	2.54E−03
	13	0.00E+00	0.00E+00	−1.21E−03
3	1	0.00E+00	0.00E+00	5.17E−04
	2	0.00E+00	0.00E+00	−2.12E−03
	3	0.00E+00	0.00E+00	2.40E−03
	4	0.00E+00	0.00E+00	9.37E−04
	5	0.00E+00	0.00E+00	−1.61E−03
	6	0.00E+00	0.00E+00	−2.12E−03
	7	0.00E+00	0.00E+00	2.99E−03
	8	0.00E+00	0.00E+00	−1.61E−03
	9	0.00E+00	0.00E+00	2.92E−03

Figure 9.11m System displacements for Example 9.12.

OUTPUT DATA

SYSTEM DISPLACEMENTS

JOINT	LOADING	X DISP	Y DISP	Z THETA
	10	0.00E+00	0.00E+00	1.45E−03
	11	0.00E+00	0.00E+00	−1.10E−03
	12	0.00E+00	0.00E+00	−1.61E−03
	13	0.00E+00	0.00E+00	3.50E−03
4	1	0.00E+00	0.00E+00	−5.74E−04
	2	0.00E+00	0.00E+00	2.03E−03
	3	0.00E+00	0.00E+00	−2.82E−03
	4	0.00E+00	0.00E+00	1.15E−03
	5	0.00E+00	0.00E+00	−2.51E−04
	6	0.00E+00	0.00E+00	2.03E−03
	7	0.00E+00	0.00E+00	−2.99E−03
	8	0.00E+00	0.00E+00	1.45E−03
	9	0.00E+00	0.00E+00	−3.39E−03
	10	0.00E+00	0.00E+00	5.80E−04
	11	0.00E+00	0.00E+00	−8.25E−04
	12	0.00E+00	0.00E+00	1.45E−03
	13	0.00E+00	0.00E+00	−3.56E−03
5	1	0.00E+00	0.00E+00	1.78E−03
	2	0.00E+00	0.00E+00	−1.01E−03
	3	0.00E+00	0.00E+00	3.90E−03
	4	0.00E+00	0.00E+00	−5.77E−04
	5	0.00E+00	0.00E+00	2.61E−03
	6	0.00E+00	0.00E+00	−1.01E−03
	7	0.00E+00	0.00E+00	3.98E−03
	8	0.00E+00	0.00E+00	7.66E−04
	9	0.00E+00	0.00E+00	5.68E−03
	10	0.00E+00	0.00E+00	1.20E−03
	11	0.00E+00	0.00E+00	4.39E−03
	12	0.00E+00	0.00E+00	7.66E−04
	13	0.00E+00	0.00E+00	5.76E−03

Figure 9.11m (*continued*)

In the solution for Example 9.12, it should be noted that loading 2 and loading 6 are identical; therefore, loading 8 and loading 12 are also identical. Consequently, only six independent loadings and five dependent loadings could have been used in solving this example.

If only the dead load case as well as the live loadings plus dead load combination cases in the above solution were of interest, the four independent loadings and six dependent loadings described on page 340 could be used to solve the problem. Also, one would only output the results for loadings 5 through 10 described on page 340 if these were the only load cases of interest.

JOINT	LOADING	FORCE X	FORCE Y	MOMENT Z
1	1	0.0	29.3	300.0
	2	0.0	58.5	630.1
	3	0.0	47.2	479.2
	4	0.0	−7.7	−102.4
	5	0.0	58.0	623.1
	6	0.0	58.5	630.1
	7	0.0	−9.8	−130.1
	8	0.0	87.8	930.1
	9	0.0	76.4	779.2
	10	0.0	21.6	197.7
	11	0.0	87.2	923.1
	12	0.0	87.8	930.1
	13	0.0	19.5	169.9
2	1	0.0	51.4	0.0
	2	0.0	31.7	0.0
	3	0.0	91.1	0.0
	4	0.0	46.8	0.0
	5	0.0	33.5	0.0
	6	0.0	31.7	0.0
	7	0.0	54.0	0.0
	8	0.0	83.1	0.0
	9	0.0	142.5	0.0
	10	0.0	98.2	0.0
	11	0.0	84.9	0.0
	12	0.0	83.1	0.0
	13	0.0	105.4	0.0
3	1	0.0	39.2	0.0
	2	0.0	29.5	0.0
	3	0.0	30.4	0.0
	4	0.0	77.2	0.0
	5	0.0	22.9	0.0
	6	0.0	29.5	0.0
	7	0.0	35.7	0.0
	8	0.0	68.7	0.0
	9	0.0	69.6	0.0
	10	0.0	116.4	0.0
	11	0.0	62.1	0.0
	12	0.0	68.7	0.0
	13	0.0	74.9	0.0
4	1	0.0	39.8	0.0
	2	0.0	33.3	0.0

Figure 9.11n   Reactions for Example 9.12.

JOINT	LOADING	FORCE X	FORCE Y	MOMENT Z
	3	0.0	34.5	0.0
	4	0.0	25.4	0.0
	5	0.0	72.7	0.0
	6	0.0	33.3	0.0
	7	0.0	33.0	0.0
	8	0.0	73.2	0.0
	9	0.0	74.3	0.0
	10	0.0	65.2	0.0
	11	0.0	112.5	0.0
	12	0.0	73.2	0.0
	13	0.0	72.8	0.0
5	1	0.0	14.4	0.0
	2	0.0	−3.1	0.0
	3	0.0	26.7	0.0
	4	0.0	−1.7	0.0
	5	0.0	22.9	0.0
	6	0.0	−3.1	0.0
	7	0.0	27.0	0.0
	8	0.0	11.3	0.0
	9	0.0	41.1	0.0
	10	0.0	12.6	0.0
	11	0.0	37.2	0.0
	12	0.0	11.3	0.0
	13	0.0	41.4	0.0

Figure 9.11n (*continued*)

FINAL MEMBER-END FORCES

MEMBER	LOADING	AT ORIGIN OF MEMBER X-AXIS			AT END OF MEMBER X-AXIS		
		FORCE X	FORCE Y	MOMENT Z	FORCE X	FORCE Y	MOMENT Z
1	1	0.0	29.3	300.0	0.0	24.7	−210.0
	2	0.0	58.5	630.1	0.0	31.5	−89.9
	3	0.0	47.2	479.2	0.0	42.8	−391.5
	4	0.0	−7.7	−102.4	0.0	7.7	−204.7
	5	0.0	58.0	623.1	0.0	32.0	−103.7
	6	0.0	58.5	630.1	0.0	31.5	−89.9
	7	0.0	−9.8	−130.1	0.0	9.8	−260.1
	8	0.0	87.8	930.1	0.0	56.2	−299.9
	9	0.0	76.4	779.2	0.0	67.6	−601.5
	10	0.0	21.6	197.7	0.0	32.4	−414.7
	11	0.0	87.2	923.1	0.0	56.8	−313.7
	12	0.0	87.8	930.1	0.0	56.2	−299.9
	13	0.0	19.5	169.9	0.0	34.5	−470.1

Figure 9.11o Final member-end forces for Example 9.12.

FINAL MEMBER-END FORCES

MEMBER	LOADING	AT ORIGIN OF MEMBER X-AXIS			AT END OF MEMBER X-AXIS		
		FORCE X	FORCE Y	MOMENT Z	FORCE X	FORCE Y	MOMENT Z
2	1	0.0	26.7	210.0	0.0	21.3	−103.9
	2	0.0	0.2	89.9	0.0	−0.2	−83.1
	3	0.0	48.3	391.5	0.0	31.7	−59.7
	4	0.0	39.1	204.7	0.0	40.9	−241.2
	5	0.0	1.5	103.7	0.0	−1.5	−45.3
	6	0.0	0.2	89.9	0.0	−0.2	−83.1
	7	0.0	44.3	260.1	0.0	35.7	−90.0
	8	0.0	26.8	299.9	0.0	21.2	−187.0
	9	0.0	75.0	601.5	0.0	53.0	−163.5
	10	0.0	65.7	414.7	0.0	62.3	−345.1
	11	0.0	28.1	313.7	0.0	19.9	−149.1
	12	0.0	26.8	299.9	0.0	21.2	−187.0
	13	0.0	70.9	470.1	0.0	57.1	−193.8
3	1	0.0	17.8	103.9	0.0	18.2	−109.0
	2	0.0	29.7	83.1	0.0	30.3	−91.7
	3	0.0	−1.3	59.7	0.0	1.3	−97.6
	4	0.0	36.3	241.2	0.0	23.7	−52.2
	5	0.0	24.4	45.3	0.0	35.6	−213.7
	6	0.0	29.7	83.1	0.0	30.3	−91.7
	7	0.0	−0.0	90.0	0.0	0.0	−90.0
	8	0.0	47.5	187.0	0.0	48.5	−200.8
	9	0.0	16.6	163.5	0.0	19.4	−206.6
	10	0.0	54.1	345.1	0.0	41.9	−161.2
	11	0.0	42.2	149.1	0.0	53.8	−322.7
	12	0.0	47.5	187.0	0.0	48.5	−200.8
	13	0.0	17.8	193.8	0.0	18.2	−199.0
4	1	0.0	21.6	109.0	0.0	14.4	−0.0
	2	0.0	3.1	91.7	0.0	−3.1	0.0
	3	0.0	33.3	97.6	0.0	26.7	−0.0
	4	0.0	1.7	52.2	0.0	−1.7	0.0
	5	0.0	37.1	213.7	0.0	22.9	−0.0
	6	0.0	3.1	91.7	0.0	−3.1	0.0
	7	0.0	33.0	90.0	0.0	27.0	−0.0
	8	0.0	24.7	200.8	0.0	11.3	−0.0
	9	0.0	54.9	206.6	0.0	41.1	−0.0
	10	0.0	23.4	161.2	0.0	12.6	0.0
	11	0.0	58.8	322.7	0.0	37.2	−0.0
	12	0.0	24.7	200.8	0.0	11.3	−0.0
	13	0.0	54.6	199.0	0.0	41.4	−0.0

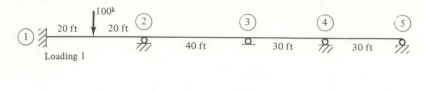

Loading 1

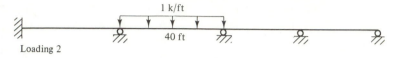

Loading 2

Loading 3

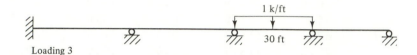

Loading 4

Figure 9.11p Alternate load cases for Example 9.12 Solution.

Independent loadings are shown in Figure 9.11p. Dependent loadings are as follows:

$$loading\,5 = 0.54 * loading\,1 + 1.2 * (loading\,2 + loading\,3 + loading\,4)$$
$$loading\,6 = loading\,5 + 0.9 * loading\,1 + 2 * loading\,3$$
$$loading\,7 = loading\,5 + 0.9 * loading\,1 + 2 * (loading\,2 + loading\,4)$$
$$loading\,8 = loading\,5 + 2 * (loading\,2 + loading\,3)$$
$$loading\,9 = loading\,6 + 2 * loading\,4$$
$$loading\,10 = loading\,5 + 2 * (loading\,2 + loading\,4)$$

9.6 NONPRISMATIC MEMBERS

As shown in Figure 9.12a, the splice points preferably do not occur at the support locations in continuous beam construction. Since the segments being spliced do not always have the same properties, the structural portion spanning between supports may not be prismatic. Linearly or parabolically varying haunches as shown in Figures 9.12 (b and c) are sometimes used in the regions of supports to improve the shear or moment capacity of a particular member. Also, if the joint size is not taken to be infinitesimal but is considered to be infinitely rigid as shown in Figure 9.12d, a nonprismatic member occurs. A structural analyst may desire to consider a member as the structural segment spanning between supports to keep the number of DOF to a minimum. These are the most common reasons why a member may be nonprismatic in a given structural analysis.

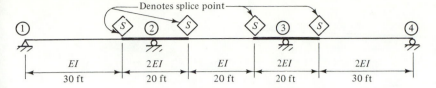

(a) Continuous beam showing field splice locations

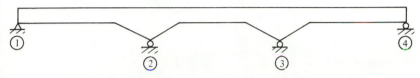

(b) Linearly varying haunches at supports 2 and 3

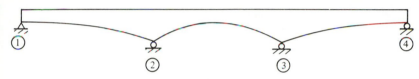

(c) Parabollically varying haunches at supports 2 and 3

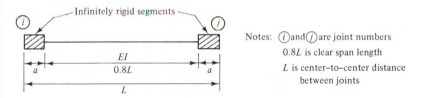

Notes: i and j are joint numbers

0.8L is clear span length

L is center–to–center distance between joints

(d) Member-end regions at joints considered as finite rigid bodies

Figure 9.12

The objective of this section is to demonstrate how to compute member-end force-deformation relations for a nonprismatic member. For graphical convenience, consider a member of constant width and a linearly varying depth as shown in Figure 9.13a; at any position x, the cross section is a rectangle. Generally, E can be treated as being constant at all points along the member length. To find member-end stiffness coefficients and fixed ended forces, the nonprismatic member can be approximated by several prismatic segments as shown in Figure 9.13b. The number of prismatic segments that should be used is a matter of engineering judgment for members involving haunches. Only members of the type shown in Figures 9.12 (a and d) are used in the numerical examples in this text. In these cases, the number of prismatic segments to use is obvious.

Figure 9.13b can be treated as an ordinary structure since all of the segments are prismatic; each prismatic segment is treated as a member spanning between joints.

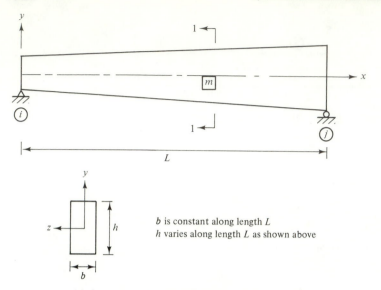

b is constant along length L
h varies along length L as shown above

(a) Original member m spanning between joints i and j

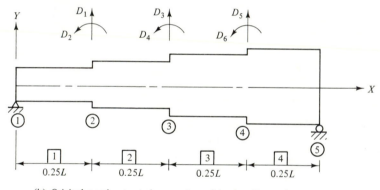

(b) Original member treated as a system of 4 prismatic members

Figure 9.13 Nonprismatic member treated as 4 prismatic members. Note: 4 segments were chosen for graphical convenience; in an actual problem more than 4 segments may be necessary.

Since axial deformations are zero in a Displacement Method analysis of the plane frame of Figure 9.13b when one is finding the member-end rotational stiffness coefficients and fixed ended moments, only 2 DOF need to be involved at each internal joint location. Therefore, in Figure 9.13b there are 6 DOF which is more than the author wishes to use in a numerical example to illustrate the calculation procedure. Consequently, examples involving fewer DOF are chosen for numerical illustration purposes.

EXAMPLE 9.13 ⎯⎯

A nonprismatic member of length L consists of two prismatic segments. Each segment length is $0.5L$ and E is constant and identical for both segments. The first segment has an area of A and a moment of inertia of I. The second segment has an area of $1.4A$ and a moment of inertia of $2.744I$. For the loadings shown in Figures 9.14a and 9.14b, find the fixed ended moments at the ends of the nonprismatic member. Also

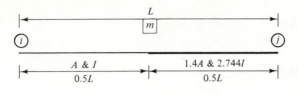

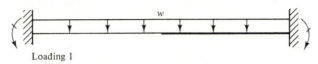

Figure 9.14a Nonprismatic member spanning between joints i and j.

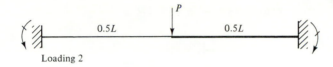

Figure 9.14b Loadings for which fixed ended moments are desired.

find the rotational member-end stiffness coefficients at the ends of the nonprismatic member.

Solution. The author chooses to solve the above described problem by the Displacement Method (Figs. 9.14c–9.14e); however, it could be solved by the Force Method.

The matrix equation of system equilibrium, $[S]\{D\} = \{P^e\}$, for the four loadings defined in Figure 9.14e has the following solution:

$$[S] = \begin{bmatrix} \dfrac{359.424EI}{L^3} & \dfrac{41.856EI}{L^2} \\[2ex] \dfrac{41.856EI}{L^2} & \dfrac{29.952EI}{L} \end{bmatrix}$$

$$\{P^e\} = \left\{ \begin{array}{cccc} \dfrac{-wL}{2} & -P & \dfrac{24EI}{L^2} & \dfrac{-65.856EI}{L^2} \\[2ex] 0 & 0 & \dfrac{-4EI}{L} & \dfrac{-10.976EI}{L} \end{array} \right\}$$

$$\{D\} = \left\{ \begin{array}{cccc} \dfrac{-0.00166wL^4}{EI} & \dfrac{-0.00332PL^3}{EI} & 0.0983L & -0.168L \\[2ex] \dfrac{0.00232wL^3}{EI} & \dfrac{0.00464PL^2}{EI} & -0.271 & -0.132 \end{array} \right\}$$

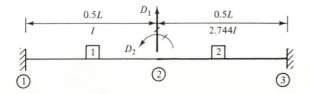

Figure 9.14c Nonprismatic member treated as a system of two prismatic members.

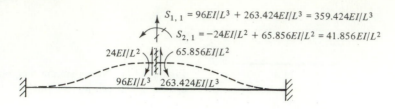

$$S_{1,1} = 96EI/L^3 + 263.424EI/L^3 = 359.424EI/L^3$$
$$S_{2,1} = -24EI/L^2 + 65.856EI/L^2 = 41.856EI/L^2$$

$6EI/(0.5L)^2 = 24EI/L^2$; $2 \star (24EI/L^2)/0.5L = 96EI/L^3$

$6E \star 2.744I/(0.5L)^2 = 65.856EI/L^2$; $2 \star (65.856EI/L^2)/0.5L = 263.424EI/L^3$

Solution for $D_1 = 1$ and $D_2 = 0$ to find column 1 of $[S]$

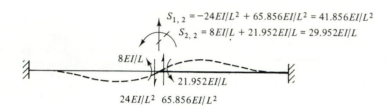

$$S_{1,2} = -24EI/L^2 + 65.856EI/L^2 = 41.856EI/L^2$$
$$S_{2,2} = 8EI/L + 21.952EI/L = 29.952EI/L$$

$4EI/0.5L = 8EI/L$; $(6EI/0.5L)/0.5L = 24EI/L^2$

$4E \star 2.744I/0.5L = 21.952EI/L$; $(6E \star 2.744I/0.5L)/0.5L = 65.856EI/L^2$

Solution for $D_2 = 1$ and $D_1 = 0$ to find column 2 of $[S]$

Figure 9.14d Solution for system stiffness matrix, $[S]$.

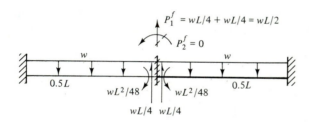

$$P_1^f = wL/4 + wL/4 = wL/2$$
$$P_2^f = 0$$

$(w \star 0.5L)/2 = wL/4$; $w \star (0.5L)^2/12 = wL^2/48$

Solution for $\{D\} = \{0\}$

$$\{P^e\} = \{P^a\} - \{P^f\} = \begin{Bmatrix} 0 \\ 0 \end{Bmatrix} - \begin{Bmatrix} wL/2 \\ 0 \end{Bmatrix} = \begin{Bmatrix} -wL/2 \\ 0 \end{Bmatrix}$$

Loading 1

$$\{P^e\} = \{P^a\} - \{P^f\} = \begin{Bmatrix} -P \\ 0 \end{Bmatrix} - \begin{Bmatrix} 0 \\ 0 \end{Bmatrix} = \begin{Bmatrix} -P \\ 0 \end{Bmatrix}$$

Loading 2

Figure 9.14e $\{P^e\}$ for Example 9.13.

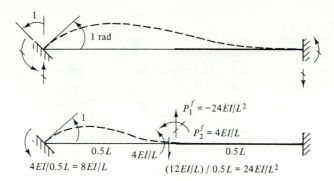

$$\{P^e\} = \{P^a\} - \{P^f\} = \begin{Bmatrix} 0 \\ 0 \end{Bmatrix} - \begin{Bmatrix} -24EI/L^2 \\ 4EI/L \end{Bmatrix} = \begin{Bmatrix} 24EI/L^2 \\ -4EI/L \end{Bmatrix}$$

Loading 3 unit member–end rotation at joint *i* of Figure 9.14a

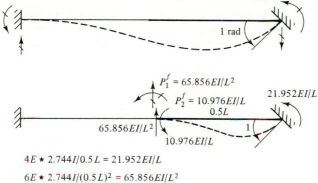

$4E \star 2.744I/0.5L = 21.952EI/L$

$6E \star 2.744I/(0.5L)^2 = 65.856EI/L^2$

$$\{P^e\} = \{P^a\} - \{P^f\} = \begin{Bmatrix} 0 \\ 0 \end{Bmatrix} - \begin{Bmatrix} 65.856EI/L^2 \\ 10.976EI/L \end{Bmatrix} = \begin{Bmatrix} -65.856EI/L^2 \\ -10.976EI/L \end{Bmatrix}$$

Loading 4 unit member–end rotation at joint *j* of Figure 9.14a

Figure 9.14e (*continued*)

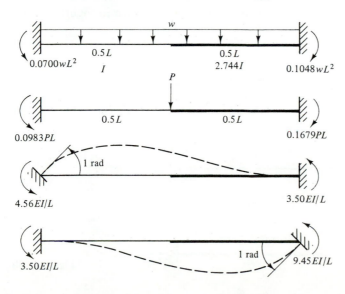

Figure 9.14f

345

The final member-end forces for the nonprismatic member of Figure 9.14a sub-jected to the loadings shown in Figure 9.14b are as shown in Figure 9.14f. To find the member-end shears, the author prefers to proceed as follows: Using known numerical values of w, P, L, and EI compute the member-end moments; then use statics (summation of moments and summation of vertical forces) to numerically find the final member-end shears.

EXAMPLE 9.14 ——

Find the axial member-end stiffness coefficients for the non-prismatic member shown on page 343 in Figure 9.14a.

Solution. Although the Displacement Method could be used in this problem, the author prefers the following approach (particularly, if there are more than two prismatic segments). See Figure 9.15a.

The nonprismatic member is treated as a system of two prismatic members in this case (in general, there could be n prismatic segments) subjected to an axial tension force P. One end of the nonprismatic member is restrained and the other end must be allowed to move parallel to the applied axial force P. Since the axial tension in each prismatic segment is P, the elongation of each segment is $(PL/EA)_i$ where the subscript i denotes the number of a particular segment 1 or 2 shown in Figure 9.15a, and $L_1 = L_2 = 0.5L$; $E_1 = E_2 = E$; $A_1 = A$; $A_2 = 1.4A$ in this example. Since point 1 in Figure 9.15a cannot move, the movement of point 3 is the sum of the elongations of the individual segments; that is,

$$\Delta = \sum_{i=1}^{2} \left(\frac{PL}{EA}\right)_i = P\left(\frac{0.5L}{EA} + \frac{0.5L}{1.4EA}\right) = 0.857143\frac{PL}{EA}$$

which can be solved for $P/\Delta = 1.16667\ EA/L$. The value of P which causes $\Delta = 1$ is defined as the stiffness coefficient; therefore, the member-end stiffness coefficients for axial deformations are as shown in Figure 9.15b.

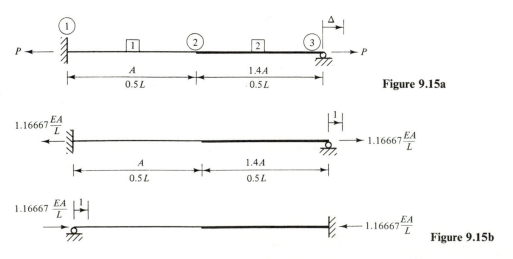

Figure 9.15a

Figure 9.15b

It should be noted that the solution for the case of axial compression stiffness is valid only if the nonprismatic member is perfectly straight and always remains straight; this means that the nonprismatic member is assumed not to buckle, otherwise a bent configuration would occur. Strictly speaking, the axial tension stiffness solution is also only valid for a perfectly straight member. If the member were initially bent to some configuration, during the application of the axial tension some of the bent configuration would be removed causing an additional axial elongation.

Therefore, if a member is bent due to any reason whatsoever, it is assumed that the bent configuration has a negligible effect on the axial stiffness coefficients. This restriction is necessary if one wishes to have a linearly elastic axial behavior for a member.

EXAMPLE 9.15 ———————————————————————————————————

In the three-span continuous beam structure loaded as shown, each exterior span is composed of two prismatic segments. Use the Displacement Method of analysis and involve only the DOF shown. See Figure 9.16a.

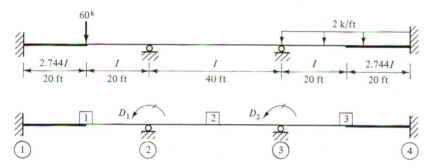

Figure 9.16a

Solution. The fixed ended moment and rotational end stiffness formulas obtained in Example 9.13 are applicable for members 1 and 3 in Figure 9.16a. See Figure 9.16b. The solution of the matrix equation of system equilibrium, $[S]\{D\} = \{P^e\}$, is

$$\begin{bmatrix} 0.214EI & 0.05EI \\ 0.05EI & 0.214EI \end{bmatrix} \begin{Bmatrix} D_1 \\ D_2 \end{Bmatrix} = \begin{Bmatrix} 235.9 \\ -224.0 \end{Bmatrix}$$

$$\begin{Bmatrix} D_1 \\ D_2 \end{Bmatrix} = \begin{Bmatrix} \dfrac{1424.67}{EI} \\ \dfrac{-1379.59}{EI} \end{Bmatrix}$$

The final member-end moments are

member 1: $F_3 = 403.0 + 0.0875EI * \dfrac{1424.67}{EI} = 527.7 \text{ ftk}$

$$F_6 = -235.9 + 0.114EI * \dfrac{1424.67}{EI} = -73.5 \text{ ftk}$$

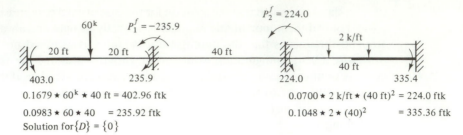

0.1679 ★ 60^k ★ 40 ft = 402.96 ftk   0.0700 ★ 2 k/ft ★ (40 ft)^2 = 224.0 ftk

0.0983 ★ 60 ★ 40   = 235.92 ftk   0.1048 ★ 2 ★ (40)^2 = 335.36 ftk

Solution for $\{D\} = \{0\}$

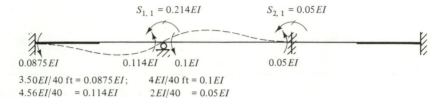

3.50EI/40 ft = 0.0875EI;   4EI/40 ft = 0.1EI

4.56EI/40   = 0.114EI 2EI/40   = 0.05EI

Solution for $D_1 = 1$ and $D_2 = 0$ to find column 1 of $[S]$

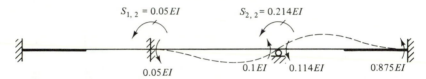

Solution for $D_2 = 1$ and $D_1 = 0$ to find column 2 of $[S]$

Figure 9.16b

$$\text{member 2:} \quad F_3 = 0 + 0.1EI * \frac{1424.67}{EI} + 0.05EI * \left(\frac{-1379.59}{EI}\right) = 73.5 \text{ ftk}$$

$$F_6 = 0 + 0.05EI * \frac{1424.67}{EI} + 0.1EI * \left(\frac{-1379.59}{EI}\right) = -66.7 \text{ ftk}$$

$$\text{member 3:} \quad F_3 = 224.0 + 0.114EI * \left(\frac{-1379.59}{EI}\right) = 66.7 \text{ ftk}$$

$$F_6 = -335.4 + 0.0875EI * \left(\frac{-1379.59}{EI}\right) = -456.1 \text{ ftk}$$

and the final member-end shears are obtained by statics for each member. The reactions for Example 9.15 are shown in Figure 9.16c.

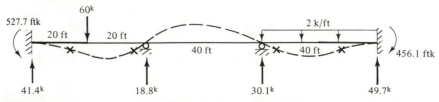

Figure 9.16c

EXAMPLE 9.16 ───────────────────────────────

A nonprismatic member consists of a deformable segment and a rigid body. For the prismatic deformable segment, $EI = 80,000$ kft²; $EA = 290,000^k$. Find:

1. The fixed ended moments for the loading shown in Figure 9.17a.
2. The axial stiffness coefficient.
3. The rotational stiffness coefficients.

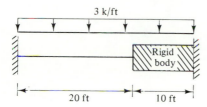

Figure 9.17a

Solution

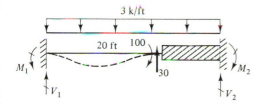

For the deformable segment

$$M_1 = \frac{(3 \text{ k/ft}) \star (20 \text{ ft})^2}{12} = 100 \text{ ftk}$$

$$V_1 = (3 \text{ k/ft}) \star (20 \text{ ft})/2 = 30^k$$

For the rigid body

$$V_2 = 30 + 10 \star 3 = 60^k$$

$$M_2 = 100 + 10 \star 30 + 5 \star 30 = 550 \text{ ftk}$$

Figure 9.17b

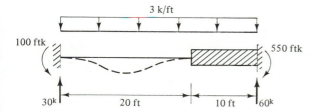

Nonprismatic member fixed ended forces

Figure 9.17c

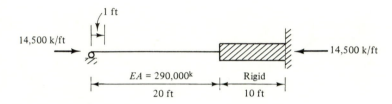

$$\Delta = \frac{P \star (20 \text{ ft})}{EA} = \frac{20P}{290,000}$$

Axial stiffness coefficient for the nonprismatic member is

$$\frac{P}{\Delta} = \frac{290,000^k}{20 \text{ ft}} = 14,500 \text{ k/ft} = 1208.33 \text{ k/in.}$$

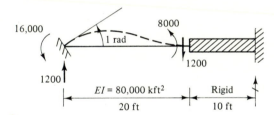

Nonprismatic member axial stiffness coefficients

Figure 9.17c (*continued*)

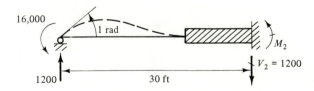

For the deformable segment

Left end moment = $\dfrac{4EI}{L} = \dfrac{4 \star 80,000}{20} = 16,000$ ftk/rad

Right end moment = $\dfrac{2EI}{L} = 8000$ ftk/rad

End shears = $(16,000 + 8000)/20 = 1200$ k/rad

M_2 is found by summation of moments = 0 at the right end

$M_2 = 30 \star 1200 - 16,000 = 20,000$ ftk/rad

Member–end stiffness coefficients for $\theta = 1$ rad induced at the left end   **Figure 9.17d**

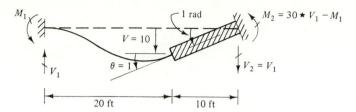

Figure 9.17e

The expressions for V_2 and M_2 were obtained by $\Sigma F_y = 0$ and $\Sigma M = 0$ at point 2. V_1 and M_1 are obtained by superposition as shown in Figure 9.17f:

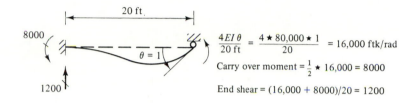

$$\frac{4EI\,\theta}{20\text{ ft}} = \frac{4 \star 80{,}000 \star 1}{20} = 16{,}000 \text{ ftk/rad}$$

Carry over moment $= \frac{1}{2} \star 16{,}000 = 8000$

End shear $= (16{,}000 + 8000)/20 = 1200$

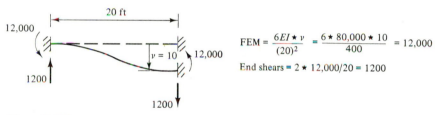

$$\text{FEM} = \frac{6EI \star v}{(20)^2} = \frac{6 \star 80{,}000 \star 10}{400} = 12{,}000$$

End shears $= 2 \star 12{,}000/20 = 1200$

Figure 9.17f

By superposition:

$$V_1 = 1200 + 1200 = 2400 \text{ k/rad}$$
$$M_1 = 8000 + 12{,}000 = 20{,}000 \text{ ftk/rad}$$

Then, $V_2 = V_1 = 2400$ k/rad and $M_2 = 30 * V_1 - M_1 = 30 * 2400 - 20{,}000 = 52{,}000$ ftk/rad. Thus, V_1, M_1, V_2, and M_2 are the nonprismatic member-end stiffness coefficients due to $\theta = 1$ rad induced at the right end of the member.

9.7 FRAMES WITH NONORTHOGONALLY INTERSECTING MEMBERS

A Displacement Method analysis involving translational DOF for a plane frame with nonorthogonally intersecting members is inherently more difficult than a plane frame with orthogonally intersecting members. The reason for the additional difficulty is because some of the member axes are not parallel to one of the system reference axes (see Figure 9.18). However, the matrix equation of system equilibrium involves member-end forces in the system reference axis directions. If a member axis is not parallel

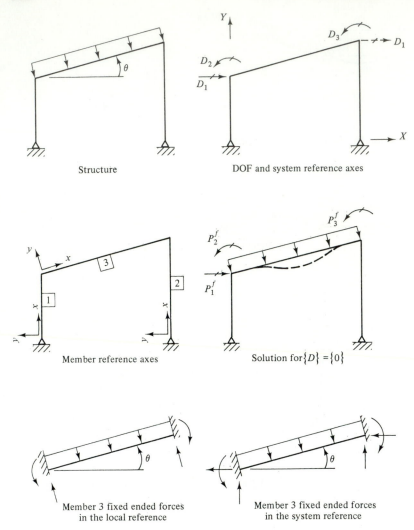

Figure 9.18 Conceptual example for resolution of member-end forces.

to one of the system reference axes, the axial and shear direction member fixed ended forces and the member-end stiffnesses have to be resolved into components in the system axes directions in order to accumulate the matrix equation of system equilibrium. Details of the resolution of the member-end forces in the local or member reference to the system reference are dealt with in this section, but there is another topic that needs our first attention if axial deformations of the members are ignored.

In a noncomputerized approach, whenever axial deformations are negligible the structural analyst chooses to ignore them in order to minimize the number of system DOF. In the following discussion axial deformations are ignored and not all of the joint translational displacements are independent. In a frame with nonorthogonally intersecting members that are treated as being axially rigid, the independent displace-

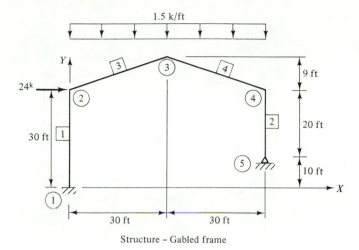

Structure – Gabled frame

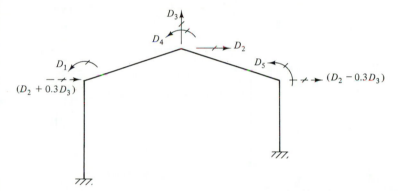

Figure 9.19 Structure involving dependent displacements, the dashed lined vectors.

ments have to be chosen and the dependent displacement relations have to be computed (see Figure 9.19). Also, it should be noted that some of the available Displacement Method computer programs allow the user to specify dependent displacement relations, but most of the available programs do not provide for this capability. The capability of specifying dependent displacement relations is particularly useful in dynamic analyses of routinely encountered structures since the dominant effects due to vibration are associated with the sidesway DOF of the structure.

For the chosen DOF (independent system displacements) in Figure 9.19, the deflected structural shapes involved in computing columns 1 through 3 of [S] are shown in Figure 9.20. The straight dashed lines in Figure 9.20 were drawn to aid in the following discussion. If frictionless pins were inserted at all interior and support joints of the structure in Figure 9.19, the resulting pin-jointed structure shown in Figure 9.21 would be unstable since it would not be able to resist any horizontally applied load. However, the unstable pin-jointed structure (for brevity purposes, the author prefers to call it a mechanism) can be used in computing the dependent displacement relations. If unit values of D_2 and D_3 as defined in Figure 9.19 are separately

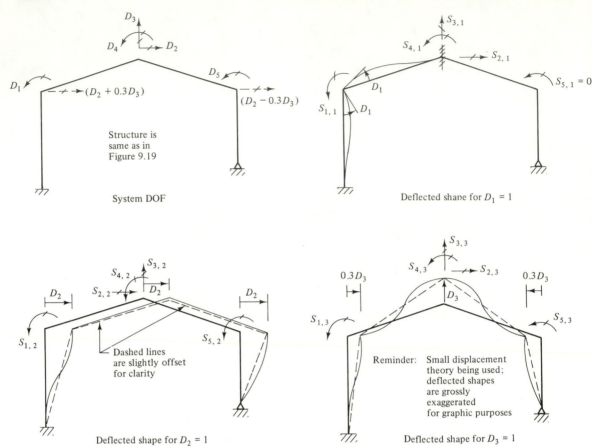

Figure 9.20 Deflected shapes for unit values of D_1, D_2, and D_3.

induced on the mechanism, the resulting deflected mechanism shapes are shown in Figure 9.21. Note that these deflected mechanism shapes correspond perfectly with the dashed straight lines drawn on the deflected shapes of the real structure in Figure 9.20. Therefore, for purposes of defining the vertical and horizontal positions of the internal joints for the deflected shapes of the real structure in Figure 9.20, we can use Figure 9.21 which is much easier to sketch and visualize. Furthermore, it will be subsequently demonstrated that the mechanism can be used to determine which of the joint translational displacements of the real structure can be chosen as the DOF in a Displacement Method of analysis involving axially rigid members. Also, since the lineal pieces of the mechanism do not deform in any manner during the inducement of the unit translational displacements, each lineal piece can be thought of as being a perfectly rigid body. Therefore, the concept of Instantaneous Center of Rotation (ICR) of a rigid body can be used in defining the mechanism relations which are identical to the joint translational displacement relations of the real structure composed of

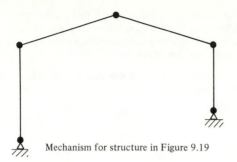

Mechanism for structure in Figure 9.19

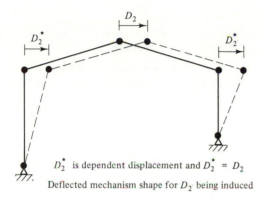

D_2^* is dependent displacement and $D_2^* = D_2$

Deflected mechanism shape for D_2 being induced

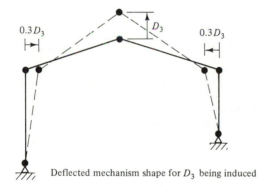

Deflected mechanism shape for D_3 being induced

Figure 9.21 Deflected mechanism shapes for unit values of D_2 and D_3.

axially rigid members. Figure 9.22 shows how an analyst is to determine the ICR of a rigid body. These ICR location definitions are used in Figure 9.24 to compute the joint displacement relations shown in Figure 9.19, but first we need to show in Figure 9.23 how an analyst can decide which of the joint translational displacements can be chosen as DOF.

In Figure 9.23 it is shown that a joint in an axially rigid membered structure cannot translate if the neighboring joints are fully restrained against translation. In

Definition: ICR (Instantaneous Center of Rotation) for a rigid body is that point in the plane of motion at which a rigid body has *or appears to have* a pure rotational type motion *for a given instant of time.*

For planar motion, the ICR locations are determined by one of the three following possible cases in which θ is the *rigid body angular motion* **at inception of motion.**

Case 1—The rigid body is constrained to rotate about a fixed axis such as a fixed hinge support joint in a structure. The ICR is at the fixed hinge location.

Note: The rotational direction of angular motion, θ, could have been shown as counterclockwise instead of clockwise. θ is a small angle such that $\sin \theta = \tan \theta = \theta$ in radians.

Case 2—Two nonparallel translational-type displacement vectors **A** and **B** are known on a rigid body. The intersection of perpendiculars to the two vectors is the ICR location. r_A and r_B are the *length of the radii* from the ICR location to the vectors.

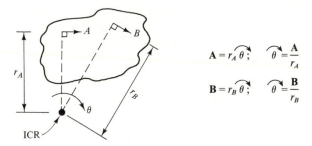

$$\mathbf{A} = r_A \, \theta \, ; \quad \theta = \frac{\mathbf{A}}{r_A}$$

$$\mathbf{B} = r_B \, \theta \, ; \quad \theta = \frac{\mathbf{B}}{r_B}$$

Note: If the directions of vectors **A** and **B** were reversed, the rotational direction of angular motion, θ, would be counterclockwise.

Case 3—Two parallel but unequal in magnitude translational-type displacement vectors are known on a rigid body. Choose a common line of origin for the two parallel vectors; connect the terminal ends of the two vectors with straight lines; the location of the ICR is defined by the intersection of the straight lines connecting the ends of the two vectors. Note that **A** is parallel to **B** and the magnitudes of **A** and **B** are not equal.

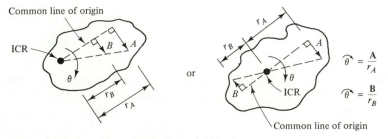

$$\theta = \frac{\mathbf{A}}{r_A}$$

$$\theta = \frac{\mathbf{B}}{r_B}$$

Figure 9.22 Location of ICR for a rigid body.

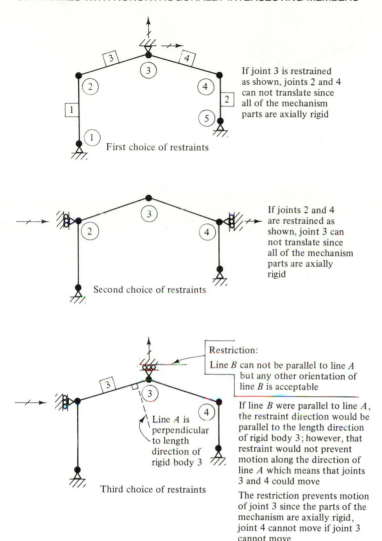

If joint 3 is restrained as shown, joints 2 and 4 can not translate since all of the mechanism parts are axially rigid

First choice of restraints

If joints 2 and 4 are restrained as shown, joint 3 can not translate since all of the mechanism parts are axially rigid

Second choice of restraints

Restriction:

Line B can not be parallel to line A but any other orientation of line B is acceptable

If line B were parallel to line A, the restraint direction would be parallel to the length direction of rigid body 3; however, that restraint would not prevent motion along the direction of line A which means that joints 3 and 4 could move

Line A is perpendicular to length direction of rigid body 3

The restriction prevents motion of joint 3 since the parts of the mechanism are axially rigid, joint 4 cannot move if joint 3 cannot move

Third choice of restraints

Figure 9.23 Choices of translational DOF for mechanism of Figure 9.21.

each of the choice of restraint sketches shown in Figure 9.23 there is at least one joint that is restrained by the neighboring joints. The author finds the first choice of restraints to be conceptually and numerically easiest to use; Figures 9.24 and 9.25 and Example 9.16 demonstrate the usage of the first choice of restraints. Example 9.17 illustrates the usage of the second choice of restraints.

In Figure 9.24 the ICR locations for rigid bodies 3 and 4 are at the intersection of the lines erected perpendicular to the known directions of the end displacements for each rigid body; that is, case 2 of Figure 9.22 is used to locate the ICR for each of these two rigid bodies.

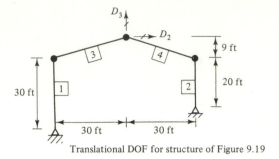

Translational DOF for structure of Figure 9.19

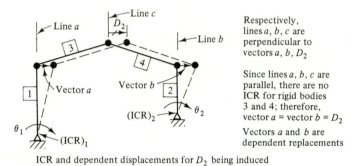

ICR and dependent displacements for D_2 being induced

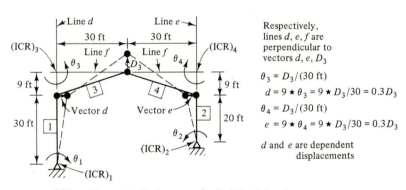

ICR and dependent displacements for D_3 being induced

Figure 9.24 Location of ICR and computation of dependent displacements.

EXAMPLE 9.17 ──

Determine the applied joint loads and fixed joint loads required in the matrix equation of system equilibrium as dictated by the choice of DOF for the structure shown in Figure 9.19.

Solution. The concept of equivalent virtual work is introduced and used to compute the values being sought at the D_2 and D_3 locations. The given applied joint loads (in this example there is only one applied joint load at joint 2) are

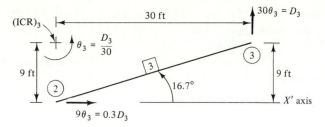

Member end displacements in system reference

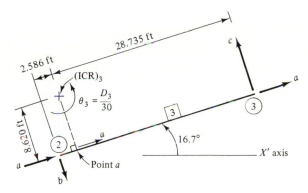

Radius from ICR to point ② is $r_2 = 9$ ft

$r_2 \sin (16.7°) = 2.586$ ft; $\quad \underline{b = 2.586\theta_3 = 0.08620D_3}$

$r_2 \cos (16.7°) = 8.620$ ft; $\quad \underline{a = 8.620\theta_3 = 0.2873D_3}$

Radius from ICR to point ③ is $r_3 = 30$ ft

$r_3 \sin (16.7°) = 8.620$ ft; $\quad a = 8.620\theta_3 = 0.2873D_3$

$r_3 \cos (16.7°) = 28.735$ ft; $\quad \underline{c = 28.735\theta_3 = 0.9578D_3}$

Radius from ICR to $\underline{\text{point } a}$ is $r_a = 8.620$ ft; $\quad a = 8.620\theta_3 = 0.2873D_3$

(note that $\underline{\text{point } a}$ has no displacement perpendicular to member 3)

Member end displacements in member reference

Figure 9.25 End displacements of member 3 in Figure 9.19.

shown on page 360 in sketch 1 and the desired applied joint loads are shown in sketch 2. Since the only given applied joint load is not coincident with an independent DOF, the applied joint loads must be computed. Note that the translational displacement at joint 2 is $(D_2 + 0.3D_3)$ which is a dependent displacement; therefore, the computational procedure using equivalent virtual work is as follows: On each of the structures in the deflected equilibrium positions in sketches 1 and 2, impose each of the following virtual displacement conditions since the dependent displacement relation coincident with the 24^k applied load in sketch 1 is a function of two displacements:

1. $\bar{D}_2$ is imposed at the location of the independent real displacement D_2 and no virtual displacements are allowed to occur at D_1, D_3, D_4, and D_5. The virtual work done by the applied loads in sketch 2 is equal to the virtual work done by the given applied loads in sketch 1. That is, $P_2^a * \bar{D}_2 = 24 * \bar{D}_2$ and since $\bar{D}_2 \neq 0$, then $P_2^a = 24^k$.

2. $\bar{D}_3$ is imposed at the location of the independent real displacement D_3 and no virtual displacements are allowed to occur at D_1, D_2, D_4, and D_5. The equivalent virtual work done is $P_3^a * \bar{D}_3 = 24 * (0.3\bar{D}_3)$; since $\bar{D}_3 \neq 0$, $P_3^a = 7.2^k$

Therefore, the applied load values that must be used for the chosen DOF are $P_1^a = P_4^a = P_5^a = 0$; $P_2^a = 24^k$; and $P_3^a = 7.2^k$.

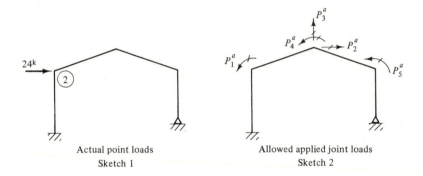

| Actual point loads | Allowed applied joint loads |
| Sketch 1 | Sketch 2 |

To facilitate the computation of the fixed joint loads using the method of equivalent virtual work, separate sketches for clarity purposes are prepared for the member-end displacements, member fixed ended forces, system displacements, and allowed fixed joint loads (see sketches 3, 3A, 4, and 4A, respectively). In sketch 3, note that the translational member-end displacements in the member reference are a function of D_2 and D_3 system DOF. Therefore, the equivalent virtual work equations are

$$P_2^f * \bar{D}_2 = 12.93 * 0.9578\bar{D}_2 + (-12.93) * 0.9578\bar{D}_2$$
$$+ 2 * 21.55 * (-0.2873\bar{D}_2) + 2 * 21.55 * 0.2873\bar{D}_2$$
$$P_3^f * \bar{D}_3 = 2 * 12.93 * 0.2873\bar{D}_3 + 2 * 21.55 * (-0.0862\bar{D}_3 + 0.9578\bar{D}_3)$$

which give $P_2^f = 0$ and $P_3^f = 45^k$. These values are obviously correct for the following reasons: (1) P_2^f must be zero since the 1.5 k/ft loading has no component in the direction of P_2^f; and (2) due to symmetry, each of the four gravity direction fixed ended forces must equally share the total gravity loading and there are two of these forces coincident with P_3^f which requires that $P_3^f = 2 * (1.5 * 60/4) = 45^k$. Since $\{P^e\} = \{P^a\} - \{P^f\}$,

$$\{P^e\} = \begin{Bmatrix} 0 \\ 24 \\ 7.2 \\ 0 \\ 0 \end{Bmatrix} - \begin{Bmatrix} 112.5 \\ 0 \\ 45 \\ 0 \\ -112.5 \end{Bmatrix} = \begin{Bmatrix} -112.5 \\ 24 \\ -37.8 \\ 0 \\ 112.5 \end{Bmatrix}$$

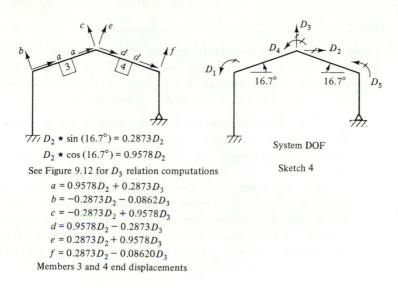

$$D_2 \star \sin(16.7°) = 0.2873 D_2$$
$$D_2 \star \cos(16.7°) = 0.9578 D_2$$

See Figure 9.12 for D_3 relation computations

$$a = 0.9578 D_2 + 0.2873 D_3$$
$$b = -0.2873 D_2 - 0.0862 D_3$$
$$c = -0.2873 D_2 + 0.9578 D_3$$
$$d = 0.9578 D_2 - 0.2873 D_3$$
$$e = 0.2873 D_2 + 0.9578 D_3$$
$$f = 0.2873 D_2 - 0.08620 D_3$$

Members 3 and 4 end displacements

Sketch 3

System DOF

Sketch 4

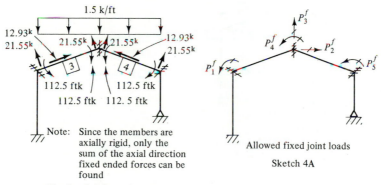

Note: Since the members are axially rigid, only the sum of the axial direction fixed ended forces can be found

Fixed ended forces in member reference

Sketch 3A

Allowed fixed joint loads

Sketch 4A

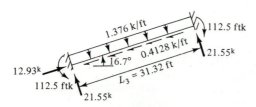

Note: Since the member is axially rigid, only the sum of the axial direction fixed ended forces can be obtained and allocated to either member end

Member 3 loading and fixed ended forces in member reference

Sketch 3B

361

EXAMPLE 9.18

Compute the dependent joint displacement relations and the member-end displacement relations in the member reference for the structure shown in Figure 9.19 and the choice of DOF shown in Figure 9.26a (note that the translational DOFs chosen are those in the second choice of restraints of Figure 9.21).

Solution. The deflected mechanism shape is shown below for $D_4 = 0$ and $D_1 = 1$. *Reminder:* Small displacement theory relations are to be computed; however, for graphical clarity the imposed D_1 must be grossly exaggerated. *At the instant of impending motion,* the direction of D_1 is perpendicular to the length direction axis of member 1 and the direction of member-end displacement *a* is perpendicular to member axis 4. Since the length of member 3 connected joints 2 and 3 in the undeflected position, this length also connects the deflected joints 2 and 3 (member is axially rigid). These pieces of information are sufficient to enable the analyst to locate the deflected positions of joints 2 and 3. The intersection of the perpendiculars to vectors *a* and D_1 define the ICR location

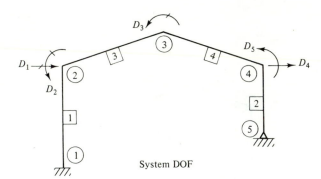

System DOF

Figure 9.26a

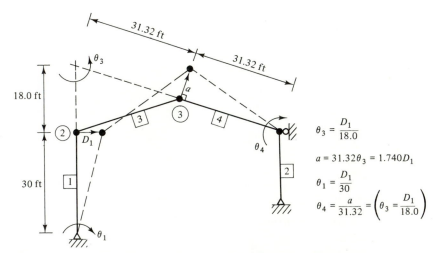

$$\theta_3 = \frac{D_1}{18.0}$$

$$a = 31.32\theta_3 = 1.740D_1$$

$$\theta_1 = \frac{D_1}{30}$$

$$\theta_4 = \frac{a}{31.32} = \left(\theta_3 = \frac{D_1}{18.0}\right)$$

Deflected mechanism shape for $D_4 = 0$ and $D_1 = 1$ being imposed   **Figure 9.26b**

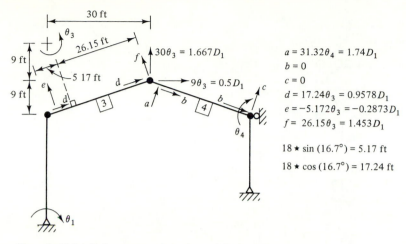

$a = 31.32\theta_4 = 1.74D_1$

$b = 0$

$c = 0$

$d = 17.24\theta_3 = 0.9578D_1$

$e = -5.172\theta_3 = -0.2873D_1$

$f = 26.15\theta_3 = 1.453D_1$

$18 \star \sin(16.7°) = 5.17$ ft

$18 \star \cos(16.7°) = 17.24$ ft

Dependent joint displacements and member end displacement relations

Figure 9.26c

for member 3. After the rigid body 3 rotation angle, θ_3, is computed as a function of D_1, the length of vector a can be related to D_1 and the dependent joint displacement relations at joint 3 as well as the member-end displacement relations in the member reference can be determined as shown in Figures 9.26b and 9.26c.

Shown in Figures 9.26d and 9.26e are the dependent joint displacements and the member-end displacement relations found from the deflected mechanism shape for $D_1 = 0$ and $D_4 = 1$.

The solution for Example 9.18 is the algebraic sum of the values obtained for each of the separately obtained solutions for the imposed unit displacements shown above. See Figures 9.26f and 9.26g.

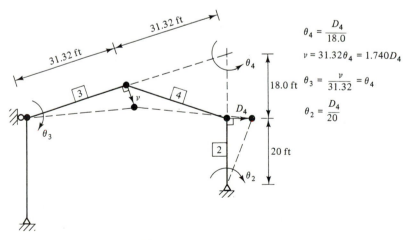

$$\theta_4 = \frac{D_4}{18.0}$$

$$v = 31.32\theta_4 = 1.740D_4$$

$$\theta_3 = \frac{v}{31.32} = \theta_4$$

$$\theta_2 = \frac{D_4}{20}$$

Deflected mechanism shape for $D_1 = 0$ and $D_4 = 1$ being imposed

Figure 9.26d

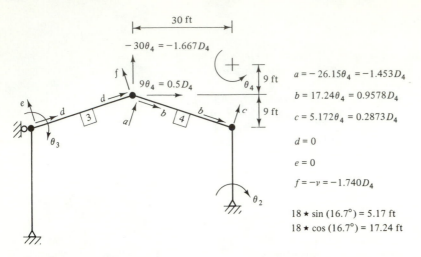

$$-30\theta_4 = -1.667D_4$$

$$9\theta_4 = 0.5D_4$$

30 ft

9 ft

9 ft

$a = -26.15\theta_4 = -1.453D_4$

$b = 17.24\theta_4 = 0.9578D_4$

$c = 5.172\theta_4 = 0.2873D_4$

$d = 0$

$e = 0$

$f = -v = -1.740D_4$

$18 \star \sin(16.7°) = 5.17$ ft
$18 \star \cos(16.7°) = 17.24$ ft

Dependent joint displacements and member end displacement relations

Figure 9.26e

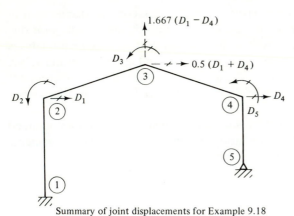

$1.667(D_1 - D_4)$

$0.5(D_1 + D_4)$

D_3

D_2 D_1

D_4

D_5

Summary of joint displacements for Example 9.18 **Figure 9.26f**

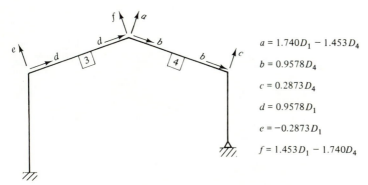

$a = 1.740D_1 - 1.453D_4$

$b = 0.9578D_4$

$c = 0.2873D_4$

$d = 0.9578D_1$

$e = -0.2873D_1$

$f = 1.453D_1 - 1.740D_4$

Summary of member end displacement relations for Example 9.18 **Figure 9.26g**

EXAMPLE 9.19

Find the reactions and final member-end forces for the axially rigid membered structure shown using the Displacement Method of analysis. For all members, $EI = 42,600$ kNm²; $L_1 = 3\sqrt{5} = 6.708$ m; $L_2 = 6$ m; $L_3 = 7.5$ m; $\tan \phi_1 = \frac{6}{3}$; $\phi_1 = 63.435°$; $\tan \phi_3 = 6/4.5$; $\phi_3 = 53.130°$; $L_4 = 3$ m. See Figure 9.27a.

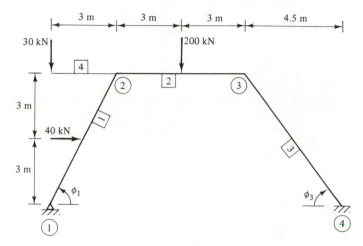

Figure 9.27a

Solution. First, remove the determinate part of the structure which reduces the problem to the solution of the structure and loading of Figure 9.27b.

Rotational DOF must be used at joints 2 and 3. Examination of the mechanism reveals that only one properly located translational restraint is needed; the author chooses the horizontal displacement at joint 2 as the other DOF. The dependent joint displacements and member-end displacement relations are computed as shown in Figure 9.27c.

$$\frac{h}{x} = \frac{6}{3}; \qquad h = 2x; \qquad \frac{6-x}{h} = \frac{4.5}{6}; \qquad (6-x) = \frac{4.5}{6}*(2x);$$

$$2.5x = 6; \qquad x = 2.4; \qquad h = 4.8$$

$$(6-x) = 3.6; \qquad r_2 = \sqrt{(2.4)^2 + (4.8)^2} = 5.37; \qquad r_3 = \sqrt{(3.6)^2 + (4.8)^2} = 6.00$$

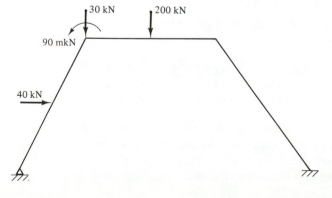

Figure 9.27b

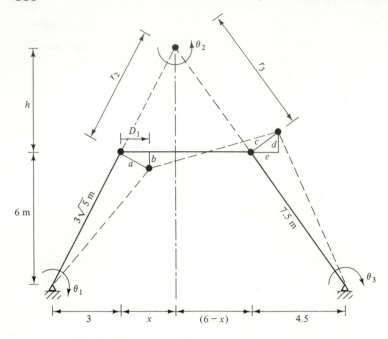

Figure 9.27c

$$6\theta_1 = 4.8\theta_2 = D_1; \qquad \theta_1 = \frac{D_1}{6}; \qquad \theta_2 = \frac{D_1}{4.8}$$

$$a = 3\sqrt{5}\theta_1 = 1.118D_1; \qquad b = 3\theta_1 = 0.5D_1;$$

$$c = 6\theta_2 = 1.25D_1; \qquad \theta_3 = \frac{c}{7.5} = \frac{D_1}{6}$$

$$d = 4.5\theta_3 = 0.75D_1; \qquad e = 6\theta_3 = D_1; \qquad (b+d) = 1.25D_1$$

The equivalent virtual work equation for imposed virtual displacements of $\bar{D}_1 = 1$ and $\bar{D}_2 = \bar{D}_3 = 0$ is $P_1^a * \bar{D}_1 = 30 * 0.5\bar{D}_1$; therefore, $P_1^a = 15$ kN, $P_2^a = 90$ mkN, and $P_3^a = 0$.

The equivalent virtual work equation for imposed virtual displacements of $D_1 = 1$ and $D_2 = D_3 = 0$ (see sketch of system DOF and dependent displacement relations, Fig. 9.27d) is $P_1^f * D_1 = -24.6 * 1.118D_1 - 100 * 0.5D_1 + 100 * 0.75D_1 + 17.89 * 0$. Therefore, $P_1^f = -2.50$ kN, $P_2^f = 105$ mkN, and $P_3^f = -150$ mkN.

$\{P^e\} = \{P^a\} - \{P^f\}$, the right-hand side values in the matrix equation of system equilibrium, are

$$\begin{Bmatrix} P_1^e \\ P_2^e \\ P_3^e \end{Bmatrix} = \begin{Bmatrix} 15 \\ 90 \\ 0 \end{Bmatrix} - \begin{Bmatrix} -2.5 \\ 105 \\ -150 \end{Bmatrix} = \begin{Bmatrix} 12.5\,\text{kN} \\ -15\,\text{mkN} \\ 150\,\text{mkN} \end{Bmatrix}$$

In order to compute the fixed-ended moments, the member-end displacement relations shown in the sketch titled system DOF and dependent displacement relations in Figure 9.27d are needed. After the fixed ended moments are computed, the member-end shears are obtained by statics; these shears must be

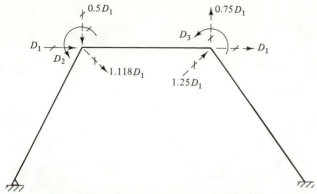

System DOF and dependent displacement relations

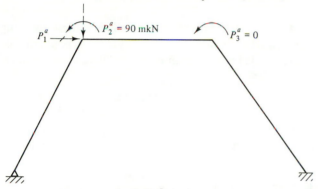

Applied joint loads

Figure 9.27d

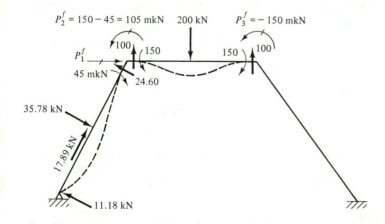

(Note: Fixed ended shears and axial members loads must be allocated to P_1^f)

Solution for $\{D\} = \{0\}$

Figure 9.27e

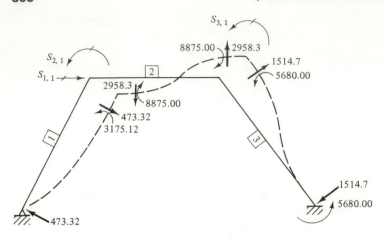

Solution for $D_1 = 1$ and $D_2 = D_3 = 0$ to find column 1 of $[S]$.   **Figure 9.27f**

allocated by computations to $S_{1,1}$ (the author chooses to allocate them by using the equivalent virtual work equation which also employs the dependent displacement relations). The fixed ended moments in Figure 9.27f and the member-end shears are

member 1: $\dfrac{3EI}{L^2}*1.118 = 3175.12\,\text{mkN};$ $\dfrac{3175.12}{L} = 473.32\,\text{kN}$

member 2: $\dfrac{6EI}{L^2}*(0.5+0.75) = 8875.00\,\text{mkN};$ $\dfrac{8875.00}{L} = 2958.3\,\text{kN}$

member 3: $\dfrac{6EI}{L^2}*1.25 = 5680.00\,\text{mkN};$ $2*\dfrac{5680}{L} = 1514.7\,\text{kN}$

For virtual displacements of $\bar{D}_1 = 1$ and $\bar{D}_2 = \bar{D}_3 = 0$ being imposed, the equivalent virtual work equation is

$$S_{1,1}*\bar{D}_1 = 473.32*1.118\bar{D}_1 + 2958.3*0.5\bar{D}_1 + 2958.3*0.75\bar{D}_1 + 1514.7*1.25\bar{D}_1$$

which gives $S_{1,1} = 6120.4$ and the algebraic summation of the member-end moments at each interior joint gives

$$S_{2,1} = 3175.12 - 8875.00 = -5699.88$$
$$S_{3,1} = 5680.00 - 8875.00 = -3195.00$$

In Figure 9.27g, the fixed ended moments and member-end shears are

member 1: $\dfrac{3EI}{L} = 19{,}051.3;$ $\dfrac{19{,}051.3}{L} = 2840.00$

member 2: $\dfrac{4EI}{L} = 28{,}400;$ $\dfrac{2EI}{L} = 14{,}200;$ $\dfrac{(28{,}400 + 14{,}200)}{L} = 7100$

For imposed virtual displacements of $\bar{D}_1 = 1$ and $\bar{D}_2 = \bar{D}_3 = 0$, the equivalent virtual work equation is

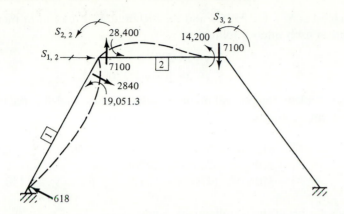

Solution for column 2 of $[S]$ **Figure 9.27g**

$$S_{1,2} * \bar{D}_1 = 2840 * 1.118\bar{D}_1 - 7100 * 0.5\bar{D}_1 - 7100 * 0.75\bar{D}_1$$

which gives $S_{1,2} = -5699.88$ and the algebraic summation of member-end moments at each interior joint gives

$$S_{2,2} = 19,051.3 + 28,400 = 47,451.3$$
$$S_{3,2} = 14,200$$

In Figure 9.27h, the fixed ended moments and member-end shears are:

member 2: $\dfrac{4EI}{L} = 28,400;$ $\dfrac{2EI}{L} = 14,200;$ $\dfrac{(28,400 + 14,200)}{L} = 7100$

member 3: $\dfrac{4EI}{L} = 22,720;$ $\dfrac{2EI}{L} = 11,360;$ $\dfrac{(22,720 + 11,360)}{L} = 4544$

For imposed virtual displacements of $\bar{D}_1 = 1$ and $\bar{D}_2 = \bar{D}_3 = 0$, the equivalent virtual work equation is

$$S_{1,3} * \bar{D}_1 = -7100 * 0.5\bar{D}_1 - 7100 * 0.75\bar{D}_1 + 4544 * 1.25\bar{D}_1$$

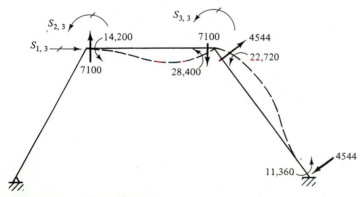

Solution for column 3 of $[S]$ **Figure 9.27h**

which gives $S_{1,3} = -3195.0$ and the algebraic summation of member-end moments at each interior joint gives

$$S_{2,3} = 14{,}200$$
$$S_{3,3} = 28{,}400 + 2{,}720 = 51{,}120$$

This completes the calculation of the values needed to define the matrix equation of system equilibrium which is

$$\begin{bmatrix} 6120.4 & -5699.9 & -3195.0 \\ -5699.9 & 47451 & 14200 \\ -3195.0 & 14200 & 51120 \end{bmatrix} \begin{Bmatrix} D_1 \\ D_2 \\ D_3 \end{Bmatrix} = \begin{Bmatrix} 12.5 \\ -15 \\ 150 \end{Bmatrix}$$

The solution of the simultaneous equations is

$$\begin{Bmatrix} D_1 \\ D_2 \\ D_3 \end{Bmatrix} = \begin{Bmatrix} 0.00289639 \text{ m} \\ -0.000982106 \text{ rad} \\ 0.00338810 \text{ rad} \end{Bmatrix}$$

The final member-end forces and the reactions are shown in Figure 9.27i.
For member 1:

$$F_6 = -45 + 3175.12*0.00289639 + 19051.3*(-0.000982106)$$
$$= -54.5 \text{ mkN}$$
$$6.708F_2 = 3*40 - 54.5; \qquad F_2 = 9.76 \text{ kN}$$

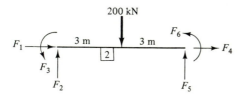

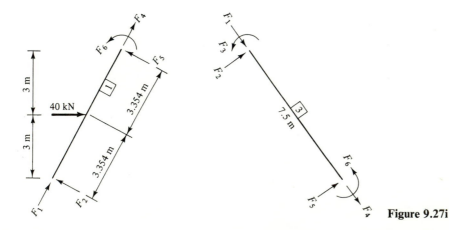

Figure 9.27i

For member 2:

$$F_3 = 150 + (-8875 * 0.00289639) + 28,400 * (-0.000982106)$$
$$+ 14,200 * 0.00338810 = 144.5 \text{ mkN}$$
$$F_6 = -150 + (-8875) * 0.00289639 + 14,200 * (-0.000982106)$$
$$+ 28,400 * 0.00338810 = -93.4 \text{ mkN}$$
$$F_2 = (3 * 200 + 144.5 - 93.4)/6 = 108.5 \text{ kN}$$
$$F_5 = 200 - 108.5 = 91.5 \text{ kN}$$

For member 3:

$$F_3 = 0 + 5680 * 0.00289639 + 22,720 * 0.00338810 = 93.4 \text{ mkN}$$
$$F_6 = 0 + 5680 * 0.00289639 + 11,360 * 0.00338810 = 54.9 \text{ mkN}$$
$$F_2 = (93.4 + 54.9)/7.5 = 19.8 \text{ kN}$$
$$F_5 = -19.8 \text{ kN}$$

Note: Since the members are axially rigid, the axially directed member-end forces have to be obtained from joint equilibrium. This can best be done after the reactions are computed.

The reactions shown in Figure 9.27j are computed as follows: Moment equilibrium at the left reaction (free body is the entire structure) gives the vertical reaction at the right support; vertical equilibrium of the entire structure gives the left reaction; moment equilibrium at the top of the rightmost inclined member as a free body gives the horizontal reaction at the right support; and horizontal equilibrium of the entire structure gives the horizontal reaction at the left support.

The reader has undoubtedly reached the conclusion that the noncomputerized analysis of frames with nonorthogonally intersecting members that are treated as being axially rigid is a lengthy endeavor. However, if axial deformations of the members in frames with nonorthogonally intersecting members are included, the noncomputerized analysis is even more lengthy. This conclusion is easily reached by considering the DOF required in Example 9.19 if axial deformations of the members are not ignored. In that case, there would be 3 DOF at each interior joint; therefore, a total of 6 DOF would be required in the solution. The solution for the applied joint loads and fixed joint loads would be much shorter than the solution for axially rigid members. However,

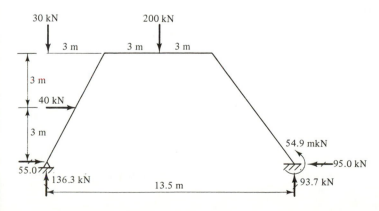

Figure 9.27j

the solution for [S] would be much longer and no less complicated than the solution for axially rigid members. Consequently, at this location in the text, the author chooses not to illustrate the solution of frames with nonorthogonally intersecting members that are axially deformable. Such illustrations are more appropriate in the computerized approach of the Displacement Method. Also, such illustrations are more appropriate after numerical examples of truss analysis by the Displacement Method have been provided. In such truss analyses, the joint displacements are due only to the axial deformations of the members. Noncomputerized Displacement Method solutions of trusses are also lengthy endeavors. However, in order for the reader to be adequately prepared to cope with the computerized Displacement Method approach, the author firmly believes that the reader must perform some lengthy and reasonably complicated noncomputerized solutions.

It will be shown in Example 10.10 that Example 9.19 can be solved much more quickly by the Moment Distribution Method. However, the moment distribution solution for Example 9.19 will require the usage of the dependent joint displacement and member-end relations that were involved in the Example 9.19 solution.

Also, it is interesting to obtain an approximate solution for Example 9.19 by treating member 2 as a rigid body as shown in Figure 9.28.

Since members 1 and 3 are axially rigid, it is impossible for rotations to occur at joints 2 and 3 unless joints 2 and 3 are allowed to translate. Therefore, if we choose the X-direction displacement of joint 2 as the DOF (to take advantage of the displacement relations found in Example 9.19), the dependent displacement relations shown in Figure 9.29 are obtained. All loads on the rigid body and at the joints must be allocated to P_1^a, whereas the loads on the deformable members are allocated to P_1^f.

For a virtual displacement, $\bar{D}_1$ (Fig. 9.30),

$$P_1^a * \bar{D}_1 = 30 * 0.5\bar{D}_1 + 90 * \left(\frac{\bar{D}_1}{4.8}\right) - 200 * \left(\frac{\bar{D}_1}{6}\right)$$

$$P_1^a = 0.417 \, \text{kN}$$

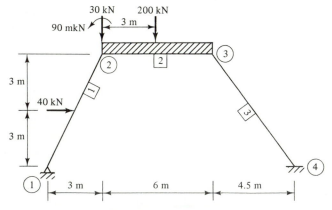

Members 1 and 3: EI = 42600 kNm² and EA = infinity

Figure 9.28 Structural portion 2 is a rigid body: EI = infinity and EA = infinity.

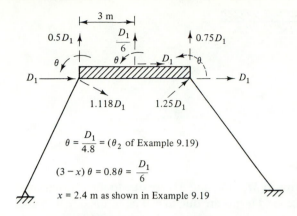

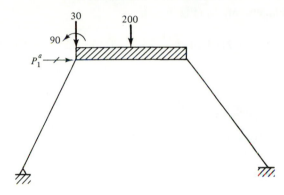

Figure 9.29 System DOF and dependent displacements needed in the solution.

Figure 9.30

For a virtual displacement, $\bar{D}_1$ (Fig. 9.31),

$$P_1^f * \bar{D}_1 = -24.60 * 1.118\bar{D}_1 - 45 * \left(\frac{\bar{D}_1}{4.8}\right)$$

$$P_1^f = -36.88\,\text{kN}$$

$$P_1^e = P_1^a - P_1^f = 0.417 - (-36.88) = 37.29\,\text{kN}$$

$$M_1 = 3175.12 + 19{,}051.3 * \left(\frac{1}{4.8}\right) = 7144.14$$

$$V_1 = \frac{M_1}{(3\sqrt{5})} = 1064.99$$

$$M_2 = 5680 + 22{,}720 * \left(\frac{1}{4.8}\right) = 10{,}413.33$$

$$M_3 = 5680 + 11{,}360 * \left(\frac{1}{4.8}\right) = 8046.67$$

$$V_2 = \frac{(M_2 + M_3)}{7.5} = 2461.33$$

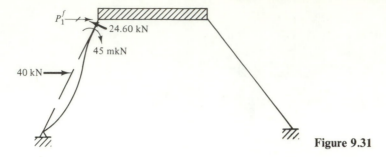

Figure 9.31

Note that the first terms in M_1, M_2, and M_3 are due to the translational displacements a and c whereas the second terms are due to θ. The terms due to a and c translational displacements were previously computed in Example 9.19.

For a virtual displacement, $\bar{D}_1$ (Fig. 9.32),

$$S_{1,1} * \bar{D}_1 = 1064.99 * 1.118\bar{D}_1 + 7144.14 * \left(\frac{\bar{D}_1}{4.8}\right) + 2461.33 * 1.25\bar{D}_1 + 10{,}413.33 * \left(\frac{\bar{D}_1}{4.8}\right)$$

$$S_{1,1} = 7925.13$$

$$[S]\{D\} = \{P^e\} \text{ gives } D_1 = \frac{37.29}{7925.13} = 0.0047059 \text{ m}$$

$$M_3 * D_1 = 8046.67 * 0.0047059 = 37.87 \text{ mkN}$$

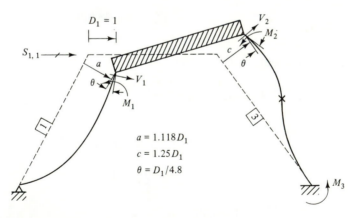

Figure 9.32

Summation of moments at joint 1 gives (Fig. 9.33)

$$13.5 Y_4 = 3 * 40 - 90 + 3 * 30 + 6 * 200 - 37.87$$
$$Y_4 = 94.97 \text{ kN}$$
$$Y_1 = 30 + 200 - Y_4 = 135.03 \text{ kN}$$

For a FBD of member 3,

$$M_2 * D_1 = 10413.33 * 0.0047059 = 49.00 \text{ mkN}$$

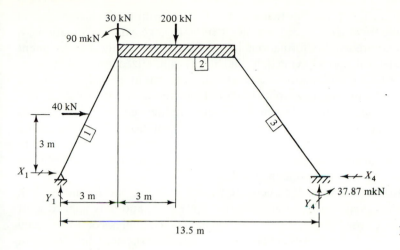

Figure 9.33

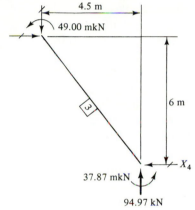

94.97 kN   **Figure 9.34**

Summation of moments at the top of member 3 (Fig. 9.34) gives

$$6X_4 = 4.5*94.97 + 3787 + 49.00$$
$$X_4 = 85.71 \text{ kN}$$

Then, using the entire structure as a FBD,

$$X_1 = X_4 - 40 = 45.71 \text{ kN}$$

At joints 1 and 4, the member-end shear and axial force can be obtained from the algebraic sum of the components of the X and Y reaction forces acting perpendicular to and parallel to the member at each joint.

9.8 TRUSSES

In all of the preceding numerical examples, axial deformations of the members have been ignored to decrease the number of system displacements and the solution time as well as to avoid the complications that may be involved with the inclusion of axial deformations of the members. Since truss members only have axial deformations, it is not possible to ignore axial deformations in a truss analysis by the Displacement Method. The reader will quickly conclude that noncomputerized truss analyses by the Displacement Method are lengthy endeavors if the truss has more than one interior joint. Therefore, in the interest of keeping the problems simple and the solution time from being excessive, only numerical examples of noncomputerized *plane truss* analysis are presented. However, the author begins with the presentation of concepts and a notation that are applicable to space truss analysis; this is essential if the reader is to progress to the stage of correctly using available computer programs for the solution of space trusses and verifying the obtained solutions.

In a truss analysis, all truss members are treated as being pinned ended (each member has only an axial tension or compressive force in it) and only applied joint loads are permitted. Consequently, in a Displacement Method analysis of trusses, only translational DOF parallel to the system coordinate axes can exist at an interior joint. If an applied joint load is not parallel to a system coordinate axis, the applied joint load must be decomposed into components parallel to the system coordinate axes. Furthermore, an equilibrium equation must be satisfied at each DOF location. Since the axial member-end forces for each member are along the member axis and at any joint in the truss not all member axes are parallel to a DOF direction, the member-end forces must be decomposed into components parallel to the system coordinate axes to write equilibrium equations at a joint. Figure 9.35 introduces the notation used in the decomposition of a space truss member-end axial tension force, T, into components parallel to the system axes. The decomposition of an applied joint load, P, acting along the direction line defined by points i and j is identical to the decomposition of T except that P is substituted for T in the decomposition definitions.

In Figure 9.35, the X', Y', Z' axes are respectively parallel to the system coordinate axes X, Y, Z (not shown), and located at the origin of the member which spans between joints i and j in a space truss. The member x-axis direction is from joint i to joint j. As shown in Figure 9.35, if numerical values (any or all of them can be negative) for the coordinates of points i and j with respect to the system axes X, Y, Z are known, the member length, direction cosines of the member x axis, and the X, Y, Z direction components of T can be computed. Figure 9.36 shows that for a plane truss lying in the XY plane, the member length, direction cosines, and components of T can be obtained from the relations shown in Figure 9.35 by deleting the Z coordinates in L and ignoring all other parameters that are a function of the Z coordinates.

In Example 9.20, a plane truss analysis is performed using sketches, logic, and the bare minimum of Displacement Method matrix symbols. Then, most of the additional definitions needed in a computerized analysis are introduced and used in Example 9.21 to solve the same problem which was solved in Example 9.20; hopefully, a thorough study of these two examples will aid the reader in the understanding of the matrix symbol definitions and convince the reader that the rigorous definitions required in a computerized approach avoid the pitfalls (wrong signs for axial forces and changes in member lengths; wrong vector components and wrong signs for them) inherent to the Example 9.20 approach. Also, the reader should conclude that if the

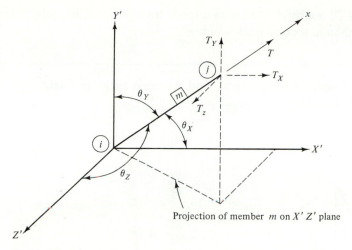

Projection of member m on $X'Z'$ plane

X', Y', Z' axes are parallel, respectively, to system axes X, Y, Z (not shown)

Joint i coordinates are: X_i, Y_i, Z_i

Joint j coordinates are: X_j, Y_j, Z_j

Length of member m is $L = \sqrt{(X_j - X_i)^2 + (Y_j - Y_i)^2 + (Z_j - Z_i)^2}$

Direction cosines of x axis of member m are:

$\cos \theta_X = (X_j - X_i)/L$
$\cos \theta_Y = (Y_j - Y_i)/L$
$\cos \theta_Z = (Z_j - Z_i)/L$

At joint j, let the X, Y, Z components of the tension force, T, in member m be respectively, T_X, T_Y, T_Z

$T_X = T \star \cos \theta_X$
$T_Y = T \star \cos \theta_Y$
$T_Z = T \star \cos \theta_Z$

Figure 9.35 Definitions of relations needed in space truss analyses.

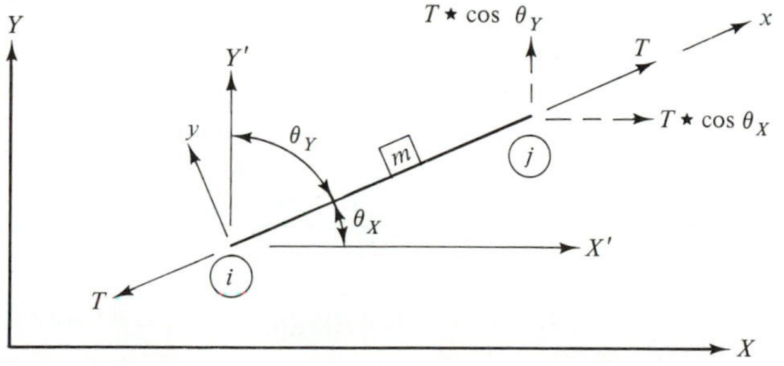

Length of member m is $L = \sqrt{(X_j - X_i)^2 + (Y_j - Y_i)^2}$

Direction cosines of the x axis for member m are:

$\cos \theta_X = (X_j - X_i)/L$
$\cos \theta_Y = (Y_j - Y_i)/L$

Note that $\cos \theta_Y \equiv \sin \theta_X$ is a special relation in plane truss analysis and some analysts prefer to use only θ_X in the definitions

Figure 9.36 Definitions of relations needed in plane truss analyses.

Example 9.20 approach is used in a space truss solution, the pitfalls are even more prevalent than in a plane truss solution.

EXAMPLE 9.20 ———————————————————————————————————————

Using the Displacement Method, find the truss member forces. For all members, $E = 29,000$ ksi, $A = 9.13$ in.^{2}, $EA = 264,770^k$.

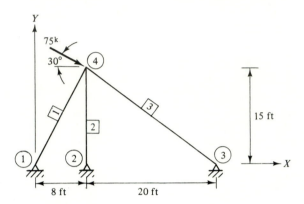

Figure 9.37a

Solution

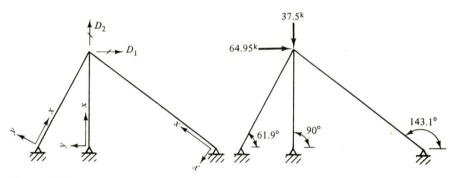

Figure 9.37b

member 1: $L = 17.0$ ft; $\dfrac{EA}{L} = \dfrac{264,770^k}{17.0\,\text{ft}} = 15,574.7$ k/ft

$\cos \theta_X = \dfrac{8}{17.0} = 0.470588;$ $\theta_X = 61.9275°$

$\cos \theta_Y = \dfrac{15}{17.0} = 0.882353;$ $\theta_Y = 28.0725°$

member 2: $L = 15$ ft; $\dfrac{EA}{L} = \dfrac{264,770^k}{15\,\text{ft}} = 17,651.3$ k/ft

$\theta_X = 90°;$ $\cos \theta_X = 0;$ $\theta_Y = 0°;$ $\cos \theta_Y = 1$

member 3: $L = 25.0 \, \text{ft};$ $\dfrac{EA}{L} = 10{,}590.8 \, \text{k/ft}$

$$\cos \theta_Y = \frac{15}{25} = 0.6; \qquad \theta_Y = 53.1301°$$

$$\theta_X = 90° + \theta_Y = 143.130°; \qquad \cos \theta_X = -0.8$$

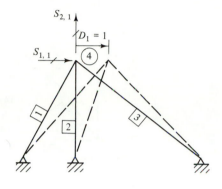

For member 1: $\theta_X = 61.9275°$

For member 3: $\theta_X = 143.130°$; $\phi = 36.8699°$

Figure 9.37c

Solution for Column 1 of [S]; *See Figure 9.37c*
  For member 1: details at joint 4.

member *elongation*, $a_1 = D_1 * \cos(61.9275°) = 0.470588$

member tension force $= \left(\dfrac{EA}{L}\right) * a_1 = 15{,}574.7 \, \text{k/ft} * 0.470588 = 7329.27 \, \text{k/ft}$

X component of member tension force
$= 7329.27 * \cos(61.9275°) = 3449.07 \, \text{k/ft} \rightarrow$
Y component of member tension force
$= 7329.27 * \sin(61.9275°) = 6467.01 \, \text{k/ft} \uparrow$

  For member 2: details at joint 4. Member neither elongates nor shortens
for small displacement theory; therefore, there is no axial force in this member
for $D_1 = 1$ being imposed.
  For member 3: details at joint 4.

member *shortening*, $a_3 = D_1 * \cos(36.8699°) = 0.8$

member *compression* force $= \left(\dfrac{EA}{L}\right) * a_3 = 10{,}590.8 * 0.8 = 8472.64 \, \text{k/ft}$

X component of compression force
$= 8472.64 * \cos(36.8699°) = 6778.11 \, \text{k/ft} \rightarrow$
Y component of compression force $= 8472.64 * \sin(36.8699°) = 5083.58 \, \text{k/ft} \downarrow$
$S_{1,1} = 3449.07 + 0 + 6778.11 = 10{,}227.18 \, \text{k/ft}$
$S_{2,1} = 6467.01 + 0 + (-5083.58) = 1383.43 \, \text{k/ft}$

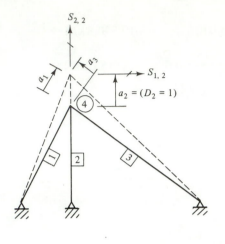

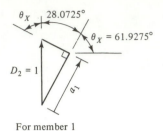

For member 1

For member 3

Figure 9.37d

Solution for Column 2 of [S]; See Figure 9.37d
   For member 1: details at joint 4.

$$\text{elongation, } a_1 = D_2 * \cos(28.0725°) = 0.882353$$

$$\text{tension force} = \left(\frac{EA}{L}\right) * a_1 = 15{,}574.7 * 0.882353 = 13{,}742.4 \text{ k/ft}$$

X component of tension force $= 13{,}742.4 * \cos(61.9275°) = 6467.01$ k/ft →
Y component of tension force $= 13{,}742.4 * \sin(61.9275°) = 12{,}125.6$ k/ft ↑

   For member 2: details at joint 4.

$$\text{elongation, } a_2 = D_2 = 1$$

$$\text{tension force} = \left(\frac{EA}{L}\right) * a_2 = 17{,}651.3 * 1 = 17{,}651.3 \text{ k/ft}$$

X component of tension force $= 0$
Y component of tension force $= 17{,}651.3$ k/ft ↑

   For member 3: details at joint 4.

$$\text{elongation, } a_3 = D_2 * \cos(53.1301°) = 0.6$$

$$\text{tension force} = \left(\frac{EA}{L}\right) * a_3 = 10{,}590.8 * 0.6 = 6354.48 \text{ k/ft}$$

X component of tension force $= 6354.48 * \cos(36.8699°) = 5083.58$ k/ft ←
Y component of tension force $= 6354.48 * \sin(36.8699°) = 3812.69$ k/ft ↑
$$S_{1,2} = 6467.01 + 0 + (-5083.58) = 1383.43 \text{ k/ft}$$
$$S_{2,2} = 12{,}125.6 + 17{,}651.3 + 3812.69 = 33{,}589.6 \text{ k/ft}$$

Therefore, the matrix equation of system equilibrium is

$$\begin{bmatrix} 10227.18 & 1383.43 \\ 1383.43 & 33589.6 \end{bmatrix} \begin{Bmatrix} D_1 \\ D_2 \end{Bmatrix} = \begin{Bmatrix} 64.95 \\ -37.5 \end{Bmatrix}$$

The solution of these simultaneous equations gives

$$\begin{Bmatrix} D_1 \\ D_2 \end{Bmatrix} = \begin{Bmatrix} 0.00653817 \, \text{ft} \\ -0.00138570 \, \text{ft} \end{Bmatrix}$$

For purposes of obtaining the axial forces in the members, let:

Tension forces be denoted as positive.

Compression forces be denoted as negative.

Member elongations be denoted as positive changes in length.

Member shortenings be denoted as negative changes in length.

For member 1:

$$a_1 = D_1 * \cos{(61.9275°)} + D_2 * \cos{(28.0725°)} = 0.00185411 \, \text{ft}$$

$$\text{axial force} = \left(\frac{EA}{L}\right) * a_1 = 15{,}574.7 * 0.00185411 = 28.88^k$$

For member 2:

$$a_2 = 0 + D_2 = -0.00138570 \, \text{ft}$$

$$\text{axial force} = \left(\frac{EA}{L}\right) * a_2 = 17{,}651.3 * (-0.00138570) = -24.46^k$$

For member 3:

$$a_3 = -D_1 * \cos{(36.8699°)} + D_2 * \cos{(53.1301°)} = -0.00606196 \, \text{ft}$$

$$\text{axial force} = \left(\frac{EA}{L}\right) * a_3 = 10{,}590.8 * (-0.00606196) = -64.20^k$$

Note: Six-digit precision was used in formulating and solving this problem since the solutions of Examples 9.20 and 9.21 are compared later in the text.

One end of each member in Example 9.20 was attached to a nonmovable support joint. If a plane truss has two or more interior joints, both ends of one or more members are attached to interior joints each of which have 2 DOF. The approach used in Example 9.20 can be used to solve trusses with two or more interior joints, but the least time-consuming and least error prone approach is the approach used in computerized solutions. Of course, this means that rigorous definitions must be made and abided by in conducting the calculations; also, for efficiency, a matrix notation is needed. The matrix notation and rigorous definitions used in a computerized approach are given in Figures 9.38 through 9.40 and used in solving two example problems.

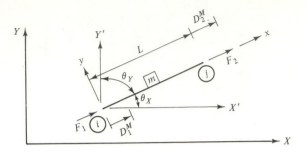

F_1 and F_2 are member-end forces in the member reference.
D_1^M and D_2^M are member-end displacements in the member reference.
For known values of D_1^M and D_2^M, F_1 and F_2 can be obtained by
superposition as shown in the following sketches.

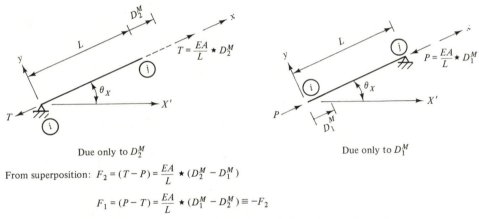

From superposition: $F_2 = (T - P) = \dfrac{EA}{L} \star (D_2^M - D_1^M)$

$$F_1 = (P - T) = \frac{EA}{L} \star (D_1^M - D_2^M) \equiv -F_2$$

Figure 9.38 Truss member-end force-deformation relations in member reference.

Since the global member-end displacements correspond to system DOF, the global member-end forces due to a unit value for any one of the system DOF are the stiffness coefficients needed for a member in calculating the member's contribution to the system stiffness matrix.

After the matrix equation of system equilibrium has been formulated and solved for the system displacements, the member-end displacements in the global reference are known since they correspond to a system displacement. Consequently, the member-end displacements in the member reference which are needed to compute the axial force in the members can be obtained as shown in Figure 9.40. Since a positive value for F_2 in the member reference (see Figure 9.38) corresponds to an axial tension force in the member and a negative value for F_2 corresponds to an axial compression force, the analyst need only compute F_2 for each member in order to determine the truss member forces.

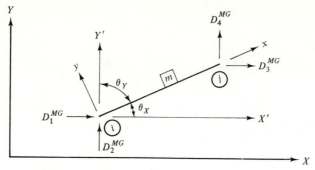

Member–end displacements in the global (system) reference

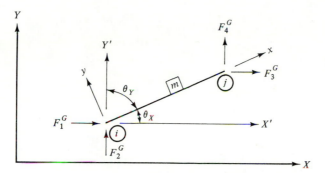

Member–end forces in the global (system) reference

For known values of the global member–end displacements, the global member–end forces can be obtained by superposition as shown in the following sketches

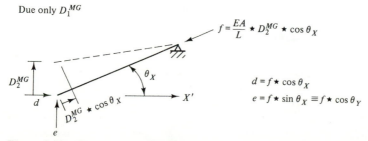

$$c = \frac{EA}{L} \star D_1^{MG} \star \cos\theta_X$$

$$a = c \star \cos\theta_X$$
$$b = c \star \sin\theta_X = c \star \cos\theta_Y$$

Due only D_1^{MG}

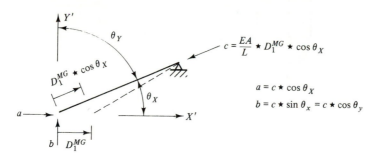

$$f = \frac{EA}{L} \star D_2^{MG} \star \cos\theta_X$$

$$d = f \star \cos\theta_X$$
$$e = f \star \sin\theta_X \equiv f \star \cos\theta_Y$$

Figure 9.39 Truss member–end force-deformation relation in global reference.

Due only to D_2^{MG}

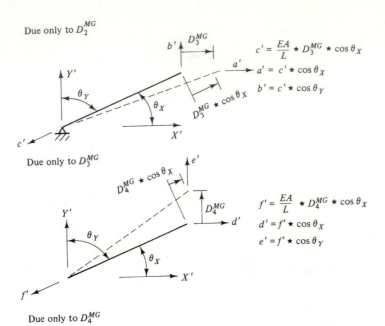

$$c' = \frac{EA}{L} \star D_3^{MG} \star \cos\theta_X$$
$$a' = c' \star \cos\theta_X$$
$$b' = c' \star \cos\theta_Y$$

Due only to D_3^{MG}

$$f' = \frac{EA}{L} \star D_4^{MG} \star \cos\theta_X$$
$$d' = f' \star \cos\theta_X$$
$$e' = f' \star \cos\theta_Y$$

Due only to D_4^{MG}

From superposition: $F_1^G = a + d - a' - d'$
$$F_2^G = b + e - b' - e'$$
$$F_3^G = -a - d + a' + d'$$
$$F_4^G = -b - e + b' + e'$$

After performing the algebraic manipulations, the force-deformation relation in matrix form becomes

$$\begin{Bmatrix} F_1^G \\ F_2^G \\ F_3^G \\ F_4^G \end{Bmatrix} = \frac{EA}{L} \begin{bmatrix} C_X^2 & C_X C_Y & -C_X^2 & -C_X C_Y \\ C_X C_Y & C_Y^2 & -C_X C_Y & -C_Y^2 \\ -C_X^2 & -C_X C_Y & C_X^2 & C_X C_Y \\ -C_X C_Y & -C_Y^2 & C_X C_Y & C_Y^2 \end{bmatrix} \begin{Bmatrix} D_1^{MG} \\ D_2^{MG} \\ D_3^{MG} \\ D_4^{MG} \end{Bmatrix}$$

where $C_X = \cos\theta_X$ and $C_Y = \cos\theta_Y$.

Figure 9.39 (*continued*)

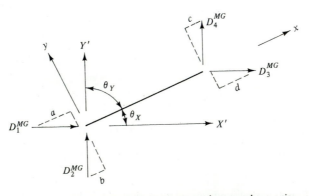

Components of global member–end displacements along member x axis

Figure 9.40 Relation between local and global member-end displacements.

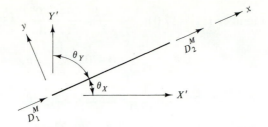

Member-end displacements in member reference

From superposition: $D_1^M = a + b = D_1^{MG} \star \cos \theta_X + D_2^{MG} \star \cos \theta_Y$

$\qquad\qquad\qquad\quad D_2^M = c + d = D_3^{MG} \star \cos \theta_X + D_4^{MG} \star \cos \theta_Y$ **Figure 9.40** (*continued*)

EXAMPLE 9.21 ——————————————————————————————————

Using the definitions shown in Figures 9.38 through 9.40, solve for the truss member forces and compare with the solution of Example 9.20; for all members, $EA = 264{,}770^k$.

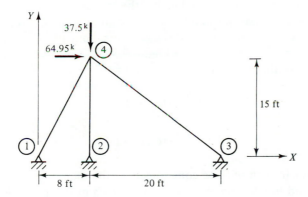

Figure 9.41a

Solution

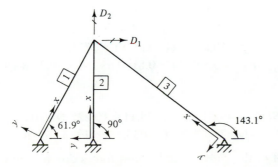

Figure 9.41b

member 1: $L = 17.0\,\text{ft}$; $\dfrac{EA}{L} = \dfrac{264{,}770^k}{17.0\,\text{ft}} = 15{,}574.7\,\text{k/ft}$

$$\cos\theta_X = \frac{8}{17.0} = 0.470588; \qquad \theta_X = 61.9275°$$

$$\cos\theta_Y = \frac{15}{17.0} = 0.882353; \qquad \theta_Y = 28.0725°$$

member 2: $L = 15\,\text{ft}$; $\dfrac{EA}{L} = \dfrac{264{,}770^k}{15\,\text{ft}} = 17{,}651.3\,\text{k/ft}$

$$\theta_X = 90°; \qquad \cos\theta_X = 0; \qquad \theta_Y = 0°; \qquad \cos\theta_Y = 1$$

member 3: $L = 25.0\,\text{ft}$; $\dfrac{EA}{L} = 10{,}590.8\,\text{k/ft}$

$$\cos\theta_Y = \frac{15}{25} = 0.6; \qquad \theta_Y = 53.1301°$$

$$\theta_X = 90° + \theta_Y = 143.130°; \qquad \cos\theta_X = -0.8$$

At joint 4, the global member-end forces due to a unit value of D_1 need to be obtained and algebraically summed to find the entries for column 1 of $[S]$. Since none of the members has the member x-axis origin at joint 4, all of the members have the same global member-end displacements, namely, $D_3^{MG} = (D_1 = 1)$ and $D_1^{MG} = D_2^{MG} = D_4^{MG} = 0$; also, from Figure 9.39, the only global member-end forces for *each* of the member ends at joint 4 are

$$F_3^G = \frac{EA}{L} * (C_X^2 * D_3^{MG})$$

$$F_4^G = \frac{EA}{L} * (C_X C_Y * D_3^{MG})$$

Note: All items in both of these equations have the member subscript number, m, attached and m takes on all values from 1 to 3 in this example. Therefore, the needed global member-end forces to find column 1 of $[S]$ are

member 1: $F_3^G = 15{,}574.7 * (0.470588)^2 * 1 = 3449.07\,\text{k/ft}$
             $F_4^G = 15{,}574.7 * 0.470588 * 0.882353 * 1 = 6467.01\,\text{k/ft}$
member 2: $F_3^G = 17{,}651.3 * (0)^2 * 1 = 0$
             $F_4^G = 17{,}651.3 * 0 * 1 * 1 = 0$
member 3: $F_3^G = 10{,}590.8 * (-0.8)^2 * 1 = 6778.11\,\text{k/ft}$
             $F_4^G = 10{,}590.8 * (-0.8) * 0.6 * 1 = -5083.58\,\text{k/ft}$

and

$$S_{1,1} = 3449.07 + 0 + 6778.11 = 10{,}227.18\,\text{k/ft}$$
$$S_{2,1} = 6467.01 + 0 + (-5083.58) = 1383.43\,\text{k/ft}$$

Similarly, at joint 4, the global member-end forces due to a unit value of D_2 need to be computed and algebraically summed to find the entries for column

2 of [S]. Again, all of the members have the same global member-end displacements which for this case are $D_4^{MG} = (D_2 = 1)$ and $D_1^{MG} = D_2^{MG} = D_3^{MG} = 0$; also, from Figure 9.39, the only global member-end forces for *each* of the member ends at joint 4 are

$$F_3^G = \frac{EA}{L} * (C_X C_Y * D_4^{MG})$$

$$F_4^G = \frac{EA}{L} * (C_Y^2 * D_4^{MG})$$

Note: Member subscript number, *m*, is attached to all items in both of these equations and *m* takes on all values from 1 to 3 in this example.

Therefore, the global member-end forces needed to find column 2 of [S] are

member 1: $F_3^G = 15{,}574.7 * 0.470588 * 0.882353 * 1 = 6467.01 \text{ k/ft}$
                $F_4^G = 15{,}574.7 * (0.882353)^2 * 1 = 12{,}125.6 \text{ k/ft}$
member 2: $F_3^G = 17{,}651.3 * 0 * 1 * 1 = 0$
                $F_4^G = 17{,}651.3 * (1)^2 * 1 = 17{,}651.3 \text{ k/ft}$
member 3: $F_3^G = 10{,}590.8 * (-0.8) * 0.6 * 1 = -5083.58 \text{ k/ft}$
                $F_4^G = 10{,}590.8 * (0.6)^2 * 1 = 3812.69 \text{ k/ft}$

and

$$S_{1,2} = 6467.01 + 0 + (-5083.58) = 1383.43 \text{ k/ft}$$
$$S_{2,2} = 12{,}125.6 + 17{,}651.3 + 3812.69 = 33{,}589.6 \text{ k/ft}$$

Therefore, the solution of the matrix equation of system equilibrium

$$\begin{bmatrix} 10227.18 & 1383.43 \\ 1383.43 & 33589.6 \end{bmatrix} \begin{Bmatrix} D_1 \\ D_2 \end{Bmatrix} = \begin{Bmatrix} 64.95 \\ -37.5 \end{Bmatrix}$$

is

$$\begin{Bmatrix} D_1 \\ D_2 \end{Bmatrix} = \begin{Bmatrix} 0.00653817 \text{ ft} \\ -0.00138570 \text{ ft} \end{Bmatrix}$$

The member axial force (+ denoting tension and − denoting compression) in each member is found by using the relations given in Figure 9.40 and the relation for F_2 in Figure 9.38. The member axial forces are

member 1: $D_1^{MG} = D_2^{MG} = 0; \qquad D_3^{MG} = D_1; \qquad D_4^{MG} = D_2$
            $D_1^M = 0 + 0 = 0$
            $D_2^M = 0.00653817 * 0.470588 + (-0.00138570) * 0.882353$
                $= 0.00185411$
            $F_2 = 15{,}574.7 * (0.00185411 - 0) = 28.88^k$
member 2: $D_1^{MG} = D_2^{MG} = 0; \qquad D_3^{MG} = D_1; \qquad D_4^{MG} = D_2$
            $D_1^M = 0 + 0 = 0$

$$D_2^M = 0.00653817*0 + (-0.00138570)*1 = -0.00138570$$
$$F_2 = 17,651.3*(-0.0013857 - 0) = -24.46^k$$

member 3: $D_1^{MG} = D_2^{MG} = 0; \qquad D_3^{MG} = D_1; \qquad D_4^{MG} = D_2$

$$D_1^M = 0 + 0 = 0$$
$$D_2^M = 0.00653817*(-0.8) + (-0.00138570)*0.6$$
$$= -0.00606196$$
$$F_2 = 10,590.8*(-0.00606196 - 0) = -64.20^k$$

The solution for Example 9.21 is identical to the solution of Example 9.20 as it should be. In performing the mathematical calculations, if the analyst has a multimemory calculator, the values of EA/L, $\cos\theta_X$, and $\cos\theta_Y$ for each member can be computed, stored, and recalled as needed. Since the calculator values have six or more digit precision, the author chose to show at least six-digit precision in the solution of the last two example problems. The author recommends that the member axial forces be recorded with at least three-digit precision which is sufficient (or more than sufficient) for subsequent calculations in an actual design situation.

EXAMPLE 9.22 ───

Using the definitions shown in Figure 9.39, formulate the matrix equation of system equilibrium. For convenience, the values of EA for the members are shown in the table of computed values for each member; this table is located in the solution of the problem. This example was devised to have nonzero global member-end displacements at both ends for some of the members; also, the origin end of some members are not attached to a stationary support. Finally, it is illustrated that a DOF must be chosen at a support that can translate in one direction (see joint 1, Fig. 9.42a); note that this circumstance was not necessary at roller supports in the noncomputerized solutions

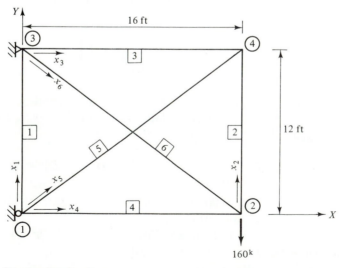

Structure is a plane truss   **Figure 9.42a**

of continuous beams since axial deformations of the members were ignored in those example problems.

Note: x_1 through x_6 arrows show origin and x-axis direction for each member. Also note that there is no joint where members 5 and 6 intersect.

Solution

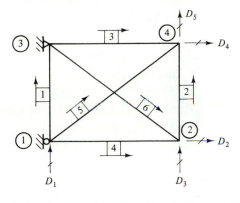

MEMBER-SYSTEM COMPATIBILITY

Member number	D_1^{MG}	D_2^{MG}	D_3^{MG}	D_4^{MG}
1	0	D_1	0	0
2	D_2	D_3	D_4	D_5
3	0	0	D_4	D_5
4	0	D_1	D_2	D_3
5	0	D_1	D_4	D_5
6	0	0	D_2	D_3

$\{D^{MG}\}$ definition is in Figure 9.39.

Member number	EA (kips)	L (ft)	EA/L (k/ft)	$\cos \theta_X$	$\cos \theta_Y$	θ_X (degrees)	θ_Y (degrees)
1	75980	12	6331.67	0	1	90	0
2	75980	12	6331.67	0	1	90	0
3	102950	16	6434.38	1	0	0	90
4	339300	16	21206.3	1	0	0	90
5	577100	20	28855.0	0.8	0.6	36.8699	53.1301
6	144130	20	7206.50	0.8	−0.6	323.130	126.870

Figure 9.42b

The calculations of the direction cosines using the definitions in Figure 9.36 are shown below for member 6 to illustrate the analyst's options.

$$\text{member 6: } \cos \theta_X = \frac{16-0}{20} = 0.8$$

$$\text{either } \theta_X = 323.130° \quad \text{or} \quad \theta_X = -36.8699° \text{ may be used}$$

$$\cos \theta_Y = \frac{0-12}{20} = -0.6$$

$$\theta_Y = 126.870° \text{ must be used}$$

For the reader's convenience, the truss member-end force-deformation relation derived in Figure 9.39 is shown below. This relation is needed in computing the system stiffness matrix coefficients. It should be noted that all items in the following relation have the member number, *m*, attached as a subscript; these items are listed in the two tables that appear in the solution part of the example problem.

$$
\begin{Bmatrix} F_1^G \\ F_2^G \\ F_3^G \\ F_4^G \end{Bmatrix}_m = \left(\frac{EA}{L}\right)_m \begin{bmatrix} C_X^2 & C_X C_Y & -C_X^2 & -C_X C_Y \\ C_X C_Y & C_Y^2 & -C_X C_Y & -C_Y^2 \\ -C_X^2 & -C_X C_Y & C_X^2 & C_X C_Y \\ -C_X C_Y & -C_Y^2 & C_X C_Y & C_Y^2 \end{bmatrix}_m \begin{Bmatrix} D_1^{MG} \\ D_2^{MG} \\ D_3^{MG} \\ D_4^{MG} \end{Bmatrix}_m
$$

where $C_X = \cos\theta_X$, $C_Y = \cos\theta_Y$, and m denotes the member number.

For column 1 of $[S]$, $D_1 = 1$ and the other system displacements are zero:

member 1: $F_2^G = 6331.67 * (1)^2 * 1 = 6331.67$

member 4: $F_2^G = 21,206.3 * (0)^2 * 1 = 0$

$F_3^G = 21,206.3 * (-1) * 1 * 0 * 1 = 0$

$F_4^G = 21,206.3 * (-1) * (0)^2 * 1 = 0$

member 5: $F_2^G = 28,855 * (0.6)^2 * 1 = 10,387.8$

$F_3^G = 28,855 * (-1) * 0.8 * 0.6 * 1 = -13,850.4$

$F_4^G = 28,855 * (-1) * (0.6)^2 * 1 = -10,387.8$

$$S_{1,1} = \sum_{m=1,4,5} (F_2^G)_m = 6331.67 + 0 + 10,387.8 = 16,719.5$$

$$S_{2,1} = \sum_{m=4} (F_3^G)_m = 0$$

$$S_{3,1} = \sum_{m=4} (F_4^G)_m = 0$$

$$S_{4,1} = \sum_{m=5} (F_3^G)_m = -13,850.4$$

$$S_{5,1} = \sum_{m=5} (F_4^G)_m = -10,387.8$$

For column 2 of $[S]$, $D_2 = 1$ and the other system displacements are zero:

member 2: $F_1^G = 6331.67 * (0)^2 * 1 = 0$

$F_2^G = 6331.67 * (0) * 1 * 1 = 0$

$F_3^G = 6331.67 * (-1) * (0)^2 * 1 = 0$

$F_4^G = 6331.67 * (-1) * 0 * 1 * 1 = 0$

member 4: $F_2^G = 21,206.3 * (-1) * 1 * 0 * 1 = 0$

$F_3^G = 21,206.3 * (1)^2 * 1 = 21,206.3$

$F_4^G = 21,206.3 * 0 * 1 * 1 = 0$

member 6: $F_3^G = 7206.5 * (0.8)^2 * 1 = 4612.16$

$F_4^G = 7206.5 * (0.8) * (-0.6) * 1 = -3459.12$

$$S_{1,2} = \sum_{m=4} (F_2^G)_m = 0$$

$$S_{2,2} = \sum_{m=2} (F_1^G)_m + \sum_{m=4,6} (F_3^G)_m = 0 + 21,206.3 + 4612.16 = 25,818.4$$

$$S_{3,2} = \sum_{m=2} (F_2^G)_m + \sum_{m=4,6} (F_4^G)_m = 0 + 0 + (-3459.12) = -3459.12$$

$$S_{4,2} = \sum_{m=2} (F_3^G)_m = 0$$

$$S_{5,2} = \sum_{m=2} (F_4^G)_m = 0$$

Since the reader needs to practice the generation of the columns of $[S]$, the other columns of $[S]$ are left for the reader to generate. The matrix equation of system equilibrium is

$$\begin{bmatrix} 16719.5 & 0 & 0 & -13850.4 & -10387.8 \\ 0 & 25818.4 & -3459.12 & 0 & 0 \\ 0 & -3459.12 & 8926.01 & 0 & -6331.67 \\ -13850.4 & 0 & 0 & 24901.5 & 13850.4 \\ -10387.8 & 0 & -6331.67 & 13850.4 & 16719.5 \end{bmatrix} \begin{Bmatrix} D_1 \\ D_2 \\ D_3 \\ D_4 \\ D_5 \end{Bmatrix}$$

$$= \begin{Bmatrix} 0 \\ 0 \\ -160 \\ 0 \\ 0 \end{Bmatrix}$$

and the system displacements are

$$\begin{Bmatrix} D_1 \\ D_2 \\ D_3 \\ D_4 \\ D_5 \end{Bmatrix} = \begin{Bmatrix} -0.0102432 \text{ ft} \\ 0.0059821 \text{ ft} \\ -0.0446495 \text{ ft} \\ 0.0134396 \text{ ft} \\ -0.0344063 \text{ ft} \end{Bmatrix}$$

and the member forces are (+ denotes tension and − denotes compression)

member 1: 64.9^k
member 2: 64.9^k
member 3: 86.5^k
member 4: -126.9^k
member 5: -108.1^k
member 6: 158.6^k

9.9 CLOSING REMARKS

Since microcomputer software for the Displacement Method is readily available, it is only natural that beginning students in structural analysis want to use the computer as soon as possible. The author has taught computer methods of structural analysis for 20 years and firmly believes that students should not be allowed to use available computer software until they can verify that the computer solutions are correct. Basically this means that the student must first thoroughly learn how to correctly formulate and solve structural analysis problems in the same manner that the computer software does it. The purpose of this chapter was to accomplish this objective for continuous beams, plane frames, and plane trusses.

Wherever possible, the author has emphasized the deflected shape of the structure. A good structural engineer is always able to determine reasonable estimates and the correct signs for the dominant system displacements before the computer software solution is performed. This capability is a gradual learning process which becomes more fully developed as the analyst learns how to design structures and matures into a good structural designer. Furthermore, a good structural designer develops and uses approximate methods to obtain reasonable estimates for the reactions and member-end forces which can be used to perform a preliminary design of an indeterminate structure. A good structural designer knows how he wants his deflected structure to respond to actual conditions and designs the structure such that it will behave that way.

At this stage, the student should have learned enough to correctly use available computer software for routine solutions of continuous beams, plane frames, plane trusses, and space trusses. However, the student should continue to study structural theory until the computerized Displacement Method has been mastered before using available computer software for the solutions of space frames. Similarly, the theory underlying finite element analysis should be mastered before attempting to perform such analyses using available computer software. The more complex the structural theory becomes, the more likely it is that a beginner will misuse an available computer software for such structural analyses. Most importantly, without an adequate knowledge of the applicable structural theory, the beginner will not know that the computer solution is incorrect and, therefore, that the computer software has been misused. Consequently, the student should not use available computer software unless the student can verify that the computer solution is correct.

PROBLEMS

If the member properties are not given, assume that all members are prismatic, axially rigid, and have the same EI value. Wherever they are applicable, use the formulas given in Figures 8.7, 8.8, 8.9, 8.11, and 8.12 to compute the member-end stiffness coefficients and fixed ended forces. Solve each problem by the noncomputerized Displacement Method involving only interior joint displacements as the DOF. Find the final member-end forces and the reactions. (Answers are not given for problems preceded by an asterisk.)

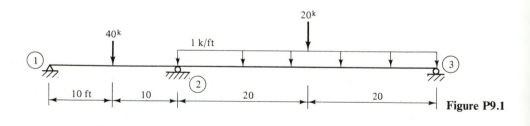

Figure P9.1

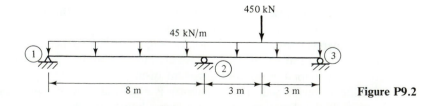

Figure P9.2

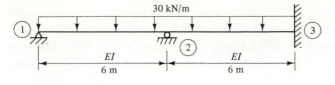

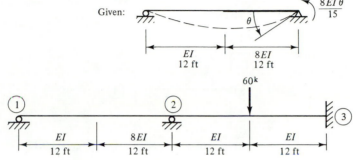

Figure P9.3

Figure P9.4

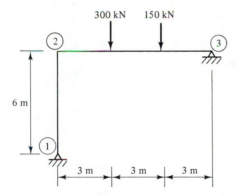

Figure P9.5

* **9.6.** $E = 29,000$ ksi and $I = 612$ in.^{4} Support 2 settles 1 in. downward. Zero settlement occurs at supports 1 and 3.

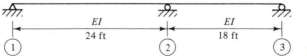

Figure P9.6

9.7. $EI = 160,000$ kNm^2; support 2 settles 15 mm downward; zero settlement occurs at supports 1 and 3.

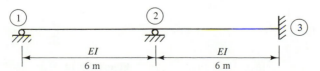

Figure P9.7

9.8. Solve Example 7.2, removal of a splice point angular gap of 0.0188315 rad at 10 ft to the right of the center support, by the Displacement Method (1 DOF).

9.9. Solve Example 7.3, a temperature change problem, by the Displacement Method (1 DOF).

9.10. Solve Example 7.4, a prescribed support displacements problem, by the Displacement Method (1 DOF).

9.11. Reminder: All members are axially rigid.

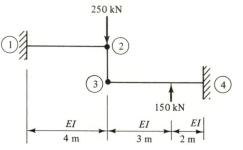

Figure P9.11

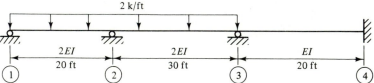

Figure P9.12

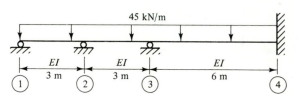

Figure P9.13

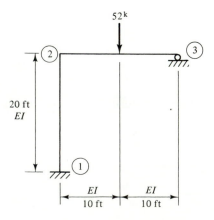

Figure P9.14

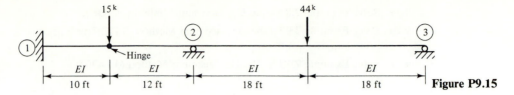

Figure P9.15

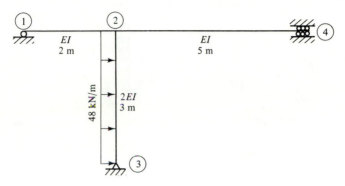

Figure P9.16

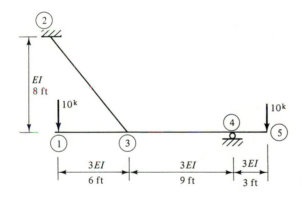

Figure P9.17

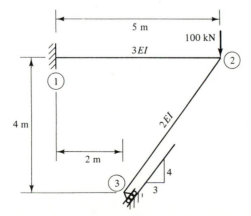

Figure P9.18

9.19. Solve Example 7.8 by the Displacement Method (3 DOF).

9.20. Solve Example 7.9 by the Displacement Method (3 DOF prescribed support movements problem).

9.21. Solve Example 7.13 by the Displacement Method (3 DOF).

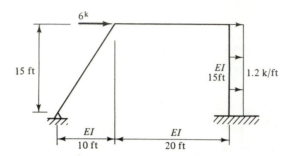

Figure P9.22

* **9.23.** *EA* is the same for all members.

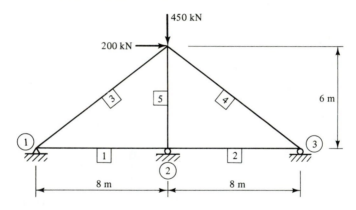

Figure P9.23

9.24. *EA* is the same for all members.

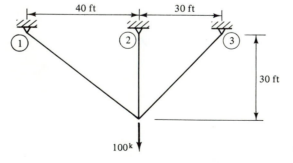

Figure P9.24

9.25. $EI = 25,000$ kft^2 for members 1 and 2. $EA = 4000^k$ and $EI = 0$ for member 3.

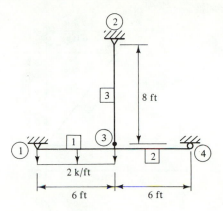

2 k/ft

6 ft

6 ft

Figure P9.25

9.26. $E = 2 \times 10^8$ kNm^2 for all members. $I_1 = I_2 = 12,000$ cm^4. $A_3 = 0.2$ cm^2; $I_3 = 0$.

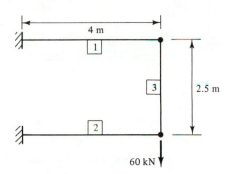

4 m

2.5 m

60 kN

Figure P9.26

9.27. $EI = 1296 \times 10^7$ kin.^{2} for members 1 and 2. $EA = 15,000^k$ for member 3.

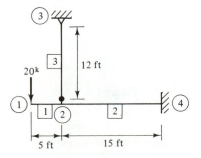

20^k

12 ft

5 ft

15 ft

Figure P9.27

* 9.28. EI = 150,000 kft^2 and EI = infinity for member 1. EA = 12,800^k and EI = 0 for member 2.

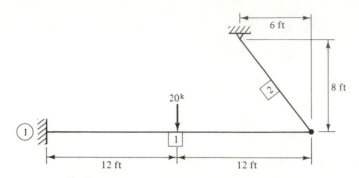

Figure P9.28

9.29. EI = 80,000 kNm^2 for members 1 through 3. EA = 2,000 kN for member 4.

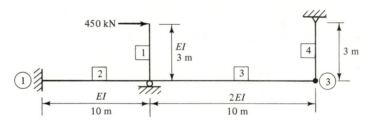

Figure P9.29

9.30. Solve Example 7.22 by the Displacement Method (1 DOF involves a cable).

9.31. Solve Example 7.23 by the Displacement Method (1 DOF involves a cable).

9.32. Members 1 and 2: EA/L = 20 k/in; EI/L = 0. Members 3 and 4: EI/L = 360,000 in.-k.

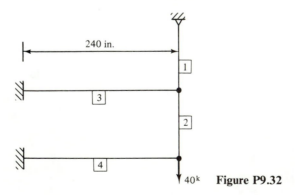

40^k **Figure P9.32**

* 9.33. Solve Problem 9.32 assuming Member 1 was fabricated 1 in. too short.

9.34. Note that due to the given loading, joint 4 does not rotate nor translate in the X direction. Choose the DOF as D_1 at joint 4 in the $+Y$ direction. For all members, $EI/L = EI/8$ ft = 640 ftk.

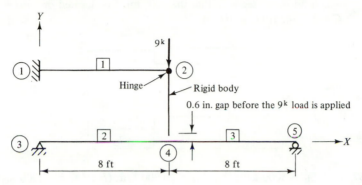

Figure P9.34

* **9.35.** $E = 29{,}000$ ksi for all members. Member 4 is a cable; $A = 4.43$ in.^{2} h is constant in all cases: h = cable length in case a.
(a) cable length is 30.00 ft
(b) cable length is 29.88 ft (0.12 ft too short)
(c) cable length is 30.12 ft (0.12 ft too long)
Member 1: $I = 640$ in.^{4} Members 2 and 3: $I = 2100$ in.^{4} Solve by the Displacement Method for each of the cable lengths stated above and the loading shown below. DOF are also shown below. Find the reactions for each of the three solutions.

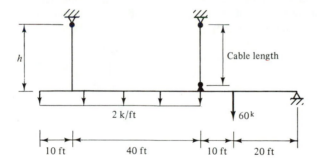

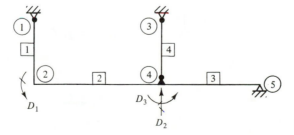

Figure P9.35

9.36. Given: The 100^k load applied at a distance C from the left support causes only axial deformations. $L_1 = L_2 = L_3 = 100$ in.; $(EA/L)_1 = 1450$ k/in.; $(EA/L)_2 = 2900$ k/in.; $(EA/L)_3 = 4350$ k/in. Find the value of C required to produce $R_1 = R_2 = 50^k$. Hint: To isolate which prismatic segment that the 100^k must be located on, first find R_1 and R_2 for (a) $C = 150$ in.; (b) $C = 100$ in.

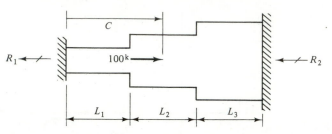

Figure P9.36

9.37. $(EA/L)_1 = 5000$ k/ft; $(EA/L)_2 = 10,000$ k/ft; $(EA/L)_3 = 15,000$ k/ft; $(EA/L)_4 = 20,000$ k/ft; the loads shown cause only axial deformations.

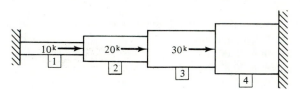

Figure P9.37

* **9.38.** For each structure shown, all members are axially rigid. Cross-hatched parts are rigid bodies.
  (a) Identify the minimum DOF required for a noncomputerized solution.
  (b) For your choice of DOF in part (a), find $\{P^e\}$. *Reminder:* Treat loads on rigid bodies as contributing to $\{P^a\}$.

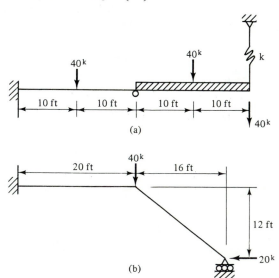

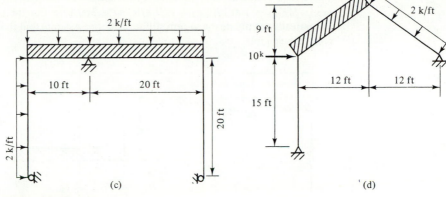

Figure P9.38

9.39. A 10-ft long rigid body weighs 20^k and is supported by two cables as shown. $EA = 3190^k$ for each cable.

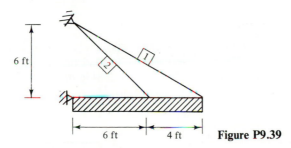

Figure P9.39

9.40. A 30-ft long rigid body weighs 120^k and is supported by four cables as shown. $EA = 7250^k$ for each cable. Choose the independent DOF at the center of the rigid body.

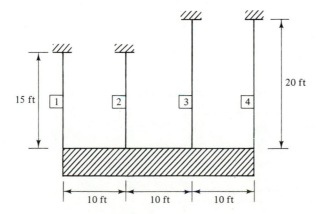

Figure P9.40

9.41. Three rigid bodies are interconnected with springs and supported as shown. Spring stiffnesses are $k_1 = 100$ k/ft; $k_2 = 200$ k/ft; $k_3 = 300$ k/ft. For the choice of DOF shown, numerically formulate the system stiffness matrix. Assume that the thickness of each rigid body is negligible.

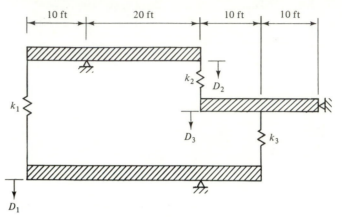

Figure P9.41

9.42. Members 1 and 2: EA = infinity and EI = 100,000 kft². Structural part 3 is a rigid body. For the loading shown, $D_1 = -0.0112611$ ft. Find the final member-end forces in the local reference at the supports. Find the reactions in the global reference.

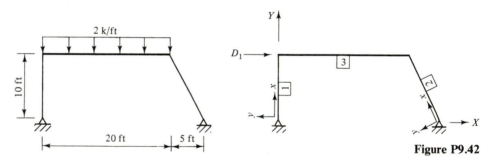

Figure P9.42

9.43. Numerically formulate the matrix equation of system equilibrium. Nothing else is required.

Member number	EA/L	EI/L
1, 2, 3	Infinity	32,000 ftk
4	870 k/ft	NA
5	Infinity	Infinity

The choice of independent DOF must be as shown in Figure P9.43.

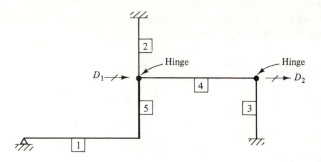

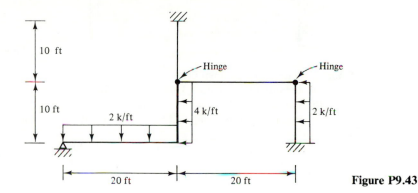

Figure P9.43

9.44. A 30-ft long rigid body weighs 4 k/ft, is supported by two members, and has a 500^k load applied as shown. The DOF of the system are to be chosen at the center of the rigid body. Member 1: $EA/L = 5000$ k/ft; $EI/L = 5000$ ftk. Member 2: $EA/L = 10,000$ k/ft; $EI/L = 10,000$ ftk.

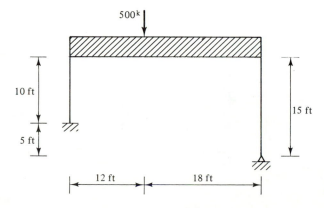

Figure P9.44

9.45. The 12.5-ft long structural part is a rigid body. For all members EA = infinity and EI is a constant.

(a) Identify the minimum DOF required for a noncomputerized solution.

(b) For your DOF choice in part (a), find $\{P^e\}$.

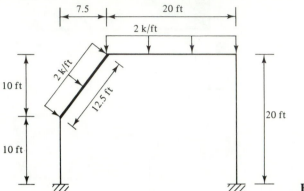

Figure P9.45

9.46. In each structure shown below, EA = infinity for all members.

(a) Identify the minimum DOF required for a noncomputerized solution

(b) For your DOF choice in part (a), using dashed lined vectors, show the dependent displacement relations in the global reference at all joints.

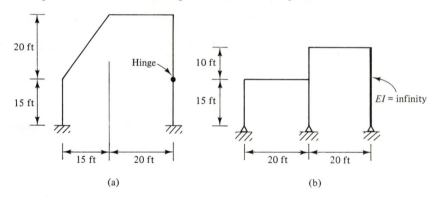

(a)   (b)

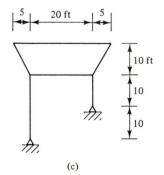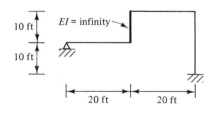

(c)   (d)   **Figure P9.46**

9.47. Given: EA = infinity for all members. Information shown in Figures P9.47a, P9.47b, P9.47c. Structure and loading shown in Figure P9.47d and DOF shown in Figure P9.47e. Find: Final member-end forces in Figure P9.47d (two-beam members—nonprismatic) (one-column member—prismatic).

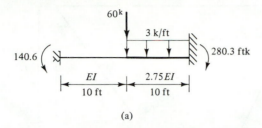

(a)

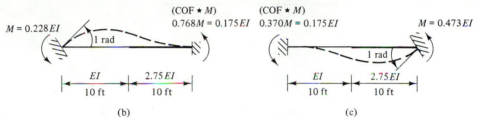

(b)   (c)

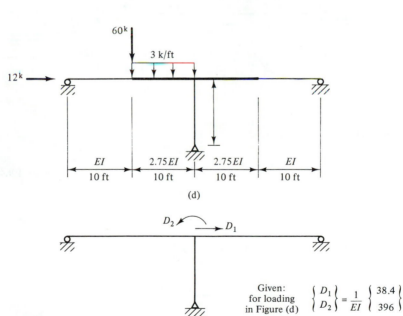

(d)

(e)

Figure P9.47

9.48. All members are axially rigid. $EI = 103,000$ kft^2 for all members.

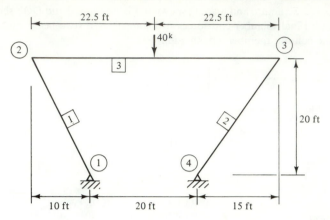

Figure P9.48

9.49. $(EI)_1 = EI;\ (EI)_2 = EI;\ (EI)_3 = 2EI.$

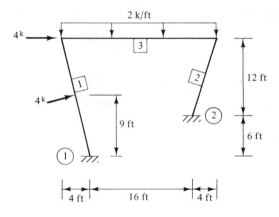

Figure P9.49

9.50. $(EI)_1 = EI;\ (EI)_2 = 1.5EI;\ (EI)_3 = 2.5EI.$

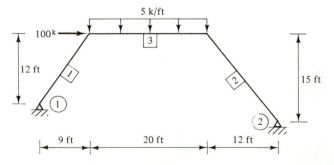

Figure P9.50

9.51. Solve Problem 9.50 with 100^k load deleted.

9.52. For both members: $E = 29{,}000$ ksi; $I = 1170$ in.^{4}

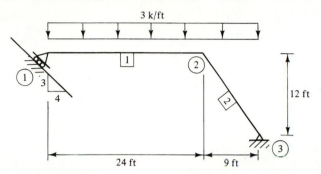

Figure P9.52

9.53. Identify the choice of independent DOF and only compute $\{P^e\}$. All members are prismatic and axially rigid.

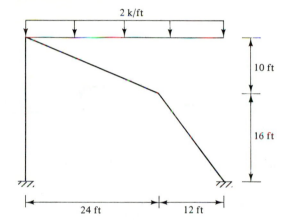

Figure P9.53

9.54. All members are prismatic and axially rigid. Identify the choice of independent DOF and only compute $\{P^e\}$.

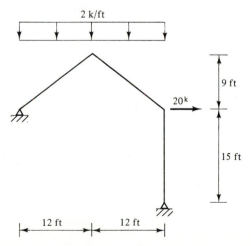

Figure P9.54

9.55. Member properties are variable for each member; however, the same member properties exist in each figure (two members in each figure). Some of the necessary calculations for a Displacement Method solution are given in the first three figures. In the last figure, find the indicated member-end forces.

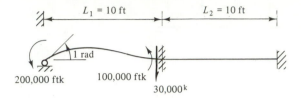

$L_1 = 10$ ft $L_2 = 10$ ft

1 rad

200,000 ftk   100,000 ftk

30,000^k

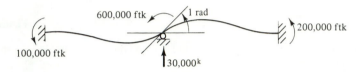

600,000 ftk   1 rad

200,000 ftk

100,000 ftk

30,000^k

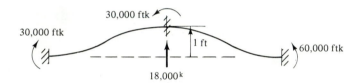

30,000 ftk

30,000 ftk

1 ft

60,000 ftk

18,000^k

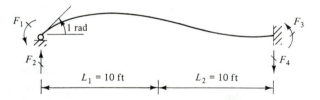

F_1 1 rad F_3

F_2 F_4

$L_1 = 10$ ft $L_2 = 10$ ft

Figure P9.55

9.56. Only numerically formulate the matrix equation of system equilibrium. The loading is an unknown moment at joint 1 due to a prescribed support rotation of 0.01 rad counterclockwise at joint 1. $EI = 180,000$ kft^2 for all members.

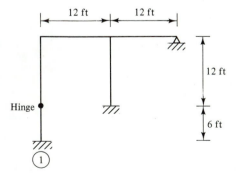

12 ft   12 ft

12 ft

Hinge

6 ft

1

Figure P9.56

9.57. $EI/L = 9000$ ftk for both members and $k = 100$ k/ft.

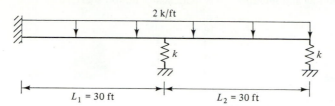

Figure P9.57

9.58. Find $\{P^e\}$ and columns 4, 5, and 6 of $[S]$. EI/L for each member is shown in parentheses on the structure in units of ftk.

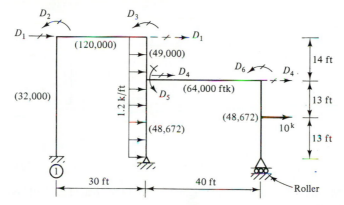

Figure P9.58

* 9.59. $E = 29,000$ ksi; $I_1 = 843$ in.⁴; $A_2 = 4.68$ in.² If member 2 were axially rigid, $M_1/\theta_1 = (3EI/L)_1$. Find C in the expression $M_1/\theta_1 = C*(EI/L)_1$ accounting for the axial deformation of member 2.

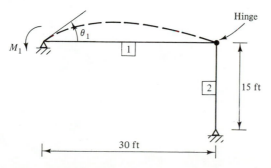

Figure P9.59

* 9.60. $E = 29,000$ ksi; $I_1 = I_2 = 500$ in.⁴; $A_3 = 0.1$ in.² The support at joint 2 was erected 1 in. too low and member 3 was fabricated 2 in. too short.

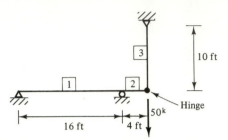

Figure P9.60

* 9.61. All members are axially rigid and DOF are not to be chosen at the supports. Only numerically formulate the matrix equation of system equilibrium for the structure and loading shown. $(EI/L)_1 = 29,000$ ftk; $(EI/L)_2 = 58,000$ ftk. $(EI/L)_3 = 81,000$ ftk; $(EI/L)_4 = 72,000$ ftk.

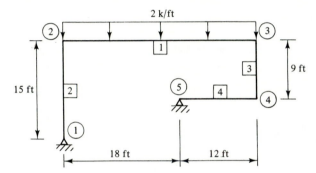

Figure P9.61

9.62. Members 2, 3, and 4 of Problem 9.61 are to be treated as rigid bodies. Find the reactions for the modified structure.

Chapter 10

Moment Distribution

10.1 INTRODUCTION

Hardy Cross [10.1] introduced the method of Moment Distribution in 1930 for the analysis of indeterminate structural plane frames and continuous beams. This method ignores the axial deformation of all members. For continuous beams and indeterminate plane frames in which no unknown joint translations can occur, this method obtains the final member-end moments by an iterative procedure that avoids the involvement of simultaneous equations. In 1932, C. T. Morris [10.2] introduced a successive corrections version of moment distribution that avoids the involvement of simultaneous equations for the analysis of indeterminate multistory frames in which joint translations are not prevented. Since the final member-end moments needed in structural design could be obtained without having to solve simultaneous equations, moment distribution quickly became immensely popular.

Recently, due to the ever increasing availability of computer software in structural design offices, the usage of Moment Distribution has naturally declined. However, in the author's opinion, the study of the Moment Distribution method is still justifiable for several reasons. There is a class of structural problems for which Moment Distribution is still an economical solution method. Also, Moment Distribution involves almost all of the concepts in the noncomputerized Displacement Method; consequently, the study of Moment Distribution enhances the understanding of these concepts. Finally, as the author emphasized in the closing remarks of Chapter 9, computerized solutions must be examined to determine whether or not they are correct; wherever applicable, Moment Distribution can be used for this purpose by the structural analyst.

10.2 BASIC CONCEPTS, DEFINITIONS, AND NOTATION

All members in Figure 10.1 are to be treated as being axially rigid (axial deformations are to be ignored). The value of M_B required to produce $\theta_B = 1$ rad is S_B, the *joint stiffness* at the rotated joint B. The *rotational member-end stiffnesses* (S_{BA}, S_{BC}, and S_{BD}) are calculated, respectively, using the relations given in Figure 8.8, Figure 8.11, and Example 9.13; note that the first subscript of the member-end stiffness is the symbol for the joint where the unit rotation is induced and the second subscript is needed to identify the opposite end of each member. Moment equilibrium of joint B requires $S_B = S_{BA} + S_{BC} + S_{BD}$; that is, the *joint stiffness* is the *sum of the member-end stiffnesses* at the rotated joint.

Now, suppose a 400 ftk moment is applied at joint B in Figure 10.1 and the member-end moments are desired—see Figure 10.2 for the posed problem. In Figure 10.1, note that the unknown moment, S_B, was found to be the sum of the individual member-end stiffnesses at joint B. Each member-end stiffness contributed a certain percentage of the sum. Therefore, the 400 ftk moment at joint B in Figure 10.2 is to be *distributed* to the member ends at joint B in the same proportions that the member-end stiffnesses contributed to S_B in Figure 10.1; these proportions are called *distribution factors* (DF$_{BA}$, DF$_{BC}$, and DF$_{BD}$). In Figure 10.2, there also are moments induced at the rotationally restrained joints A and D; these restraining moments are called *carry over moments*. Each carry over moment is the member's *carry over factor* (COF$_{BA}$ or COF$_{BD}$) times the member-end moment distributed at joint B. Each carry over factor is found from Figure 10.1 by dividing the restraining end moment by the member-end stiffness.

The process shown in Figure 10.2 (distribution of an applied moment at an interior joint to the member ends at that joint and the calculation of the carry over

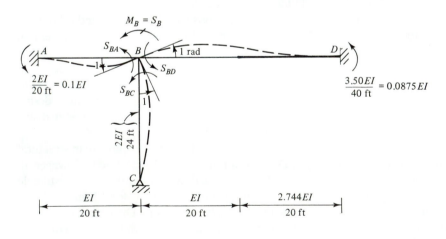

$S_{BA} = 4EI/20 \quad\quad = 0.2EI$

$S_{BC} = 3 \star 2EI/24 = 0.25EI$

$S_{BD} = 4.56EI/40 = 0.114EI$
_____

$S_B = 0.564EI = S_{BA} + S_{BC} + S_{BD}$

Figure 10.1 Joint stiffness at interior joint B.

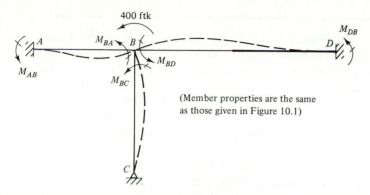

(Member properties are the same as those given in Figure 10.1)

See Figure 10.1 for the member-end stiffnesses, joint stiffness, and restraining end moments used in the following calculations.

Distribution factors

$$DF_{BA} = \frac{S_{BA}}{S_B} = \frac{0.2EI}{0.564EI} = 0.355$$

$$DF_{BC} = \frac{S_{BC}}{S_B} = \frac{0.25EI}{0.564EI} = 0.443$$

$$DF_{BD} = \frac{S_{BD}}{S_B} = \frac{0.114EI}{0.564EI} = 0.202$$

sum = 1.000

Distributed moments

$$M_{BA} = DF_{BA} * 400 = 142.0 \text{ ftk}$$

$$M_{BC} = DF_{BC} * 400 = 177.2 \text{ ftk}$$

$$M_{BD} = DF_{BD} * 400 = 80.8 \text{ ftk}$$

sum = 400.0 ftk

Carry over factors

$$COF_{BA} = \frac{0.1EI}{S_{BA}} = 0.5$$

$$COF_{BC} = \frac{0}{S_{BC}} = 0$$

$$COF_{BD} = \frac{0.0875EI}{S_{BD}} = 0.768$$

Carry over moments

$$M_{AB} = COF_{BA} * M_{BA} = 71.0 \text{ ftk}$$

$$M_{CB} = COF_{BC} * M_{BC} = 0$$

$$M_{DB} = COF_{BD} * M_{BD} = 62.1 \text{ ftk}$$

Figure 10.2 Distributed and carry over moments.

moments) is a fundamental aspect of the Moment Distribution Method of indeterminate structural analysis.

10.3 MOMENT DISTRIBUTION—JOINT TRANSLATIONS NOT PERMITTED

In the following conceptual example, the detailed procedural steps of the Moment Distribution Method are stated and graphically illustrated. Also, the solution is shown in the customary tabular form which is used for brevity purposes.

CONCEPTUAL EXAMPLE ——————————————————————————————

For the structure shown in Figure 10.3, use the Moment Distribution Method to obtain the final member-end moments. Also, find the reactions. All of the members are prismatic and each joint is labeled with an alphabetic character. Using an alphabetic character instead of a number to identify a joint is helpful in keeping the notations of the definitions as simple as possible. For a noncomputerized Displacement Method analysis, 2 rotational DOF ($D_1 = \theta_B$ and $D_2 = \theta_C$ with counterclockwise D_1 and D_2 being positive) would have to be chosen. Each rotational DOF would have to be dealt with independently to obtain the matrix equation of system equilibrium. Similarly, in the Moment Distribution Method, each of the two rotational DOF must be dealt with independently, but in an alternating, iterative manner.

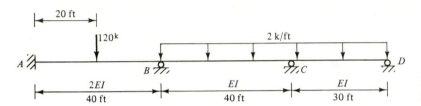

Figure 10.3 Structure for the conceptual example.

Solution. The conceptual steps in the moment distribution solution for the structure shown in Figure 10.3 are:

1. Remove all loads and restrain each interior joint against rotational motion. Note: Do not restrain any hinged boundary joints against rotation.

Restrained structure

2. At each interior joint, compute the following:
    a. *Joint stiffness*—Independently rotate each interior joint by the amount of 1 rad for each member end at the rotated joint and compute the rotational member-end stiffnesses—the sum of these stiffnesses is the *joint stiffness.*
    b. *Distribution factors* for all member ends at each interior joint are obtained by dividing each member-end stiffness by the joint stiffness.
    c. *Carry over factors*—When an interior joint is rotated 1 rad, the member ends at that joint also rotate 1 rad. If the opposite ends of these members are rotationally restrained, *carry over moments* are induced at these locations. For each member, the carry over moment divided by the member-end stiffness at the rotated end is the *carry over factor* from the rotated end to the opposite end.

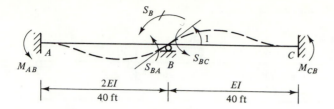

From Figure 8.8:

$$S_{BA} = 4*(2EI)/40 = 0.2EI \qquad DF_{BA} = S_{BA}/S_B = 0.2EI/0.3EI = 0.667$$
$$S_{BC} = 4*(EI)/40 = 0.1EI \qquad DF_{BC} = S_{BC}/S_B = 0.1EI/0.3EI = 0.333$$

$$S_B = 0.3EI \qquad \text{sum} = 1.000$$

$$M_{AB} = 2*(2EI)/40 = 0.1EI \qquad COF_{BA} = M_{AB}/S_{BA} = 0.1EI/0.2EI = 0.5$$
$$M_{CB} = 2*(EI)/40 = 0.05EI \qquad COF_{BC} = M_{CB}/S_{BC} = 0.05EI/0.1EI = 0.5$$

Joint B of restrained structure is rotated 1 rad.

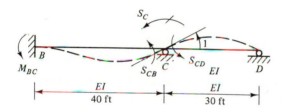

From Figures 8.8 and 8.11:

$$S_{CB} = 4EI/40 = 0.1EI \qquad DF_{CB} = S_{CB}/S_C = 0.1EI/0.2EI = 0.500$$
$$S_{CD} = 3EI/30 = 0.1EI \qquad DF_{CD} = S_{CD}/S_C = 0.1EI/0.2EI = 0.500$$

$$S_C = 0.2EI \qquad \text{sum} = 1.000$$

$$M_{BC} = 2EI/40 = 0.05EI \qquad COF_{CB} = M_{BC}/S_{CB} = 0.05EI/0.1EI = 0.5$$
$$M_{DC} = 0 \qquad COF_{CD} = M_{DC}/S_{CD} = 0$$

Joint C of restrained structure is rotated 1 rad.

3. Place the member loads on the structure that has all of the interior joints restrained against rotation and compute the member-end moments at the fixed ends—counterclockwise member-end moments are positive. At each interior joint, obtain the algebraic sum of all member fixed ended moments; this algebraic sum is the *restraining joint moment* or the *unbalanced joint moment* (the amount that the joint lacks being in moment equilibrium).

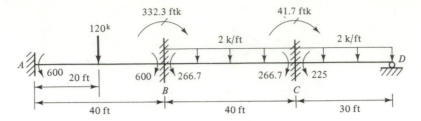

The correct vector directions of the fixed ended moments and joint restraining moments are shown in the above sketch; however, in the tabular form that is customarily used in moment distribution, signs must be used for these moments. The sign convention is counterclockwise fixed ended moments and joint restraining moments are positive. From Figures 8.7 and 8.12:

$$M^f_{AB} = 120*40/8 = 600 \text{ ftk}; \qquad M^f_{BA} = -120*40/8 = -600 \text{ ftk}$$
$$M^f_{BC} = 2*(40)^2/12 = 266.7 \text{ ftk}; \qquad M^f_{CB} = -2*(40)^2/12 = -266.7 \text{ ftk}$$
$$M^f_{CD} = 2*(30)^2/8 = 225 \text{ ftk}$$
$$\text{joint } B \text{ restraining moment} = M^f_{BA} + M^f_{BC}$$
$$= -600 + 266.7 = -332.3 \text{ ftk (it is CW)}$$
$$\text{joint } C \text{ restraining moment} = M^f_{CB} = M^f_{CD}$$
$$= -266.7 + 225 = -41.7 \text{ ftk (it is CW)}$$

4. Release only the interior joint that is most out of balance; it will rotate opposite to the direction that the joint-restraining moment was acting. Consequently, change the sign of the *unbalanced joint moment* and distribute it to the member ends at the released joint using the distribution factors at that joint. Multiply each distributed member-end moment by its carry over factor and record these carry over moments at the opposite ends of these members. Now, the released joint is balanced, but it must be restrained before the next step begins.*

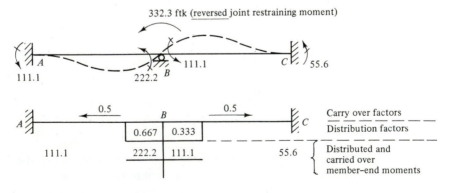

* Notes:
1. Horizontal line beneath the distributed moments at *B* indicates that the joint has been balanced and once again restrained.
2. Carry over moment at *C* must be algebraically added to the previously obtained joint *C* restraining moment; that is, the current joint *C* restraining moment = −41.7 + 55.6 = 13.9.

5. Repeat step 4 for the joint that is currently most out of balance. Step 5 is to be executed until all unbalanced joint moments are deemed to be negligible or until all interior joints are in balance.

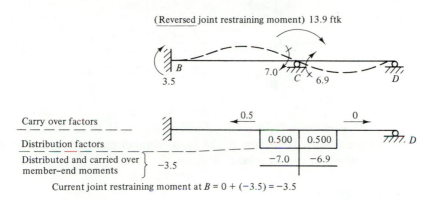

(Reversed joint restraining moment) 13.9 ftk

Carry over factors				0.5		0	
Distribution factors					0.500	0.500	777. D
Distributed and carried over member-end moments		−3.5			−7.0	−6.9	

Current joint restraining moment at $B = 0 + (-3.5) = -3.5$

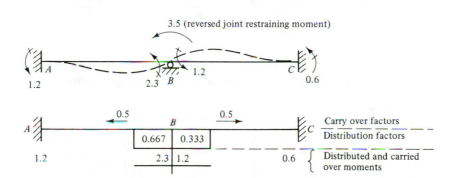

3.5 (reversed joint restraining moment)

1.2   2.3 B   1.2   0.6

	0.5	B	0.5	C	Carry over factors
A		0.667	0.333		Distribution factors
1.2		2.3	1.2	0.6	Distributed and carried over moments

Current joint restraining moment at $C = 0 + 0.6 = 0.6$
(It is considered to be negligible and 0.6 carry over moment is deleted)

6. After step 5 is completed, each final member-end moment is obtained from the algebraic summation of the fixed ended moment, distributed moments, and carried over moments at each member end. These algebraic summations are most easily done if the numerical calculations are recorded in a tabular form as shown below.†

† Notes pertaining to tabular form used in moment distribution:
1. Abbreviations in right margin—COF (carry over factors), DF (distribution factors), FEM (fixed ended moments), #1 D&CO (first distribution and carry over).
2. A single line was drawn beneath the distributed moments each time a joint was balanced. This helps to identify the joint restraining moment which is to be reversed and distributed the next time that joint is released. After a joint has been balanced once, the sum of the carry over moments immediately beneath the single line is the joint restraining moment. If a joint has never been balanced, the sum of all existing moments at that joint is the joint restraining moment.
3. A double line was drawn to signify that the moment distribution process was being stopped. Any carry over moment values beneath the last single lines are deleted prior to the summation of any column of figures, otherwise all joints will not be perfectly in balance.

A	0.5	B	0.5		0.5	C	0	D COF
		0.667	0.333			0.500	0.500	DF
600		−600	266.7			−266.7	225	FEM
111.1		222.2	111.1			55.6		#1 D & CO
			−3.5			−7.0	−6.9	#2 D & CO
1.2		2.3	1.2			0.6		#3 D & CO
712.3		−375.5	375.5			−218.1	218.1	Sum

7. Draw a free body diagram of each member. Since the member-end moments are known from step 6, the final member-end shears can be obtained by statics for each free body diagram.

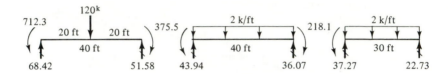

8. The reactions and any axial forces in the members are obtained either from equilibrium of the joints or from system equilibrium.

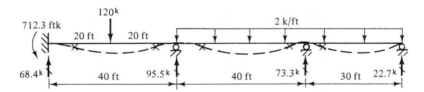

It should be noted that the rotational DOF are never computed at any stage in the Moment Distribution Method although each unbalanced interior joint is independently released and allowed to rotate to a state of temporary moment equilibrium before it is restrained again.

10.4 EXAMPLES FOR JOINT TRANSLATIONS NOT PERMITTED

The tabular form introduced in the conceptual example is used hereafter for brevity reasons.

EXAMPLE 10.1 ————————————————————————————————————

Determine the final member-end moments by Moment Distribution Method. Compare them to Example 9.3 final member-end moments. See Figure 10.4a.

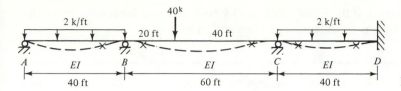

Figure 10.4a

Solution. Fixed ended moments are found from Figures 8.5 and 8.10:

span *AB*: $2\,\text{k/ft} * (40\,\text{ft})^2/8 = 400\,\text{ftk}$

span *BC*: $40^k * 20\,\text{ft} * (40\,\text{ft})^2/(60\,\text{ft})^2 = 356\,\text{ftk}$

$40^k * 40\,\text{ft} * (20\,\text{ft})^2/(60\,\text{ft})^2 = 178\,\text{ftk}$

span *CD*: $2\,\text{k/ft} * (40\,\text{ft})^2/12 = 267\,\text{ftk}$

Joint stiffnesses, distribution factors, and carry over factors are found as follows using Figures 8.6 and 8.9:

joint B: $S_{BA} = 3EI/40 = 0.075EI;$ $DF_{BA} = S_{BA}/S_B = 0.529$

$S_{BC} = 4EI/60 = 0.0667EI;$ $DF_{BC} = S_{BC}/S_B = 0.471$

$S_B = 0.1417EI;$ sum = 1.000 (check)

joint C: $S_{CB} = 4EI/60 = 0.0667EI;$ $DF_{CB} = S_{CB}/S_C = 0.400$

$S_{CD} = 4EI/40 = 0.100EI;$ $DF_{CD} = S_{CD}/S_C = 0.600$

$S_C = 0.1667EI$ sum = 1.000 (check)

$COF_{BA} = 0$ and all other carry over factors are 0.5; see Figures 8.10 and 8.5, respectively.

The final member-end moments obtained by moment distribution are shown on the line marked Sum and those obtained by the Displacement Method are shown on the line marked Example 9.3 for ease of comparison (Fig. 10.4b). Note that the results begin to disagree in the fourth digit. This is because the moment distribution process was stopped with joint *C* having an unbalanced moment of 0.7 which was deleted because it was deemed negligible. Final member-end moments need not be obtained to more than 3-digit precision by any

	0	0.5		0.5		0.5		COF
A		*B*			*C*		*D*	
	0.529	0.471		0.400	0.600			DF
	−400	356		−178	267		−267	FEM
		−17.8		−35.6	−53.4		−26.7	#1 D & CO
	32.7	29.1		14.6				#2 D & CO
		−2.9		−5.8	−8.8		−4.4	#3 D & CO
	1.5	1.4		0.7				#4 D & CO
	−365.8	365.8		−204.8	204.8		−298.1	Sum
	−365.7	365.7		−204.4	204.4		−298.3	Example 11.3

Figure 10.4b

analysis method. If the results obtained by both methods are rounded to 3-digit precision, there is no difference in the solution results.

EXAMPLE 10.2 ──

Determine the final member-end moments by Moment Distribution Method. Note that spans *BC* and *CD* are nonprismatic members (Fig. 10.5a).

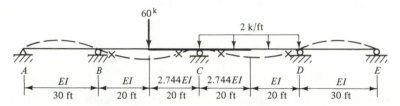

Figure 10.5a

Solution. The fixed ended moments, member-end stiffnesses, and carry over factors for spans *BC* and *CD* are obtained from the solution of Example 9.13; for the two exterior spans, the member-end stiffness is given in Figure 8.9.
   Fixed ended moments are

$$\text{span } BC: \quad 0.0938 * 60^k * 40 \text{ ft} = 225.1 \text{ ftk}$$
$$0.1679 * 60^k * 40 \text{ ft} = 403.0 \text{ ftk}$$
$$\text{span } CD: \quad 0.1048 * 2 \text{ k/ft} * (40 \text{ ft})^2 = 335.4 \text{ ftk}$$
$$0.0700 * 2 \text{ k/ft} * (40 \text{ ft})^2 = 224.0 \text{ ftk}$$

Joint stiffnesses and distribution factors are

joint B: $\quad S_{BA} = 3EI/30 \quad = 0.100EI; \qquad DF_{BA} = S_{BA}/S_B = 0.467$
$\qquad\qquad S_{BC} = 4.56EI/40 = 0.114EI; \qquad DF_{BC} = S_{BC}/S_B = 0.533$

$$\overline{\qquad\qquad\qquad S_B = 0.214EI \qquad\qquad\qquad}$$

joint C: $\quad S_{CB} = 9.45EI/40 = 0.236EI; \qquad DF_{CB} = S_{CB}/S_C = 0.500$
$\qquad\qquad S_{CD} = 9.45EI/40 = 0.236EI; \qquad DF_{CD} = S_{CD}/S_C = 0.500$

$$\overline{\qquad\qquad\qquad S_C = 0.472EI \qquad\qquad\qquad}$$

joint D: $\quad S_{DC} = 4.56EI/40 = 0.114EI; \qquad DF_{DC} = S_{DC}/S_D = 0.533$
$\qquad\qquad S_{DE} = 3EI/30 \quad = 0.100EI; \qquad DF_{DE} = S_{DE}/S_D = 0.467$

$$\overline{\qquad\qquad\qquad S_D = 0.214EI \qquad\qquad\qquad}$$

From Example 9.13, the carry over factors are

$$\text{COF}_{BC} = \text{COF}_{DC} = \frac{(3.50EI/40)}{(4.56EI/40)} = 0.768$$

$$\text{COF}_{CB} = \text{COF}_{CD} = \frac{(3.50EI/40)}{(9.45EI/40)} = 0.370$$

Due to round off, joint *C* is not perfectly in balance; but, as noted previously, these final member-end moments can be rounded to 3-digit precision and then

A	B		C		D		E	
0		0.768	0.370		0.370	0.768	0	COF
	0.467	0.533	0.500	0.500		0.533	0.467	DF
	0	225.1	−403.0	335.4	−224.0		0	FEM
	−105.1	−120.0	−92.2					#1 D & CO
				91.7	119.4	104.6		#2 D & CO
		12.6	34.0	34.0	12.6			#3 D & CO
	−5.9	−6.7	−5.1					#4 D & CO
				−5.1	−6.7	−5.9		#5 D & CO
		1.9	5.1	5.1	1.9			#6 D & CO
	−0.9	−1.0	−0.4					#7 D & CO
				−0.4	−1.0	−0.9		#8 D & CO
	−111.9	111.9	−461.2	461.2	−97.8	97.8		Sum

Figure 10.5b

all joints are perfectly balanced (Fig. 10.5b). Provided fixed ended moments and member-end stiffnesses can be obtained for nonprismatic members, the moment distribution process is easily performed. The only item that causes beginners a problem is in remembering that the carry over factors are not 0.5 for nonprismatic members that are not symmetric with respect to midspan of the nonprismatic member.

EXAMPLE 10.3

Determine the final member-end moments by Moment Distribution Method. Also determine the reactions. See Figure 10.6a.

Solution. Fixed ended moments are

$$\text{span } AB: \quad (30\,\text{kN/m})*(12\,\text{m})^2/12 = 360\,\text{mkN}$$
$$\text{span } CE: \quad (30\,\text{kN/m})*(9\,\text{m})^2/12 = 202.5\,\text{mkN}$$

Joint stiffnesses and distribution factors are

joint B:
$$S_{BA} = 4*(2EI)/12 = 0.667EI; \qquad DF_{BA} = S_{BA}/S_B = 0.500$$
$$S_{BC} = 4EI/6 \quad\quad = 0.667EI; \qquad DF_{BC} = S_{BC}/S_B = 0.500$$

$$S_B = 1.334EI$$

joint C:
$$S_{CB} = 4EI/6 \quad\quad\quad = 0.667EI; \qquad DF_{CB} = S_{CB}/S_C = 0.258$$
$$S_{CD} = 4*(1.38EI)/4.5 = 1.227EI; \qquad DF_{CD} = S_{CD}/S_C = 0.474$$
$$S_{CE} = 4*(1.56EI)/9 \quad = 0.693EI; \qquad DF_{CB} = S_{CE}/S_C = 0.268$$

$$S_C = 2.587EI$$

joint E:
$$S_{EC} = 4*(1.56EI)/9 \quad = 0.693EI; \qquad DF_{EC} = S_{EC}/S_E = 0.400$$
$$S_{EF} = 3*(1.56EI)/4.5 = 1.04EI; \qquad DF_{EF} = S_{EF}/S_E = 0.600$$

$$S_E = 1.733EI$$

Final member-end shears are obtained by drawing free body diagrams of each member, summing moments at one end of the member, and summing forces in the direction of the member-end shears (Fig. 10.6b). After the shears at D and

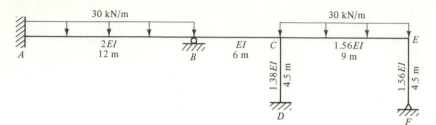

Figure 10.6a

F are known, using the entire structure as a free body diagram and summing forces in the horizontal direction gives the horizontal reaction at A. The vertical reaction at B, axial force in CD at C, and axial force in EF at F are obtained by summing the member-end shears at joints B, C, and E, respectively. See Figure 10.6c.

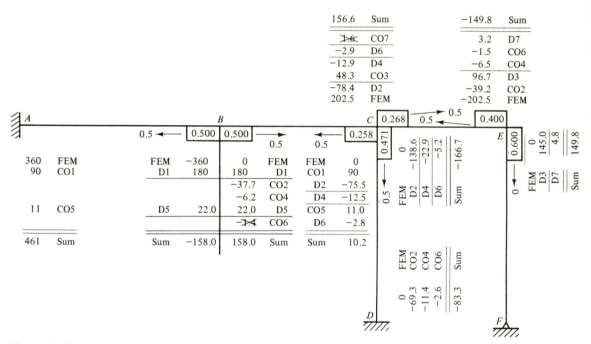

Figure 10.6b

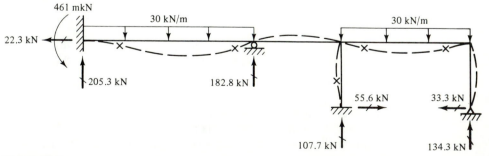

Figure 10.6c

10.5 EXAMPLES OF PRESCRIBED SUPPORT MOVEMENTS

For the reasons pointed out in Section 9.3, it is customary to obtain separate solutions for the effects due to prescribed support movements and superimpose these solutions on the solutions for no support movements. In the Moment Distribution Method, the prescribed support movements are imposed on the *restrained structure* to obtain the member fixed ended moments. Since all of the support movement(s) is(are) imposed to obtain the fixed ended moments, no movements of the supports are permitted thereafter. Since the prescribed support movements are small, the geometry of the undeformed structure can be used to obtain the distribution factors, carry over factors, final member-end shears, axial forces, and reactions. Therefore, the procedure described in Section 10.3 is applicable after the fixed ended moments due to prescribed support movements have been obtained.

EXAMPLE 10.4 ———————————————————————————————

The structure shown in Figure 10.7a is the same one as in Examples 10.1 and 9.9; all members have the same $EI = 312,153$ kft^2. Determine the final member-end moments due only to a prescribed downward movement of 1.5 in. = 0.125 ft at joint B.

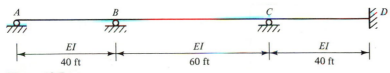

Figure 10.7a

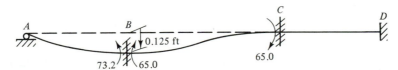

Figure 10.7b

Solution. Fixed ended moments are (Fig. 10.7b)

$$\text{span } AB: \qquad \frac{3EI}{L^2}*0.125 \text{ ft} = 73.2 \text{ ftk}$$

$$\text{span } BC: \qquad \frac{6EI}{L^2}*0.125 \text{ ft} = 65.0 \text{ ftk}$$

Joint stiffnesses, distribution factors, and carry over factors are exactly the same as in Example 10.1 since the structure is the same as it was in that example (Fig. 10.7c).

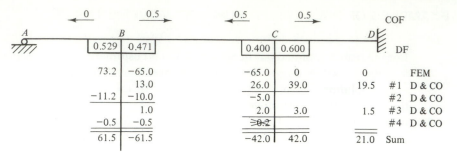

Figure 10.7c

EXAMPLE 10.5 ───

For the structure shown in Example 9.5, find the fixed ended moments due to the following prescribed support movements:

1. 0.002 rad clockwise at the left support.
2. 0.03 ft relative separation horizontally between the supports.
3. 0.024 ft relative separation vertically between the supports.

Solution for Part 1

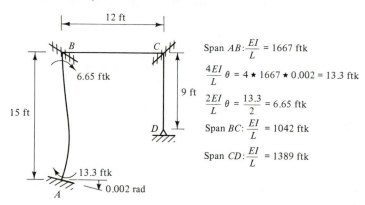

Solution for Part 2. All of the 0.03-ft separation can be applied at either point *A* or point *D*. Alternatively, a portion of the 0.03-ft separation can be applied at *A* and the remainder applied at point *D*. In the following illustrative calculations, all of the 0.03-ft separation is applied at point *A*.

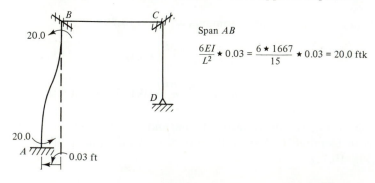

Solution for Part 3. All of the 0.024 ft can be applied downward at point A or upward at point D. Alternatively, a portion of the 0.024 ft can be applied downward at point A and the remainder applied upward at point D. In the following illustrative calculations, all of the 0.024-ft separation is applied downward at point A.

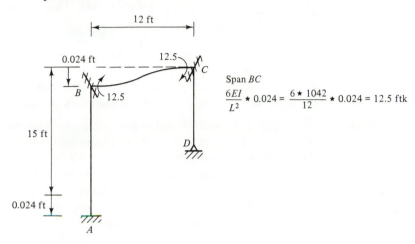

Span BC

$$\frac{6EI}{L^2} \star 0.024 = \frac{6 \star 1042}{12} \star 0.024 = 12.5 \text{ ftk}$$

Note. In Example 10.9, the final member-end moments are found by the Moment Distribution Method for the prescribed support movements and fixed ended moments of Example 10.5 part 2.

10.6 PROCEDURE FOR ONE TRANSLATIONAL DOF

The reader is reminded that in the Moment Distribution Method all members are treated as being axially rigid (axial deformations of all members are ignored). As a consequence of this assumption and the location of the supports in the examples of Section 10.4, no joint translations could occur. However, a structural analyst frequently encounters plane frame problems which involve unknown joint translations at the interior joints.

Two methods are available for computing the effects of interior joint translations in a moment distribution analysis. One method, known as the method of successive shear corrections, was presented originally by C. T. Morris. The other method, known as the direct method, involves the superposition for the effects of separate translational DOF as described below.

In the interest of simplicity, the following description of the direct method is made for only one translational DOF. However, the direct method can be applied to frames involving more than one translational DOF as shown in Section 10.9 for multistory frames.

Consider the structural problem shown in Figure 10.8a (this problem was solved in Example 9.4 by the Displacement Method). The objective is to describe and numerically illustrate the direct method of moment distribution for a structure involving only one translational DOF. Therefore, treat the structural system as two prismatic members which requires that point B must be treated as an interior joint. Obviously

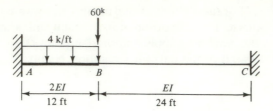

Figure 10.8a Original structure and loading.

joint B must rotate and translate downward an unknown amount due to the loading shown. However, in the moment distribution procedure presented in Section 10.3 no unknown joint translation was permitted. Therefore, the procedure must be modified as described below to account for the effects of the unknown translation at joint B:

Step 1. Temporarily make joint B a roller supported joint to prevent translation in the vertical direction at joint B (see Figure 10.8b). Using the moment distribution procedure defined in Section 10.3 for no joint translations, find the final member-end moments due to the original loading in the structure having a temporary roller support at joint B. Also compute the reaction, R_B, that the temporary roller support provided. See Figure 10.8b which numerically illustrates step 1.

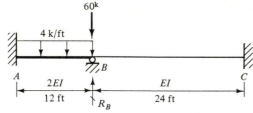

The fixed ended moments for span AB are $\dfrac{4 \star (12)^2}{12} = 48$ ftk

$$DF_{BA} = \frac{4 \star (2EI)/12}{4 \star (2EI)/12 + 4EI/24} = 0.800; \quad DF_{BC} = 1 - DF_{BA} = 0.200$$

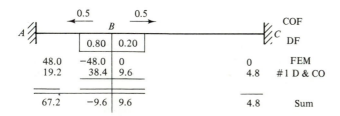

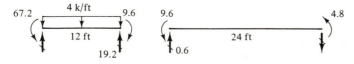

$R_B = 60 + (19.2 + 0.6) = 79.8^k$

Figure 10.8b Solution for joint translation prevented at B.

Step 2. For the original structure with all loads removed, restrain joint B against rotation and induce a translational displacement with an arbitrary magnitude, Δ, at but opposite in direction to the temporary roller support force found in step 1. The fixed ended moments due to the induced Δ are

$$\text{span } AB: \qquad \frac{6*(2EI)\Delta}{(12)^2} = \frac{EI\Delta}{12}$$

$$\text{span } BC: \qquad \frac{6EI\Delta}{(24)^2} = \frac{EI\Delta}{96}$$

Since Δ is unknown, choose an arbitrary value for one of the fixed ended moment (FEM)—for example, choose $EI\Delta/96 = 10$ ftk which is the same as choosing $\Delta = 960/EI$ and this choice requires that the other FEM be $EI\Delta/12 = EI/12*960/EI = 80$ ftk. Using these compatibly chosen FEM due to an arbitrarily prescribed Δ, perform a moment dis-

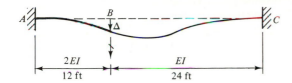

Choose $M_{BC}^f = M_{CB}^f = \dfrac{6EI\Delta}{(24)^2} = 10$ ftk which gives $\Delta = \dfrac{960}{EI}$ and requires

that $M_{AB}^f = M_{BA}^f = \dfrac{6(2EI)\Delta}{(12)^2} = \dfrac{12EI}{144} \star \dfrac{960}{EI} = 80$ ftk

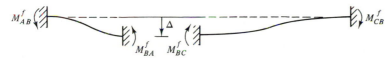

	0.5	B	0.5			COF
	A				C	DF
		0.80	0.20			
80		80	−10		−10	FEM
−28		−56	−14		−7	# 1 D & CO
52		24	−24		−17	Sum

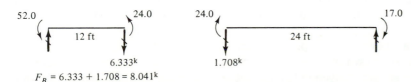

$F_B = 6.333 + 1.708 = 8.041^k$

Figure 10.8c Solution for arbitrary Δ induced at joint B.

tribution analysis to balance the interior joints in the usual manner. Also compute the force required at B, F_B, to induce the arbitrarily chosen Δ. (Instead of choosing an arbitrary value for one of the FEM and compatibly computing the other FEM, the analyst can choose $\Delta = 1$ in. (or 1 ft), substitute actual values of E, I, and L with compatible units, and compute all FEM. For this approach, the computed force at B is the stiffness coefficient, S_B. Also, for this approach the correction factor is the actual value of Δ which eventually is needed anyway to check the control of drift. However, if only relative values of the member EI are known, the analyst probably will want to use scaled values of the FEM which corresponds to the previously introduced approach.) See Figure 10.8c which numerically illustrates step 2.

Step 3. To obtain the final member-end moments for the original structure and loading (Figure 10.8a), multiply the final member-end moments computed in step 2 by the ratio R_B/F_B and algebraically add them to the final member-end moments computed in step 1. See Figure 10.8d for the final numerical results. From step 1, $R_B = 79.8^k$ and from step 2, $F_B = 8.041^k$; therefore, the final member-end moments obtained in step 2 need to be multiplied by the ratio $R_B/F_B = 79.8/8.041 = 9.924$ and algebraically added to the final member-end moments obtained in step 1.

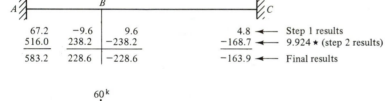

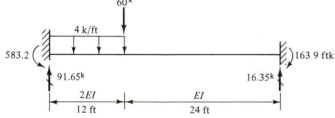

Figure 10.8d Solution for structure and loading in Figure 10.8a.

10.7 EXAMPLES FOR ONE TRANSLATIONAL DOF

Single story frames, similar to the ones shown in the following example problems, are usually so constructed that they have one translational DOF. If the structure is symmetrical but the loading is not symmetrical, the structure will sway (interior joints will translate) as shown in Example 10.6—see Section 10.8 for examples of structural symmetry and loading symmetry or antisymmetry. Example 10.9 illustrates the solution for a prescribed support movement. The structure shown in Example 10.10 is a frame with sloping legs (nonorthogonally intersecting members); this structure was solved previously by the Displacement Method (Example 9.19). Section 10.9 deals with frames having more than one translational DOF.

EXAMPLE 10.6

Find the final member-end moments by Moment Distribution Method. All members have the same EI and length $L = 9$ m. See Figure 10.9a.

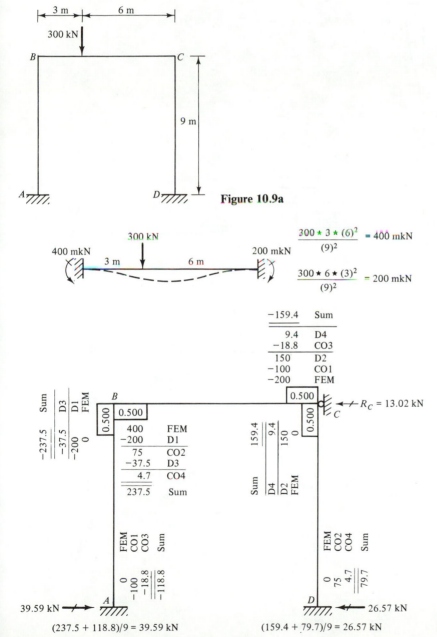

Figure 10.9a

$$\frac{300 \star 3 \star (6)^2}{(9)^2} = 400 \text{ mkN}$$

$$\frac{300 \star 6 \star (3)^2}{(9)^2} = 200 \text{ mkN}$$

Figure 10.9b

$(237.5 + 118.8)/9 = 39.59$ kN

$(159.4 + 79.7)/9 = 26.57$ kN

Solution. Provide a temporary roller support at *C* to prevent joint translations, compute the member fixed ended moments due to the loading, and distribute them as shown in Figure 10.9b.

The fixed ended moments due to joint translations are computed and distributed as shown in Figure 10.9c.

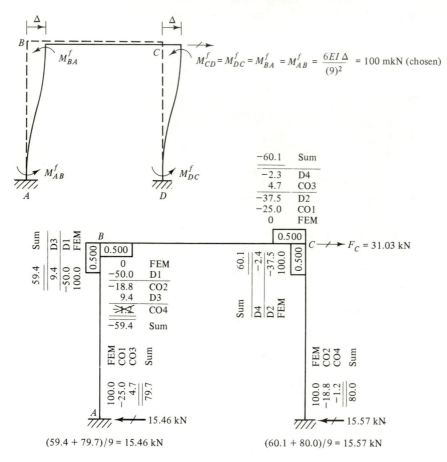

$$M_{CD}^f = M_{DC}^f = M_{BA}^f = M_{AB}^f = \frac{6EI\,\Delta}{(9)^2} = 100 \text{ mkN (chosen)}$$

$(59.4 + 79.7)/9 = 15.46 \text{ kN}$ $(60.1 + 80.0)/9 = 15.57 \text{ kN}$ **Figure 10.9c**

Figure 10.9d

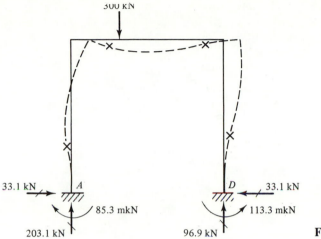

300 kN

33.1 kN   A   D   33.1 kN

85.3 mkN   113.3 mkN

203.1 kN   96.9 kN   **Figure 10.9e**

Since $R_C/F_C = 13.02/31.30 = 0.420$, the values in Figure 10.9c must be multiplied by 0.420 and added to the values in the joint translation prevented solution to obtain the final member-end moments shown in Figure 10.9d. The reactions are shown in Figure 10.9e.

EXAMPLE 10.7 ──

Find the final member-end moments by moment distribution in Figure 10.10a.

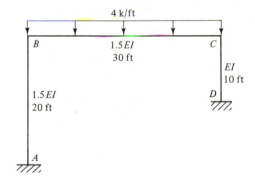

4 k/ft

B   1.5EI C
      30 ft

                         EI
                         10 ft

1.5EI D
20 ft

A

Figure 10.10a

Solution

$$M^f_{BC} = M^f_{CB} = \frac{4*(30)^2}{12} = 300 \text{ ftk with proper signs attached}$$

$S_{BA} = 4*1.5EI/20 = 0.300EI$ $DF_{BA} = 0.300EI/0.500EI = 0.6$
$S_{BC} = 4*1.5EI/30 = 0.200EI$ $DF_{BC} = 0.200EI/0.500EI = 0.4$
────────────────────────────────────
            $S_B = 0.500EI$

$S_{CB} = 4*1.5EI/30 = 0.2EI$ $DF_{CB} = 0.2/0.6 = 0.333$
     $S_{CD} = 4EI/10 = 0.4EI$ $DF_{CD} = 0.4/0.6 = 0.667$
────────────────────────────────────
            $S_C = 0.6EI$

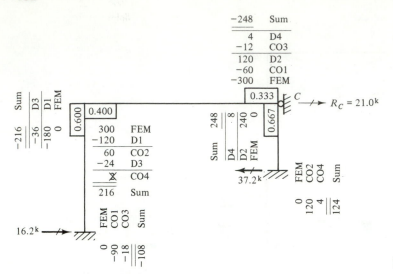

Figure 10.10b

With a temporary roller support provided at C to prevent joint translation and due to the FEM for the applied loads, moment distribution gives Figure 10.10b. The FEM due to joint translations are computed as shown in Figure 10.10c.

$$M^f_{AB} = M^f_{BA} = \frac{-6*1.5EI\Delta}{(20)^2} = -0.0225EI\Delta$$

$$M^f_{CD} = M^f_{DC} = \frac{-6EI\Delta}{(10)^2} = -0.060EI\Delta = 2.67M^f_{AB}$$

Choose $M^f_{AB} = M^f_{BA} = -100$ ftk and compute $M^f_{CD} = M^f_{DC} = -267$ ftk. Distribution of these FEM gives Figure 10.10d. The values in Figure 10.10d must be multiplied by $R_C/F_C = 21.0/35.3 = 0.595$ and added to the solution for no joint translations to obtain the final member-end moments shown in Figure 10.10e.

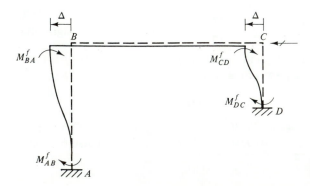

Figure 10.10c

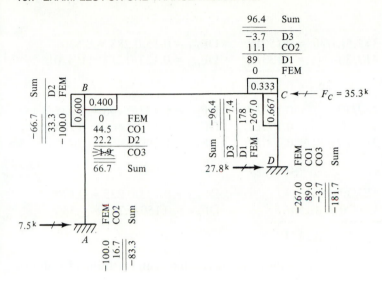

Figure 10.10d

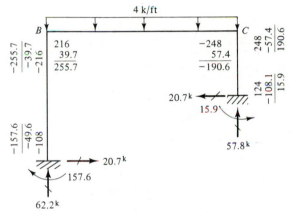

Figure 10.10e

EXAMPLE 10.8

Find the final member-end moments by moment distribution (Fig. 10.11a).

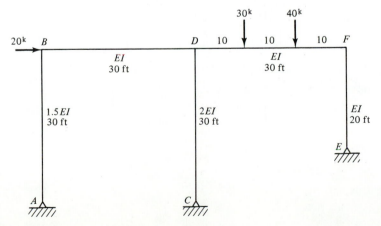

Figure 10.11a

Solution

$$S_{BA} = 3*1.5EI/30 = 0.150EI \qquad DF_{BA} = 0.15/0.283 = 0.530$$
$$S_{BD} = 4EI/30 \qquad\ \ = 0.133EI \qquad DF_{BD} = 0.133/0.283 = 0.470$$

$$S_B = 0.283EI$$

$$S_{DB} = 4EI/30 \qquad\ \ = 0.133EI \qquad DF_{DB} = 0.133/0.466 = 0.285$$
$$S_{DF} = 4EI/30 \qquad\ \ = 0.133EI \qquad DF_{DF} = 0.133/0.466 = 0.285$$
$$S_{DC} = 3*(2EI)/30 = 0.200EI \qquad DF_{DC} = 0.200/0.466 = 0.429$$

$$S_D = 0.466EI$$

$$S_{FD} = 4EI/30 = 0.133EI \qquad DF_{FD} = 0.133/0.285 = 0.470$$
$$S_{FE} = 3EI/20 = 0.150EI \qquad DF_{FE} = 0.150/0.285 = 0.530$$

$$S_F = 0.285EI$$

With a temporary roller support at F, the FEM are computed and distributed as shown below (Fig. 10.11b):

$$M^f_{DF} = 30*10*(20)^2/(30)^2 + 40*20*(10)^2/(30)^2 = 222.2 \text{ ftk}$$
$$M^f_{FD} = -30*20*(10)^2/(30)^2 - 40*10*(20)^2/(30)^2 = -244.4 \text{ ftk}$$

The FEM due to joint translations are computed and distributed as shown in Figures 10.11c and 10.11d.

$$M^f_{BA} = 3*(1.5EI)\Delta/(30)^2 = 0.00500EI\Delta$$
$$M^f_{DC} = 3*(2EI)\Delta/(30)^2\ \ = 0.00667EI\Delta = 1.334M^f_{BA}$$
$$M^f_{FE} = 3EI\Delta/(20)^2 \qquad\ = 0.00750EI\Delta = 1.500M^f_{BA}$$

Choose $M^f_{BA} = 100$ ftk.

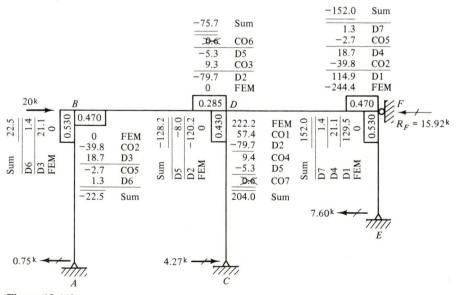

Figure 10.11b

The values in Figure 10.11d must be multiplied by the correction factor $= R_F/F_F = 15.92/8.86 = 1.80$ and added to the solution for no joint translations to obtain the final member-end moments shown in Figure 10.11e.

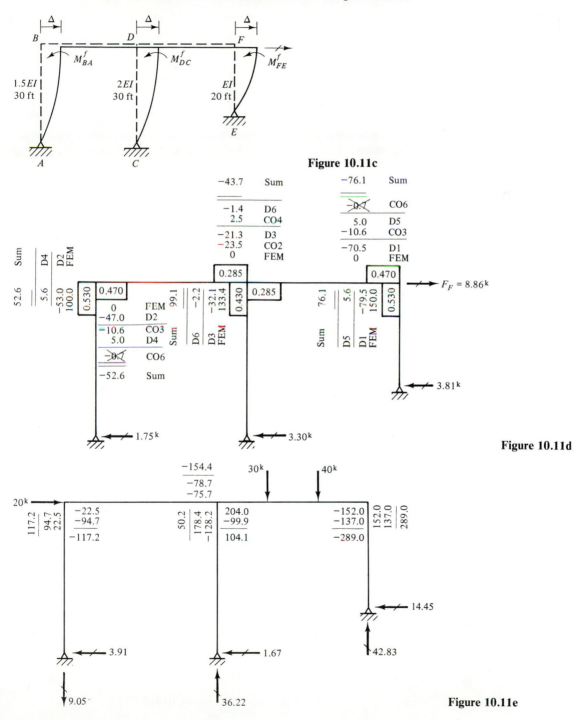

Figure 10.11c

Figure 10.11d

Figure 10.11e

EXAMPLE 10.9

Find the final member-end moments due to a prescribed 0.03-ft movement of support A to the left. From Example 10.5, $M^f_{AB} = M^f_{BA} = 20.0$ ftk.

$$\text{member } AB: \qquad \frac{EI}{L} = 1667 \text{ ftk}$$

$$BC: \qquad \frac{EI}{L} = 1042 \text{ ftk}$$

$$CD: \qquad \frac{EI}{L} = 1389 \text{ ftk}$$

Solution. Due to the prescribed support movement and no interior joint translations, the moment distribution analysis is as shown in Figure 10.12b. The

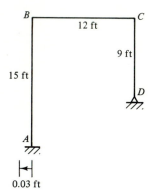

Figure 10.12a

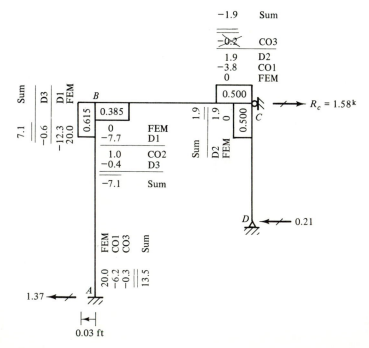

Figure 10.12b

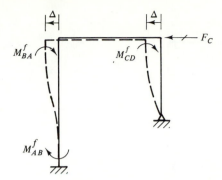

Figure 10.12c

solution part to cancel the effects of having restrained interior joint translations in the preceding solution part are as shown in Figure 10.12c.

$$M^f_{AB} = M^f_{BA} = \frac{6*1667*\Delta}{15} = 666.8\Delta$$

$$M^f_{CD} = \frac{3*1389*\Delta}{9} = 463\Delta$$

$$\frac{666.8\Delta}{463\Delta} = 1.44$$

Choose $463\Delta = 10$ ftk; $666.8\Delta = 14.4$ ftk.

The final results, shown in Figure 10.12e, are the results from Figure 10.12d times $R_C/F_C = 1.58/1.84 = 0.859$ plus the results from Figure 10.12b.

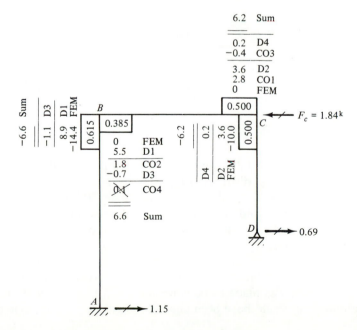

Figure 10.12d

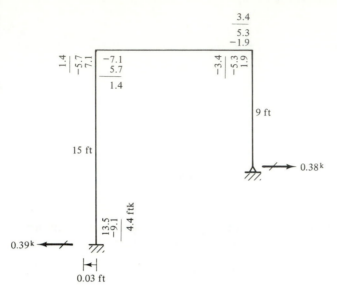

Figure 10.12e

EXAMPLE 10.10 ──

Find the final member-end moments for the structure shown in Figure 10.13a which was previously solved in Example 9.19 by the Displacement Method. EI = 42,600 kNm^2 for all members. Member lengths are 6.708 m for AC; 7.5 m for DE. ϕ_A = 63.435° and ϕ_E = 53.130°.

Solution. From Example 9.19, M^f_{CA} = 45 mkN; M^f_{CD} = M^f_{DC} = 150 mkN; M_{CB} = 90 mkN.

$$S_{CA} = 19051.3 \text{ mkN} \qquad \text{DF}_{CA} = 19{,}051.3/47{,}451.3 = 0.401$$
$$S_{CD} = 28{,}400 \qquad\qquad \text{DF}_{CD} = 1 - 0.401 = 0.599$$

$$S_C = 47{,}451.3$$

$$S_{DC} = 28{,}400 \qquad\qquad \text{DF}_{DC} = 28{,}400/51{,}120 = 0.556$$
$$S_{DE} = 22{,}720 \qquad\qquad \text{DF}_{DE} = 1 - 0.556 = 0.444$$

$$S_D = 51{,}120$$

In the following moment distribution analysis, the first time that joint C is released to balance it, we must distribute M_{CB} = 90 mkN in order to introduce it into the ends of members CA and CD. Since member CB is determinate it provides no stiffness at joint C and never receives any of the distributed moment at joint C. Due to the applied loads and no joint translations permitted, the moment distribution analysis is as shown in Figure 10.13b.

The location of the force required to prevent joint translations in the preceding moment distribution is chosen as the horizontal force at D and named R_D. This choice was made to be consistent with Example 9.19; however, the restraining force could have been chosen coincident with any one of the inde-

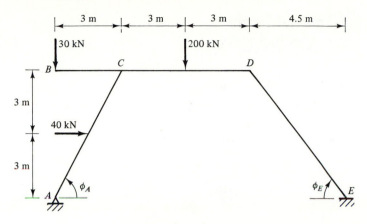

Figure 10.13a

pendent or dependent displacement vectors shown in Example 9.19 for the translational DOF. R_D is most conveniently computed by summing moments at the instantaneous center of rotation of member CD which is named point i in Figure 10.13c. The location of point i was obtained from the solution of Example 9.19.

Since summation of moments is at point i, the axial force direction reactions at A and E are not needed (they pass through point i). The shear direction reactions at A and E are computed from free body diagrams of members AC and DE by summing moments at points C and D, respectively:

$$\text{shear at } A = \frac{3*40 - 69.6}{6.708} = 7.51 \text{ kN}$$

$$\text{shear at } E = \frac{74.1 + 37.1}{7.5} = 14.8 \text{ kN}$$

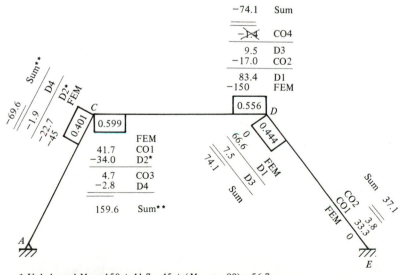

* Unbalanced $M_C = 150 + 41.7 - 45 + (M_{CB} = -90) = 56.7$

** $M_{CA} + M_{CD} + M_{CB} = -69.6 + 159.6 - 90 = 0$ as it should be

Figure 10.13b

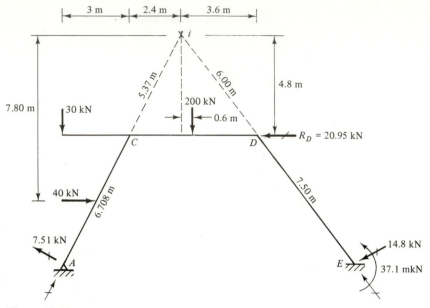

Figure 10.13c

Summation of moments = 0 at point i requires

$$4.8R_D = 37.1 + 5.4*30 + 7.8*40 - 12.08*7.51 - 13.5*14.8 - 0.6*200$$
$$R_D = 20.95 \text{ kN}$$

Information needed from Example 9.19 to compute the fixed ended moments due to joint translation is shown in Figure 10.13d.

$$M^f_{CA} = 3175\Delta$$
$$M^f_{CD} = M^f_{DC} = 8875\Delta$$
$$M^f_{DE} = M^f_{ED} = 5680\Delta$$

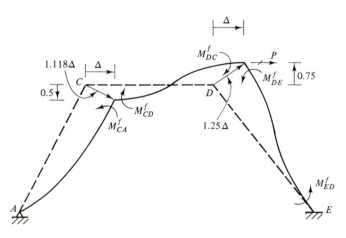

Figure 10.13d

Choose $M_{CA}^f = 100$ mkN and this choice requires

$$M_{CD}^f = M_{DC}^f = 279.5\,\text{mkN}$$
$$M_{DE}^f = M_{ED}^f = 178.9\,\text{mkN}$$

Moment distribution for these fixed ended moments is as shown in Figure 10.13e.

$$\text{shear at } A = \frac{166.8}{6.708} = 24.87\,\text{kN}$$

$$\text{shear at } E = \frac{(201.4 + 190.2)}{7.5} = 52.21\,\text{kN}$$

Summation of moments = 0 at instantaneous center of rotation for member CD requires

$$4.8F_D = 13.5*52.21 + 12.08*24.87 - 190.2$$
$$F_D = 169.8\,\text{kN}$$

$R_D/F_D = 20.95/169.8 = 0.123$ is the correction factor for the values shown in Figure 10.13f. Final member-end moments are shown in Figure 10.13f.

It should be noted that perfect agreement of the final member-end moments in the moment distribution solution and the Displacement Method solution, Example 9.18, is not possible. The reason for the small disagreements in final member-end moment values is that the moment distribution process was terminated with a carry over moment which was deemed negligible in each part of the solution. Therefore, the moment distribution solution is an approximate but very nearly exact solution and additional cycles of distribution could have been performed to get even closer agreement provided more than three-digit precision was used in all calculated values.

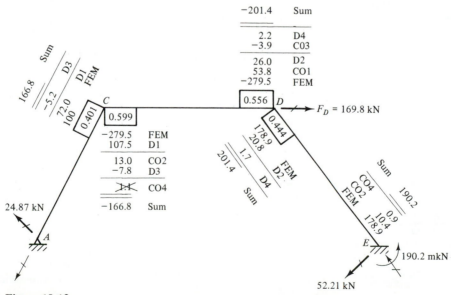

Figure 10.13e

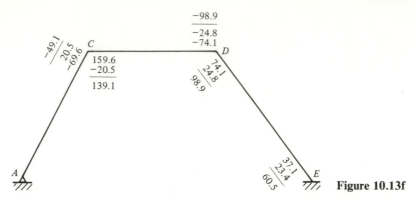

Figure 10.13f

10.8 STRUCTURAL SYMMETRY AND LOADING SYMMETRY OR ANTISYMMETRY

If a structure has an axis of symmetry, the analyst can take advantage of structural symmetry (and loading symmetry or antisymmetry) to decrease the time required to perform a moment distribution analysis. Each of the structures shown in Figure 10.14 is structurally symmetric with respect to the vertical line marked by the symbol $. For the structures shown, if the reader carefully and completely folded the page along the line marked $, the two structural halves would exactly overlay one another in regard to geometry, member properties, and support conditions. Alternately stated, if the reader imagined cutting the structure along the line marked $ and rotated the left half of the structure 180° about the axis marked $, the right half of the structure would be perfectly generated in regard to geometry, member properties, and support conditions.

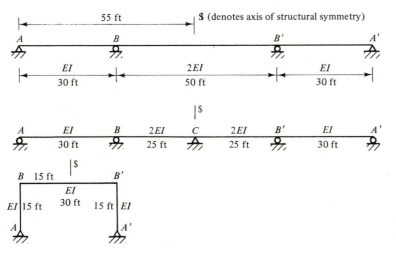

Figure 10.14 Examples of structurally symmetric structures.

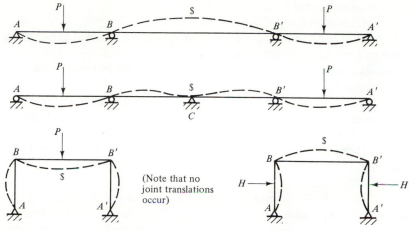

Examples of Symmetric Structures Subjected to Symmetric Loadings

(Note that no joint translations occur)

Figure 10.15 Examples of symmetric structures subjected to symmetric loadings.

In the examples of symmetric structures subjected to symmetric loadings shown in Figure 10.15, note that:

1. On the axis of structural symmetry:
   a. The slope of the deflected shape is zero.
   b. The horizontal deflection is zero; therefore, no joint translations occur since the members are axially rigid.
   c. The vertical deflection is not zero unless there is a support at this location.
2. For a pair of joints, B and B' (or A and A'), θ_B and θ'_B have opposite rotational directions.
3. If the left half of the loading and deflected shape are rotated 180° about the axis marked $, the right half of the loading and deflected shape are perfectly generated.

In Figure 10.16, deflected shapes are shown for symmetric structures subjected to antisymmetric loading. Note that:

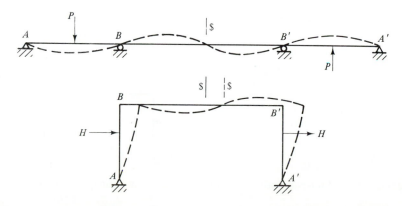

Figure 10.16 Examples of symmetric structures subjected to antisymmetric loadings.

1. On the axis of structural symmetry:
   a. The horizontal deflection is not zero.
   b. The vertical deflection is zero since all members are assumed to be axially rigid in a moment distribution analysis.
   c. There is a point of inflection which means that the internal moment is zero at this location.
2. For a pair of joints, B and B', θ_B and θ'_B have the same rotational directions.
3. If the left half of the loading and deflected shape are rotated 180° about the axis marked $, the vector directions of all loads and deflections are exactly opposite of those on the right half of the structure, but the vector magnitudes are identical.

As shown in Figure 10.17, any unsymmetrical loading can be decomposed into a symmetrical loading and an antisymmetrical loading. However, it is only profitable to do so when the structure is symmetric. Since superposition is involved in the concepts of Figure 10.17, it is assumed that the symmetric structure behaves in a linearly elastic manner.

If there is no joint located on the axis of structural symmetry, the member whose midspan coincides with the axis of structural symmetry bends as shown in Figure 10.18 when there are no member loads on that member. To maintain symmetry in a moment distribution analysis, both ends of this member must rotate simultaneously. Therefore, the definition of member-end stiffness computed by the Conjugate Beam Method for this member in Figure 10.18 is $2EI/L$. Also, the carry over factor is zero since both joints are rotated simultaneously.

Figure 10.19 shows a member whose midspan coincides with the axis of structural symmetry, has no member loads on it, but is subjected to antisymmetric bending. Therefore, the definition of member-end stiffness for both ends of the member rotating simultaneously due to antisymmetric bending is $3EI/0.5L = 6EI/L$. Also, the carry over factor is zero since both ends of the member are rotated simultaneously.

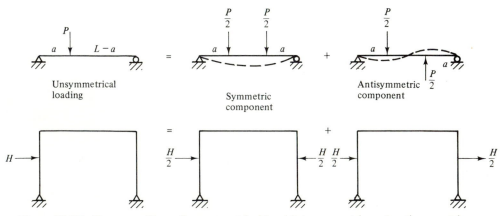

Figure 10.17 Decomposition of unsymmetrical load into symmetric and antisymmetric components.

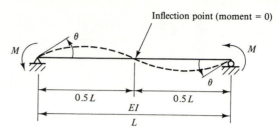

Inflection point (moment = 0)

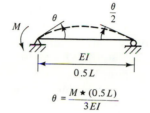

0.5 L   0.5 L

EI

L

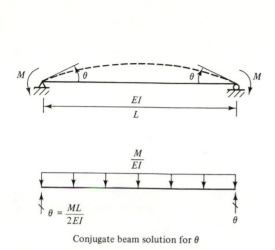

$$\theta = \frac{ML}{2EI}$$

Conjugate beam solution for θ

The definition of member-end stiffness is $\dfrac{M}{\theta} = \dfrac{2EI}{L}$

Figure 10.18 Member-end stiffness for symmetric bending.

$$\theta = \frac{M \star (0.5L)}{3EI}$$

The definition of member-end stiffness is $\dfrac{M}{\theta} = \dfrac{3EI}{0.5L} = \dfrac{6EI}{L}$

Figure 10.19 Member-end stiffness for antisymmetric bending.

EXAMPLE 10.11

Find the final member-end moments by moment distribution. Take advantage of the fact that the structure and loading are symmetric. Note that no joint translations can occur since the *loading and structure* are symmetric.

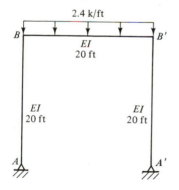

2.4 k/ft

B   EI   20 ft   B'

EI 20 ft   EI 20 ft

A   A'

Figure 10.20a

Solution

$$S_{BA} = 3EI/20 = 0.15EI \qquad DF_{BA} = 0.15/0.25 = 0.600$$
$$S_{BB'} = 2EI/20 = 0.10EI \qquad DF_{BB'} = 1 - 0.6 = 0.400$$

$$S_B = 0.25EI$$

Note that $COF_{BA} = 0$ since point A is hinged and $COF_{BB'} = 0$ since $S_{BB'}$ was computed based on the stiffness derivation for joints B and B' rotating simul-

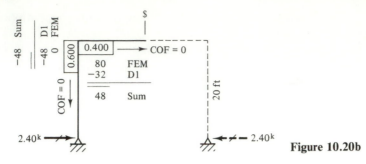

Figure 10.20b

taneously. The moment distribution solution for the fixed ended moment, $M^f_{BB'} = 2.4(20)^2/12 = 80$ ftk, is shown in Figure 10.20b. If symmetry of structure and loading had not been accounted for, it would have taken four cycles of moment distribution to achieve the same solution.

EXAMPLE 10.12 ──

Find the final member-end moments by moment distribution accounting for structural symmetry and decomposing the loading into symmetrical and antisymmetrical components. See Figure 10.21a.

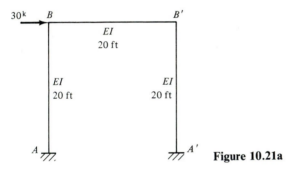

Figure 10.21a

Solution. Since the loading is neither symmetric nor antisymmetric, it must be decomposed into its symmetric and antisymmetric parts (if one is to take full advantage of structural symmetry) as shown in Figure 10.21b. Note that there are no fixed ended moments for the symmetric part of the loading; therefore, the final member-end moments for this part are zero since the joints do not translate due to this loading.

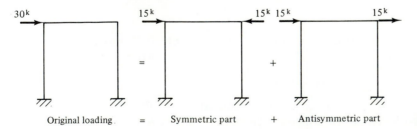

Original loading   =   Symmetric part   +   Antisymmetric part   Figure 10.21b

For the antisymmetric part of the loading, either half of the structure can be analyzed; recall that a point of inflection (moment = 0) and a zero vertical deflection occurs at the axis of symmetry for antisymmetric bending. This is equivalent to having a roller at this location constrained to roll on a horizontal surface as shown in Figure 10.21c.

$$S_{BA} = 4EI/20 = 0.200EI \qquad DF_{BA} = 0.2/0.5 = 0.400$$
$$S_{BC} = EI/10 \ \ = 0.300EI \qquad DF_{BC} = 1 - 0.4 = 0.600$$
$$\overline{\phantom{S_{BA} = 4EI/20 = 0.200EI}}$$
$$S_B = 0.500EI$$

$$COF_{BC} = 0 \,(\text{end } C \text{ is hinged})$$
$$COF_{BA} = 0.5 \,(\text{prismatic member with fixed end})$$

Fixed ended moments due to joint translations are

$$M^f_{AB} = M^f_{BA} = 6EI\Delta/(20)^2 = 100 \,\text{ftk (chosen)}$$

Distribution of these FEM gives Figure 10.21d. The chosen FEMs when distributed only provide an applied force $F_B = 7.00^k$ and an applied force of 15^k was needed. Therefore, the values in Figure 10.21d must be multiplied by the correction factor of $15.00/7.00 = 2.14$ to obtain the correct final member-end moments shown in Figure 10.21e.

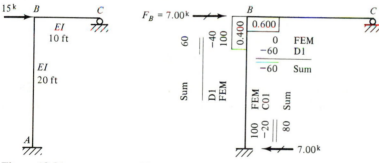

Figure 10.21c   **Figure 10.21d**

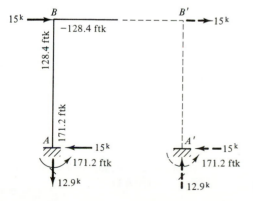

Figure 10.21e

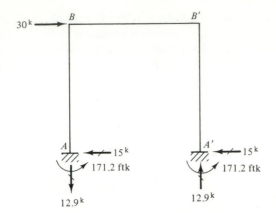

Figure 10.21f

The solution for the antisymmetric part of the loading is shown in Figure 10.21e. Since the symmetric part of the loading causes no bending and no reactions, when the antisymmetric solution part is added to the symmetric solution part the final solution is obtained as shown in Figure 10.21f.

Additional examples of loading symmetry and antisymmetry solutions are shown in the multistory buildings section.

10.9 MULTISTORY BUILDINGS

When there are two or more translational DOF, the direct method presented in Section 10.6 for one translational DOF must be modified as shown in the following discussion to account for two or more translational DOF.

For purposes of presenting the direct method for two or more translational DOF, consider Figure 10.22 which depicts a multistory frame of n stories and n translational

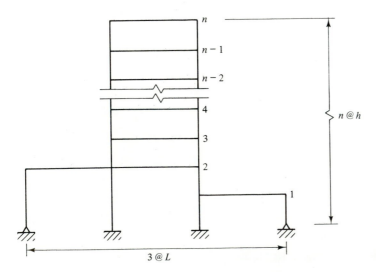

Figure 10.22 Multistory building with n translational DOF.

DOF. At each floor or roof level there is a translational DOF. For convenience in the discussion, the translational DOF are numbered from 1 through n. The direct method involves the superposition for the effects due to each of the n translational DOF as described below.

The direct method for two or more translational DOF is:

Step 1. At each of the levels numbered 1 through n, temporarily provide a roller support at the joints on one of the exterior faces of the building. This roller restrained structure has no translational DOF; therefore, it can be analyzed using the moment distribution procedure defined in Section 10.3 for no joint translations to find the final member-end moments for the original loading and no joint translations at levels 1 through n. Also compute the reactions, R_1 through R_n, provided by the temporary roller supports. Let the positive direction of these reactions be from right to left.

Step 2. For the original structure with all loads removed and roller supports at levels 2 through n, induce $\Delta_1 = 1$ (positive Δ_1 is to the right). Compute the fixed ended moments for all columns in the story below and story above the induced Δ. Due to these FEMs which must have the same units as those computed in Step 1, perform a moment distribution analysis to balance all joints that were not supports in the original structure. Also compute the inducing and restraining forces at levels 1 through n; these are stiffness coefficients, $S_{1,1}$ through $S_{n,1}$.

Step 3. Repeat step 2 for each of the levels $j = 2$ through n. For example, when $\Delta_2 = 1$ is induced, roller supports are provided at levels 1 and 3 through n; the stiffness coefficients to be computed are $S_{1,2}$ through $S_{n,2}$.

Step 4. Solve the following matrix equation of system equilibrium for the translational displacements $\{\Delta\}$: $[S]\{\Delta\} = \{R\}$.

Step 5. To obtain the final member-end moments for the original structure and loading:

    a. Multiply the final moments in step 2 by the value of Δ_1 obtained in step 4; also multiply the final moments in each of the cases $j = 2$ through n of step 3 by the value of Δ_j (for $j = 2$ through n) obtained in step 4.

    b. Algebraically add the final moments computed in step 5a for each of the n cases to the final moments in step 1.

EXAMPLE 10.13 ──

Find the final member-end moments by moment distribution using the Direct Method. All members have the same length, $L = 20$ ft, and $EI = 268,000$ kft^2. See Figure 10.23a.

Solution. Since the structure is symmetric, decompose the loading into its symmetric part and antisymmetric part in order to shorten the time required to perform the moment distribution analyses. See Figure 10.23b. For the symmetric loading there are no joint translations and the left half of the structure is chosen

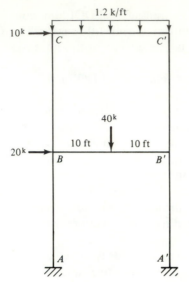

Figure 10.23a

for purposes of analysis. The fixed ended moments are $M^f_{CC'} = 1.2(20)^2/12 = 40$ ftk and $M^f_{BB'} = 40*20/8 = 100$ ftk. See Figure 10.23c.

$$S_{BA} = 4EI/20 = 0.200EI \qquad DF_{BA} = 0.2/0.5 = 0.400 \qquad COF_{BA} = 0.5$$
$$S_{BC} = 4EI/20 = 0.200EI \qquad DF_{BC} = 0.2/0.5 = 0.400 \qquad COF_{BC} = 0.5$$
$$S_{BB'} = 2EI/20 = 0.100EI \qquad DF_{BB'} = 0.1/0.5 = 0.200 \qquad COF_{BB'} = 0$$
$$\overline{\qquad\qquad\qquad\quad S_B = 0.500EI \qquad\qquad\qquad\qquad\qquad\qquad\qquad}$$

$$S_{CB} = 4EI/20 = 0.200EI \qquad DF_{CB} = 0.2/0.3 = 0.667 \qquad COF_{CB} = 0.5$$
$$S_{CC'} = 2EI/20 = 0.100EI \qquad DF_{CC'} = 0.1/0.3 = 0.333 \qquad COF_{CC'} = 0$$
$$\overline{\qquad\qquad\qquad\quad S_C = 0.300EI \qquad\qquad\qquad\qquad\qquad\qquad\qquad}$$

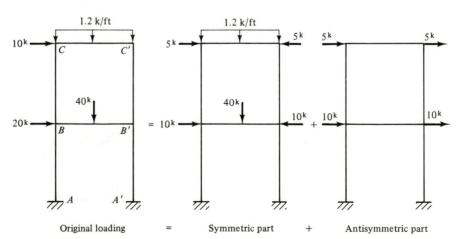

Original loading   =   Symmetric part   +   Antisymmetric part

Figure 10.23b

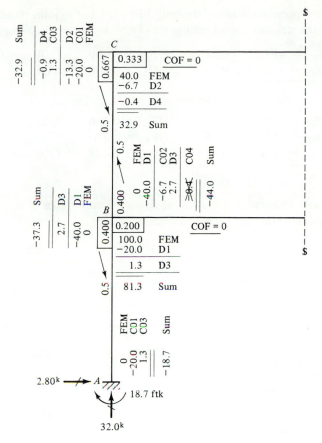

Figure 10.23c

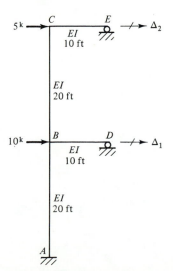

Figure 10.23d

For the antisymmetric loading, the only FEM are due to joint translations. The left half of the structure, the loading, and DOF are shown in Figure 10.23d.

$$S_{CB} = 4EI/20 = 0.2EI \qquad DF_{CB} = 0.400 \qquad COF_{CB} = 0.5$$
$$\underline{S_{CE} = 3EI/10 = 0.3EI \qquad DF_{CE} = 0.600 \qquad COF_{CB} = 0}$$

$$S_C = 0.5EI$$

$$S_{BC} = 4EI/20 = 0.2EI \qquad DF_{BC} = 0.286 \qquad COF_{BC} = 0.5$$
$$S_{BA} = 4EI/20 = 0.2EI \qquad DF_{BA} = 0.286 \qquad COF_{BA} = 0.5$$
$$\underline{S_{BD} = 3EI/10 = 0.3EI \qquad DF_{BD} = 0.428 \qquad COF_{BD} = 0}$$

$$S_B = 0.7EI$$

The FEM and moment distribution analyses for unit values of Δ_1 and Δ_2 being induced one at a time are shown in Figure 10.23e.

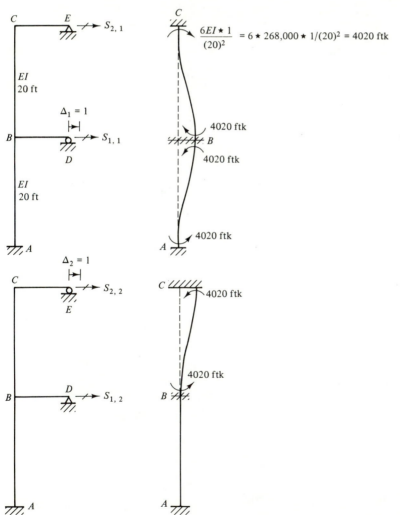

$$\frac{6EI \star 1}{(20)^2} = 6 \star 268{,}000 \star 1/(20)^2 = 4020 \text{ ftk}$$

Figure 10.23e

Moment distribution analysis for the FEM due to $\Delta_1 = 1$ is shown in Figure 10.23f. Moment distribution analysis for the FEM due to $\Delta_2 = 1$ is shown in Figure 10.23g. For the antisymmetric loading, the matrix equation of equilibrium is

$$\begin{bmatrix} 681.2 & -295.9 \\ -296.4 & 224.9 \end{bmatrix} \begin{Bmatrix} \Delta_1 \\ \Delta_2 \end{Bmatrix} = \begin{Bmatrix} 10^k \\ 5^k \end{Bmatrix}$$

$$\begin{Bmatrix} \Delta_1 \\ \Delta_2 \end{Bmatrix} = \begin{Bmatrix} 0.0570\,\text{ft} \\ 0.0972\,\text{ft} \end{Bmatrix}$$

The values in the figure due to a unit value of Δ_1 and Δ_2 are to be multiplied, respectively, by the calculated values of Δ_1 and Δ_2 and algebraically summed to obtain the final member-end moments for the antisymmetric loading component shown in Figure 10.23h.

The algebraic sum of the values obtained for the symmetric loading component and the antisymmetric loading component give the final member-end moments and reactions for the original loading as shown in Figure 10.23i.

Note that one benefit of the Direct Method is that the correct values of the translational displacements are obtained as a byproduct.

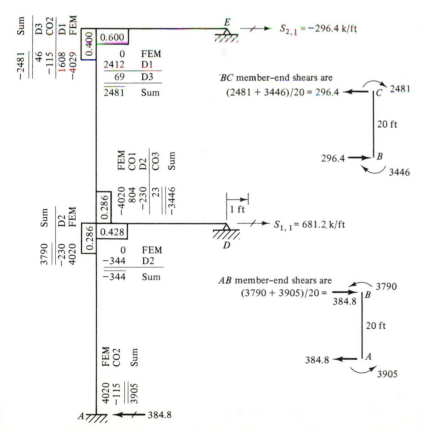

Figure 10.23f

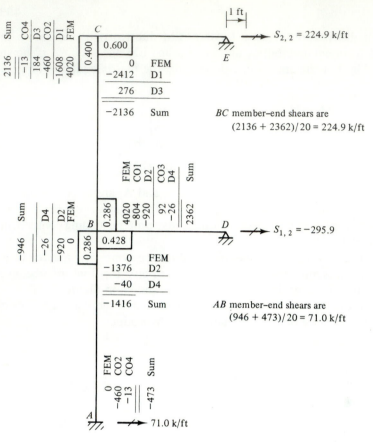

$1\ ft$

$S_{2,\,2} = 224.9\ \text{k/ft}$

Sum	CO4	D3	CO2	D1	FEM
2136	−13	184	−460	−1608	4020

C 0.400 | 0.600

	FEM
0	FEM
−2412	D1
276	D3
−2136	Sum

E

BC member-end shears are
$(2136 + 2362)/20 = 224.9\ \text{k/ft}$

FEM	CO1	D2	CO3	D4	Sum
4020	−804	−920	92	−26	2362

Sum	D4	D2	FEM
−946	−26	−920	0

B 0.286 | 0.286   0.428

D $S_{1,\,2} = -295.9$

	FEM
0	FEM
−1376	D2
−40	D4
−1416	Sum

AB member-end shears are
$(946 + 473)/20 = 71.0\ \text{k/ft}$

FEM	CO2	CO4	Sum
0	−460	−13	−473

A $71.0\ \text{k/ft}$

Figure 10.23g

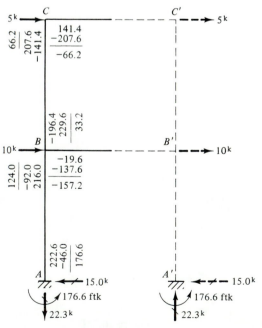

5^k C C' 5^k

66.2	207.6	−141.4

| 141.4 |
| −207.6 |
| −66.2 |

−196.4	229.6	33.2

10^k B B' 10^k

| −19.6 |
| −137.6 |
| −157.2 |

124.0	−92.0	216.0

222.6	−46.0	176.6

A 15.0^k A' 15.0^k

$176.6\ \text{ftk}$ $176.6\ \text{ftk}$

22.3^k 22.3^k

Figure 10.23h

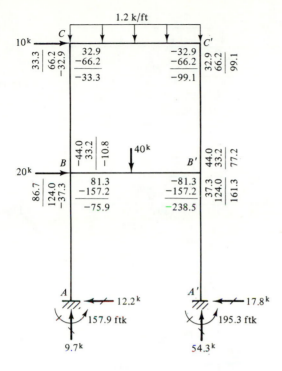

Figure 10.23i

PROBLEMS

If the member properties are not given, assume that all members are prismatic, axially rigid, and have the same *EI* value. Wherever they are applicable, use the formulas given in Figures 8.7, 8.8, 8.9, 8.11, and 8.12 to compute the member-end stiffness coefficients and fixed ended moments. Solve each problem by the Moment Distribution Method. Find the final member-end forces and the reactions. In addition to the problems shown below, Problems 9.1 through 9.22 and 9.47 through 9.54 are suggested problems. (Answers are not given for problems preceded by an asterisk.)

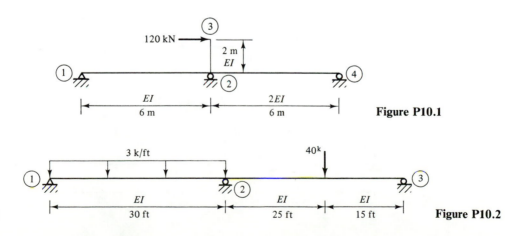

Figure P10.1

Figure P10.2

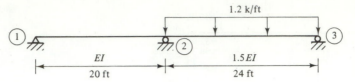

Figure P10.3

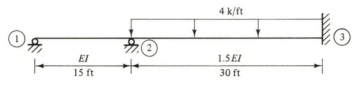

Figure P10.4

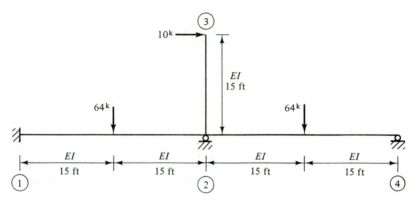

Figure P10.5

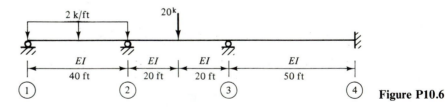

Figure P10.6

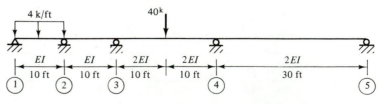

Figure P10.7

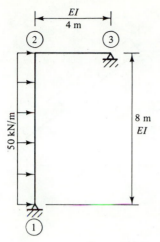

Figure P10.8

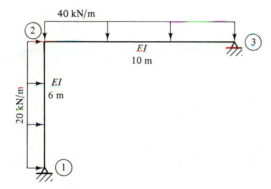

Figure P10.9

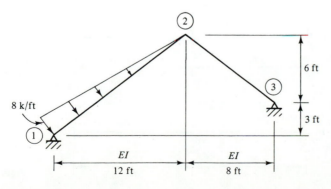

Figure P10.10

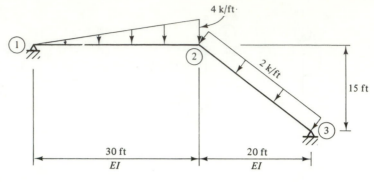

Figure P10.11

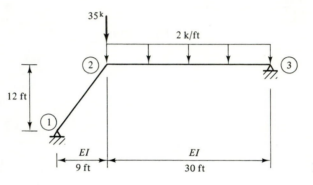

Figure P10.12

10.13. $EI = 400,000$ kft^2. Support 2 settles 0.06 ft downward. Support 3 settles 0.0524 rad CCW. No other support settlements occur.

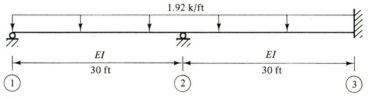

Figure P10.13

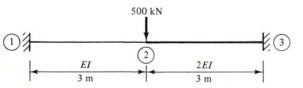

Figure P10.14

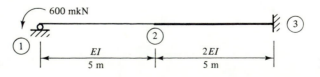

Figure P10.15

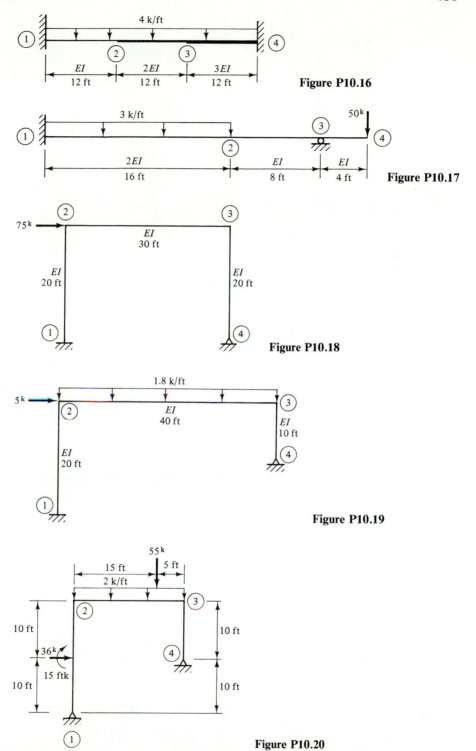

Figure P10.16

Figure P10.17

Figure P10.18

Figure P10.19

Figure P10.20

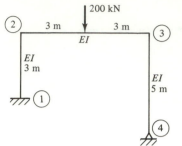

Figure P10.21

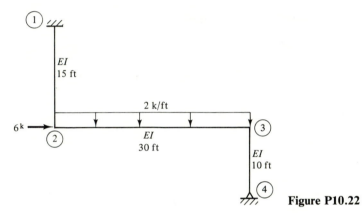

Figure P10.22

Solve the following problems using symmetry and antisymmetry concepts of moment distribution.

10.23. All members: $L = 30$ ft; $EI = 210,000$ kft^2.

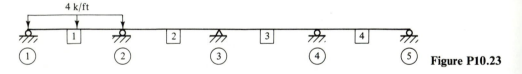

Figure P10.23

* 10.24. All members: $L = 30$ ft; $EI = 210,000$ kft^2.

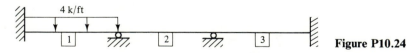

Figure P10.24

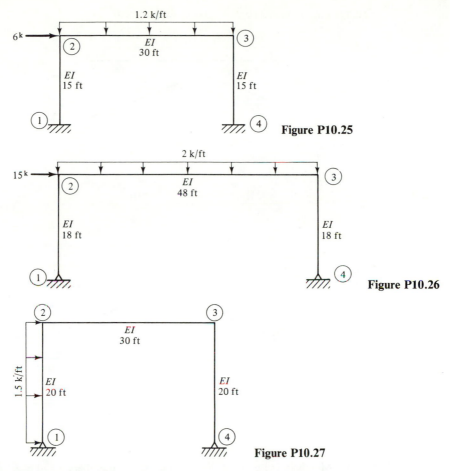

Figure P10.25

Figure P10.26

Figure P10.27

* 10.28. $EI = 276,458$ kft^2 and $EA = $ infinity for members 1 through 3. $EA = 60,000^k$ and $EI = 0$ for member 4. Comments: The solution for member 4 being deleted (inactive) and joint 4 being a fixed hinge is a symmetric structure. The solution for the effect of the elongation of member 4 can be treated as being symmetric with respect to the plane at the midlength of member 4.

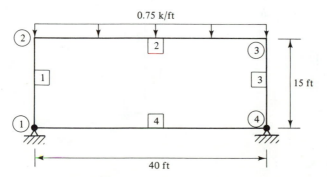

Figure P10.28

* **10.29.** For all members: $E = 29,000$ ksi; $I = 2850$ in.^{4}

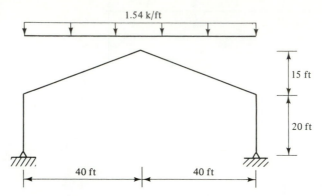

Figure P10.29

* **10.30.** Same structure as in Problem 10.29.

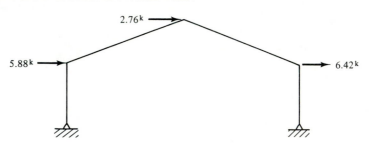

Figure P10.30

* **10.31.** Members 1 through 4: $EI/L = 13,500$ ftk; members 5 and 6: $EI/L = 10,800$ ftk.

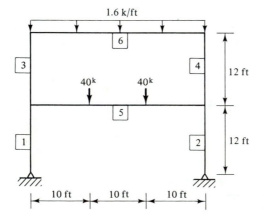

Figure P10.31

* 10.32. See Problem 10.31 for member properties and member lengths.

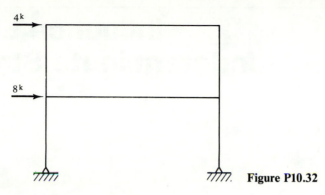

Figure P10.32

Chapter 11

Influence Lines for Indeterminate Structures

11.1 INTRODUCTION

A review of the following definitions is appropriate.

An *influence line* is the plot of a structural response function (a support reaction; an internal bending moment or shear of a beam or frame; an axial force in a truss member) due to a unit concentrated load moving along the surface of the structure where moving loads can occur after the structure is completed. *Ordinates of an influence line* are the values of the response function when the unit concentrated load is positioned on the structure's loading platform at the abscissa values of the influence line.

Influence lines aid the designer in deciding where to position the live load(s) to cause the magnitude of the structural response function to be a maximum. For determinate structures, it was found that the influence lines were either one straight line or a series of straight lines. To plot influence lines for a determinate structure, only the ordinates for the end points on each straight line segment had to be computed and then each straight line segment was drawn. However, influence lines for indeterminate structures are either curved lines or a series of chord lines. Chord-shaped influence lines occur only for trusses and floor systems in which the floor loads are transferred from the deck to stringers to floor beams to girders; the transfer of the load between the stringers and floor beams as well as between the floor beams and the girders are at finite intervals. Otherwise, the influence lines for indeterminate structures are curved lines and the influence line ordinates need to be computed at the quarter-span (or, preferably, the tenth span) points of beams to have a sufficient number of points on the curve to draw a usable quantitative influence line. The computation of ordinates on an influence line for a structural response function of an indeterminate structure involves the solution of an indeterminate problem. Consequently, to obtain influence lines for an indeterminate structure, the designer must have performed a preliminary design of the structure since the member properties are needed in solving the indeterminate problem. If the analyst has access to an indeterminate structural analysis software package and a computer to process it, generally the analyst proceeds as follows:

1. If they are needed to aid in the decision of where to place the live loads, qualitative influence lines are obtained by the Muller-Breslau principle.
2. If the live loads are of the uniformly distributed type, the qualitative influence lines generally are sufficient to enable the designer to decide where to place the live loads to maximize each structural response function of interest. As many different live load cases are devised as necessary to maximize the pertinent structural response functions.
3. If the live loads are concentrated loads, the qualitative influence lines are used to choose a neighborhood in which to search for the worst loading position of a particular configuration of concentrated live loads. The computer is given the boundaries of the chosen neighborhood, told where to start the search, and informed to step the loading configuration forward by a certain interval of distance until the neighborhood boundary is reached. Computerized solutions are obtained for the starting position and for each stepped interval. From these computerized solutions, the analyst can determine the loading configuration that causes each structural response function of interest to be a maximum.

Since the emphasis in this text is to gradually progress toward the usage of computerized solutions, only a few quantitative influence lines are obtained via noncomputerized solutions to illustrate the general procedure. Either the Force Method or the Displacement Method (or moment distribution for beams and frames) can be used to obtain quantitative influence lines. As explained in the preceding paragraph, if computerized solutions are obtainable generally only the qualitative influence lines are needed.

11.2 MULLER-BRESLAU PRINCIPLE AND QUALITATIVE INFLUENCE LINES

The Muller-Breslau principle can be stated as follows: The influence line for any structural response function is identical to the deflection curve of the structure's surface where moving loads can be applied when the response function's capability is removed or destroyed and a unit change in displacement is induced at and in the direction of the response function.

Qualitative influence lines are directly obtainable via the Muller-Breslau principle. For the structure shown in Figure 11.1, obtain qualitative influence lines for the following structural response functions: the Y-direction reactions at A and C, Y_A and Y_C; the positive bending moment at point 1, M_1; the negative bending moment at points 2 and C, $-M_2$ and $-M_C$; and the shear at points 1 and 2, V_1 and V_2.

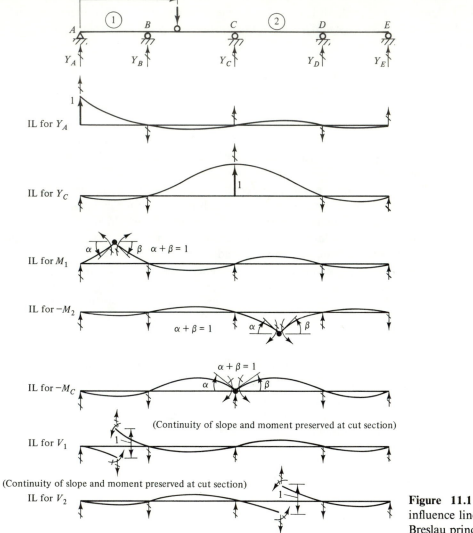

Figure 11.1 Qualitative influence lines via Muller-Breslau principle.

The qualitative influence lines for the Y-direction reactions are the easiest to describe and they are done first. By the Muller-Breslau principle, only the Y-direction reaction capability at point A, for example, is destroyed and an unit change in displacement is induced at and in the positive direction of the destroyed Y-direction reaction. The deflected shape is the qualitative influence line. To emphasize that unknown forces are required to induce the unit change in displacement, the unknown forces that are required are shown in Figure 11.1 as vectors with slashes on them. However, it must be emphasized that the unknown forces do not have to be obtained in order to sketch the qualitative influence lines.

To obtain the qualitative influence line for the bending moment at point C, for example, only the capability of the structure to resist bending moment is destroyed at point C. Therefore, an internal hinge must be inserted at point C to maintain continuity of shear, axial force, and u, v displacements. According to the Muller-Breslau principle, a unit change in displacement is induced at and in the direction of the structural function that was destroyed. Since the structural function that was destroyed was a rotational force, angle changes whose absolute values must sum to 1 rad are induced at points 1, 2, and C as shown in Figure 11.1 on the influence lines for M_1, $-M_2$, and $-M_C$, respectively. A positive moment causes compression in the topmost fiber of the beam's cross section. On the right end of a FBD a counterclockwise moment is positive, and on the left end of a FBD a clockwise moment is positive. Positive unknown moments are applied on the segment ends each side of the internal hinge which has to be inserted at the point of the destroyed structural function. The segment ends each side of the internal hinge rotate in the same direction as the unknown applied moments and the deflection curve must conform to this requirement.

The qualitative influence line for shear at point 1, for example, is obtained by destroying only the capability of the member to resist shear at point 1 and inducing changes in displacement in the directions of positive shears on the member ends each side of point 1 where shear capability was destroyed; the absolute sum of the induced changes in displacement must be unity. On the right end of a FBD a shear in the negative Y direction is positive, and on the left end of a FBD a shear in the positive Y direction is positive. Unknown positive shear forces must be applied each side of the section where the shear capability was destroyed.

In Figures 11.2 through 11.4, three examples of qualitative influence lines for a plane frame from a multistory building are shown. Generally, the structural response functions of practical importance in a design application are either near midspan of a member or at the member's ends. In general it is not possible to correctly draw the qualitative influence lines for shear or bending moment at or in the near vicinity of the quarter-span points without some computed ordinate information. The form of the influence lines for shear and bending moment changes in shape somewhere in the near proximity of the quarter-span points.

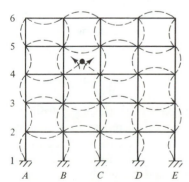

Figure 11.2 Unknown moments applied at midspan of span $B4$–$C4$.

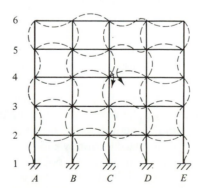

Figure 11.3 Unknown moments applied at dx to right of joint $C4$.

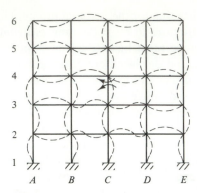

Figure 11.4 Unknown moments applied at *dy* below joint *C*4.

Qualitative influence lines for frames are generally of interest only for the purpose of deciding where to place gravity-type live loads. Consequently, in drawing qualitative influence lines for frames the joint translations are ignored and this simplifies the graphics considerably. In drawing the influence line for the positive moment at the midspan of the member between joints *B*4 and *C*4, Figure 11.2, the author began at the point of the unknown applied moments and sketched the influence line in that span. This established the rotational directions of joints *B*4 and *C*4. Joints that are neighbors of joint *B*4 rotate in the opposite direction of joint *B*4. Similarly, the opposite rotational direction process is repeated for the neighbors of joint *C*4, the neighbors of the neighboring joints for *B*4 and *C*4, and so on. To cause the positive moment at midspan of span *B*4–*C*4 to be a maximum, a uniformly distributed live load should be placed on the members that have a deflection curve with positive *Y*-direction ordinates.

Also, the author began at the locations of the unknown applied moments in sketching Figures 11.3 (IL for the negative moment at the left end of span *C*4–*D*4) and Figure 11.4 (IL for the moment that causes tension on the right side of the column ends at joint *C*4). In those figures, not all of the neighbors of a joint named *i* rotate in the opposite direction to the rotational direction of joint *i*. Note that in the *Y* direction along each column line, the rotational joint directions alternate in direction from joint to joint, but some neighboring joints in the *X* direction along a beam line do not alternate in their rotational directions. To cause the moment at the location of the unknown applied moments in Figure 11.3 or in Figure 11.4 to be a maximum, the live load should be placed only on the portion of the member spans that have a deflection curve with positive *Y*-direction ordinates. However, for all practical purposes the effect of a load on a member that is two or three bays or stories away from the point of the structural function being sought is negligible. Therefore, in a practical analysis the loading is placed entirely along all members for which the dominate portion of the deflection curve is in the positive *Y* direction. Thus, for example, members at levels 3, 4, and 5 in bay *CD* of Figure 11.3 would be entirely loaded with uniformly distributed live loads in a practical analysis, whereas levels 2 and 6 in bay *CD* would be entirely unloaded.

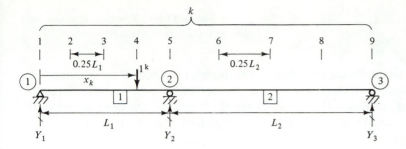

Figure 11.5 Notation and sign convention.

Vertical reactions (Y-direction components of the reactions) are positive as shown

11.3 INFLUENCE LINES FOR BEAMS

A notation and sign convention needs to be adopted as shown in Figure 11.5 for usage in the following recommended procedure of computing the ordinates of a quantitative influence line. If the influence line is to be used to make numerical calculations in an actual design problem, the ordinates of the influence line need to be computed at the quarter-span points or, preferably, at the tenth span points of each span. As shown in Figure 11.5, let k denote an arbitrary span point at which an ordinate is to be computed. Member numbers are enclosed in boxes and joint numbers are enclosed in circles. The recommended procedure is more difficult to describe for a general situation than it is to illustrate for a specific problem. Therefore, the recommended procedure is illustrated in a few example problems using either the Force Method or the Displacement Method to solve the indeterminate part of the problem.

EXAMPLE 11.1 ──

In Figure 11.6a, a unit load successively moves from point $k = 1$ to point $k = 7$. Compute and plot the ordinates of the influence lines at $k = 1$ through 7 for the vertical reaction components Y_1, Y_2, and Y_3. Members 1 and 2 are prismatic and have the same EI.

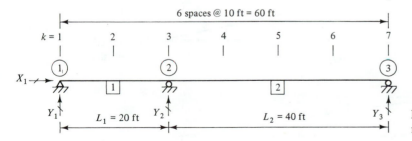

Figure 11.6a Problem statement sketch.

Solution. For a solution of this structure by the Force Method, one redundant must be chosen. As shown in Chapter 7, support moments of continuous beams

Figure 11.6b Released structure and positive redundant choice.

are computationally the best choice of redundants. Choose the moment at joint ② as the redundant, R_1 (Fig. 11.6b). In Figure 11.6a for the unit load located at any k point, note that if any one of the three vertical reactions is known the other two vertical reactions can be computed from the equations of statics:

1. Summation of moments $= 0$.
2. Summation of vertical forces $= 0$.

Therefore, the author chooses to find the ordinates of the IL for Y_2 using the Force Method and then uses statics to find the ordinates of the IL for Y_1 and IL for Y_3.

According to the Muller-Breslau principle, the deflection curve of the structure in Figure 11.6a due to a unit change in displacement induced at and in the direction of Y_2 is the IL for Y_2 (see Fig. 11.6c). Since the induced unit change in displacement in Figure 11.6c is not coincident with the chosen redundant, R_1, the Force Method solution for R_1 requires the following approach. The compatibility equation at the location of R_1 in Figure 11.6c is

$$f_{1,1} * R_1 + D_1 = D_1^s$$

$$\frac{20}{EI} * R_1 + \left(-\frac{3}{40}\right) = 0$$

$$R_1 = \frac{3EI}{800}$$

Figure 11.6c is the algebraic sum of Figure 11.6d and R_1 times Figure 11.6e; ordinates at $k = 1$ through 7 in Figure 11.6d are computed from similar triangle relations; and, ordinates at $k = 1$ through 7 for R_1 times Figure 11.6e are computed

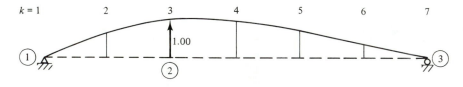

Figure 11.6c Qualitative IL for Y_2 obtained by Muller-Breslau principle.

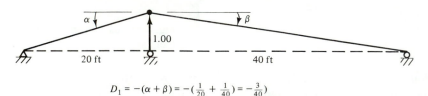

$$D_1 = -(\alpha + \beta) = -\left(\tfrac{1}{20} + \tfrac{1}{40}\right) = -\tfrac{3}{40}$$

Figure 11.6d Deformation due to induced unit change in displacement at Y_2.

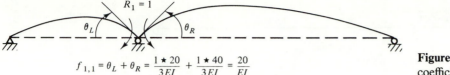

$$f_{1,1} = \theta_L + \theta_R = \frac{1 \star 20}{3EI} + \frac{1 \star 40}{3EI} = \frac{20}{EI}$$

Figure 11.6e Flexibility coefficient.

from the following formulas obtained from a conjugate beam solution. See Figure 11.6f. Therefore, the ordinates in Figure 11.6c (and in Fig. 11.6g) are

at $k = 2$:
$$0.5 + \frac{\left(\dfrac{3EI}{800}\right) \star 10 \star [(20)^2 - (10)^2]}{6EI \star 20} = \frac{19}{32} = 0.594$$

at $k = 4$:
$$0.75 + \frac{\left(\dfrac{3EI}{800}\right) \star 10 \star (40 - 10) \star (80 - 10)}{6EI \star 40} = \frac{69}{64} = 1.078$$

at $k = 5$:
$$0.5 + \frac{\left(\dfrac{3EI}{800}\right) \star 20 \star (40 - 20) \star (80 - 20)}{6EI \star 40} = \frac{7}{8} = 0.875$$

at $k = 6$:
$$0.25 + \frac{\left(\dfrac{3EI}{800}\right) \star 30 \star (40 - 30) \star (80 - 30)}{6EI \star 40} = \frac{31}{64} = 0.484$$

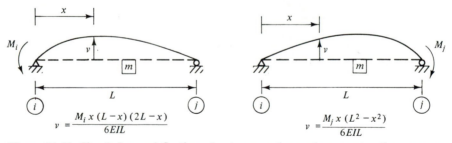

$$v = \frac{M_i x (L - x)(2L - x)}{6EIL} \qquad v = \frac{M_j x (L^2 - x^2)}{6EIL}$$

Figure 11.6f Simple beam deflections due to a member-end moment acting on member m.

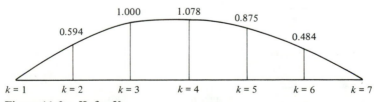

Figure 11.6g IL for Y_2.

$$\overset{\curvearrowright}{\Sigma M_{\text{③}}} = 60Y_1 + 40 \star 0.594 - 50 \star 1 = 0; \quad Y_1 = 0.437^{\text{k}}$$

$$\overset{\text{+}}{\Sigma F_Y} = Y_3 + 0.437 + 0.594 - 1 = 0; \quad Y_3 = -0.031^{\text{k}}$$

Figure 11.6h Free body diagram and equilibrium when the unit load is at $k = 2$.

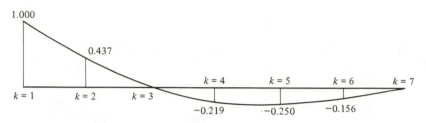

Figure 11.6i IL for Y_1.

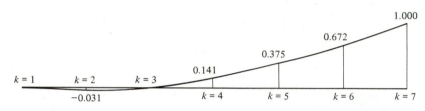

Figure 11.6j IL for Y_3.

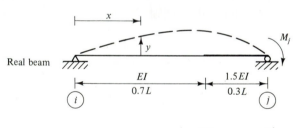

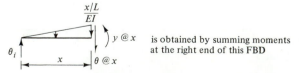

is obtained by summing moments at the right end of this FBD

Figure 11.6k

472

The ordinates of the IL for Y_1 and the IL for Y_3 are obtained by using the Muller-Breslau principle and the following procedure illustrated for $k = 2$. When the unit load is positioned at $k = 2$, for example, $Y_2 = 0.594$ is obtained from Figure 11.6g and the ordinates at $k = 2$ in Figures 11.6i and 11.6j are computed from the equations of statics as shown in Figure 11.6h. Similarly, the equations of statics are used to compute the ordinates at $k = 4$ through 6 in Figures 11.6i and 11.6j when the unit load is positioned at $k = 4$ through 6.

It should be emphasized that the formulas in Figure 11.6f are not valid unless EI is constant within each span of a continuous beam structure. If EI is not constant within a span, it is recommended that the Conjugate Beam Method be used either to numerically obtain the specifically desired ordinates or to derive formulas that are applicable for the actual variation in EI. The initial stages of the recommended approach are illustrated in Figure 11.6k.

EXAMPLE 11.2

In Figure 11.7a members 1 through 3 are prismatic and have the same EI. A unit load successively moves from point $k = 1$ to point $k = 10$. The following itemized tasks are to be performed in this example:

1. Choose the internal moments at support joints ② and ③ as the redundants (see Fig. 11.7b) to compute the ordinates at points $k = 1$ through 10 of the IL for the vertical, interior reaction components Y_2 and Y_3.
2. Use statics to compute the ordinates of the IL for the exterior, vertical reaction components Y_1 and Y_4.
3. Use statics to compute the ordinates of the IL for M_6 (internal moment at $k = 6$), V_7 (internal shear at $k = 7$), and M_8 (internal moment at $k = 8$).

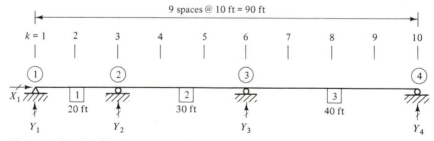

Figure 11.7a Problem statement sketch.

Solution. The matrix equation of compatibility is

$$[F]\{R\} + \{D\} = \{D^s\}$$

$$\frac{1}{3EI}\begin{bmatrix} 50 & 15 \\ 15 & 70 \end{bmatrix}\begin{Bmatrix} R_1 \\ R_2 \end{Bmatrix} + \begin{Bmatrix} -\frac{1}{12} & \frac{1}{30} \\ \frac{1}{30} & -\frac{7}{120} \end{Bmatrix} = \begin{Bmatrix} 0 & 0 \\ 0 & 0 \end{Bmatrix}$$

For loading 1 $\downarrow$ $\uparrow$ For loading 2

$$\begin{Bmatrix} R_1 \\ R_2 \end{Bmatrix} = \frac{EI}{3275}\begin{Bmatrix} 19 & -9.625 \\ -8.75 & 10.25 \end{Bmatrix}$$

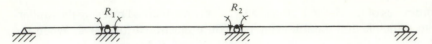

Figure 11.7b Released structure and positive rendundants.

The IL for Y_2 (and Y_3) is the algebraic sum of Figure 11.7c (Figure 11.7d for Y_3), R_1 times Figure 11.7e, and R_2 times Figure 11.7f. Ordinates at $k = 1$ through 10 in Figures 11.7c and 11.7d are computed from similar triangle relations. Ordinates at $k = 1$ through 10 in Figures 11.7e and 11.7f are computed from the formulas given in Figure 11.6f; for example, the ordinate at $k = 4$ of the IL for Y_2 and the IL for Y_3 is:

1. For Y_2:
$$y_4 = \frac{2}{3} + \frac{\left(\dfrac{19EI}{3275}\right) * 10 * (30 - 10) * (60 - 10)}{6EI * 30}$$
$$+ \frac{\left(\dfrac{-8.75EI}{3275}\right) * 10 * [(30)^2 - (10)^2]}{6EI * 30} = 0.870$$

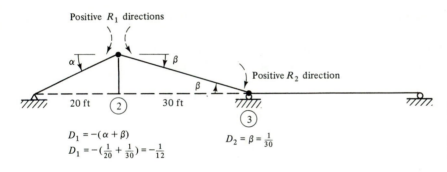

$$D_1 = -(\alpha + \beta)$$
$$D_1 = -\left(\tfrac{1}{20} + \tfrac{1}{30}\right) = -\tfrac{1}{12}$$

$$D_2 = \beta = \tfrac{1}{30}$$

Figure 11.7c Loading 1—induced unit displacement at Y_2 in the released structure.

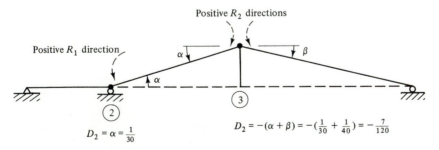

$$D_2 = \alpha = \tfrac{1}{30}$$

$$D_2 = -(\alpha + \beta) = -\left(\tfrac{1}{30} + \tfrac{1}{40}\right) = -\tfrac{7}{120}$$

Figure 11.7d Loading 2—induced unit displacement at Y_3 in the released structure.

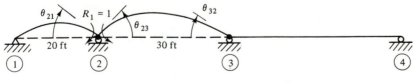

$$f_{1,1} = \theta_{21} + \theta_{23} = \frac{1 * 20}{3EI} + \frac{1 * 30}{3EI} = \frac{50}{3EI}; \quad f_{2,1} = \theta_{32} = \frac{1 * 30}{6EI} = \frac{15}{3EI}$$

Figure 11.7e Column 1 of $[F]$, the flexibility matrix.

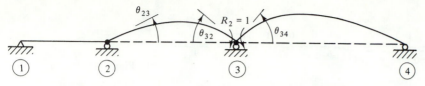

$$f_{1,2} = \theta_{23} = f_{2,1} = \frac{15}{3EI} \; ; \quad f_{2,2} = \theta_{32} + \theta_{34} = \frac{1 * 30}{3EI} + \frac{1 * 40}{3EI} = \frac{70}{3EI}$$

Figure 11.7f Column 2 of $[F]$.

2. For Y_3: $\quad y_4 = \dfrac{1}{3} + \dfrac{\left(\dfrac{-9.625EI}{3275}\right)*10*(30-10)*(60-10)}{6EI*30}$

$$+ \frac{\dfrac{10.25EI}{3275}*10*[(30)^2-(10)^2]}{6EI*30} = 0.309$$

The IL for Y_1 and Y_4 are obtained by statics as shown in Figures 11.7i–11.7k; the procedure is illustrated for the ordinates at $k = 5$. The IL for M_6, V_7, and

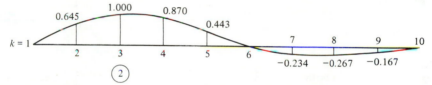

Figure 11.7g IL for Y_2.

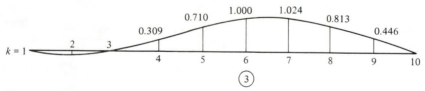

Figure 11.7h IL for Y_3.

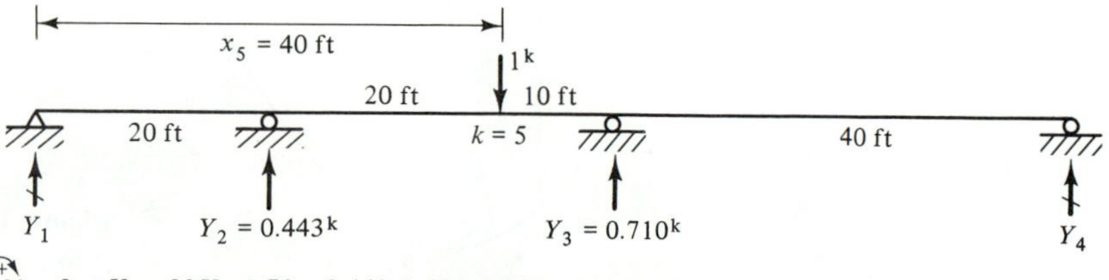

$x_5 = 40$ ft

20 ft $\quad$ 10 ft

$k = 5$

20 ft $\quad\quad$ 40 ft

$Y_1 \quad\quad Y_2 = 0.443^k \quad\quad Y_3 = 0.710^k \quad\quad Y_4$

$\overset{(+)}{\Sigma M} = 0$ at Y_4: $\quad 90 Y_1 + 70 * 0.443 + 40 * 0.710 - 50 * 1 = 0$

$\qquad Y_1 = -0.105^k$ (-0.104326^k using 6–digit precision for Y_2, Y_3, etc)

$\overset{+\uparrow}{\Sigma F_Y} = 0$: $\quad Y_4 + 0.710 + 0.443 + (-0.105) - 1 = 0$

$\qquad Y_4 = -0.0480^k$ (-0.0483460^k using 6–digit precision for Y_2, Y_3, etc)

Figure 11.7i

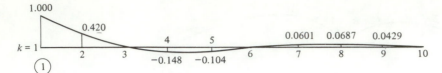

Figure 11.7j

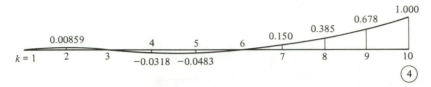

Figure 11.7k

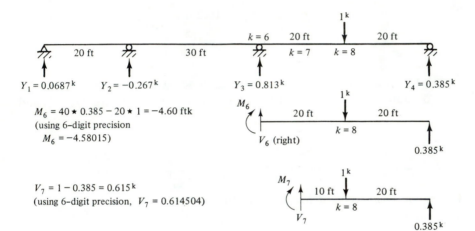

$M_6 = 40 \star 0.385 - 20 \star 1 = -4.60$ ftk
(using 6–digit precision
$M_6 = -4.58015$)

$V_7 = 1 - 0.385 = 0.615^k$
(using 6–digit precision, $V_7 = 0.614504$)

$M_8 = 20 \star 0.385 = 7.70$ ftk
(using 6–digit precision, $M_8 = 7.70992$)

Figure 11.7l Free body diagrams and equilibrium for unit load at $k = 8$.

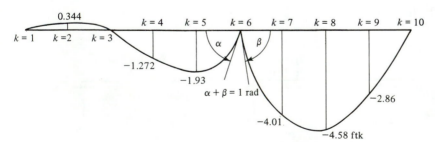

Figure 11.7m IL for M_6 (moment at $k = 6$).

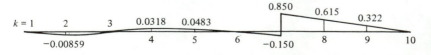

Figure 11.7n IL for V_7 (shear at $k = 7$).

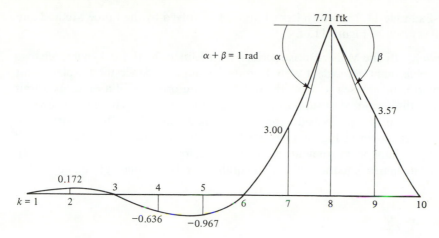

Figure 11.7o IL for M_8 (moment at $k = 8$).

M_8 are obtained by statics; the procedure is illustrated for the ordinates at $k = 8$. See Figures 11.7l–11.7o.

EXAMPLE 11.3 _____

In Figure 11.8a members 1 through 3 are prismatic and have the same EI. A unit load successively moves from point $k = 1$ to point $k = 10$. The objectives of this example are to:

1. Use the Displacement Method and the formulas given in Figure 11.6f to compute and plot the influence lines for the interior vertical reaction components, Y_2 and Y_3.
2. Use statics to compute and plot the IL for Y_1 and Y_4.
3. Use statics to compute and plot the IL for M_6 (internal moment at $k = 6$), V_7 (internal shear at $k = 7$), and M_8 (internal moment at $k = 8$).

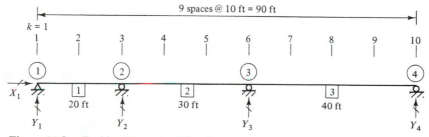

Figure 11.8a Problem statement sketch.

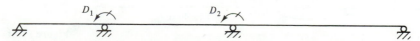

Figure 11.8b DOF (positive system displacements).

Note: See Example 11.2 in which item 1 above was solved by the Force Method and the formulas given in Figure 11.6f.

Solution. By the Muller-Breslau principle, to obtain the IL for Y_2 a unit change in displacement coincident with Y_2 must be induced. Since the Displacement Method is being used to solve the indeterminate problem of a prescribed unit vertical displacement at support 2 and since the DOFs are chosen as shown in Figure 11.8b, the prescribed support movement must be induced with $D_1 = D_2 = 0$ as shown in Figure 11.8c for loading 1. Loading 2 (Fig. 11.8d) is the prescribed support movement to obtain the IL for Y_3.

The matrix equation of system equilibrium is (see Figs. 11.8e and 11.8f)

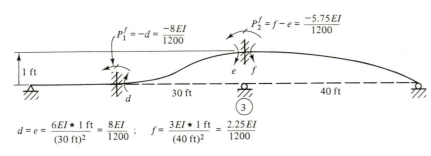

$$a = \frac{3EI \star 1 \text{ ft}}{(20 \text{ ft})^2} = \frac{3EI}{400}; \quad b = c = \frac{6EI \star 1 \text{ ft}}{(30 \text{ ft})^2} = \frac{6EI}{900} = \frac{8EI}{1200}$$

Figure 11.8c Loading 1—fixed joint loads due to induced unit displacement at reaction 2.

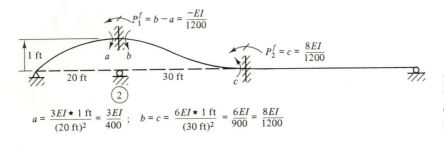

$$d = e = \frac{6EI \star 1 \text{ ft}}{(30 \text{ ft})^2} = \frac{8EI}{1200}; \quad f = \frac{3EI \star 1 \text{ ft}}{(40 \text{ ft})^2} = \frac{2.25EI}{1200}$$

Figure 11.8d Loading 2—fixed joint loads due to induced unit displacement at reaction 3.

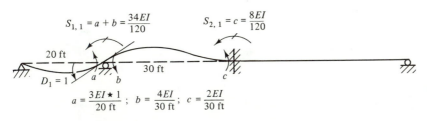

$$a = \frac{3EI \star 1}{20 \text{ ft}}; \quad b = \frac{4EI}{30 \text{ ft}}; \quad c = \frac{2EI}{30 \text{ ft}}$$

Figure 11.8e Column 1 of $[S]$.

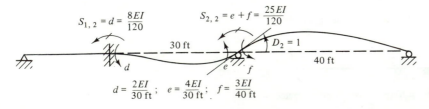

$$d = \frac{2EI}{30 \text{ ft}}; \quad e = \frac{4EI}{30 \text{ ft}}; \quad f = \frac{3EI}{40 \text{ ft}}$$

Figure 11.8f Column 2 of $[S]$.

$$[S]\{D\} = \{P^e\} \quad \overset{\text{loading 1}}{\underset{\text{loading 2}}{\downarrow \downarrow}}$$

$$\frac{EI}{120}\begin{bmatrix} 34 & 8 \\ 8 & 25 \end{bmatrix}\begin{Bmatrix} D_1 \\ D_2 \end{Bmatrix} = \frac{EI}{1200}\begin{Bmatrix} 1 & 8 \\ -8 & 5.75 \end{Bmatrix}$$

$$\begin{Bmatrix} D_1 \\ D_2 \end{Bmatrix} = \begin{Bmatrix} 0.0113232 & 0.0195929 \\ -0.0356234 & 0.0167303 \end{Bmatrix}$$

The final member-end moments are

member 1:   loading 1: $F_6 = \dfrac{-3EI}{400} + \dfrac{3EI}{20}*0.0113232 = -0.0058015EI$

loading 2: $F_6 = 0 + \dfrac{3EI}{20}*0.0195929 = 0.0029389EI$

member 3:   loading 1: $F_3 = 0 + \dfrac{3EI}{40}*(-0.0356234) = -0.0026718EI$

loading 2: $F_3 = \dfrac{3EI}{1600} + \dfrac{3EI}{40}*0.0167303 = 0.0031298EI$

member 2:   loading 1: $F_3 = 0.0058015EI$
$F_6 = 0.0026718EI$

loading 2: $F_3 = -0.0029389EI$
$F_6 = -0.0031298EI$

The influence lines for Y_2 and Y_3 are obtained by superposition as shown in Figure 11.8g.

IL for Y_2:

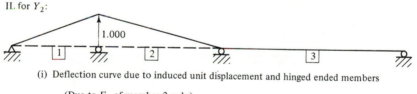

1.000

(i) Deflection curve due to induced unit displacement and hinged ended members

(Due to F_3 of member 2 only)

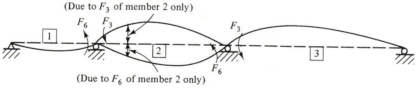

F_6 F_3 F_3

F_6

(Due to F_6 of member 2 only)

(ii) Deflection curves due to the loading 1 final member-end moments on hinged ended members

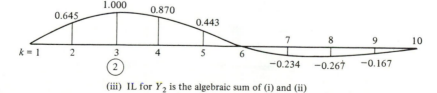

0.645   1.000   0.870   0.443

$k = 1$ 2   3   4   5   6   7   8   9   10

(2)

−0.234   −0.267   −0.167

(iii) IL for Y_2 is the algebraic sum of (i) and (ii)

Figure 11.8g Superposition method of obtaining IL for Y_2.

Ordinates of Figure 11.8g at $k = 1$ through 10 are obtained for each member-end moment from the formulas given in Figure 11.6f; for example, the ordinate at $k = 4$ of the IL for Y_2 is

$$y_4 = \frac{2}{3} + \frac{0.0058015EI * 10 * (30 - 10) * (60 - 10)}{6EI * 30}$$
$$+ \frac{0.0026718EI * 10 * [(30)^2 - (10)^2]}{6EI * 30} = 0.870$$

Similarly, the IL for Y_3 (Fig. 11.8h) can be obtained by superposition; for example, the ordinate at $k = 4$ of the IL for Y_3 is

$$y_4 = \frac{1}{3} + \frac{(-0.0029389EI) * 10 * (30 - 10) * (60 - 10)}{6EI * 30}$$
$$+ \frac{(-0.0031298EI) * 10 * [(30)^2 - (10)^2]}{6EI * 30} = 0.309$$

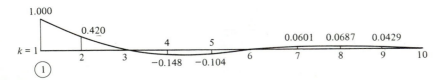

Figure 11.8h IL for Y_3.

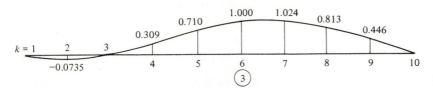

$\stackrel{\curvearrowright}{\Sigma M} = 0$ at Y_4: $90 Y_1 + 70 \star 0.443 + 40 \star 0.710 - 50 \star 1 = 0$

$Y_1 = -0.105^k$ (−0.104326^k using 6-digit precision for Y_2, Y_3, etc)

$\stackrel{+\uparrow}{\Sigma F_Y} = 0$: $Y_4 + 0.710 + 0.443 + (-0.105) - 1 = 0$

$Y_4 = -0.0480^k$ (−0.0483460^k using 6-digit precision for Y_2, Y_3, etc)

Figure 11.8i Free body diagrams and equilibrium for unit load at $k = 8$.

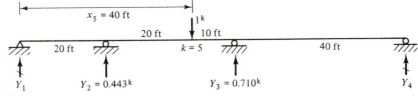

Figure 11.8j IL for M_6 (moment at $k = 6$).

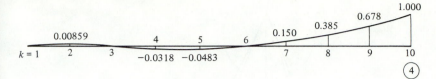

Figure 11.8k IL for V_7 (shear at $k = 7$).

The IL for Y_1 and Y_4 are obtained by statics as shown in Figures 11.8l, 11.8m, 11.8n; the procedure is illustrated for the ordinates at $k = 5$. The IL for M_6, V_7, and M_8 are obtained by statics; the procedure is illustrated for the ordinates at $k = 8$ (Fig. 11.8o).

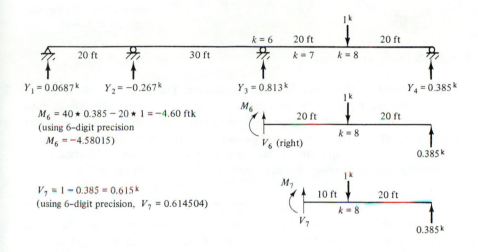

$M_6 = 40 \star 0.385 - 20 \star 1 = -4.60$ ftk
(using 6-digit precision
  $M_6 = -4.58015$)

$V_7 = 1 - 0.385 = 0.615^k$
(using 6-digit precision, $V_7 = 0.614504$)

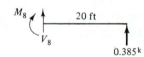

$M_8 = 20 \star 0.385 = 7.70$ ftk
(using 6-digit precision, $M_8 = 7.70992$)

Figure 11.8l Free body diagram and equilibrium equations.

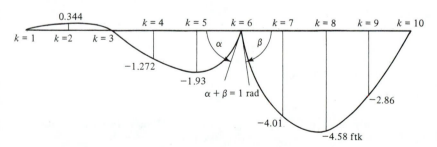

Figure 11.8m IL for Y_1.

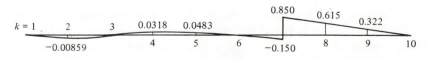

Figure 11.8n IL for Y_4.

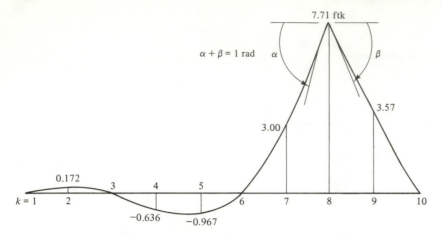

Figure 11.8o IL for M_8 (moment at $k = 8$).

11.4 INFLUENCE LINES FOR TRUSSES

Two examples solved by the Force Method are shown. The first example is for an externally indeterminate truss and the second example is for an internally indeterminate truss. The author chose to use the Force Method to solve the indeterminate problem in the illustrated solutions. However, these two examples appear in the list of suggested practice problems with the requirement that the author's Displacement Method computer program for trusses be used to obtain the same influence lines.

EXAMPLE 11.4 _____

Successively position a unit load at the bottom panel points of the truss shown in Figure 11.9a; compute and plot the IL for Y_4, then use statics to obtain the IL for Y_1, Y_8, F_1, F_2, and F_{10} (F_1, F_2, F_{10} are the forces in members 1, 2, and 10, respectively). For members 1 through 8, $A = 4$ in.^{2} and for members 9 through 13, $A = 3$ in.^{2} $E = 29,000$ ksi for all members.

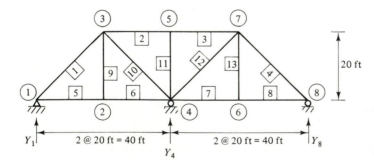

Figure 11.9a Problem statement sketch.

Solution. Choose Y_4 as the redundant, R_1. By the Muller-Breslau principle, the IL for Y_4 is the deflection curve of the panel points at which the unit load is successively positioned when a unit change in displacement is induced coincident with Y_4 (and R_1) (Fig. 11.9b).

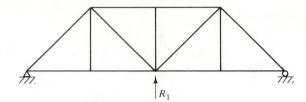

Figure 11.9b Released structure and positive redundant.

The vertical deflection at and in the direction of R_1 in Figure 11.9c is the flexibility coefficient, $f_{1,1}$. Since $f_{1,1}$ is computed by virtual work, a unit virtual load must be applied at and in the direction of $f_{1,1}$ which is where $R_1 = 1$ is applied in Figure 11.9c. Therefore, the member forces in Figure 11.9c are also the virtual member forces, $\bar{F}_i$, due to the unit virtual load which needs to be applied to compute

$$f_{1,1} = \sum_{i=1}^{13} \bar{F}_i * \left(\frac{FL}{EA}\right)_i$$

Member i	L (in.)	A (in.^{2})	F_i	$\bar{F}_i$	$\bar{F}_i * \left(\Delta L = \frac{FL}{EA}\right)_i$
1	339.4	4	0.707	0.707	0.001463
2	240	4	1	1	0.002068
3	240	4	1	1	0.002068
4	339.4	4	0.707	0.707	0.001463
5	240	4	−0.5	−0.5	0.000517
6	240	4	−0.5	−0.5	0.000517
7	240	4	−0.5	−0.5	0.000517
8	240	4	−0.5	−0.5	0.000517
9	240	3	0	0	0
10	339.4	3	−0.707	−0.707	0.001951
11	240	3	0	0	0
12	339.4	3	−0.707	−0.707	0.001951
13	240	3	0	0	0
					$f_{1,1} = 0.013032$

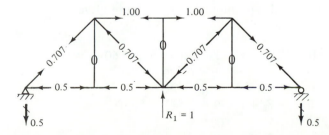

Figure 11.9c Real member forces, F_i, due to $R_1 = 1$.

The compatibility equation is

$$f_{1,1} * R_1 + D_1 = D_1^s$$
$$0.013032 * R_1 + 0 = 1.000$$
$$R_1 = 76.734^k$$

Therefore, if $Y_4 = 76.734^k$ is applied to the released structure, the vertical deflection at Y_4 is 1.000 in. upward and the member forces, F_i, are 76.734 times those in Figure 11.9c. To obtain the vertical deflection at ②, a unit virtual load must be applied at ② on the released structure.

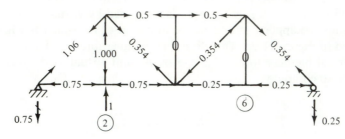

Figure 11.9d $\bar{F}_i$ due to unit virtual load applied at ②.

Member i	L (in.)	A (in.^{2})	F_i	$\bar{F}_i$	$\bar{F}_i * \left(\Delta L = \dfrac{FL}{EA} \right)_i$
1	339.4	4	54.3	1.06	0.16839
2	240	4	76.7	0.5	0.07938
3	240	4	76.7	0.5	0.07938
4	339.4	4	54.3	0.354	0.05613
5	240	4	−38.4	−0.75	0.05954
6	240	4	−38.4	−0.75	0.05954
7	240	4	−38.4	−0.25	0.01985
8	240	4	−38.4	−0.25	0.01985
9	240	3	0	−1	0
10	339.4	3	−54.3	0.354	−0.07484
11	240	3	0	0	0
12	339.4	3	−54.3	−0.354	0.07484
13	240	3	0	0	0
					$y_2 = 0.542$

Since the structure in Figure 11.9d is symmetric with respect to midspan, if the virtual unit load were applied at ⑥ instead of at ②, $y_6 = y_2$ (Fig. 11.9e).

The IL for Y_1 and Y_8 (Fig. 11.9f) are obtained by statics. For example, when the unit load is positioned at ② in Figure 11.9a:

1. $\overset{\curvearrowright +}{\sum M} = 0$ @ Y_8: $80 * Y_1 + 40 * 0.542 - 60 * 1 = 0$
$$Y_1 = 0.479$$

2. $\overset{+\uparrow}{\sum F_Y} = 0$: $Y_8 + 0.479 + 0.542 - 1 = 0$
$$Y_8 = -0.021$$

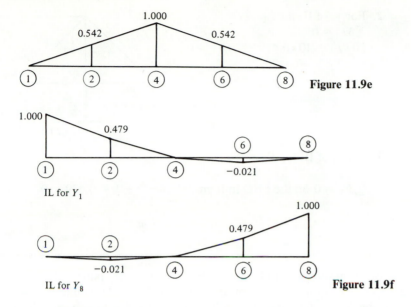

Figure 11.9e

IL for Y_1

IL for Y_8

Figure 11.9f

Influence lines for the force in members 1, 2, and 10 (IL for F_1, F_2, and F_{10}) are obtained by statics (Fig. 11.9g). The procedure is illustrated for the unit load positioned at ② in Figure 11.9a:

1. For y_2 of IL for F_1

$$\Sigma F_y = 0: \quad \frac{F_1}{\sqrt{2}} + 0.479 = 0$$

$$F_1 = -0.479\sqrt{2} = -0.677$$

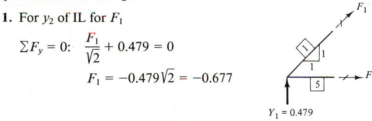

$Y_1 = 0.479$

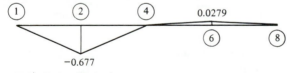

IL for F_1 (tension is +)

IL for F_2 (tension is +)

IL for F_{10} (tension is +)

Figure 11.9g

2. For y_2 of IL for F_2

$\Sigma M_\circledcirc = 0$:

$20 * F_2 + 40 * 0.479 - 20 * 1 = 0$

$\quad F_2 = -0.0420$

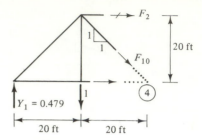

3. For y_2 of IL for F_{10}

$\overset{+\downarrow}{\Sigma F_Y} = 0$ on the FBD in item 2: $\dfrac{F_{10}}{\sqrt{2}} + 1 - 0.479 = 0$

$$F_{10} = -0.737$$

EXAMPLE 11.5 ———————————————————————————————————

Successively position a unit load at the bottom panel points of the truss shown in Figure 11.10a; compute and plot the IL for the force in member 8, then use statics to obtain the IL for the force in members 5 and 10. For members 1 through 6, $A = 2$ in.² and for members 7 through 9, $A = 1$ in.² $E = 29{,}000$ ksi for all members.

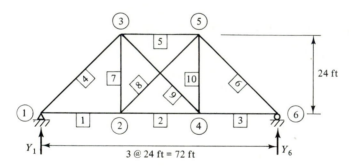

Figure 11.10a   Problem statement sketch.

Solution. Choose the force in member 8 as the redundant, R_1. By the Muller-Breslau principle, the IL for F_8 (the force in member 8) is the deflection curve of the panel points at which the unit load is successively positioned when a unit change in displacement is induced coincident with F_8 (and R_1). For the Force Method solution of R_1, the induced unit change in displacement corresponds to $D_1^s = 1$ (Fig. 11.10b).

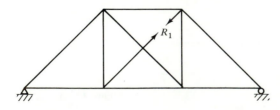

Figure 11.10b   Released structure and positive redundant.

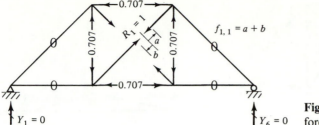

Figure 11.10c Real member forces, F_i, due to $R_1 = 1$.

To compute $f_{1,1}$ a pair of unit virtual forces must be applied coincident with R_1. Therefore, the member forces in Figure 11.10c are also the virtual member forces, $\bar{F}_i$.

Member i	L (in.)	A (in.²)	F_i	$\bar{F}_i$	$\bar{F}_i * \left(\Delta L = \dfrac{FL}{EA} \right)_i$
1	288	2	0	0	0.002483
2	288	2	−0.707	−0.707	0
3	288	2	0	0	0
4	288	2	0	0	0
5	288	2	−0.707	−0.707	0.002483
6	288	2	0	0	0
7	288	1	−0.707	−0.707	0.004966
8	407.3	1	1	1	0.014045
9	407.3	1	1	1	0.014045
10	288	1	−0.707	−0.707	0.004966
					$f_{1,1} = 0.04299$

The compatibility equation is

$$f_{1,1} * R_1 + D_1 = D_1^s$$
$$0.04299 * R_1 + 0 = 1.000 \text{ in.}$$
$$R_1 = 23.26$$

To obtain the vertical deflection in Figure 11.10b at ② and ④ when $R_1 = 23.26$ is applied to induce the unit change in displacement coincident with R_1, as shown in Figures 11.10d and 11.10e, a unit virtual load must be applied separately

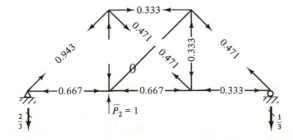

Figure 11.10d $\bar{F}_i$ due to $\bar{P}_2 = 1$.

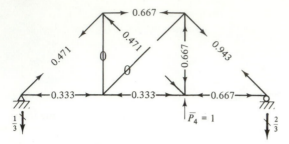

Figure 11.10e $\bar{F}_i$ due to $\bar{P}_4 = 1$.

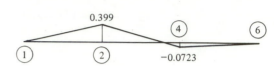

Figure 11.10f IL for F_8 (tension is + in member 8).

at ② and ④ in Figure 11.10b (note that the real member forces are 23.26 times those in Figure 11.10c). These computed displacements, y_2 and y_4, are the ordinates of the IL for F_8 (see Fig. 11.10f).

Member i	L (in.)	A (in.²)	F_i	For $\bar{P}_2 = 1$ (Fig. 11.10d) $\bar{F}_i$	$\bar{F}_{i*}\left(\dfrac{FL}{EA}\right)_i$	For $\bar{P}_4 = 1$ (Fig. 11.10e) $\bar{F}_i$	$\bar{F}_{i*}\left(\dfrac{FL}{EA}\right)_i$
1	288	2	0	−0.667	0	−0.333	0
2	288	2	−16.45	−0.667	0.05446	−0.333	0.02723
3	288	2	0	−0.333	0	−0.667	0
4	288	2	0	0.943	0	0.471	0
5	288	2	−16.45	0.333	−0.02723	0.667	−0.05446
6	288	2	0	0.471	0	0.943	0
7	288	1	−16.45	−1	0.16337	0	0
8	407.3	1	23.26	0	0	0	0
9	407.3	1	23.26	0.471	0.15400	−0.471	−0.15400
10	288	1	−16.45	−0.333	0.05446	−0.667	0.10891
					$y_2 = 0.39906$		$y_4 = -0.07232$

IL for F_5 (tension is + in member 5)

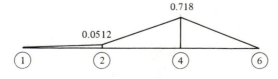

IL for F_{10} (tension is + in member 10)

Figure 11.10g Influence lines for force in members 5 and 10.

Influence lines for the force in members 5 and 10 are obtained by statics (Fig. 11.10g). The procedure is illustrated for the unit load positioned at ② in Figure 11.10a:

1. For y_2 of IL for F_5 (force in member 5).

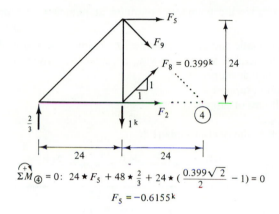

$$\overset{+}{\Sigma M}_{④} = 0: \quad 24 \star F_5 + 48 \star \frac{2}{3} + 24 \star \left(\frac{0.399\sqrt{2}}{2} - 1\right) = 0$$

$$F_5 = -0.6155^k$$

2. For y_2 of IL for F_{10} (force in member 10).

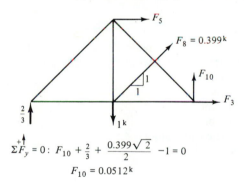

$$\overset{+\uparrow}{\Sigma F}_y = 0: \quad F_{10} + \frac{2}{3} + \frac{0.399\sqrt{2}}{2} - 1 = 0$$

$$F_{10} = 0.0512^k$$

PROBLEMS

In Problems 11.1 through 11.5, use the method of indeterminate analysis specified by the instructor and the formulas in Figure 11.6f to obtain the indicated ordinates of influence lines in each problem.

11.1. All members are prismatic and have the same *EI*. At intervals of 7.5 ft, determine the ordinates of the influence line for the following structural response functions:
   (a) Reaction at support 2.
   (b) Reaction at support 3.
   (c) Shear at the left face of support 2.
   (d) Moment at 15 ft from support 1.
   (e) Moment at 15 ft from support 3.
   (f) Moment at support 2.
   (g) Moment at support 3.

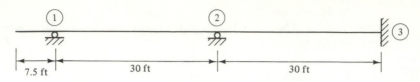

Figure P11.1

11.2. In Problem 11.1, change the boundary condition at joint 3 to a hinge and delete item (g).

11.3. All members are prismatic and have the same *EI*. At intervals of 6 ft, determine the ordinates of the influence line for the following structural response functions:
  (a) Horizontal reaction at support 1.
  (b) Vertical reaction at support 1.
  (c) Shear at the left face of support 3.
  (d) Moment at support 3.
  (e) Moment at 12 ft to the right of support 2.

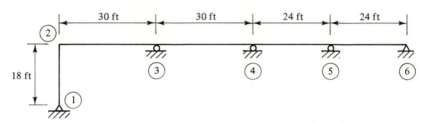

Figure P11.3

11.4. All members are prismatic and have the same *EI*. At intervals of 6 ft, determine the ordinates of the influence line for the following structural response functions:
  (a) Horizontal reaction at support 3.
  (b) Vertical reaction at support 3.
  (c) Shear at the left face of joint 4.
  (d) Moment at the right face of joint 4.
  (e) Moment at the top of the column at joint 4.
  (f) Moment at 12 ft to the right of joint 4.

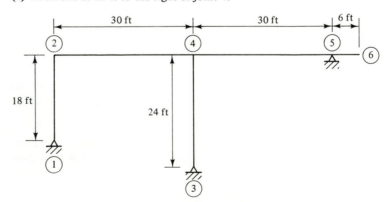

Figure P11.4

11.5. Change the hinge at joint 5 of Problem 11.4 to a roller.

In the following problems, use the computerized method of indeterminate analysis specified by the instructor and the formulas in Figure 11.6f where applicable to obtain the indicated ordinates of the influence lines in each problem.

11.6. Solve Problem 11.1.

11.7. Solve Problem 11.2.

11.8. Solve Problem 11.3.

11.9. Solve Problem 11.4.

11.10. Solve Problem 11.5.

11.11. See Example 7.19 and delete the center reaction. The platform on which the wheel loads roll is attached to the bottom panel points of the truss. Determine the ordinates of the influence line at the panel points for the following structural response functions:
   (a) Force in member 2.
   (b) Force in member 12.
   (c) Force in member 6.

11.12. See Example 7.21. The platform on which the wheel loads roll is attached to the bottom panel points of the truss. Determine the ordinates of the influence line at the panel points for the following structural response functions:
   (a) Force in member 12.
   (b) Force in member 13.

11.13. Solve Example 11.4 (verify the influence line ordinates of Example 11.4).

11.14. Solve Example 11.5 (verify the influence line ordinates of Example 11.5).

11.15. See Example 7.24 in which the wheel loads roll on members 1 through 3. At intervals of 5 ft, obtain the ordinates of the following influence lines: (a) Y_C; (b) M_B; (c) moment at midspan of member 2.

Chapter 12

Approximate Methods of Indeterminate Analysis

12.1 INTRODUCTION

In the strictest sense, the analysis of every structure is approximate since it is necessary to make certain assumptions to enable an analysis to be made. For example, in a truss analysis it is assumed that the joints are frictionless pins and only axial forces are caused in members. For small displacement theory solutions of continuous beams and frames, the Bernoulli-Euler beam theory is generally accepted as giving the exact solution. However, the boundary conditions of continuous beams and frames are a matter of judgment. The true boundary conditions are somewhere between hinged and fixed in every structure. A structural designer assesses the interaction of the supports and the structure, chooses the boundary conditions deemed to be the most appropriate, analyzes the structure for this choice of boundary conditions, and designs the structural boundary conditions to behave as closely as possible to the chosen assumptions.

In this text, the phrase "approximate methods of indeterminate analysis" means that assumptions in addition to the ones made in the exact analyses are made to enable an indeterminate structure to be analyzed as a determinate structure. The purposes of an approximate analysis are:

1. To obtain results needed to perform a preliminary design of the structural members.
2. To provide results which are, say, within + or −15% of the exact solution to aid the analyst in deciding whether or not the exact computerized solutions are correct.

In the initial stages of design, the designer may consider several structural framing schemes or/and types of material. An approximate cost for each of the alternative preliminary designs is needed in order to decide which alternative is to be chosen for the final design. Approximate analysis results which are quickly and easily obtained are needed to provide the structural and cost data for the alternative preliminary design studies. After the designer has decided which design alternative is to be used, reasonably good estimates of the moments of inertia and/or areas of the members are needed as starting values for input to a computerized structural analysis program which obtains exact indeterminate solutions. If the initial member property estimates are close enough, only one computer run is needed. Otherwise, one or more subsequent computer runs are necessary; in each subsequent run, changes are made for the member properties that were poorly estimated in the previous computer run until a converged computerized solution is obtained and used to perform the final design checks.

If any data entries to the computer software package are incorrect, the computerized solution is incorrect. If one or more data entries involving the boundary conditions are incorrect, usually the computerized solution is totally unusable. Input errors involving the structural geometry or/and member properties can lead to either grossly incorrect results or, perhaps, results that are usable except in the neighborhood of each data error.

Prior to about 1960, final design checks were often made using only the results obtained from approximate methods of analysis since exact solutions were expensive, time consuming, error prone, and computerized solutions were not readily available. Some structural analysis textbooks [12.1, 12.2] which were written prior to about 1970 gave exhaustive reviews of the most popularly used approximate methods of analysis. Iterative computerized solutions are now available to perform not only the analysis but also the design checks. Thus, the current objectives of the approximate methods of analysis are quite different than they were prior to about 1960. Consequently, the author chooses to discuss only the approximate methods of analysis that he uses in checking exact computerized results for gross errors and in sizing members during the preliminary design phase.

12.2 BEAMS AND FRAMES SUBJECTED TO GRAVITY LOADS

Part 2 of the 1980 AISC Manual [12.3] gives exact solutions for shears, bending moments, and maximum deflections for continuous beams of one to four spans, provided all spans have the same EI and span length. These solutions are for various uniformly distributed live load configurations and for a dead load configuration. Since these solutions should be readily accessible to the reader, they are not completely shown here. (The author does not discuss approximate methods of analysis in the classroom until the students need them to obtain results for the preliminary design of a structure.) However, some parts of these solutions are shown in the discussion below. The maximum values for the four span case solutions can be used as the approximate solutions for more than four equal continuous spans having the same constant EI.

Chapter 8 of the 1983 ACI Code [12.4] for reinforced concrete structures gives approximate maximum values for shears, positive bending moments, and negative bending moments for unequal spans in continuous beams and frames subjected to uniformly distributed loads. The inherent conditions that must be satisfied for these maximum values to be valid approximate solutions for continuous indeterminate structures are clearly stated in the ACI Code. Although these maximum values are given in a reinforced concrete design code, they can be used for structures of any elastic material provided the inherent conditions stated in the ACI Code are not violated. Since these solutions and inherent conditions should be readily accessible to the reader, they are not completely shown here. However, some of the maximum bending moment values are given in the discussion below.

Good sketches of the deflected structure were emphasized in the example problem solutions of indeterminate beams and frames by the Force and Displacement Methods of analysis. Approximate methods of indeterminate analysis are usually based on the structural engineer's knowledge of the deflected structure. At points of inflection on the deflected structure, the curvature is zero and the bending moment is zero. Thus, if the locations of the points of inflection ($M = 0$ points) can be reasonably estimated as shown in Figure 12.1, this information enables an approximate moment diagram to be obtained.

The assumption of $a = 0.1L = 0.1*40 = 4$ ft in Figure 12.1b is recommended unless a better assumption is available from previously designed structures similar to this one. The basis for the assumption of $a = 0.1L$ can be justified as shown in the

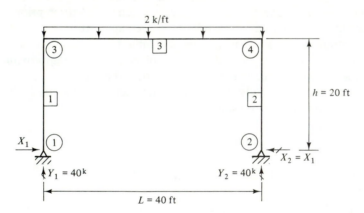

All members have the same E. $I_2 = I_1$ and $I_3 \neq I_1$

Y_1, Y_2, and $X_2 = X_1$ were obtained from the equations of statics.

(a) Indeterminate structure

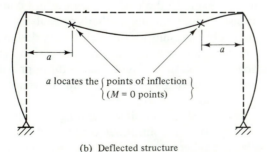

a locates the $\left\{ \begin{array}{c} \text{points of inflection} \\ (M = 0 \text{ points}) \end{array} \right\}$

(b) Deflected structure

Figure 12.1 Approximate method of analysis.

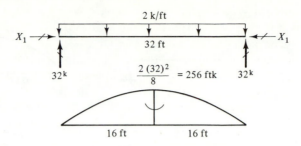

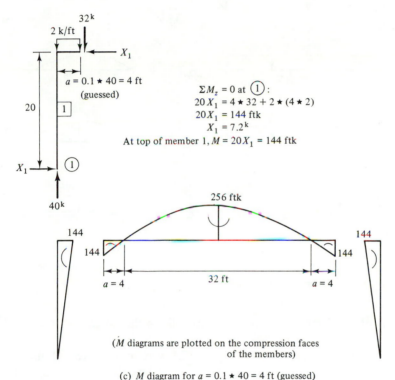

(M diagrams are plotted on the compression faces
of the members)

(c) M diagram for a = 0.1 ★ 40 = 4 ft (guessed)

following discussion. First, consider the moment diagrams shown in Figure 12.2; they were obtained from Part 2 of the 1980 AISC Manual [12.3].

The information shown in Figures 12.2a and 12.2b can be used to justify the assumption of $a = 0.1L$ in Figure 12.1b as follows: In Figure 12.1a:

1. If $I_2 = I_1 = \infty$, joints 3 and 4 would not be able to rotate and for this extreme, limiting mathematical case (see Fig. 12.2b) $a = 0.2113L$.
2. If $I_2 = I_1 = 0$, the ends of member 3 would be completely free to rotate and $a = 0$ for this case (see Fig. 12.2a).

Thus, if the structure in Figure 12.1 is properly designed (the rotational stiffness of members 1 and 2 is approximately equal to the rotational stiffness of member 3), the

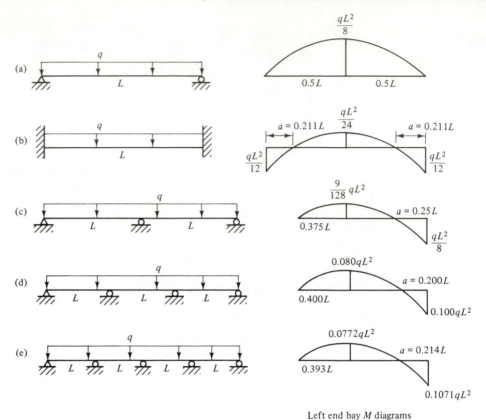

Left end bay M diagrams

Figure 12.2 All members have same EI.

points of inflection in Figure 12.2b must lie in the range $0 < a < 0.2113L$ and $a = 0.1L$ is at about the middle of this range.

Suppose joints 1 and 2 in Figure 12.1 were fixed instead of hinged. In that case, if $a = 0.1L = 4$ ft were also assumed, Figure 12.3 would show the approximate moment diagram which was obtained as follows:

1. From the FBD shown in Figure 12.3b, at the left end of member 3 (and at the top end of member 1), $M = 4*32 + 2*(4*2) = 144$ ftk.
2. From the Displacement Method or Moment Distribution Method concepts, if a prismatic member is fixed at one end, hinged at the other end, and subjected to an applied moment M at the hinged end, the carry over moment is half of M. Therefore, in Figure 12.3b, $M_1 = 0.5*144 = 72$ ftk.
3. In Figure 12.3b, $20X_1 = 4*32 + 2*(4*2) + 72 = 216$, $X_1 = 10.8^k$.

In Figure 12.2 it was shown that the $M = 0$ points depend on the number of spans and the boundary conditions. Also, not shown in Figure 12.2, the $M = 0$ points are significantly different for the exterior and the interior bays of a multispan structure. Other parameters that affect the location of the $M = 0$ points for a multibay structure

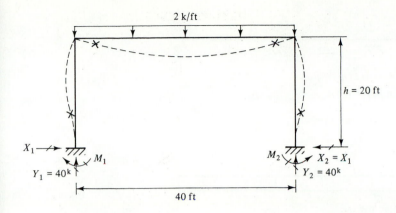

(a) Same structure as Figure 12.1 (a) except for boundary conditions

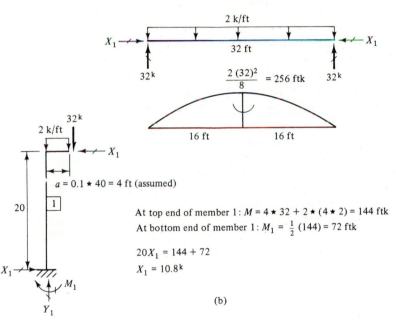

At top end of member 1: $M = 4 \star 32 + 2 \star (4 \star 2) = 144$ ftk

At bottom end of member 1: $M_1 = \frac{1}{2} (144) = 72$ ftk

$20X_1 = 144 + 72$

$X_1 = 10.8^k$

(b)

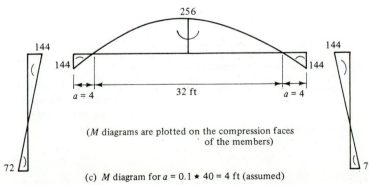

(*M* diagrams are plotted on the compression faces of the members)

(c) *M* diagram for $a = 0.1 \star 40 = 4$ ft (assumed)

Figure 12.3 Approximate method of analysis.

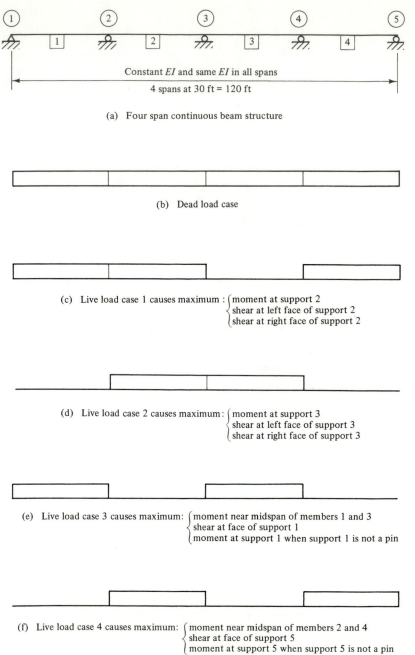

(a) Four span continuous beam structure

(b) Dead load case

(c) Live load case 1 causes maximum : {moment at support 2
shear at left face of support 2
shear at right face of support 2

(d) Live load case 2 causes maximum: {moment at support 3
shear at left face of support 3
shear at right face of support 3

(e) Live load case 3 causes maximum: {moment near midspan of members 1 and 3
shear at face of support 1
moment at support 1 when support 1 is not a pin

(f) Live load case 4 causes maximum: {moment near midspan of members 2 and 4
shear at face of support 5
moment at support 5 when support 5 is not a pin

Structure and loading Configurations to Cause Maximum Shears and Moments

Figure 12.4 Structure and loading configurations to cause maximum shears and moments.

are: the structural properties; the live load configuration; and the live load to dead load ratio. The solutions of several multibay structures are shown in Figures 12.4 through 12.20 to illustrate that estimating the $M = 0$ points for the beams in a multibay structure subjected only to gravity loads is considerably more difficult than estimating the $M = 0$ points in the beam of a one-story and one-bay plane frame structure. For multibay structures, the author prefers the approach recommended in ACI 318-83 [7.2] which gives maximum moment coefficients at the beam ends and near midspan of the beams for exterior spans and interior spans. In order to compare the following multispan solutions to the ACI moment coefficients, the author's solutions are cast in the form of the ACI coefficients. Also, the author's solutions are provided to show the locations of the $M = 0$ points.

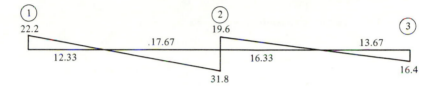

(a) Dead load + live load case 3 (causes maximum shear at support 1)

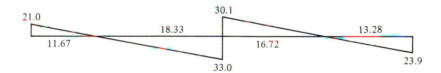

(b) Dead load + live load case 1 (causes maximum shear at support 2)

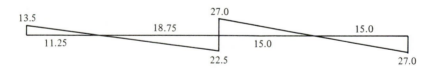

(c) Dead load + live load case 2 (causes maximum shear at support 3)

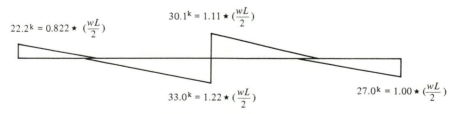

(d) Maximum shear curves at each member end

Figure 12.5 Shear curves for Structure of Figure 12.4 when live load/dead load = 0.6 k/ft/1.2 k/ft = 0.5.

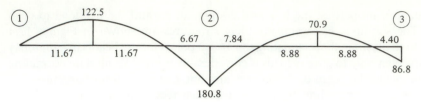

(a) Dead load + live load case 1 (causes maximum moment at support 2)

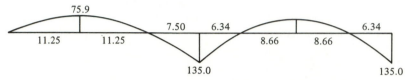

(b) Dead load + live load case 2 (causes maximum moment at support 3)

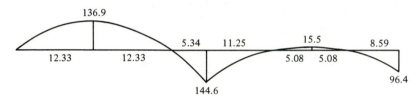

(c) Dead load + live load case 3 (causes maximum moment near midspan of member 1)

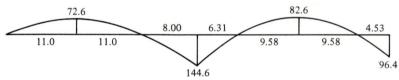

(d) Dead load + live load case 4 (causes maximum moment near midspan of member 2)

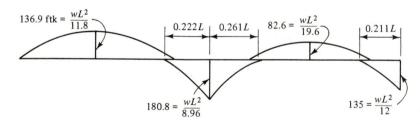

(e) Maximum moment curves

Figure 12.6 Moment curves for structure of Figure 12.4 when LL/DL = 0.5.

500

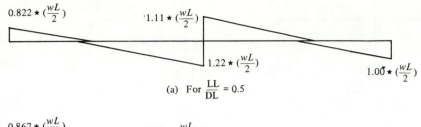

$0.822 \star \left(\frac{wL}{2}\right)$

$1.11 \star \left(\frac{wL}{2}\right)$

$1.22 \star \left(\frac{wL}{2}\right)$

$1.00 \star \left(\frac{wL}{2}\right)$

(a) For $\dfrac{LL}{DL} = 0.5$

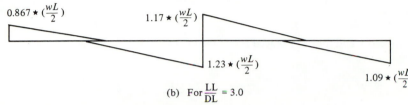

$0.867 \star \left(\frac{wL}{2}\right)$

$1.17 \star \left(\frac{wL}{2}\right)$

$1.23 \star \left(\frac{wL}{2}\right)$

$1.09 \star \left(\frac{wL}{2}\right)$

(b) For $\dfrac{LL}{DL} = 3.0$

Figure 12.7 Maximum shear curves for structure of Figure 12.4.

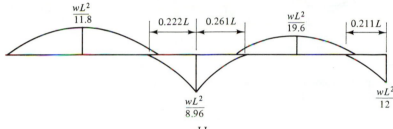

$\dfrac{wL^2}{11.8}$

$0.222L$ $0.261L$

$\dfrac{wL^2}{19.6}$

$0.211L$

$\dfrac{wL^2}{8.96}$

$\dfrac{wL^2}{12}$

(a) For $\dfrac{LL}{DL} = 0.5$

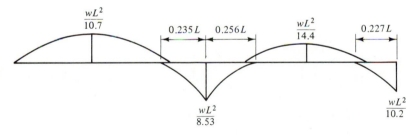

$\dfrac{wL^2}{10.7}$

$0.235L$ $0.256L$

$\dfrac{wL^2}{14.4}$

$0.227L$

$\dfrac{wL^2}{8.53}$

$\dfrac{wL^2}{10.2}$

(b) For $\dfrac{LL}{DL} = 3.0$

Figure 12.8 Maximum moment curves for structure of Figure 12.4.

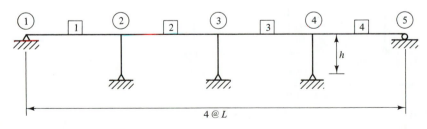

①   1   ②   2   ③   3   ④   4   ⑤

h

$4 @ L$

Beams. $L = 30$ ft; 4 64I
Columns: $h = 12.5$ ft; I

Figure 12.9 One-story and four-bay braced frame for LL/DL = 0.5.

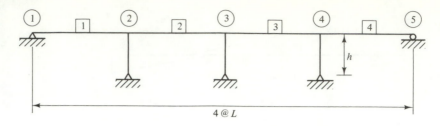

Beams: $L = 30$ ft; $2.86I$
Columns: $h = 12.5$ ft; I

Figure 12.10 One-story and four-bay braced frame for LL/DL = 3.0.

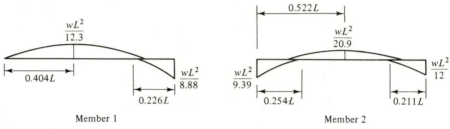

Member 1 Member 2

Figure 12.11 Maximum moment curves for structure of Figure 12.9 and LL/DL = 0.6 k/ft/1.2 k/ft = 0.5 (see Figure 12.4).

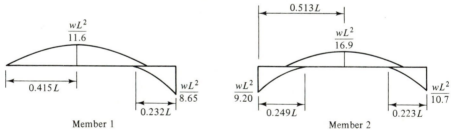

Member 1 Member 2

Figure 12.12 Maximum moment curves for structure of Figure 12.10 and LL/DL = 3.6/1.2 = 3.0 (see Figure 12.4).

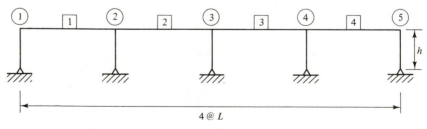

Beams: $L = 30$ ft; $1.65I$
Columns: $h = 12.5$ ft; I

Figure 12.13 One-story and four-bay unbraced frame for LL/DL = 0.5.

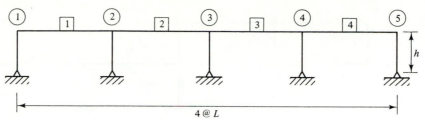

Figure 12.14 One-story and four-bay unbraced frame for LL/DL = 3.0.

Beams: L = 30 ft; 2.86I
Columns: h = 12.5 ft; I

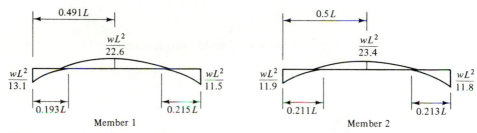

Figure 12.15 Maximum moment curves for structure of Figure 12.13 and LL/DL = 0.6 k/ft/1.2 k/ft = 0.5 (see Figure 12.4).

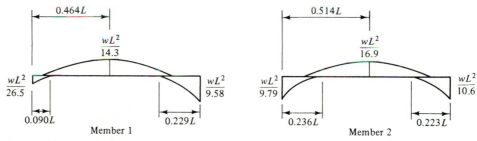

Figure 12.16 Maximum moment curves for structure of Figure 12.14 and LL/DL = 3.6/1.2 = 3.0 (see Figure 12.4).

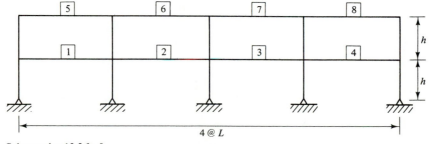

Columns: h = 12.5 ft; I

Members 1 through 4: L = 30 ft; 3.20I; $\dfrac{\text{LL}}{\text{DL}} = \dfrac{1.8 \text{ k/ft}}{1.8 \text{ k/ft}} = 1.0$

Members 5 through 8: L = 30 ft; 1.26I; $\dfrac{\text{LL}}{\text{DL}} = \dfrac{0.6}{1.2} = 0.5$

Figure 12.17 Two-story and four-bay structure for floor LL/DL = 1.0.

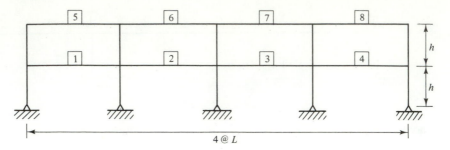

Columns: $h = 12.5$ ft; $1.63I$

Members 1 through 4: $L = 30$ ft; $5.34I$; $\dfrac{\text{LL}}{\text{DL}} = \dfrac{5.4 \text{ k/ft}}{1.8 \text{ k/ft}} = 3.0$

Members 5 through 8: $L = 30$ ft; I; $\dfrac{\text{LL}}{\text{DL}} = \dfrac{0.6}{1.2} = 0.5$

Figure 12.18 Two-story and four-bay structure for floor LL/DL = 3.0.

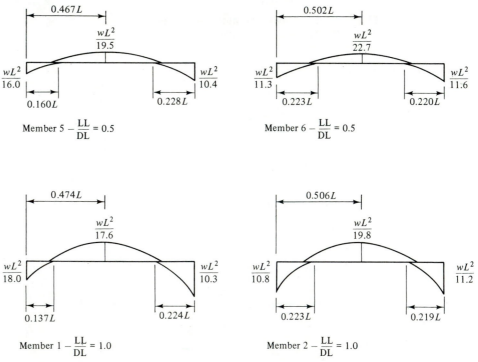

Figure 12.19 Maximum moment curves for structure of Figure 12.17 and load configurations of Figure 12.4.

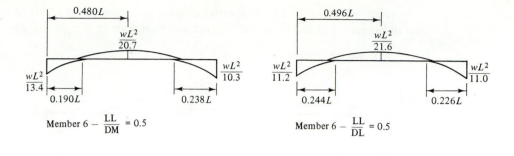

Member 6 $-\dfrac{\mathrm{LL}}{\mathrm{DM}} = 0.5$ Member 6 $-\dfrac{\mathrm{LL}}{\mathrm{DL}} = 0.5$

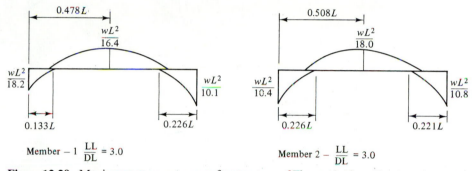

Member $-1 \dfrac{\mathrm{LL}}{\mathrm{DL}} = 3.0$ Member 2 $-\dfrac{\mathrm{LL}}{\mathrm{DL}} = 3.0$

Figure 12.20 Maximum moment curves for structure of Figure 12.18 and load configurations of Figure 12.4.

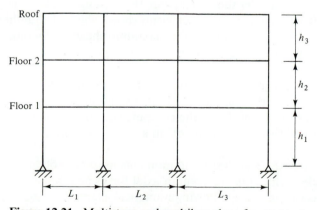

Figure 12.21 Multistory and multibay plane frame structure.

According to the ACI Code approximate analysis coefficients [12.4], at any girder level in Figure 12.21 the maximum shears and moments in the girders are:

Exterior bays – shear and moment envelopes

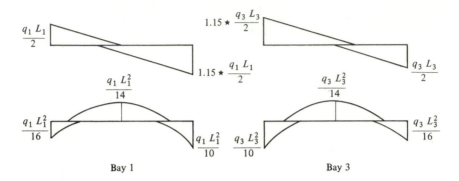

Bay 1   Bay 3

Interior bays – shear and moment envelopes

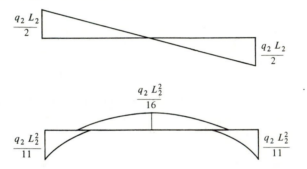

The moment envelopes are plotted on the compression faces of the member along the member length. The reader should compare the author's results shown in Figures 12.4 through 12.20 with the ACI Code coefficients shown above. The author personally uses the ACI Code coefficients to obtain the maximum shears and moments due only to gravity loads for the following purposes:

1. Preliminary design of the girders—to obtain the first estimate of the girder sizes for input to the computerized exact solution.
2. To perform rough checks on the computerized plane frame analysis results for girder end shears and moments in a multibay frame.

After the maximum girder end shears and moments have been obtained numerically from the ACI Code coefficients, the column axial forces and member-end moments can be obtained as shown in the following numerical example.

EXAMPLE 12.1 ———

Perform an approximate analysis of the plane frame shown in Figure 12.22a using the ACI Code coefficients for the girder-end shears and moments.

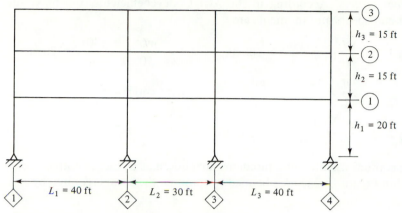

Distributed loads are: 1 k/ft on the roof girders (level 3)
2 k/ft on the floor girder (levels 1 and 2)

Figure 12.22a Multistory and multibay plane frame.

Solution. Since the structure and loading in Figure 12.22a are symmetric with respect to the vertical line through the midpoints of the girders in bay 2, we only need to obtain the girder-end moments and shears in bays 1 and 2. Also, we only need to obtain the column axial forces and member-end moments on column lines 1 and 2.

Roof girders: According to the ACI Code coefficients, the maximum member-end shears and moments are

bay 1: $\quad \dfrac{wL^2}{16} = \dfrac{1*(40)^2}{16} = 100 \, \text{ftk};$ $\qquad \dfrac{wL^2}{10} = \dfrac{1*(40)^2}{10} = 160 \, \text{ftk}$

$\quad \dfrac{wL}{2} = \dfrac{1*40}{2} = 20^k;$ $\qquad 1.15*20 = 23^k$

bay 2: $\quad \dfrac{wL^2}{11} = \dfrac{1*(30)^2}{11} = 81.8 \, \text{ftk};$ $\qquad \dfrac{wL}{2} = \dfrac{1*30}{2} = 15^k$

If the analyst chooses to accept the member-end moments as being the correct values and wants the member-end shears that satisfy equilibrium, the FBD in Figure 12.22b are obtained. The axial forces in the roof girders are obtained, if desired, after the member-end shears for the third story columns have been estimated.

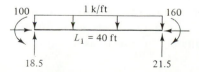

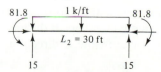

Figure 12.22b

Floor girders: According to the ACI Code coefficients, the maximum member-end shears and moments are

bay 1: $\dfrac{wL^2}{16} = \dfrac{2*(40)^2}{16} = 200 \text{ ftk};$ $\dfrac{wL^2}{10} = \dfrac{2*(40)^2}{10} = 320 \text{ ftk}$

$\dfrac{wL}{2} = \dfrac{2*40}{2} = 40^k;$ $1.15*40 = 46^k$

bay 2: $\dfrac{wL^2}{11} = \dfrac{2*(30)^2}{11} = 163.6 \text{ ftk};$ $\dfrac{wL}{2} = \dfrac{2*30}{2} = 30^k$

If the maximum member-end moments are chosen as being the correct values, the FBD in Figure 12.22c are obtained to compute the shears required by equilibrium.

Figure 12.22c

Columns: The girder-end shears are reversed and applied to the joints to obtain the axial compression forces in the columns, starting at the roof level (Fig. 12.22d). For the columns on column line 1, the reversed girder-end moments applied to the joints are as shown in Figure 12.22e. Since the axial forces in the girders are not known, statics cannot be used to find the moment at the bottom of each story, starting at the third story and moving down to the foundation. At each floor level, the joint moment shown must be distributed to the column ends at that floor level. If the column properties are not known, allocate half of the joint moment to each column end at any floor level. If the column properties are known, obtain the sum of the column rotational stiffnesses at each floor level and allocate the joint moments in proportion to the column rotational stiffness divided by the sum of the column rotational stiffnesses. For example, at floor level 2,

$$M_{\text{bot(story 3)}} = \frac{(4EI/h)_3}{(4EI/h)_2 + (4EI/h)_3} * 200 \text{ ftk}$$

Since the bottom end of the first story column is pinned, the rotational stiffness at the top end of that column is $(3EI/h)_1$. If the bottom end of the first story column were fixed against rotation instead of being pinned, there would be a moment to be assigned to that column end. For a symmetric structure and a symmetric gravity loading there is no X-direction movement of the joints. For this case the moment to be assigned to a fixed end of a column at the foundation is half of the moment at the top end of that column; that is, the carry-over factor is 0.5 for prismatic members. If the loading due to gravity only is not symmetric, the joints translate in the X direction. Fortunately, in most cases, these joint

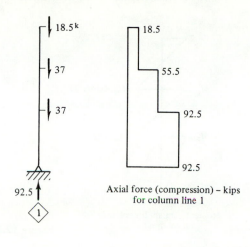

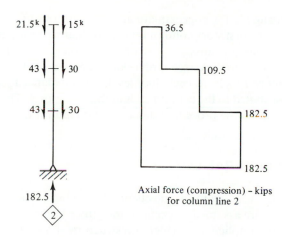

Figure 12.22d

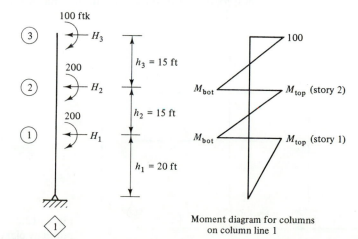

Moment diagram for columns
on column line 1

Figure 12.22e

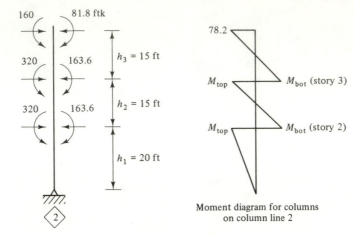

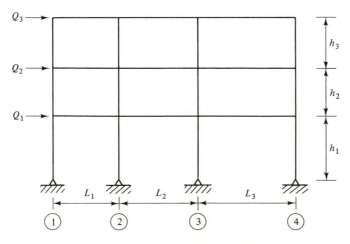

Moment diagram for columns
on column line 2

Figure 12.22f

translations are negligible for approximate analysis purposes. Consequently, a fixed bottom end of the first story column is assigned half of the top end moment even if the gravity loading is unsymmetric or if the structure is unsymmetric.

On column line 2, the reversed girder-end moments applied to the joints are as shown in Figure 12.22f. At each floor level, the unbalanced joint moment is obtained and distributed to the column ends at that floor level. The distribution procedure is identical to the procedure described for column line 1.

12.3 FRAMES SUBJECTED TO LATERAL LOADS

Generally, the lateral loads are due to the effects of either wind or earthquakes and, as shown in Figure 12.23, are assumed to occur at the girder levels.

First, for reasons of simplicity, consider the structure shown in Figure 12.24 which is subjected only to lateral loads. From the approximate analysis due to gravity

Figure 12.23 Multistory, multibay plane frame subjected to lateral loads.

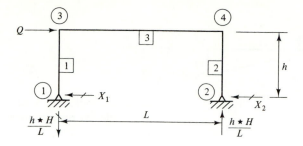

From statics, the Y–direction reactions are as shown, and $X_1 + X_2 = Q$

Figure 12.24 One-bay, one-story plane frame.

loads only, a preliminary design can be performed to size the members. That is, tentative areas and moments of inertia for all members are assumed to be known at this stage of the approximate analysis process. Consequently, in Figure 12.24 Q can be allocated to X_1 and X_2 in proportion to the column shear stiffnesses (see Figure 12.25). Therefore, in Figure 12.24, Q is allocated as follows:

$$X_1 = \frac{H_1}{H_1 + H_2} * Q \quad \text{and} \quad X_2 = \frac{H_2}{H_1 + H_2} * Q$$

It is easier to continue the discussion using numbers instead of symbols. Suppose the parameters in Figure 12.24 are chosen as follows: $Q = 10^k$; $h = 15$ ft; $L = 40$ ft; $I_2 = I_1$; and $I_3 \neq I_1$. Since members 1 and 2 have the same shear stiffness, the 10^k lateral load is allocated equally to the X-direction reactions (see Figure 12.26). Now, all member-end forces due to lateral loads only can be found using member FBDs and

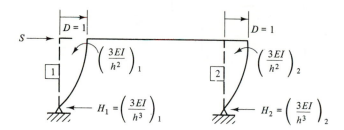

$S = H_1 + H_2$

Figure 12.25

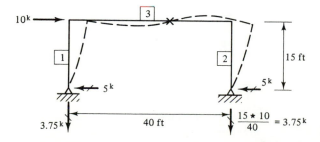

Figure 12.26

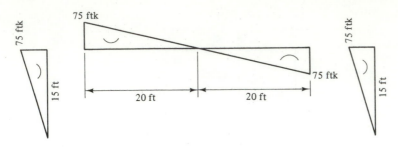

Figure 12.27 Moment diagrams plotted on compression faces of the members.

statics. Figure 12.27 shows the moment diagram plotted on the compression faces of the members. Superposition is valid and these results can be combined with the results from an approximate gravity load analysis.

Now, let us consider the structure in Figure 12.28 and discuss approximate methods of analysis for it. In each story, the total lateral load that the column shears must sum up to can be allocated to the columns in proportion to their shear stiffnesses. For example, in the second story (see Figure 12.28): Each column in the second story is prismatic and has the shear stiffness shown in Figure 12.29. Let

$$S_2 = \sum_{k=1}^{4} \frac{12EI_k}{h_2^3} = \frac{12E}{h_2^3} \sum_{k=1}^{4} I_k$$

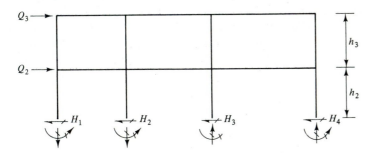

$(H_1 + H_2 + H_3 + H_4) = (Q_2 + Q_3)$ is required by statics

Figure 12.28

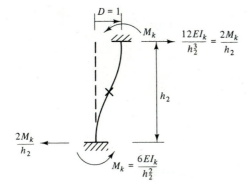

Figure 12.29 Shear stiffness for column k in second story.

Then, $(Q_2 + Q_3)$ is allocated in Figure 12.28 as follows:

$$H_k = \frac{(12EI_k/h_2^3)}{S_2} * (Q_2 + Q_3) = \frac{I_k}{\sum_{m=1}^{4} I_m} * (Q_2 + Q_3) \tag{12.1}$$

for $k = 1$ to 4.

If $M = 0$ points are assumed to occur at midheight of the columns in the third story, the following FBD is obtained:

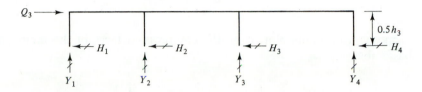

Note that there are eight unknown reactions and only three equations of statics are available for this substructure. Therefore, this substructure is statically indeterminate to the fifth degree. If precisely five force conditions are assumed, this substructure becomes determinate. For example, if Q_3 is distributed to the column shears in the manner described in Eq. (12.1), this constitutes assuming three force conditions (the other column shear satisfies one of the three equations of statics, namely, $\Sigma F_X = 0$). Consequently, only two more conditions can be assumed and retain a stable structure. Therefore, if $M = 0$ points are assumed in the roof girders, only two such $M = 0$ points can be assumed, otherwise the substructure is unstable.

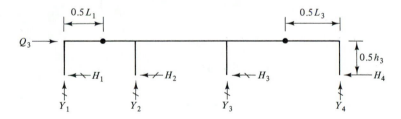

Based on the preceding discussion for the substructure of the third story, it can be concluded that in addition to the assumed distribution of story shears in each story as described in Eq. 12.1, to render Figure 12.23 into a determinate structure it is only possible to assume:

1. $M = 0$ points in all but one girder of each floor and roof level.
2. $M = 0$ points in each column except for the bottom story columns which already have known $M = 0$ points at the foundation.

To find the column and girder-end moments, any subassemblage from Figure 12.30 can be treated as an FBD after Eq. 12.1 has been used to define the column shears. For example, consider the subassemblage at girder level 2 on column line 4:

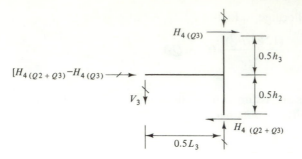

The moment diagrams plotted on the compression faces of the members for this subassemblage are:

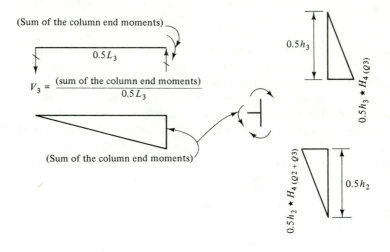

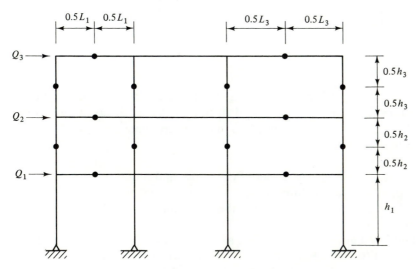

Figure 12.30

Since this approach is applicable for each subassemblage, the shears at the mid-span of each girder having an assumed $M = 0$ point are obtainable. Starting with the subassemblages at the roof girder level, summation of Y-direction forces equal to zero on each subassemblage enables the axial force to be obtained in each column below the roof girder level. Then, the analyst moves down one level, uses the same procedure, and finds the axial force in each column of another story, and so on. Of course the analyst must first solve at each girder level for the forces on the exterior subassemblages of the building and then solve for the forces on the neighboring interior subassemblages until all subassemblages at each girder level have been solved.

EXAMPLE 12.2 _____

Obtain the column shears in each story assuming the story shear is distributed to the columns in proportion to the column shear stiffness. The interior columns have 1.2 times as much moment of inertia as the exterior columns in each story. $M = 0$ points are assumed to occur at midheight of the columns in the top two stories and at midspan of the exterior span girders at each girder level as shown in Figure 12.31a.

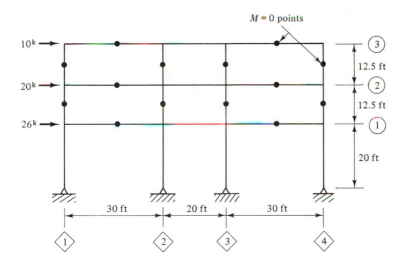

Figure 12.31a Plane frame subjected to lateral loads.

Solution. The column shear stiffness is $(12EI/h^3)_m$ for any member m. In each story of Figure 12.31, all columns have the same length; consequently, the only variable in the column shear stiffness in each story is the moment of inertia of the columns. Therefore, the story shear is distributed to the column shears as follows:

exterior columns: $\quad$ shear $= \dfrac{I}{4.4I} * (\text{story shear}) = 0.227 * (\text{story shear})$

interior columns: $\quad$ shear $= \dfrac{1.2I}{4.4I} * (\text{story shear}) = 0.273 * (\text{story shear})$

third story shear $= 10^k$

exterior columns: $\quad$ shear $= 0.227 * 10 = 2.27^k$

interior columns: shear $= 0.273 * 10 = 2.73^k$
second story shear $= 10 + 20 = 30^k$
exterior columns: shear $= 0.227 * 30 = 6.81^k$
interior columns: shear $= 0.273 * 30 = 8.19^k$
first story shear $= 30 + 26 = 56^k$
exterior columns: shear $= 0.227 * 56 = 12.7^k$
interior columns: shear $= 0.273 * 56 = 15.3^k$

A few of the subassemblages are solved for the other forces to illustrate the procedure. See Figure 12.31b.

$$\text{at 2, } \Sigma M = 0: \qquad 20 Y_3 = 6.25 * (7.73 - 2.27) - 15 * 0.946 - 35 * 0.946$$
$$Y_3 = -0.659$$
$$\Sigma F_Y = 0: \qquad Y_2 = 0.946 - 0.946 - Y_3 = 0.659$$

The preceding approximate method based on an assumed distribution of column shears in each story is not the only possible approximate method of analysis for a multistory, multibay plane frame. However, in the author's opinion it is the best assumption to make when the lateral loads do not govern the member sizes.

If the lateral loading governs the member sizes for the structure shown in Figure 12.32, the following approximate method is preferred. $M = 0$ points are assumed at the midspan of each girder and at midheight of the columns in each story except in the first story when the columns are pinned to the foundation. In each story at the $M = 0$ points in the columns, the column axial forces are assumed to be distributed in the manner that bending stresses are distributed in a cantilever beam subjected to lateral loads. Internal bending stresses in a cantilever beam subjected to lateral loads are linearly varying across the depth of the cantilever cross section and proportional to the distance from the neutral axis of the cross section. This approach is illustrated in Example 12.3.

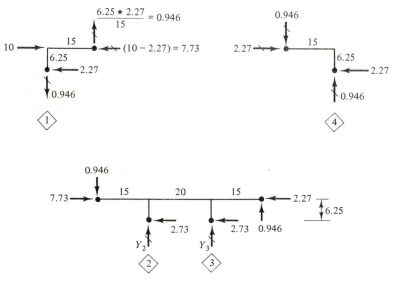

Figure 12.31b

EXAMPLE 12.3

Obtain the column axial forces in the bottom two stories of the structure of Figure 12.32a assuming that the column axial forces vary linearly in proportion to the distance from the centroid of the column areas. The exterior column areas are A and the interior column areas are $1.2A$.

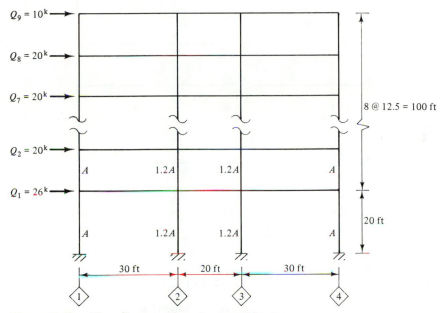

Figure 12.32a Plane frame subjected to lateral loads.

Solution. The centroid of the column areas in the bottom two stories is at 40 ft from column line 1. Since there are not any externally applied loads in the Y direction on the FBD shown in Figure 12.32b, some of the column axial forces must be tensile and the others must be compressive (their algebraic sum must be zero). With respect to the bottom of the FBD shown in Figure 12.32b, the externally applied loads have a clockwise external moment of

$$M_{ext} = 10*26 + (10+4*12.5)*(7*20) + (10+8*12.5)*10$$
$$= 9760 \text{ ftk (clockwise)}$$

The internal or resisting moment provided by the axial forces in the columns must be 9760 ftk (counterclockwise). Consequently, the axial forces on column line 1 must be tensile and on column line 2 they must be compressive. Since column lines 2 and 3 are 10 ft from the centroid of the column areas and column lines 1 and 4 are 40 ft from the centroid of the column areas, if F is the force in the columns on column lines 2 and 3 then the force in the columns on column lines 1 and 2 is $(40 \text{ ft}/10 \text{ ft})*F = 4F$. Summation of moments for these column axial forces with respect to the centroid of the column areas on the $M = 0$ line must be 9760 ftk (CCW). Therefore, $80*(4F) + 20*(F) = 9760$; $F = 28.71^k$;

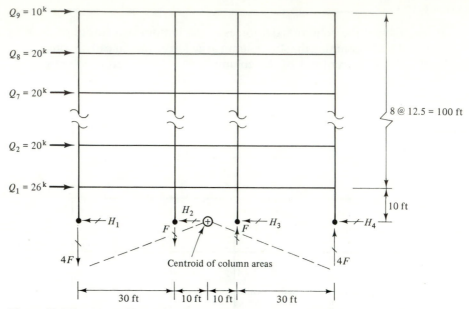

Figure 12.32b   Note: The solid dots at the bottom of the FBD are $M = 0$ points in the columns of Figure 12.32a in the first story.

$4F = 114.82^k$. In a similar manner, the column axial forces in the second story columns are found using the FBD shown in Figure 12.32c.

$$M_{\text{int}} = 80 * (4F) + 20 * (F) = 340F\,(\text{CCW})$$
$$M_{\text{ext}} = (6.25 + 3*12.5)*(7*20) + (6.25 + 7*12.5)*10 = 7062.5\,\text{ftk}\,(\text{CW})$$
$$340F = 6062.5; \qquad F = 20.77^k; \qquad 4F = 83.09^k$$

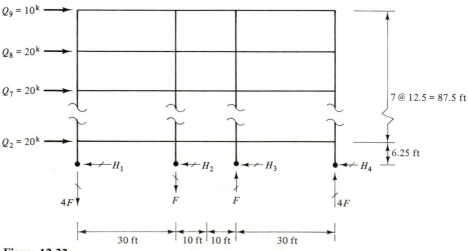

Figure 12.32c

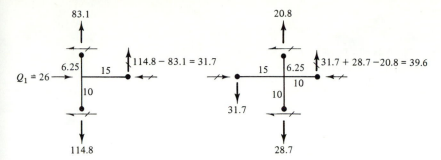

Figure 12.32d

The subassemblages between the $M = 0$ points in the first and second story columns on column lines 1 and 2 are as shown in Figure 12.32d. Note that if the analyst starts finding the forces for the subassemblages at the bottom of the structure, only the girder shears and girder-end moments can be obtained when only the axial forces in the columns are known. Therefore, the analyst must begin with the subassemblages at the roof level in order to find all of the forces on each subassemblage (Figs. 12.32e and 12.32f).

In the topmost stories of a highrise building, where the column axial forces due to lateral loads are less than those due to gravity loads, the author prefers the method of distributing the story shear to the column shears. After the column axial forces due to lateral loads become greater than those due to gravity loads, the author prefers the method of distributing the column axial forces in a "cantilever-beam-type action." However, it should be noted that only one of the two approximate methods (distribute the shears or distribute the axial forces) can be used in any story; otherwise, inconsistent results will be obtained for the other forces on the subassemblages. Since the top two stories in Example 12.2 and Example 12.3 have the same dimensions and lateral loads, a comparison of the results obtained by the two different assumptions is readily made for the subassemblage forces in the top story of both examples (Fig. 12.32g).

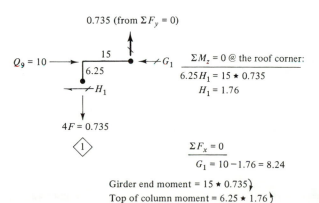

Figure 12.32e

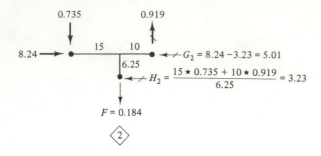

$G_2 = 8.24 - 3.23 = 5.01$

$H_2 = \dfrac{15 \star 0.735 + 10 \star 0.919}{6.25} = 3.23$

$F = 0.184$

⟨2⟩

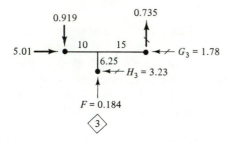

$G_3 = 1.78$

$H_3 = 3.23$

$F = 0.184$

⟨3⟩

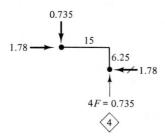

$4F = 0.735$

⟨4⟩

Figure 12.32f

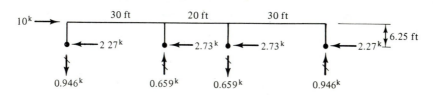

From Example 12.2 — column shears assumed

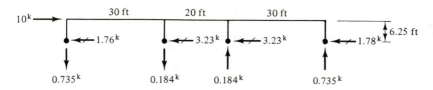

From Example 12.3 — column axial forces assumed

Figure 12.32g

520

12.4 TRUSSES

Externally or internally indeterminate, parallel chord trusses as shown in Figure 12.33 occur frequently in bridges and buildings. A truss is a "deep beam with large holes in it." The truss chord members provide the "beam bending moment," the diagonal members provide the "beam shear," and the vertical members maintain the parallel chord geometry (they keep the top and bottom chord members separated at a constant distance, h).

If the reactions in Figure 12.33a are known, the truss is then determinate and the Method of Joints or/and the Method of Sections can be used to find all of the forces in the truss members. Therefore, in the preliminary design phase the primary objective of an approximate method of analysis for Figure 12.33a is only to obtain a reasonable estimate of the reactions. The truss can be treated as a beam of constant EI to obtain the estimated reactions from either a Displacement Method or Force Method indeterminate beam analysis. After the individual truss member sizes have been determined in the preliminary design phase, the following approach can be used to obtain an estimate of the maximum deflection:

1. Let A_c be the average or weighted average of the areas of the top and bottom chord members of the truss and $I_c = 2 * A_c(h/2)^2 = 0.5 * A_c h^2$ be the moment of inertia of the chord areas.

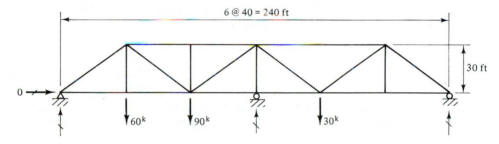

(a) Externally indeterminate truss

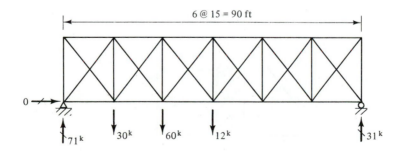

(b) Internally indeterminate truss

Figure 12.33 Indeterminate trusses.

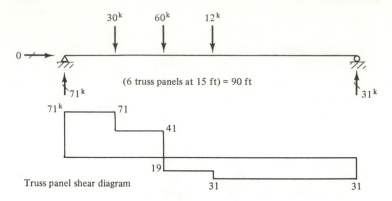

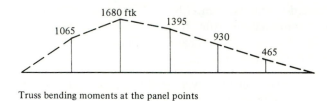

Figure 12.34 Truss treated as a beam.

2. Use $I = 0.75 * I_c$ as the constant moment of inertia of an equivalent beam to estimate deflections. The 0.75 factor accounts for the deformation of the truss diagonal members.

Consider the approximate analysis of Figure 12.34 assuming that all diagonal members can resist either compression or tension. In each panel it is assumed that the panel shear is equally shared by these two diagonals. For example, Figure 12.35 shows the approximate analysis in the second panel from the left support.

$$\sum F_Y = 0 \text{ (assuming each diagonal equally shares the panel shear)}$$
$$2V = 71 - 30 = 41^k$$
$$H = \tfrac{3}{4}V = 15.375^k$$

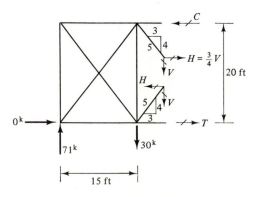

Figure 12.35 Approximate analysis in second truss panel.

Force in each diagonal is $\frac{5}{4}V = 25.625^k$. One diagonal is in tension and the other one is in compression.

$$\Sigma F_X = 0: \qquad T + H - H + C = 0; \qquad C = T$$
$$[\Sigma M_Z \text{ at } X = 15\,\text{ft}] = 0: \qquad 20*(C-H) = 15*71$$
$$(C-H) = 53.25^k$$
$$C = 53.25 + 15.375 = 68.625^k$$

After the forces in all of the diagonals have been allocated by the equal sharing assumption, all of the other truss member forces can be obtained by the Method of Joints.

If the diagonal members in Figure 12.33b are designed to resist only tension, the following approximate method of analysis is used. The analyst justifiably assumes that if a diagonal member is called upon to resist compression, the member buckles and the buckling force is negligible. Consider the second panel from the left support in Figure 12.33b assuming the diagonal members can resist only tension axial forces (see Fig. 12.36).

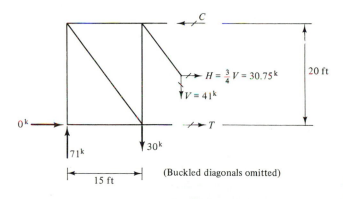

(Buckled diagonals omitted)

$\Sigma F_x = 0: \ T + H - C = 0; \ C \neq T$ in this case.

Figure 12.36 Approximate analysis for tension only diagonals.

12.5 INDUSTRIAL BUILDINGS

An approximate analysis of the structure shown in Figure 12.37 can be made by assuming that:

1. In the columns, points of inflection occur halfway between the foundations and the kneebraces.
2. The X-direction reactions are equal. That is, they are half of the total, applied, X-direction loads.

The reaction moments are the column shears times the distance from the foundation to the assumed $M = 0$ point in the columns. After the reaction moments are computed, the Y-direction reactions are obtained by statics. After the reactions are obtained, each

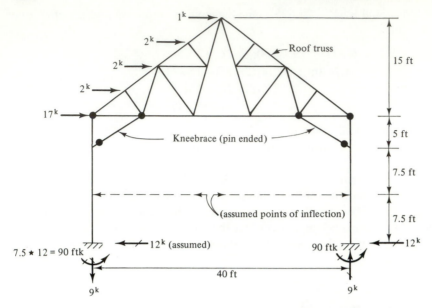

Figure 12.37

column can be considered as an FBD. Summation of moments equal to zero at the top end of the column FBD gives the force in each kneebrace. Then, a routine truss analysis by the Method of Joints can be performed.

PROBLEMS

12.1–12.3. Using the ACI Code coefficients, compute:
  (a) The shear envelope values for the beams.
  (b) The moment envelope values for the beams.
  (c) The maximum axial forces and moments in the columns.
  Uniformly distributed loads are:

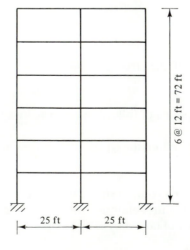

Figure P12.1

roof beams: dead load = 3.14 k/ft;   live load = 0.578 k/ft;   (exterior bays)
                dead load = 3.04 k/ft;   live load = 0.538 k/ft;   (interior bays)
floor beams: dead load = 3 k/ft;   live load = 1.56 k/ft;   (exterior bays)
                dead load = 2.9 k/ft;   live load = 2.93 k/ft;   (interior bays)

Note: Take advantage of symmetry and repetition (of floor loads) to minimize the number of calculations required in each problem.

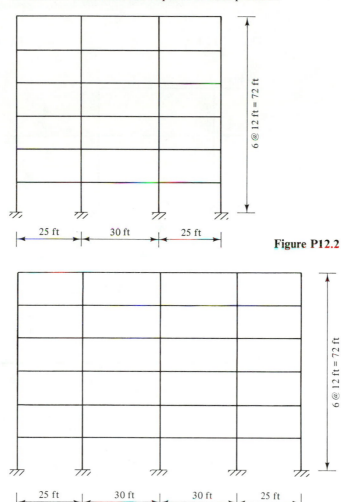

Figure P12.2

Figure P12.3

12.4–12.10. In each story, assume all columns have the same properties (A and I). Wherever applicable, take advantage of structural symmetry to minimize the number of required calculations. Compute the shears, moments, and axial forces in all members using:

(a) The Cantilever Method.

(b) The method of distributing the story shear to the columns and assuming a sufficient number of midspan and midheight $M = 0$ points to render the structure determinate.

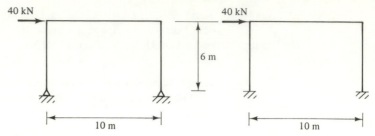

Figure P12.4   Figure P12.5

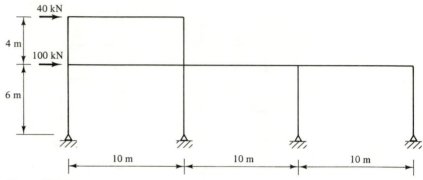

Figure P12.6

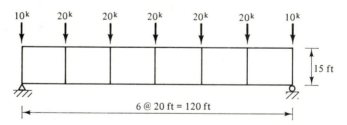

The left half of this structure subjected to the symmetric loading shown
can be analyzed as the following structure can be analyzed

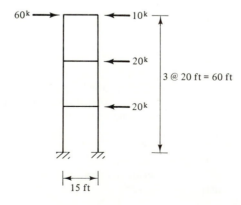

Figure P12.7

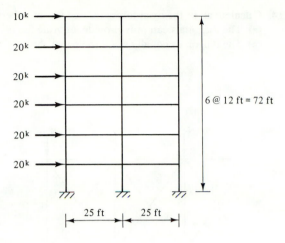

10^k
20^k
20^k
20^k
20^k
20^k

6 @ 12 ft = 72 ft

25 ft 25 ft

Figure P12.8

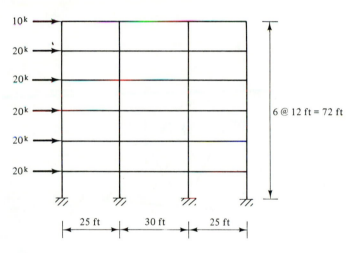

10^k
20^k
20^k
20^k
20^k
20^k

6 @ 12 ft = 72 ft

25 ft 30 ft 25 ft

Figure P12.9

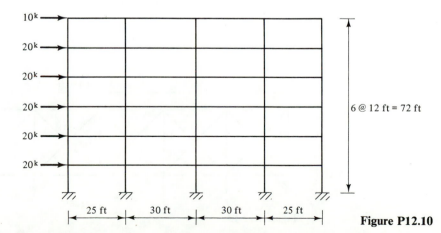

10^k
20^k
20^k
20^k
20^k
20^k

6 @ 12 ft = 72 ft

25 ft 30 ft 30 ft 25 ft

Figure P12.10

12.11–12.14. Calculate all of the axial forces in the truss members assuming:
  (a) The diagonals can only provide a tensile force.
  (b) The diagonals equally share the panel shear.

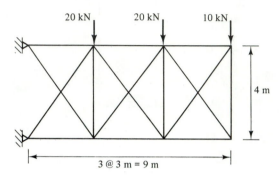

3 @ 3 m = 9 m

Figure P12.11

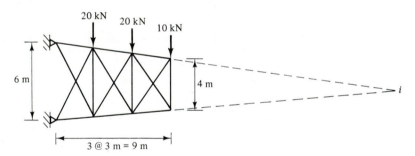

Invoke the assumptions for the diagonal in each panel when summing
moments equal to zero at point *i*

Figure P12.12

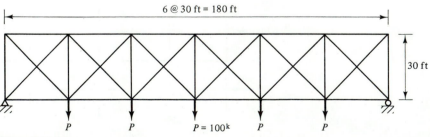

Figure P12.13

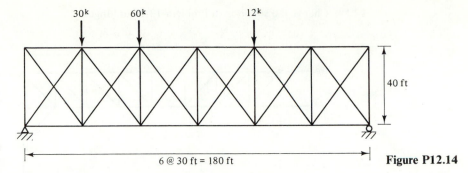

Figure P12.14

In the following problems, use the method described in Section 12.5 to obtain the forces in the members framing into the columns and to obtain the axial force, shear, and moment diagrams of the columns. Assume all knee braces are pinned ended.

12.15. Finish the solution of Figure 12.37.

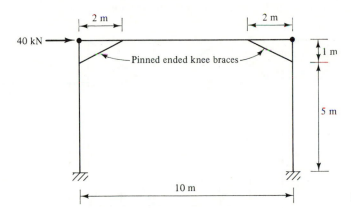

Figure P12.16

12.17. Change the supports in Problem 12.16 to hinges.

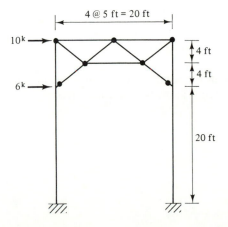

Figure P12.18

12.19. Change the supports in Problem 12.18 to hinges.

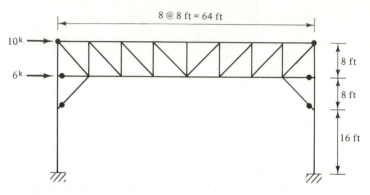

Figure P12.20

12.21. Delete the kneebraces in Problem 12.20.

12.22. Change the supports in Problem 12.20 to hinges.

12.23. Assume all diagonals can resist compression.

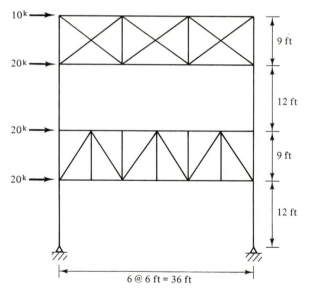

Figure P12.23

12.24. In Problem 12.23, shift the bottom truss down to the foundation level and delete the two horizontal loads acting at the bottom truss level.

Chapter 13

Displacement Method— Computerized Approach

13.1 INTRODUCTION

It is assumed that the reader has adequately studied Chapters 8 through 10 before embarking on the study of this chapter. Chapters 8 and 9 dealt with some of the fundamentals involved in computerizing the Displacement Method and chapter 10 dealt with Moment Distribution. In this chapter the remaining fundamentals needed in computerizing the Displacement Method to solve plane frames are presented. The efficient means of mathematical organization for a computerized solution is to use matrix notation in the formulation.

A structural engineer needs a good understanding of mechanics and structural behavior as well as good judgment in idealizing a structure to be analyzed. This is true whether the analyst chooses to use either a classical or matrix approach. During the study of Chapters 8 and 9, the reader undoubtedly observed that the matrix approach which is to be computerized in this chapter required a rigorous and meticulous order in performing a structural analysis. In either a classical or matrix approach, careful attention to signs is necessary and structural analysts have to think rationally and carefully. In computerizing the Displacement Method, one must define and rigorously obey a routine that is tedious to a human but is very efficient for a computer. For example, in the noncomputerized solutions of Chapter 9, the matrix equation of system equilibrium was assembled by looking at sketches to find the needed values (and their signs) which were superimposed to obtain $[S]$ and $\{P^e\}$. In this chapter a scheme is presented to have the computer to generate and examine a matrix of coded numbers defining the compatibility of member-end displacements in the global reference and the system DOF at the joints in order to obtain $[S]$ and $\{P^e\}$ by superposition. Using such a number coded scheme is tedious for a human, but it is very efficient for a computer.

13.2 SEQUENCING TIPS FOR THE SYSTEM DISPLACEMENTS (DOF)

Available Displacement Method computer programs either have a fixed order for the DOF (their order usually is tied to the joint ordering scheme) or allow the user to specify the DOF order independently of the joint and member ordering schemes. Some of the available programs also have the capability of reordering the DOF for maximum computational efficiency. The DOF order is important for two reasons: (1) conservation of computer memory; and (2) precision in the solution for the system displacements.

Consider Figure 13.1 which illustrates the best and worst choices for the order of the DOF. The $\times$ symbol in the system stiffness matrices denotes the positions of all nonzero entries. The best DOF order has all of the nonzero entries located in a narrow band centered on the main diagonal of $[S]$. It will be subsequently shown that $[S]$ always is a symmetric matrix for a linearly elastic, first order, small displacement theory solution. Consequently, only the nonzero values above the main diagonal and the main diagonal values need to be appropriately stored in a computer in order to completely define $[S]$. Simultaneous equation solvers which deal with $[S]$ as a banded symmetric matrix are available. By appropriately storing only the nonzero upper banded part of $[S]$, computer memory is kept to a minimum; this is important for two reasons: (1) it enables the analyst to solve structural problems with more DOF before the main memory of the computer is used up; and (2) computer costs are a function of computer memory used. Note that the nonzero entries in $[S]$ for the worst DOF order are situated such that the band width of $[S]$ is a maximum, whereas the best DOF order gives the minimum possible band width for $[S]$.

The author's recommended rule-of-thumb procedure for ordering the DOF is to consider separate horizontal and vertical cuts through the structure such that a minimum number of joints is involved on a typical cut. For example, in the highrise (multistory) building with only a few bays shown in Figure 13.2, a typical horizontal cut involves only three joints, whereas a typical vertical cut involves five joints. Consecutively number the joints along each horizontal cut starting at the foundation cut and proceed floor by floor to the roof (or start with the horizontal cut at the roof level and work downward). Beginning at the first joint number at which there is a DOF, begin to consecutively number the DOF in X translation, Y translation, and Z rotation

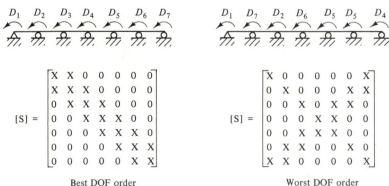

Best DOF order   Worst DOF order   **Figure 13.1**

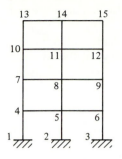

Best joint order − high rise

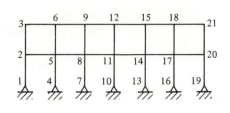

Best joint order − low rise multibay

Figure 13.2

order at each joint. This DOF order gives the minimum band width of [S] for the highrise building shown. Likewise, in the lowrise, multibay building shown in Figure 13.2 a typical horizontal cut involves seven joints, whereas a typical vertical cut only involves three joints. If the joints are numbered from bottom to top (or top to bottom) along each vertical cut starting at the left side of the building and proceeding from column line to column line to the right side of the building, and if the DOF are numbered in X, Y, Z displacement vector order at each consecutively encountered joint, the band width of [S] is a minimum.

For structures having a radial geometry, the trial typical cuts should be radial and circumferential to determine which cut has the minimum number of joints. If the radial geometry defines a complete circle in plan view, the concepts of symmetry and antisymmetry of loading for a symmetric structure or the concepts of substructuring if the structure is not symmetric are needed in order to minimize the bandwidth of [S].

For structures having complex or irregular joint geometry, the best joint ordering scheme is not always obvious and the rule-of-thumb procedure can only be used to get some idea of how to begin to think about the joint and DOF ordering scheme which gives the minimum bandwidth of [S].

13.3 MAXWELL-BETTI RECIPROCAL THEOREM

In Chapter 9, the author chose not to take advantage of the fact that [S] was a symmetric matrix. In the early stages of learning the Displacement Method, all nonzero entries in [S] should be computed for practice to learn which ones are more easily computed. If [S] does not turn out to be symmetric, then the analyst knows there is one or more errors in the mathematical calculations since [S] must be a symmetric matrix as we will shortly prove. For computerized solutions, a member stiffness matrix instead of a sketch is used to contain the member-end stiffness coefficients due to separate member-end displacements. Every member stiffness matrix is also a symmetric matrix and we need to prove this.

Wherever possible in the computer, we take advantage of the fact that a matrix must be symmetric. The author prefers to use the Maxwell-Betti reciprocal theorem in proving that the system stiffness matrix and each member stiffness matrix are symmetric matrices.

The proof of the Maxwell-Betti reciprocal theorem was made in Section 9.3; also, some simple examples were given to illustrate the usefulness of the Maxwell-Betti reciprocal theorem. For the case of a structure subjected to only concentrated loads, the Maxwell-Betti reciprocal theorem (MBRT) is:

> The external virtual work done by a set of m forces, F, during the deformation caused by a set of n forces, P, is equal to the external virtual work done by the P forces during the deformation caused by the F forces.

EXAMPLE 13.1 ——

The objective of this example is to prove that the system stiffness matrix is symmetric; that is, we want to prove that $S_{i,j} = S_{j,i}$ for all i and j except for $i = j$. In the interest of familiarity, we choose to use the results of Example 9.5 in the proof. The necessary deflected structures from Example 9.5 are repeated in Figure 13.3.

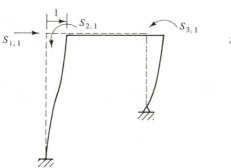
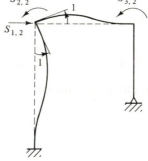

System 1 solution for $D_1 = 1, D_2 = D_3 = 0$ System 2 solution for $D_2 = 1, D_1 = D_3 = 0$

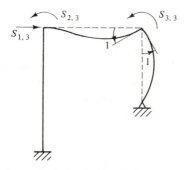

System 3 solution for $D_3 = 1, D_1 = D_2 = 0$ **Figure 13.3**

Solution

Apply MBRT for Systems 1 and 2: $S_{1,2} * 1 = S_{2,1} * 1$

Apply MBRT for Systems 1 and 3: $S_{1,3} * 1 = S_{3,1} * 1$

Apply MBRT for Systems 2 and 3: $S_{2,3} * 1 = S_{3,2} * 1$

The member stiffness matrix, $[S^M]_m$, for a prismatic member was derived in Section 8.4; by inspection it is symmetric and need not be proved to be symmetric. The proof that $[S^M]_m$ is symmetric for a nonprismatic member is given in Section 13.4 where the derivation of $[S^M]_m$ is shown.

13.4 MEMBER STIFFNESS MATRIX FOR NONPRISMATIC MEMBERS

For a nonprismatic, plane frame member, EA and EI vary along the member length. It was shown in Section 9.6 that member fixed ended forces due to loads and member-end stiffness coefficients can be obtained for such members provided the variation in EA and EI is known. The reader should review Section 9.6 before proceeding to study the following material.

In Figure 13.4, a typical nonprismatic member for a plane frame analysis is shown. Let EA and EI be the values at the *origin* of the *local* (or member oriented) *axes*, x and y, and accept the fact that we must already know the variation in EA and EI at all points along the member length. The digits 1 through 6 on the member-end vectors are the subscript numbers defining the positions in the arrays to be created in a computer program to contain the entries for the member-end displacements ($\{D^M\}$), forces ($\{F\}$ and $\{F^f\}$), and stiffness coefficients ($[S^M]$) in the local reference for member m. Also the vector directions shown in Figure 13.4 define the positive directions for the member-end displacements, forces, and stiffness coefficients. For member m, the member-end force-deformation relation is $\{F\}_m = \{F^f\}_m + [S^M]_m\{D^M\}_m$ where $\{F\}_m$ are the total member-end forces, $\{F^f\}_m$ are the member fixed ended forces due to member loads, and via matrix multiplication $[S^M]_m\{D^M\}_m$ gives the member-end forces due to member-end displacements after the system displacements have been

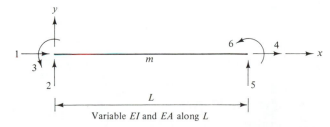

Note: m is the member number which in the computer is an additional
    subscript on all items shown in the above sketch; for example, L_m,
    A_m, I_m, x_m, y_m and $\{D^M\}_m$ where the entries in $\{D^M\}_m$ are $D^M_{1,m}, ..., D^M_{6,m}$.

Figure 13.4   Typical nonprismatic, plane frame member.

computed. Our objective here is to define the expressions from which the numerical entries in $[S^M]_m$ are to be computed.

For a first order, linear displacement method of analysis, $[S^M]_m$ is derived assuming that member m is perfectly straight and does not buckle when loads are applied. In other words, the assumptions are that (1) axial stiffness for axial shortening is the same as it is for axial elongation; (2) no member bending occurs due to axial tension or compression member-end forces; and (3) equations of statics involving the member-end forces are written in terms of the member geometry which existed before the member-end displacements occurred. Expressions for the entries in $[S^M]_m$ are obtained one column at a time for member m. That is, we induce $D_i^M = 1$ for each value of $i = 1, 2, \ldots, 6$ and we require that all of the other member-end displacements be zero while the unit valued member-end displacement is being induced. In Figure 13.5, only the nonzero member-end stiffness coefficients are shown and a slash is placed on all unknown vectors. The known force vectors are obtained as described in Section 9.6 for a nonprismatic member. Example 9.14 numerically illustrated how to determine the axial stiffness value for a member composed of two prismatic pieces. The axial stiffness was $1.6667EA/L$ and can be expressed as $C_1 EA/L$ which means that for

(a) Nonzero entries in column 1 of $[S^M]_m$

(b) Nonzero entries in column 4 of $[S^M]_m$

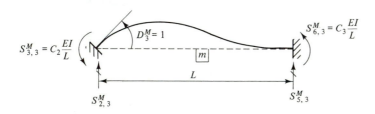

Summation of moments at the origin of member m gives $S_{5,3}^M = -(C_2 + C_3)\dfrac{EI}{L^2}$.

For convenience, let $C_5 = C_2 + C_3$ and $S_{5,3}^M = -C_5 \dfrac{EI}{L^2}$. Summation of vertical

forces gives $S_{2,3}^M = C_5 \dfrac{EI}{L^2}$

(c) Nonzero entries in column 3 of $[S^M]_m$

Figure 13.5 Nonzero entries in $[S^M]_m$.

$$S_{3,6}^M = C_3 \frac{EI}{L} \qquad \boxed{m} \qquad S_{6,6}^M = C_4 \frac{EI}{L}$$

$$S_{2,6}^M = C_6 \frac{EI}{L^2} \qquad D_6^M = 1 \qquad S_{5,6}^M = -C_6 \frac{EI}{L^2}$$

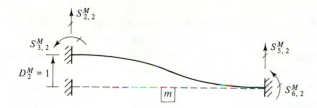

where $C_6 = C_3 + C_4$

Maxwell-Betti reciprocal theorem for Figures (c) and (d) gives $S_{3,6}^M \star D_3^M = S_{6,3}^M \star D_6^M$, since $D_3^M = D_6^M = 1$, then $S_{3,6}^M = S_{6,3}^M$.

(d) Nonzero entries in column 6 of $[S^M]_m$

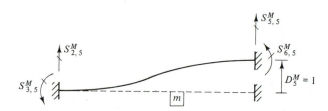

MBRT for Figures (c) and (e) gives $S_{3,2}^M \star D_3^M = S_{2,3}^M \star D_2^M$; since $D_2^M = D_3^M = 1$, then $S_{3,2}^M = S_{2,3}^M = C_5 \frac{EI}{L^2}$. MBRT for Figures (d) and (e) gives

$S_{6,2}^M \star D_6^M = S_{2,6}^M \star D_2^M$; since $D_2^M = D_6^M = 1$, then $S_{6,2}^M = S_{2,6}^M = C_6 \frac{EI}{L^2}$. Summation of

moments at the origin of member m gives $S_{5,2}^M = -C_7 \frac{EI}{L^3}$ where $C_7 = (C_5 + C_6)$.

Summation of vertical forces gives $S_{2,2}^M = C_7 \frac{EI}{L^3}$.

(e) Nonzero entries in column 2 of $[S^M]_m$

MBRT for Figures (c) and (f) gives $S_{3,5}^M \star D_3^M = S_{5,3}^M \star D_5^M$; since $D_3^M = D_5^M = 1$,

then $S_{3,5}^M = S_{5,3}^M = -C_5 \frac{EI}{L^2}$. MBRT for Figures (d) and (f) gives

$S_{6,5}^M \star D_6^M = S_{5,6}^M \star D_5^M$; since $D_5^M = D_6^M = 1$, then $S_{6,5}^M = S_{5,6}^M = -C_6 \frac{EI}{L^2}$. Summation of

moments at the origin of member m gives $S_{5,5}^M = C_7 \frac{EI}{L^3}$. Summation of

vertical forces gives $S_{2,5}^M = -C_7 \frac{EI}{L^3}$.

(f) Nonzero entries in column 5 of $[S^M]_m$

Example 9.14, $C_1 = 1.6667$. Similarly, the rotationally oriented member-end stiffness coefficients can be obtained for the member of Example 9.14 as shown in Example 9.13 and in expression form they are $C_2 EI/L = 4.56EI/L$; $C_3 EI/L = 3.50EI/L$; and $C_4 EI/L = 9.54EI/L$.

By expressing the axial and rotational direction member-end stiffness coefficients in terms of the symbols C_1, C_2, C_3, and C_4 as well as EA/L and EI/L where EA and EI are values at the origin of member m, this enables a computer user to input C_1 through C_4, EA, EI, and L which may be different for each nonprismatic member. The computer is programmed to evaluate the expressions for each member using the numerical input values and stores the computed stiffness coefficients in the locations specified in the arithmetic assignment statements of the computer program. For prismatic members, the computer program automatically sets $C_1 = 1$, $C_2 = 4$, $C_3 = 2$, and $C_4 = 2$.

In Figures 13.5 and 13.6, all values defining the member stiffness matrix coefficients (C_1 through C_7, E, A, I, and L) have the member number m attached as a subscript by implication. The author chose to show this implication by placing the member number m on the member and on the outside of the matrix symbols instead of placing m on each symbol that is potentially different for each member. Although the author prefers to use the values of EA and EI at the origin of the member as the reference values, any other position along L can be chosen as the point at which EA and EI values are used to compute the values of C_1 through C_4.

The member-end force-deformation relation, $\{F\}_m = \{F^f\}_m + [S^M]_m\{D^M\}_m$, stated in nonmathematical form is (total member-end forces) = (fixed-ended forces) + (member-end forces due to member-end displacements). After the matrix equation of system equilibrium, $[S]\{D\} = \{P^e\}$, has been formulated and solved for the system displacements, $\{D\}$, the member-end displacements can be determined in the local (or member) reference. Consequently, the member-end force-deformation relation

$$[S^M]_m = \begin{bmatrix} C_1\dfrac{EA}{L} & 0 & 0 & -C_1\dfrac{EA}{L} & 0 & 0 \\[2ex] 0 & C_7\dfrac{EI}{L^3} & C_5\dfrac{EI}{L^2} & 0 & -C_7\dfrac{EI}{L^3} & C_6\dfrac{EI}{L^2} \\[2ex] 0 & C_5\dfrac{EI}{L^2} & C_2\dfrac{EI}{L} & 0 & -C_5\dfrac{EI}{L^2} & C_3\dfrac{EI}{L} \\[2ex] -C_1\dfrac{EA}{L} & 0 & 0 & C_1\dfrac{EA}{L} & 0 & 0 \\[2ex] 0 & -C_7\dfrac{EI}{L^3} & -C_5\dfrac{EI}{L^2} & 0 & C_7\dfrac{EI}{L^3} & -C_6\dfrac{EI}{L^2} \\[2ex] 0 & C_6\dfrac{EI}{L^2} & C_3\dfrac{EI}{L} & 0 & -C_6\dfrac{EI}{L^2} & C_4\dfrac{EI}{L} \end{bmatrix}_m$$

where EA and EI are reference values at the origin of the member; $C_5 = C_2 + C_3$; $C_6 = C_3 + C_4$, and $C_7 = C_5 + C_6$. For a prismatic member, note that $C_1 = 1$, $C_2 = C_4 = 4$, $C_3 = 2$, $C_5 = C_6 = 6$, and $C_7 = 12$.

Figure 13.6 Expanded form of $[S^M]_m$ for a plane frame member.

can be evaluated for each member to obtain the total member-end forces for each loading. Note that the system displacements are parallel to the global axes and the member-end displacements are needed parallel to the local axes of each member. Global member-end displacements can be obtained by inspection from $\{D\}$. If the member's local x axis is not parallel to the global X axis, the global member-end displacements must be resolved into components parallel to the member's local axes and algebraically summed to obtain the local member-end displacements. In a computerized solution terminology, a transformation must be made from global member-end displacements to local member-end displacements in order to use the member-end force-deformation relation shown above which is for local member-end forces and displacements. In Section 13.5, this transformation relation is derived and stated in the form of a transformation matrix.

In this chapter, the emphasis is on information needed in a plane frame analysis. See Section 9.8 for information pertaining to plane and space trusses. The member stiffness matrix for a space frame analysis is given in Appendix B.

13.5 TRANSFORMATION MATRICES

Global member-end forces are needed in an analysis to create the matrix equation of system equilibrium and to check equilibrium of the system. Local member-end forces are needed in design to check the member's design code requirements for (1) shear and (2) axial compression or tension plus bending moment. The objective in this section is to derive the relations to transform (1) the global member-end forces to local member-end forces, and (2) the local member-end forces to global member-end forces.

In Figure 13.7 the plane frame member shown is not parallel to the global X axis. Assume that we know or are given numerical values for the global member-end forces as well as the angle θ in Figure 13.7a and desire the local member-end forces in Figure 13.7b. The translational-type member-end forces in Figure 13.7a have components parallel to the x and y axes in Figure 13.7b. Therefore, at the origin of the member, the relation of the local member-end forces to the global member-end forces is as shown in Figure 13.7b. At the $x = L$ end of the member, the relation of the local member-end forces to the global member-end forces is obtained by inspection from the relationship at the origin of the member. Consequently, the local member-end forces are related to the global member-end forces as shown in Figure 13.7c.

Since the member-end forces in the local and global references are located at and have the same positive direction definitions, respectively, as the local and global member-end displacements, it is obvious that the same transformation matrix that transforms global member-end forces to local member-end forces also transforms global member-end displacements, $\{D^{MG}\}_m$, to local member-end displacements, $\{D^M\}_m$. That is, $\{D^M\}_m = [T]_m\{D^{MG}\}_m$.

Since the elements in matrix $[T]$ are as shown in Figure 13.7c, the transpose of $[T]$, denoted as $[T]^T$, can be obtained by taking each of the elements in row i of $[T]$ and storing them in the same order in column i of $[T]^T$ for all values of $i = 1$ through 6 in this example. Therefore, the definition of $[T]_m^T$ is as shown below.

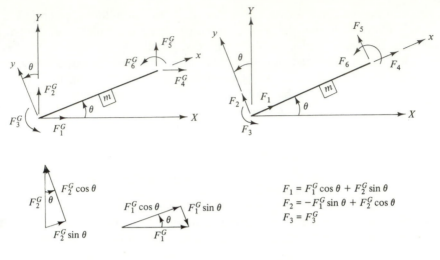

$$F_1 = F_1^G \cos\theta + F_2^G \sin\theta$$
$$F_2 = -F_1^G \sin\theta + F_2^G \cos\theta$$
$$F_3 = F_3^G$$

(a) Global member-end forces   (b) Local member-end forces

$$
\left\{ \begin{array}{c} F_1 \\ F_2 \\ F_3 \\ F_4 \\ F_5 \\ F_6 \end{array} \right\}_m
=
\begin{bmatrix}
\cos\theta & \sin\theta & 0 & 0 & 0 & 0 \\
-\sin\theta & \cos\theta & 0 & 0 & 0 & 0 \\
0 & 0 & 1 & 0 & 0 & 0 \\
0 & 0 & 0 & \cos\theta & \sin\theta & 0 \\
0 & 0 & 0 & -\sin\theta & \cos\theta & 0 \\
0 & 0 & 0 & 0 & 0 & 1
\end{bmatrix}_m
\left\{ \begin{array}{c} F_1^G \\ F_2^G \\ F_3^G \\ F_4^G \\ F_5^G \\ F_6^G \end{array} \right\}_m
$$

and in condensed form $\left\{ F \right\}_m = [T]_m \left\{ F^G \right\}_m$

(c) Relation of local to global member-end forces

Figure 13.7   Transformation of global to local member-end forces.

$$
[T]_m^T =
\begin{bmatrix}
\cos\theta & -\sin\theta & 0 & 0 & 0 & 0 \\
\sin\theta & \cos\theta & 0 & 0 & 0 & 0 \\
0 & 0 & 1 & 0 & 0 & 0 \\
0 & 0 & 0 & \cos\theta & -\sin\theta & 0 \\
0 & 0 & 0 & \sin\theta & \cos\theta & 0 \\
0 & 0 & 0 & 0 & 0 & 1
\end{bmatrix}_m
$$

For convenience of discussion, let $C = \cos\theta$, $S = \sin\theta$, and $[X] = [T]^T[T]$ where matrix multiplication is implicitly understood. After the indicated matrix multiplications have been performed and the trigonometric expressions have been evaluated, we find that all of the diagonal elements in $[X]$ are unity and the other elements in $[X]$ are zero (for example, $X_{1,1} = C^2 + S^2 = 1$). This tells us that $[T]^T[T] = [I]$

where $[I]$ is the identity matrix in which all of the diagonal elements are unity and the other elements are zero. Let $[T]^{-1}$ denote the inverse matrix of $[T]$. The definition of an inverse matrix is: if the original matrix is premultiplied by its inverse matrix, the resulting matrix is the identity matrix, $[I]$. Therefore, $[T]^T = [T]^{-1}$ since $[T]^T[T] = [I]$.

If we premultiply both sides of $\{D^M\}_m = [T]_m\{D^{MG}\}_m$ by $[T]_m^T$, we obtain $\{D^{MG}\}_m = [T]_m^T\{D^M\}_m$. Similarly, $\{F^G\}_m = [T]_m^T\{F\}_m$. That is, the transpose of the transformation matrix transforms local member-end displacements or forces to global member-end displacements or forces.

See Appendix B for the transformation matrix derivation for a space frame member.

13.6 GLOBAL MEMBER STIFFNESS MATRIX

In order to computerize the generation of the matrix equation of system equilibrium, $[S]\{D\} = \{P^e\}$, the member fixed ended forces due to loads and member-end stiffness coefficients in the global reference are needed. Assume that the member fixed ended moments in the local reference, $\{F^f\}_m$, are to be input as data to the computer program. After the member properties have been input as data, the computer program can generate the member stiffness matrix in the local reference, $[S^M]_m$. Also, the relation of the local member-end displacements to global member-end displacements is known; namely, $\{D^M\}_m = [T]_m\{D^{MG}\}_m$. The member-end force-deformation relation in the global reference can be obtained as follows.

Premultiply both sides of $\{F\}_m = \{F^f\}_m + [S^M]_m\{D^M\}_m$ by $[T]_m^T$ and substitute $\{D^M\}_m = [T]_m\{D^{MG}\}_m$ to obtain

$$[T]_m^T\{F\}_m = [T]_m^T\{F^f\}_m + [T]_m^T[S^M]_m[T]_m\{D^{MG}\}_m$$

The first two matrix multiplications, respectively, transform the total member-end forces and member fixed ended forces in the local reference to total member-end forces and member fixed ended forces in the global reference. If we properly rename these two matrix products and let

$$[S^{MG}]_m = [T]_m^T[S^M]_m[T]_m$$

we have obtained the member-end force-deformation relation in the global reference:

$$\{F^G\}_m = \{F^{fG}\}_m + [S^{MG}]_m\{D^{MG}\}_m$$

$[S^{MG}]_m = [T]_m^T[S^M]_m[T]_m$ is the member stiffness matrix in the global reference. In the interest of simplicity, the author prefers to say that $[S^{MG}]_m$ is the global member stiffness matrix. If a member's local x axis is parallel to the global X axis, $[S^{MG}]_m = [S^M]_m$; otherwise $[S^{MG}]_m$ must be obtained from the matrix triple product definition shown above. It should be noted that the matrix triple product using the definition of matrix multiplication must be performed in the following sequence: (1) the first two matrices are multiplied to obtain a matrix, $[X]$; (2) the last matrix $[T]$ is premultiplied by $[X]$. These two matrix multiplications can be performed in the computer. Since there are a lot of zero values in each of the three matrices, to speed up the computer computations the matrix triple product can be performed in algebraic terms to obtain algebraic expressions for the elements in $[S^{MG}]_m$. In computer programming terminology, these algebraic expressions become the arithmetic assignment statements which define the elements in $[S^{MG}]_m$. The author chooses not to show these algebraic expressions in the text, but they were obtained and used in the computer program which is available to teachers who adopt this book as a textbook requirement for a structural analysis course.

13.7 MEMBER-SYSTEM-COMPATIBILITY MATRIX

In the noncomputerized solutions of Chapter 11, the author chose to make extensive usage of the deformed structure sketches to communicate with the reader in regard to proper signs as well as the values for fixed ended forces, member-end stiffness coefficients, fixed joint loads, and system stiffness matrix coefficients. This approach enabled the reader to focus directly on what absolutely had to be done as a minimum in regard to computing global member-end forces, for example. Also, usage of the deformed structure sketches made it obvious which member ends of which members had to be moved a unit amount in a global direction when the entries, for example, in the jth column of the system stiffness matrix were being computed. At that time, the analyst did not have to worry about the member number nor the global member stiffness matrix name and subscript values in visually selecting either an axial-type, or a shear-type, or moment-type member-end force value to be used in the algebraic summation process to compute each of the elements in column j of the system stiffness matrix. However, as pointed out in the next paragraph, efficiency of decisions in a computerized solution is achieved best via a nonvisual process.

In an efficient computerized Displacement Method solution, the needed global member-end stiffnesses and fixed ended forces are stored in arrays (in computer programming terminology, a matrix is an array with two subscripts) instead of on sketches as was done in the noncomputerized solutions of Chapter 11. Suppose we examine a computer program involving arrays A and B which have two subscripts (assume each subscript ranges from 1 to 6) and a scalar variable named C, and find the statement $C = A(3,5) + B(4,2)$. This assignment statement and the contained arithmetic expression instruct the computer to (1) compute the sum of $A_{3,5}$ and $B_{4,2}$ and (2) store the sum in C. Note that recall of an element in an array is achieved by specifying the array name and a value for each of the array's subscripts. Similarly, in a decision statement that involves two arithmetic expressions, the computer is instructed to obtain and compare the numerical values of the arithmetic expressions. That is, the computer can compare no more than two numbers at a time in order to make a decision posed in terms of arithmetic expressions. Furthermore, all possibilities that can exist in a multifaceted decision must be sequentially programmed and sequentially examined by the computer until the unique answer is found. In the visual decision-making process of the noncomputerized Displacement Method approach, the analyst could readily determine which of the members and their global member-end stiffnesses were needed in computing the entries in the jth column of the system stiffness matrix; the unneeded global member-end stiffnesses were not computed. In the computerized

Displacement Method approach, each global member stiffness matrix is computed and a scheme must be devised to define which entries in each global member stiffness matrix contribute via superposition to the generation of certain entries in the system stiffness matrix. The devised scheme must ensure compatibility of global member-end displacements and system DOF. Therefore, the matrix that contains the information for the computer to use in selecting the proper global member stiffness matrix elements to superimpose in obtaining the system stiffness matrix elements is named the member-system-compatibility matrix.

Before we can define the member-system-compatibility matrix and show how it is used, there are a few preliminary details that must be introduced. Consider Figure 13.8 which shows the following information: global (system) axes; joint numbers; local (member) axes; member numbers; the system DOF required in a computerized solution; and the reaction direction numbers. Note that the DOF are numbered sequentially in X-translation, Y-translation, Z-rotation order at each joint in sequential joint order beginning at the first joint which has a DOF. Also, note that the reaction direction numbers are similarly assigned beginning at the lowest joint number at which there are reaction components. If we denote the last DOF number as n, the

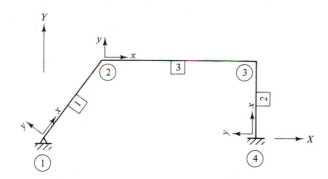

(a) Global axes, joint numbers, member numbers, and local axes

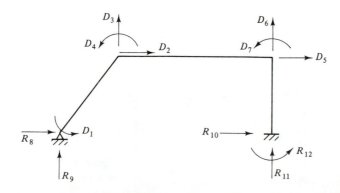

(b) DOF and reaction direction numbers

Figure 13.8 Example of information needed to generate [MSC] entries.

reaction direction numbers are sequentially numbered beginning with $n + 1$. The reaction direction numbers are used in the author's Plane Frame Program to obtain the reactions via superposition from the final global member-end forces of members connected to support joints.

Data from Figure 13.8 which must be known in the author's computer program before the program can generate the system DOF and reaction direction numbers are total number of all joints; total number of support joints; the support joint number and support release data for each support joint; number of members; and member incidences for each member (member incidences are the origin joint number and the end joint number of each member's local x axis). After the needed data are available, the computer program generates the DOF numbers and reaction direction numbers at the appropriate joints in the manner described in the previous paragraph (any joint for which no support release data were given automatically has 3 DOF).

From the computer input data for the example shown in Figure 13.8, the computer has the information needed to establish the following relations between the global member-end displacements and the system DOF (note: before proceeding, the reader may need to review the definitions given in Figure 13.7a to establish that the first three and last three global member-end displacements, respectively, are always located at the origin joint and end joint of each member's local x axis):

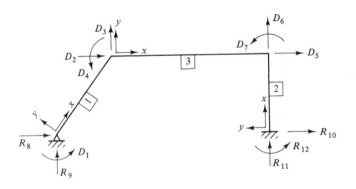

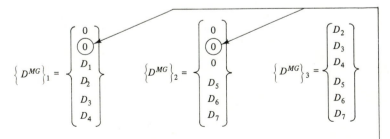

$$\{D^{MG}\}_1 = \begin{Bmatrix} 0 \\ 0 \\ D_1 \\ D_2 \\ D_3 \\ D_4 \end{Bmatrix} \qquad \{D^{MG}\}_2 = \begin{Bmatrix} 0 \\ 0 \\ 0 \\ D_5 \\ D_6 \\ D_7 \end{Bmatrix} \qquad \{D^{MG}\}_3 = \begin{Bmatrix} D_2 \\ D_3 \\ D_4 \\ D_5 \\ D_6 \\ D_7 \end{Bmatrix}$$

(A zero entry signifies that there is no system DOF at this global member-end displacement location)

Figure 13.9 Relation of global member-end displacements to the system DOF.

An entry in any column of the member-system-compatibility matrix [MSC], is either a system DOF subscript number or, where there is no system DOF, a negative number whose absolute value is a reaction direction number. Note that [MSC] is generated one column at a time (or, in structural engineering terminology, [MSC] is generated on a member by member basis). For the example shown in Figure 13.9, [MSC] is:

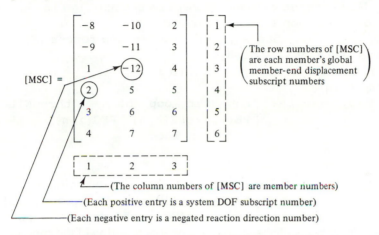

Note that [MSC] is a 6 by number of members matrix—there are 6 entries for every member. Let i denote an arbitrary matrix row number and m denote an arbitrary matrix column number. For $i = 4$ and $m = 1$, the entry in [MSC] is 2; this tells the computer that $D_4^{MG} = D_2$ for member 1. For $i = 3$ and $m = 2$ the entry in [MSC] is -12; since this entry is negative, this tells the computer that there is not a system DOF at D_3^{MG} of member 2 and that reaction component R_{12} is coincident with D_3^{MG} of member 2. For each member, m, the computer can scan all entries in the mth column of [MSC] using an index or counting parameter, i, for $i = 1$ through 6 and either determine which system DOF or determine, if the entry is negative, which reaction component is coincident with a global member-end displacement direction.

In Section 13.2, it was pointed out that the system stiffness matrix is symmetric and only the upper, banded part of the system stiffness matrix is needed in solving the matrix equation of system equilibrium for the system displacements. Let MBW denote the matrix bandwidth of the system stiffness matrix. In a computer program, MBW can be determined from the positive entries in [MSC] as follows:

1. For each column in [MSC], subtract the smallest positive entry from the largest positive entry and add 1 which gives MBW for each structural member—for the above example, the values obtained are $(4 - 1) + 1 = 4$, $(7 - 5) + 1 = 3$, and $(7 - 2) + 1 = 6$.
2. The largest of all these values is the MBW of [S], MBW = 6 for the above example. See Figure 13.14 for an example of a banded system stiffness matrix.

Also, the positive entries in [MSC] can be used in a computer program to find the fixed joint loads, $\{P^f\}$, needed in the matrix relation $\{P^e\} = \{P^a\} - \{P^f\}$. To conserve computer memory, the applied joint loads, $\{P^a\}$, are stored directly into [PE] which is a NSD by NILC array where NSD denotes the number of system displacements (number of DOF) and NILC denotes the number of independent load cases. After all of the data needed to fully define a structure and the independent load cases have been input, [PE] is updated as shown in the following pseudocomputer program:

```
reminder: current values in [PE] are the applied loads
in a computer loop for m = 1 thru NM
   compute m-th member's global fixed-ended forces [FFG]
   which is a 6 by NILC array.
   in a computer loop for i = 1 thru 6
   k=MSC(i,m)
   if k>0, execute:   ┌ computer loop for n=1 thru NILC
                      │ PE(k,n)=PE(k,n)-FFG(i,n)
                      └ end of n loop
   end of i loop
end of m loop
```

That is, each member's global fixed ended force coinciding with the kth system DOF is subtracted from the current kth value of [PE] for all independent load cases.

Similarly, the positive entries in [MSC] can be used in a computer program to generate the system stiffness matrix, [S], by superposition of the appropriate elements of the global member stiffness matrices, $[S^{MG}]_m$ for $m = 1$ through NM. In the author's program, [S] is a NSD by MBW array; that is, [S] is generated and stored in banded matrix form for computer efficiency. To simplify the discussion, the author chooses to illustrate how the positive entries in [MSC] can be used to generate [S] as a NSD by NSD array. First, all elements in [S] are set to zero and [S] is updated as shown in the following pseudocomputer program where [SMG] is a 6 by 6 array:

```
in a computer loop for m = 1 thru NM
   compute m-th global member stiffness matrix [SMG]
   in a computer loop for n = 1 thru 6
   kk=MSC(n,m)
   if kk>0 execute:   ┌ computer loop for j=1 thru 6
                      │ jj=MSC(j,m)
                      │ if jj>0, S(jj,kk)=S(jj,kk)+SMG(j,n)
                      └ end of j loop
   end of n loop
end of m loop
```

That is, the element in the jth row and nth column of $[S^{MG}]_m$ is added to the element in the jjth row and kkth column of [S].

After the matrix equation of system equilibrium has been solved for the system displacements, the computer uses the positive entries in [MSC] to pick out the global member-end displacements $\{D^{MG}\}_m$ for each member m from the array of computed system displacements.

After the final local member-end forces have been computed, they are transformed to obtain the final global member-end forces. Finally, the negative entries in

[MSC] are used in the author's computer program to obtain the reactions by super-position of the final global member-end forces from the contributing members.

Therefore, the member-system-compatibility matrix, [MSC], plays a very important role in the author's computer program.

13.8 ILLUSTRATIVE EXAMPLE AND COMPUTERIZED PROCEDURAL STEPS

In all of the examples in Chapter 9, axial deformations of all members were ignored to keep the number of DOF to a minimum. However, almost all available Displacement Method computer programs automatically account for axial deformations of all mem-

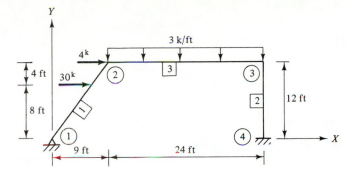

Members 1 and 2: $EA = 298,000^k$; $EI = 103,312.5$ kft^2
    Member 3: $EA = 469,800^k$; $EI = 229\,583.3$ kft^2

Loading 1 is shown above

Loading 2 is prescribed 0.036 ft horizontal movement to the right
        at joint 4

Loading 3 is prescribed 0.099 ft vertical movement downward at joint 4

Loading 4 is loading 1 + loading 2

Loading 5 is loading 1 + loading 3

Loading 6 is loading 1 + loading 2 + loading 3

(a) Structural geometry, properties, and loadings

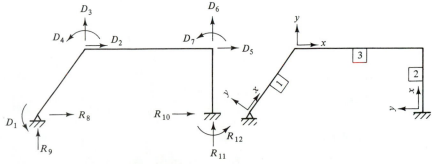

(b) DOF and reaction direction numbers   (c) Local axes

Figure 13.10 Illustrative problem description.

bers. If axial deformations are automatically accounted for, the computer program is easier to write and solving a few more simultaneous equations is no big deal in the computer.

The structure shown in Figure 13.10 is chosen as the illustrative example in the following discussion. It was chosen such that it contains all of the concepts involved in the computerized solution of a plane frame composed of all prismatic members. All of the solution pieces needed in a noncomputerized solution to generate the matrix equation of system equilibrium for this example are shown in addition to the results of the computerized solution. The computer program was written and debugged in small, manageable portions and the noncomputerized solution pieces were used as an example in debugging the computer program. In the debugging phase, all of the intermediate calculations prior to the solution of the matrix equation of system equilibrium were printed out and compared to the noncomputerized solution. After this

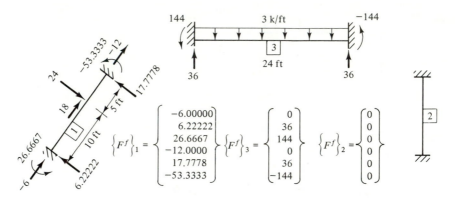

Local fixed ended forces (only the nonzero ones are shown)

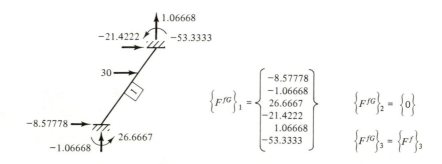

Global fixed ended forces

(a) Fixed ended forces for loading 1

Figure 13.11   Fixed ended forces in local and global references.

major portion of the program was fully debugged, the print statements for the intermediate results were deleted. Similarly, although they are not shown, the noncomputerized solution pieces involved after the solution of the matrix equation of system equilibrium were obtained and used in debugging the remainder of the computer program. It should be noted that the noncomputerized solution pieces were obtained by looking at sketches and by resolving the force vectors into components instead of following the definitions that the computerized solution used. However, the noncomputerized solution pieces were done in such a manner that all of the intermediate calculations in the computerized solution could be fully checked. Furthermore, the noncomputerized solution pieces were done before the computer program writing process was started.

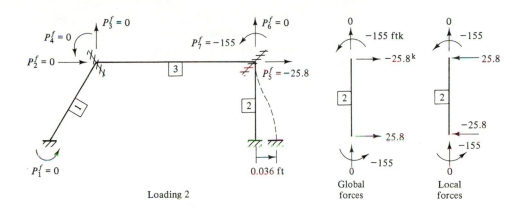

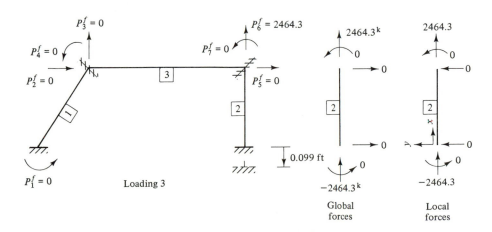

(b) Fixed ended forces for loadings 2 and 3

$$\{P^a\} = \begin{Bmatrix} 0 & 0 & 0 \\ 4 & 0 & 0 \\ 0 & 0 & 0 \\ 0 & 0 & 0 \\ 0 & 0 & 0 \\ 0 & 0 & 0 \\ 0 & 0 & 0 \end{Bmatrix} \quad \{P^f\} = \begin{Bmatrix} 26.6667 & 0 & 0 \\ -21.4222 & 0 & 0 \\ 37.0668 & 0 & 0 \\ 90.6667 & 0 & 0 \\ 0 & -25.8 & 0 \\ 36 & 0 & 2464.3 \\ -144 & -155 & 0 \end{Bmatrix}$$

$$\{P^e\} = \{P^a\} - \{P^f\} = \begin{Bmatrix} -26.6667 & 0 & 0 \\ 25.4222 & 0 & 0 \\ -37.0668 & 0 & 0 \\ -90.6667 & 0 & 0 \\ 0 & 25.8 & 0 \\ -36 & 0 & -2464.3 \\ 144 & 155 & 0 \end{Bmatrix}$$

Figure 13.12 Applied, fixed, and effective joint loads.

The author's FRAME program which solves either plane frame or space frame problems was written in FORTRAN 77 for execution on IBM compatible microcomputers. FRAME was written for microcomputers that have two floppy disk drives and no hard disk. A User's Manual accompanies FRAME which is available on a floppy disk from the publisher to those professors who adopt this book as a textbook requirement for a course. Also on this disk is an interactive, free-format program which creates the input data file in the sequence required by FRAME.

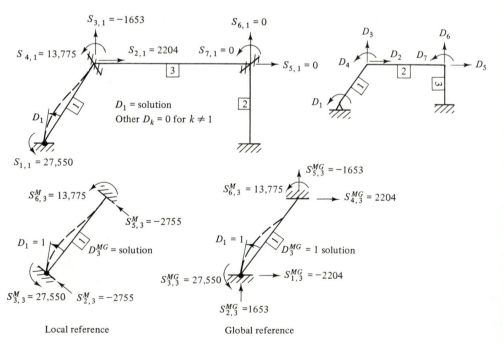

Member stiffness coefficients needed to obtain column 1 of system stiffness matrix

(a) Column 1 of system stiffness matrix

Figure 13.13

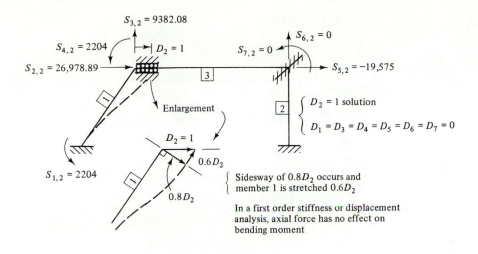

$S_{3,2} = 9382.08$

$S_{4,2} = 2204$

$S_{2,2} = 26,978.89$

$D_2 = 1$

$S_{6,2} = 0$

$S_{7,2} = 0$

$S_{5,2} = -19,575$

3

Enlargement

$\begin{cases} D_2 = 1 \text{ solution} \\ D_1 = D_3 = D_4 = D_5 = D_6 = D_7 = 0 \end{cases}$

2

$S_{1,2} = 2204$

$D_2 = 1$

$0.6D_2$

$0.8D_2$

$\begin{cases} \text{Sidesway of } 0.8D_2 \text{ occurs and} \\ \text{member 1 is stretched } 0.6D_2 \end{cases}$

In a first order stiffness or displacement analysis, axial force has no effect on bending moment

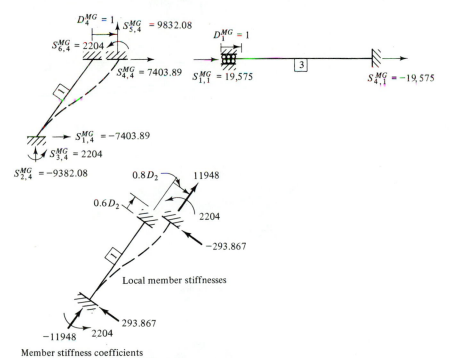

$D_4^{MG} = 1$ $S_{5,4}^{MG} = 9832.08$

$S_{6,4}^{MG} = 2204$

$S_{4,4}^{MG} = 7403.89$

$D_1^{MG} = 1$

$S_{1,1}^{MG} = 19,575$

3

$S_{4,1}^{MG} = -19,575$

$S_{1,4}^{MG} = -7403.89$

$S_{3,4}^{MG} = 2204$

$S_{2,4}^{MG} = -9382.08$

$0.8D_2$ 11948

$0.6D_2$ 2204

-293.867

Local member stiffnesses

293.867

-11948 2204

Member stiffness coefficients

(b) Column 2 of system stiffness matrix

551

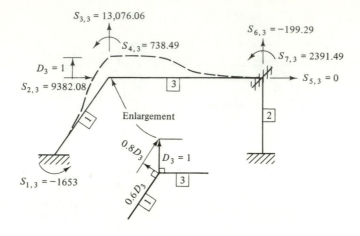

Enlargement

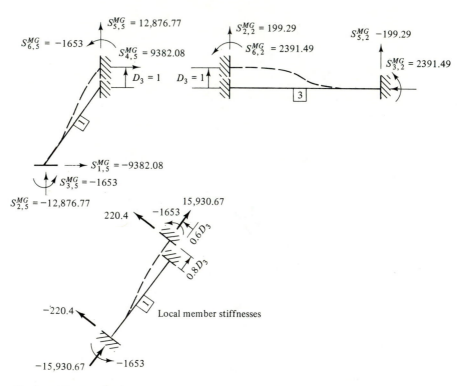

Local member stiffnesses

Member stiffness coefficients

(c) Column 3 of system stiffness matrix

Figure 13.13 (*continued*)

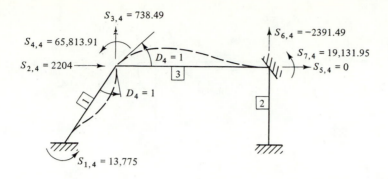

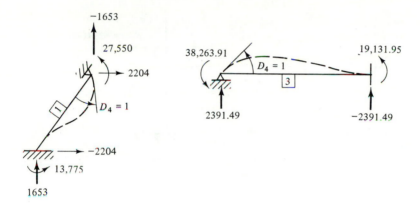

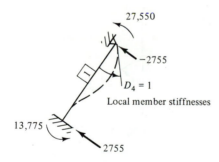

Member stiffness coefficients

(d) Column 4 of system stiffness matrix

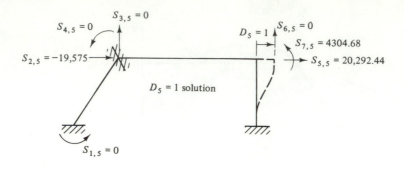

$S_{3,5} = 0$

$S_{4,5} = 0$

$D_5 = 1$ $S_{6,5} = 0$

$S_{7,5} = 4304.68$

$S_{2,5} = -19,575$

$D_5 = 1$ solution

$S_{5,5} = 20,292.44$

$S_{1,5} = 0$

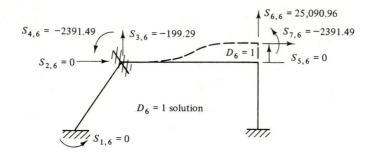

$S_{6,6} = 25,090.96$

$S_{4,6} = -2391.49$ $S_{3,6} = -199.29$

$S_{7,6} = -2391.49$

$S_{2,6} = 0$

$D_6 = 1$ $S_{5,6} = 0$

$D_6 = 1$ solution

$S_{1,6} = 0$

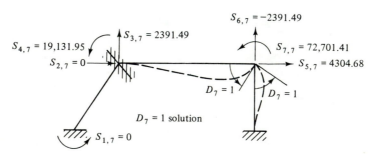

$S_{6,7} = -2391.49$

$S_{3,7} = 2391.49$

$S_{4,7} = 19,131.95$

$S_{7,7} = 72,701.41$

$S_{2,7} = 0$

$S_{5,7} = 4304.68$

$D_7 = 1$ $D_7 = 1$

$D_7 = 1$ solution

$S_{1,7} = 0$

(e) Columns 5, 6, and 7 of system stiffness matrix

Figure 13.13 (*continued*)

554

The matrix equation of system equilibrium, $[S]\{D\} = \{P^e\}$, is:

$$
\begin{bmatrix}
27550 & 2204 & -1653 & 13775 & 0 & 0 & 0 \\
2204 & 26978.88 & 9382.08 & 2204 & -19575 & 0 & 0 \\
-1653 & 9382.08 & 13076.06 & 738.49 & 0 & -199.29 & 2391.49 \\
13775 & 2204 & 738.49 & 65813.91 & 0 & -2391.49 & 19131.95 \\
0 & -19575 & 0 & 0 & 20292.44 & 0 & 4304.68 \\
0 & 0 & -199.29 & -2391.49 & 0 & 25090.96 & -2391.49 \\
0 & 0 & 2391.49 & 19131.95 & 4304.68 & -2391.49 & 72701.41
\end{bmatrix}
\begin{Bmatrix}
D_1 \\ D_2 \\ D_3 \\ D_4 \\ D_5 \\ D_6 \\ D_7
\end{Bmatrix}
$$

$$
= \begin{Bmatrix}
26.6667 & 0 & 0 \\
25.4222 & 0 & 0 \\
-37.0668 & 0 & 0 \\
-90.6667 & 0 & 0 \\
0 & 25.8 & 0 \\
-36 & 0 & -2464.3 \\
144 & 155 & 0
\end{Bmatrix}
$$

This matrix equation is solved for the three independent loadings to find the system displacements. Recall that loadings 4, 5, 6 solutions:

$$
\begin{aligned}
\text{loading } 4 &= \text{loading } 1 + \text{loading } 2 \\
\text{loading } 5 &= \text{loading } 1 + \text{loading } 3 \\
\text{loading } 6 &= \text{loading } 1 + \text{loading } 2 + \text{loading } 3
\end{aligned}
$$

Results for loadings 4, 5, and 6 are dependent; they are obtainable by superposition (straightforward addition) once the independent solutions are available.

The banded system stiffness matrix form is shown below.

$$
\begin{bmatrix}
27550 & 2204 & -1653 & 13775 & 0 & 0 \\
26978.88 & 9382.08 & 2204 & -19575 & 0 & 0 \\
13076.06 & 738.49 & 0 & -199.29 & 2391.49 & 0 \\
65813.91 & 0 & -2391.49 & 19131.95 & 0 & 0 \\
20292.44 & 0 & 4304.68 & 0 & 0 & 0 \\
25090.96 & -2391.49 & 0 & 0 & 0 & 0 \\
72701.41 & 0 & 0 & 0 & 0 & 0
\end{bmatrix}
$$

Note: If member 2 and member 3 did not orthogonally intersect at joint 3, there would be a nonzero entry in the last column of the banded system stiffness matrix.

Figure 13.14   Matrix equation of system equilibrium and banded system stiffness matrix.

The procedural steps in the program named FRAME are:

1. The user is asked to give the data file name from which FRAME is to read the input data.
2. The user is asked to choose the output destination from the choices: (a) output file on disk; and (b) printer.
3. Input the following from the data file:
   a. Number of problems to be solved.
   b. Problem title of 1 to 72 characters.
   c. Type of analysis code number (0 or 1 which designates either plane frame or space frame).

    d. Units output message choice (1 or 2 or 3 for FEET AND KIPS or INCHES AND KIPS or METERS AND KILONEWTONS).

    e. Number of members.

    f. Number of joints.

    g. Number of support joints.

    h. Number of independent load cases.

    i. Number of dependent load cases.

    j. Modulus of elasticity.

    k. Joint coordinates in sequential joint order and X, Y order for each joint.

    l. Support joint number and force release code numbers for each support joint.

    m. Member incidences for each member in sequential member order.

    n. Member properties for each member.

4. Generate DOF and reaction direction numbers needed to define the MSC array of member-system-compatibility.

5. Determine matrix bandwidth of system stiffness matrix and determine whether or not the computer memory is large enough to solve the current problem. If the computer memory is inadequate, program prints out a message and stops; otherwise, program continues to perform the following tasks.

6. For each member:

    a. Compute member length.

    b. Compute direction cosines.

    c. Compute global member stiffness matrix, $[S^{MG}]_m$.

    d. Using MSC and global member stiffness matrix, update system stiffness matrix, $[S]$.

7. For each independent load case, input applied joint loads into PE array.

8. For each independent load case, input member fixed ended forces, $\{F^f\}_m$, into F array (eventually F becomes the final member-end forces in the local reference).

9. For each member:

    a. Compute global fixed ended forces, $\{F^{fG}\}_m = [T]_m^T \{F^f\}_m$.

    b. Update PE array (using MSC, subtract fixed joint loads contributions).

10. Solve matrix equation of system equilibrium, $[S]\{D\} = \{P^e\}$, for the system displacements for all independent load cases.

11. For each member:

    a. Using MSC, pick out the global member-end displacements, $\{D^{MG}\}_m$, from the system displacements solution for all independent load cases.

    b. Transform the global member-end displacements to local member-end displacements, $\{D^M\}_m = [T]_m \{D^{MG}\}_m$.

    c. Update F array (compute and add on local member-end forces due to local member-end displacements, $[S^M]_m \{D^M\}_m$).

    d. Transform local member-end forces to global member-end forces, $\{F^G\}_m = [T]_m^T \{F\}_m$, and update the reactions array using MSC.

12. Create dependent loading solutions if there are any.

13. At each joint, output system displacements for all loadings.

14. At each support joint, output reactions for all loadings.
15. For each member, output final member-end forces in local reference for all loadings.
16. If this is the last problem to be solved, stop; otherwise, repeat items 3b through 15 of the list.

The printout generated by FRAME for the illustrative problem is shown below.

PROBLEM TITLE: ILLUSTRATIVE EXAMPLE—CHAPTER 13.8

PLANE FRAME ANALYSIS

UNITS CHOSEN WERE—FEET, KIPS, RADIANS

MATERIAL PROPERTIES
MODULUS OF ELASTICITY: E = 4176000.

JOINT NUMBER	JOINT COORDINATES X	Y	Z
1	.000	.000	.000
2	9.000	12.000	.000
3	33.000	12.000	.000
4	33.000	.000	.000

MEMBER DATA FOR PLANE FRAME ANALYSIS

MEMBER NUMBER	INCIDENCES JOINT #1	JOINT #2	AREA	IZ	C1	C2	C3	C4
1	1	2	.0715	.025	1.00	4.00	2.00	4.00
2	4	3	.0715	.025	1.00	4.00	2.00	4.00
3	2	3	.1125	.055	1.00	4.00	2.00	4.00

C O N T R O L   I N F O R M A T I O N

NUMBER OF JOINTS	= 4
NUMBER OF SUPPORT JOINTS	= 2
NUMBER OF MEMBERS	= 3
NUMBER OF INDEPENDENT LOAD CASES	= 3
NUMBER OF DEPENDENT LOAD CASES	= 3
TOTAL NUMBER OF SYSTEM DOF	= 7
STORAGE AVAILABLE	= 8000
STORAGE REQUIRED	= 723
MATRIX SEMI-BANDWIDTH	= 6

APPLIED JOINT LOADS

JOINT	LOADING	X	Y	ZZ
1				
	1	.0	.0	.0
	2	.0	.0	.0
	3	.0	.0	.0
2				
	1	4.0	.0	.0
	2	.0	.0	.0
	3	.0	.0	.0
3				
	1	.0	.0	.0
	2	.0	.0	.0
	3	.0	.0	.0
4				
	1	.0	.0	.0
	2	.0	.0	.0
	3	.0	.0	.0

MEMBER FIXED ENDED FORCES

MEMBER NUMBER	LOADING NUMBER	JOINT NUMBER	X	Y	ZZ
1					
	1	1	−6.0	6.2	26.7
		2	−12.0	17.8	−53.3
	2	1	.0	.0	.0
		2	.0	.0	.0
	3	1	.0	.0	.0
		2	.0	.0	.0
2					
	1	4	.0	.0	.0
		3	.0	.0	.0
	2	4	.0	−25.8	−155.0
		3	.0	25.8	−155.0
	3	4	−2464.3	.0	.0
		3	2464.3	.0	.0
3					
	1	2	.0	36.0	144.0
		3	.0	36.0	−144.0
	2	2	.0	.0	.0
		3	.0	.0	.0
	3	2	.0	.0	.0
		3	.0	.0	.0

SYSTEM DISPLACEMENTS

NOTE: All SUPPORT DISPLACEMENTS are shown as zero. If any
PRESCRIBED SUPPORT DISPLACEMENT was not zero for any
loading, the user will have to pencil in the non-zero
PRESCRIBED SUPPORT DISPLACEMENTS.

JOINT	LOADING	X	Y	ZZ
1				
	1	.0000	.0000	−.0077
	2	.0000	.0000	−.0026
	3	.0000	.0000	−.0015
	4	.0000	.0000	−.0103
	5	.0000	.0000	−.0092
	6	.0000	.0000	−.0118
2				
	1	.0583	−.0457	−.0014
	2	.0227	−.0169	−.0004
	3	.0235	−.0178	−.0029
	4	.0810	−.0625	−.0019
	5	.0819	−.0635	−.0043
	6	.1045	−.0804	−.0048
3				
	1	.0562	−.0019	.0005
	2	.0229	.0000	.0015
	3	.0234	−.0990	−.0033
	4	.0790	−.0019	.0019
	5	.0796	−.1009	−.0028
	6	.1024	−.1009	−.0014
4				
	1	.0000	.0000	.0000
	2	.0000	.0000	.0000
	3	.0000	.0000	.0000
	4	.0000	.0000	.0000
	5	.0000	.0000	.0000
	6	.0000	.0000	.0000

REACTIONS

JOINT	LOADING	X	Y	ZZ
1				
	1	8.3	25.0	.0
	2	−3.2	−1.0	.0
	3	2.6	1.3	.0
	4	5.1	24.1	.0
	5	10.9	26.4	.0
	6	7.7	25.4	.0

REACTIONS

JOINT	LOADING	X	Y	ZZ
4				
	1	−42.3	47.0	249.5
	2	3.2	1.0	−31.8
	3	−2.6	−1.3	44.1
	4	−39.1	47.9	217.7
	5	−44.9	45.6	293.6
	6	−41.7	46.6	261.8

MEMBER-END FORCES

MEMBER NUMBER	LOADING NUMBER	JOINT NUMBER	X	Y	ZZ
1					
	1	1	25.0	8.4	.0
		2	−43.0	15.6	5.9
	2	1	−2.7	2.0	.0
		2	2.7	−2.0	29.6
	3	1	2.6	−1.3	.0
		2	−2.6	1.3	−19.6
	4	1	22.3	10.4	.0
		2	−40.3	13.6	35.5
	5	1	27.6	7.1	.0
		2	−45.6	16.9	−13.6
	6	1	25.0	9.1	.0
		2	−43.0	14.9	16.0
2					
	1	4	47.0	42.3	249.5
		3	−47.0	−42.3	257.7
	2	4	1.0	−3.2	−31.8
		3	−1.0	3.2	−6.8
	3	4	−1.3	2.6	44.1
		3	1.3	−2.6	−12.5
	4	4	47.9	39.1	217.7
		3	−47.9	−39.1	250.9
	5	4	45.6	44.9	293.6
		3	−45.6	−44.9	245.2
	6	4	46.6	41.7	261.8
		3	−46.6	−41.7	238.4
3					
	1	2	42.3	25.0	−5.9
		3	−42.3	47.0	−257.7
	2	2	−3.2	−1.0	−29.6
		3	3.2	1.0	6.8

MEMBER-END FORCES

MEMBER NUMBER	LOADING NUMBER	JOINT NUMBER	X	Y	ZZ
	3	2	2.6	1.3	19.6
		3	−2.6	−1.3	12.5
	4	2	39.1	24.1	−35.5
		3	−39.1	47.9	−250.9
	5	2	44.9	26.4	13.6
		3	−44.9	45.6	−245.2
	6	2	41.7	25.4	−16.0
		3	−41.7	46.6	−238.4

13.9 USAGE OF AVAILABLE STRUCTURAL ANALYSIS SOFTWARE

At this stage, each teacher needs to devise and implement a plan that provides the students an opportunity to obtain an adequate understanding of the computerized displacement method. There are several different ways to accomplish this objective. Some teachers give their students a few pieces of the software and require the students to write and interface the remainder of the computer program. Alternatively, the students can be given a completed plane frame program and required to revise it to create a plane grid program. Other teachers like to have their students to use CAL [13.1] or programs devised by the teacher at this stage of the learning process. Regardless of which way the teacher chooses to proceed, the author firmly believes that the students should either know the correct solution or be required to estimate a solution *before* they are required to obtain computerized solutions using available computer software.

PROBLEMS

Solve the following problems for the indicated unknowns using only the Maxwell-Betti reciprocal theorem and basic Displacement Method concepts. The member properties are not needed if they are not given. However, the same structure is used in each system shown.

13.1. The same member of unknown length and unknown variation in EI is used in each of the following figures. Find M_1 and M_2.

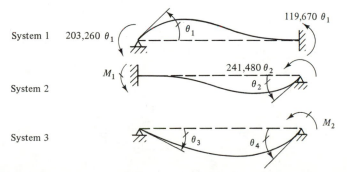

Figure P13.1

13.2. Find P, ϕ, and D

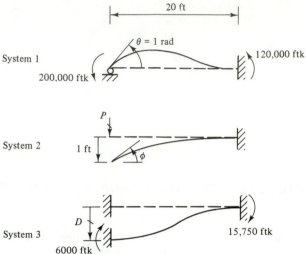

Figure P13.2

13.3. Given $\theta_{1(1)} = 0.0016$ rad; $\theta_{2(2)} = 0.0064$ rad; $\theta_{3(3)} = 0.0144$ rad. Find deflections $v_{1(2)}$, $v_{1(3)}$, and $v_{2(3)}$; reactions $R_{3(4)}$ and $R_{3(5)}$.

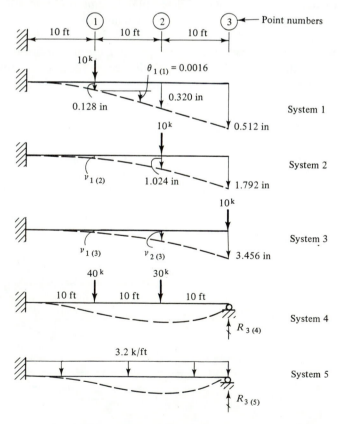

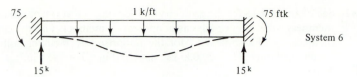

System 6

Figure P13.3

13.4. Given $\theta_{1(1)} = 2900/EI_1$ CW; $\theta_{3(1)} = 1000/EI_1$ CCW; $\theta_{1(2)} = 1500/EI_1$ CW; $\theta_{3(2)} = 1000/EI_1$ CCW. Find $v_{2(1)}$, $v_{2(2)}$, and $M_{1(3)}$.

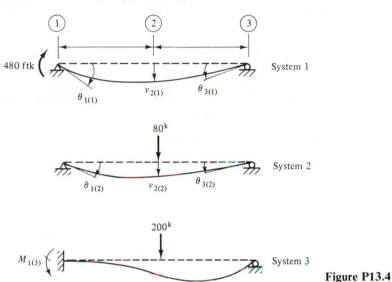

Figure P13.4

The following problems are to be solved using the computer program specified by the instructor. Problems 13.5 through 13.17 are recommended for solution to provide the student with opportunities to become familiar with the usage of the specified computer program for known solutions. Thereafter, the student should be required to estimate the solution answers using the methods of Chapter 12 or assumed $M = 0$ points before the computerized solutions are obtained.

13.5. See Example 7.13 for the geometry and loading. For all members $E = 29,000$ ksi, $A = 27.7$ in.^{2}, $I = 3270$ in.^{4}

13.6. See Example 7.14 for the geometry and loading. For all members $E = 29,000$ ksi. Member AB: $A = 13.0$ in.^{2}; $I = 843$ in.^{4} Member CD: $A = 16.2$ in.^{2}; $I = 1265$ in.^{4} Member BC: $A = 20.1$ in.^{2}; $I = 1686$ in.^{4}

13.7. See Examples 7.8 and 7.9 for the geometry and loadings. For all members $E = 29,000$ ksi, $A = 20.1$ in.^{2}, $I = 1830$ in.^{4} Treat the solution as a multiple-loading problem.

13.8. See Examples 7.1 through 7.4 for the geometry and loadings. For all members $E = 29,000$ ksi, $A = 20.1$ in.^{2}, $I = 1830$ in.^{4} Treat the solution as a multiple-loading problem.

13.9. See Problem 7.35 for the geometry and loading. For all members, $E = 20,000$ kN/cm^2. Inclined member: $A = 105$ cm^2; $I = 25400$ cm^4 ($2I = 50,800$ cm^4). Other member: $A = 130$ cm^2; $I = 25,400$ cm^4 ($3I = 76,200$ cm^4). If the author's FRAME program is specified by the instructor, the roller (support 3) must be replaced with a pinned ended member perpendicular to the inclined member—choose $A = 10,500$ cm^2 and $L = 5$ m.

13.10. See Example 9.19 for the geometry and loading. For all members $E = 20,000$ kN/cm^2, $A = 76.1$ cm^2, $I = 25,500$ cm^4.

13.11. See Example 9.15 for the geometry and loading. Treat the problem as a system of three members—the exterior bays as nonprismatic members and the center bay as prismatic member. See Example 9.15 for the nonprismatic member stiffness coefficients and fixed ended forces. $E = 29,000$ ksi, $A = 20.0$ in.^{2}, $I = 1454$ in.^{4}; $A = 29.1$ in.^{2}, $2.744I = 3990$ in.^{4}

13.12. See Figure 8.13 (conceptual example) for the geometry and loading. See Figure 8.13d for the member properties. Use $A = 18.3$ in.^{2} for members 1 through 3.

13.13. See Example 7.23 for the geometry, loading, and member properties.

13.14. See Example 7.24 for the geometry, loading, and member properties.

13.15. See Example 7.21 for the geometry, loading, and member properties.

13.16. See Examples 7.19 and 7.20 for the geometry, loadings, and member properties. Treat as a multiple-loading problem.

13.17. See Examples 7.15 through 7.18 for the geometry, loadings, and member properties. Treat as a multiple-loading problem.

13.18. For all members $E = 29,000$ ksi, $A = 7.08$ in.^{2}, $I = 82.8$ in.^{4}
    (a) Analyze as a plane truss.
    (b) Analyze as a plane frame and interior joints not pins.

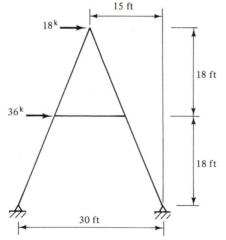

Figure P13.18

13.19. Same member properties as in Problem 13.18. Analyze as a plane frame and compare to Problem 13.18(b) solution.

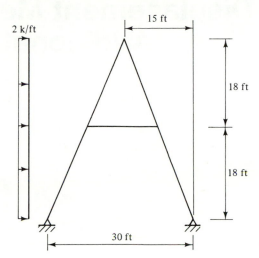

2 k/ft

15 ft

18 ft

18 ft

30 ft

Figure P13.19

13.20. See Problem 10.28 for the geometry, loading, and member properties. Use $EA = 469,800$ kft^2 for members 1 through 3.

13.21. See Figure 9.16 for the geometry and loading. For all members $E = 29,000$ ksi, $A = 14.7$ in.^{2}, $I = 984$ in.^{4}

13.22. For all members $E = 29,000$ ksi. Members 1 through 4: $A = 20.0$ in.^{2}; $I = 723$ in.^{4} Members 5 and 6: $A = 16.2$ in.^{2}; $I = 1350$ in.^{4}

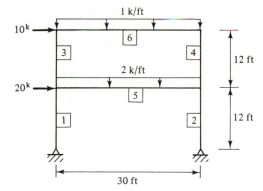

1 k/ft

10^k

6

3

4

12 ft

2 k/ft

20^k

5

1

2

12 ft

30 ft

Figure P13.22

13.23. See Problem 9.57 for the geometry, loading, and member properties. Use $A_1 = A_2 = 16.2$ in.^{2} If the author's FRAME program is specified by the instructor, the translational springs will have to be replaced by pinned ended members such that $EA/L = 100$ k/ft.

13.24. See Problem 9.33 for the geometry, loading, and member properties.

Displacement Method—
Additional Topics

14.1 INTRODUCTION

While his students are acquiring familiarity and skill in coping with some available computer programs to solve problems dealing with the material presented in Chapter 13, the author presents the additional theoretical material in the same sequence as it appears in this chapter. The order of presentation was chosen to continue with the topics that involve matrix operations to simplify the learning process. Consequently, substructuring is dealt with before symmetric and antisymmetric loading components are discussed even though an analyst naturally takes advantage of structural symmetry when it exists rather than using substructuring if the bandwidth of [S] for the entire system becomes too large.

14.2 CONDENSATION

Using an available Displacement Method computer program, a routine solution of the problem shown in Figure 14.1a typically involves 3 DOF at each of joints 3 through 6 which gives a total of 12 DOF. As shown below, it is possible to solve the problem in Figure 14.1a by using only the 2 DOF in Figure 14.1b, but this approach is not preferred in statical analyses. However, in dynamical analyses of this structure it is highly desirable *to condense out* the gravity direction and rotational direction DOF which leaves only the 2 DOF shown in Figure 14.1b. The natural frequencies of free vibration and the effects of horizontally directed forced vibrations are predominately a function of these two translational DOF for the structure shown. Furthermore, removal of the rotational DOF eliminates the need for rotary masses. This text does not deal with dynamical analyses, but the system stiffness matrix is the same in statical

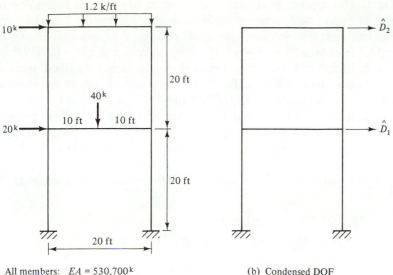

All members: $EA = 530,700^k$
$EI = 268,000 \text{ kft}^2$

(a) Structure and loading

(b) Condensed DOF

Figure 14.1 Structure, loading, and condensed DOF.

and dynamical analyses for a chosen set of DOF. Therefore, the primary objective of the following example is to illustrate how the system stiffness matrix can be obtained in terms of the 2 DOF shown in Figure 14.1b. The structure shown in Figure 14.1a was solved in Example 10.13 by the direct method of moment distribution which ignored axial deformations, condensed out the rotational DOF, and gave the system stiffness matrix for the left half of the structure shown in Figure 14.1b.

Numerical values for the symbols shown in Figure 14.2 were obtained by using the author's computer program as follows:

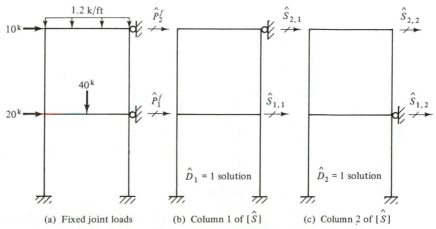

(a) Fixed joint loads   (b) Column 1 of $[\hat{S}]$ (c) Column 2 of $[\hat{S}]$

Figure 14.2 Fixed joint loads and $[\hat{S}]$ for condensed DOF.

1. Loading 1: The applied joint loads and member fixed ended forces due to the loads in Figure 14.2a were input. Solutions for loadings 1, 2, and 3 were made in one run; they involved the DOF and reaction numbers shown in Figure 14.3 to obtain $\hat{P}_1^f = R_{17} = -19.6784^k$ and $\hat{P}_2^f = R_{18} = -10.0894^k$.
2. Loading 2: Member fixed ended forces due to a unit prescribed support movement coincident with R_{17} in Figure 14.3 were input to obtain the solution for Figure 14.2b: $\hat{S}_{1,1} = R_{17} = 1307.0760$ and $\hat{S}_{2,1} = R_{18} = -561.6031$.
3. Loading 3: Member fixed ended forces due to a unit prescribed support movement coincident with R_{18} in Figure 14.3 were input to obtain the solution for Figure 14.2c: $\hat{S}_{1,2} = R_{17} = -561.6031$ and $\hat{S}_{2,2} = R_{18} = 423.8984$.

Since $\hat{P}_1^a = \hat{P}_2^a = 0$, the matrix equation of system equilibrium for the condensed DOF of Figure 14.1b is

$$\begin{bmatrix} 1307.0760 & -561.6031 \\ -561.6031 & 423.8984 \end{bmatrix} \begin{Bmatrix} \hat{D}_1 \\ \hat{D}_2 \end{Bmatrix} = \begin{Bmatrix} 19.6784 \\ 10.0894 \end{Bmatrix}$$

which gives $\hat{D}_1 = 0.0586916$ ft and $\hat{D}_2 = 0.1015591$ ft.

This completes the objective of finding the system stiffness matrix, $[\hat{S}]$, for the condensed DOF of Figure 14.1b and no further results are shown. If an analyst wished to continue the calculations to find the final member-end forces and reactions for Figure 14.1a, the following steps would have to be performed in using the author's computer program:

1. The member fixed ended forces of loading 1 are the same as in the previous run; however, the member fixed ended forces of loading 2 and loading 3 in

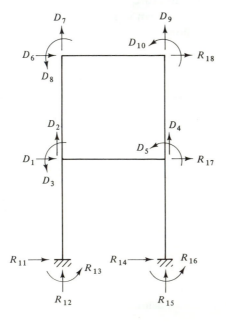

Figure 14.3 DOF and reaction numbers involved in solutions of Figure 14.2.

the current run are 0.0586916 and 0.1015591, respectively, times those of loadings 2 and 3 in the previous run.

2. A dependent loading is needed:

$$\text{loading } 4 = \text{loading } 1 + \text{loading } 2 + \text{loading } 3$$

The final member-end forces and reactions are the output for loading 4.

Some computer programs compute the reactions by using a subdivided system stiffness matrix approach which is described below for the structure shown in Figure 14.3. The advantage of this approach is that the input by the user is simpler in accounting for prescribed support movements than in the author's computer program. However, this approach has a larger bandwidth for the system stiffness matrix than in the author's computer program. In the subdivided system stiffness matrix approach to computing the reactions, the free DOF (those that are computed when the matrix equation of system equilibrium is solved) are allocated first in the computer and the support DOF are allocated last. For simplicity in the discussion, the following description deals with the system stiffness matrix of the complete structure as a square matrix (18 rows and 18 columns for the example in Figure 14.2). The matrix equation of system equilibrium including the reactions DOF is

$$
\begin{bmatrix}
S_{1,1} & \cdots & S_{1,10} & \vdots & S_{1,11} & \cdots & S_{1,18} \\
\vdots & & \vdots & \vdots & \vdots & & \vdots \\
S_{10,1} & \cdots & S_{10,10} & \vdots & S_{10,11} & \cdots & S_{10,18} \\
\hdashline
S_{11,1} & \cdots & S_{11,10} & \vdots & S_{11,11} & \cdots & S_{11,18} \\
\vdots & & \vdots & \vdots & \vdots & & \vdots \\
S_{18,1} & \cdots & S_{18,10} & \vdots & S_{18,11} & \cdots & S_{18,18}
\end{bmatrix}
\begin{Bmatrix}
D_1 \\ \vdots \\ D_{10} \\ -- \\ D_{11} \\ \vdots \\ D_{18}
\end{Bmatrix}
=
\begin{Bmatrix}
P_1^e \\ \vdots \\ P_{10}^e \\ -- \\ P_{11}^e \\ \vdots \\ P_{18}^e
\end{Bmatrix}
$$

Note that D_{11} through D_{18} are prescribed support movements (they are known) and P_{11}^e through P_{18}^e are the unknown reactions to be computed. For the loadings shown in Figure 14.2 (a through c), the prescribed support movements are

loading 1—D_{11} through D_{18} are zero
loading 2—$D_{17} = 1$ and D_{11} through D_{16} are zero and $D_{18} = 0$
loading 3—$D_{18} = 1$ and D_{11} through D_{17} are zero

First, the matrix equation of system equilibrium is solved for the unknown system displacements:

$$
\begin{bmatrix}
S_{1,1} & \cdots & S_{1,10} \\
\vdots & & \vdots \\
S_{10,1} & \cdots & S_{10,10}
\end{bmatrix}
\begin{Bmatrix}
D_1 \\ \vdots \\ D_{10}
\end{Bmatrix}
=
\begin{Bmatrix}
P_1^e \\ \vdots \\ P_{10}^e
\end{Bmatrix}
-
\begin{bmatrix}
S_{1,11} & \cdots & S_{1,18} \\
\vdots & & \vdots \\
S_{10,11} & \cdots & S_{10,18}
\end{bmatrix}
\begin{Bmatrix}
D_{11} \\ \vdots \\ D_{18}
\end{Bmatrix}
$$

Then, the reactions are computed:

$$
\begin{Bmatrix}
P_{11}^e \\ \vdots \\ P_{18}^e
\end{Bmatrix}
=
\begin{bmatrix}
S_{11,1} & \cdots & S_{11,10} \\
\vdots & & \vdots \\
S_{18,1} & \cdots & S_{18,10}
\end{bmatrix}
\begin{Bmatrix}
D_1 \\ \vdots \\ D_{10}
\end{Bmatrix}
+
\begin{bmatrix}
S_{11,11} & \cdots & S_{11,18} \\
\vdots & & \vdots \\
S_{18,11} & \cdots & S_{18,18}
\end{bmatrix}
\begin{Bmatrix}
D_{11} \\ \vdots \\ D_{18}
\end{Bmatrix}
$$

The preceding matrix equations can be stated in condensed matrix form by giving each matrix in the subdivided matrices a name. In the names chosen below, the subscript symbols f and s denote free and support DOF, respectively.

$$\begin{bmatrix} [S_{ff}] & \vdots & [S_{fs}] \\ \hdashline [S_{sf}] & \vdots & [S_{ss}] \end{bmatrix} \begin{Bmatrix} \{D_f\} \\ \{D_s\} \end{Bmatrix} \begin{Bmatrix} \{P_f^e\} \\ \{P_s^e\} \end{Bmatrix}$$

First, solve $[S_{ff}]\{D_f\} = \{P_f^e\} - [S_{fs}]\{D_s\}$ for $\{D_f\}$. Then, solve for the reactions

$$\{P_s^e\} = [S_{sf}]\{D_f\} + [S_{ss}]\{D_s\}$$

where $\{D_s\}$ are the support displacements (they are prescribed—known)

$\{D_f\}$ are the free (unknown) displacements

$[S_{ff}]$ contains the system stiffness coefficients at the free DOF locations due to unit induced displacements one at a time at the free DOF locations

$[S_{sf}]$ contains the system stiffness coefficients at the support DOF locations due to the unit induced displacements at the free DOF locations

$[S_{fs}]$ contains the system stiffness coefficients at the free DOF locations due to the prescribed support movements

$[S_{ss}]$ contains the system stiffness coefficients at the support DOF locations due to the prescribed support movements

14.3 SUBSTRUCTURING

If the DOF of a structural system are efficiently ordered to minimize the bandwidth of the banded system stiffness matrix and the bandwidth is larger than the computer memory can tolerate, and if the structure has no axis of structural symmetry, an analyst has to resort to substructuring the structural system. That is, the structural system is decomposed or subdivided into two or more substructures each of which has a system stiffness matrix bandwidth less than the maximum size that the computer memory can tolerate. It should be emphasized that if a structure has one or more axes of symmetry, the approach described in Section 14.4 is computationally and economically superior to the method of substructuring described below.

The author's previously described computerized approach proceeds member by member to formulate the matrix equation of system equilibrium and to calculate the final member-end forces as well as the reactions. Therefore, it is convenient to think of each substructure as a "member with a variable number of boundary-end displacements" which are located at the actual support joints (if there are any) and the joints where the substructures interface in the structural system. Except for the fact that each substructure does not necessarily have the same number of boundary-end DOF (analogous to a member's global member-end DOF), all of the other member global reference concepts involved in the author's previously described computerized solution are applicable to a substructure, namely:

1. Member fixed ended forces in the global reference were needed and substructure fixed-boundary-interface-DOF forces are needed in the global reference.

2. Global member stiffness matrices were needed and substructure stiffness matrices in the global reference are needed.

3. Items 1 and 2 are needed to define the force-deformation relations in the global reference.

4. Member-system-compatibility had to be ensured and substructure-system-compatibility has to be ensured.

5. The fixed joint loads are obtained by superposition of the contributing values available from item 1.

6. The banded system stiffness matrix is obtained by superposition of the contributing values available from item 2.

7. Final forces at the boundary ends are the fixed ended forces plus those due to the end displacements as defined in item 3; (fortunately, at the same time that the final boundary-interface-joint forces of a substructure are obtained, the final member-end forces in the local reference are also obtained for each member in a substructure).

The structure shown in Figure 14.4 is chosen to illustrate how an analyst obtains the solution by the method of substructuring. It should be noted that in actual practice this structure would not be solved by substructuring since the bandwidth of the system

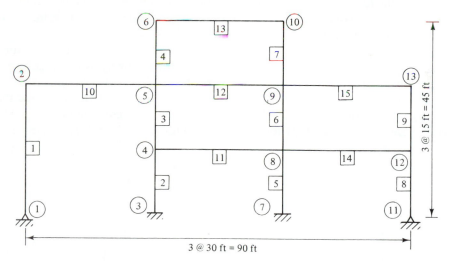

3 @ 30 ft = 90 ft

All columns (members 1 through 9) : EA = 817,800^k ; EI = 167,757 kft^2
All beams (members 10 through 15) : EA = 484,300^k ; EI = 235,625 kft^2
Loading 1 : 1.5 k/ft downward on members 10, 13, and 15
          3 k/ft downward on members 11, 12, and 14
Loading 2 : wind loads : 0.3 k/ft upward on members 10, 13, and 15
          Horizontal to right joint loads : 6^k at joint 2; 3^k at joint 6
                       2^k at joint 10
                       4^k at joints 12 and 13
Loading 3 : 0.75★ (loading 1 + loading 2)
Substructures interface at joints 5, 8, and 9

Figure 14.4 Structure to be solved by substructuring.

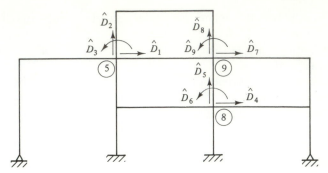

Figure 14.5 Substructure interface DOF (at joints 5, 8, and 9).

stiffness matrix is not too large for a microcomputer. However, if the author were to choose a realistic structure for which substructuring is a necessity as the example problem to illustrate substructuring, the detailed calculations would overshadow the descriptions of the needed concepts. The author recommends that before an analyst attempts to use an available computer program and the procedure described below to perform the substructuring of a realistic structure, a simple example like the one shown in Figure 14.4 should be routinely solved and also solved by substructuring. Unless the analyst can demonstrate for a simple example that he or she fully comprehends the substructuring concepts and that the available computer program is properly working, it is futile to try to solve a realistic structure that must be broken into substructures.

The objective is to solve the structure shown in Figure 14.5 for the loadings defined in Figure 14.4 by the method of substructuring using the substructures shown in Figure 14.6. Loading 1 is the gravity loading and loading 2 is the wind loading given in Figure 14.4; also needed are the loadings to individually prescribe unit valued DOF shown in Figure 14.5. For convenience, the prescribed unit valued DOF loadings are named as D1EQ1 for $\hat{D}_1 = 1$ and D9EQ1 for $\hat{D}_9 = 1$, for example.

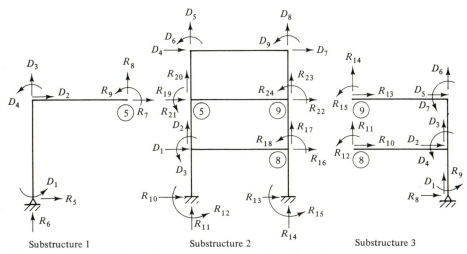

Figure 14.6 Substructures showing DOF and reaction direction numbers.

The only output needed from the computerized solution of the first run of each substructure are the reactions at the interface joints of the substructures—reactions at joints 5, 8, and 9 are needed. The reactions for loading 1 and 2 are needed to define the fixed joint loads contributions from each substructure. The reactions for the other loadings (D1EQ1, ..., D9EQ1) are needed to define the contributions of $[\hat{S}]$ from each substructure. Fortunately, the computerized solution of each substructure gives the joint number and the *X*-direction, *Y*-direction, and *Z*-rotational-direction reactions at the joint. The purpose of showing the DOF and reaction direction numbers in Figure 14.6 was to clarify the number of unknown DOF and reactions involved in the solution of each substructure.

Interface DOF Prescribed as (a) Zero (Loadings 1, 2); (b) Unity One at a Time

Substructure 1 Reactions Are: (Units Are Feet and Kips)

JOINT	LOADING	X FORCE	Y FORCE	Z MOMENT
5	1	−1.2918358	26.2052612	−149.9135284
	2	−5.7371206	−5.2342920	29.9152374
	D1EQ1	12.1299324	18.1774292	−181.4250031
	D2EQ1	18.1774292	53.3954010	−1056.5388184
	D3EQ1	−181.4250488	−1056.5388184	26253.4062500

Substructure 2 Reactions Are: (Units Are Feet and Kips)

JOINT	LOADING	X FORCE	Y FORCE	Z MOMENT
5	1	−0.0454731	88.5218048	142.5701141
	2	−4.1557531	−5.5094948	30.5828247
	D1EQ1	16685.3085937	−29.0424500	2094.1711426
	D2EQ1	−29.0424500	27405.3359375	1726.6872559
	D3EQ1	2094.1728516	1726.6877441	93480.9375000
	D4EQ1	−555.4240723	0.0000006	−4165.6796875
	D5EQ1	58.0848999	−42.1226044	290.4240723
	D6EQ1	−580.5695801	682.8291016	−2902.8442383
	D7EQ1	−16274.9375000	−0.0000044	1397.5410156
	D8EQ1	−0.0000044	−124.7932892	−1871.8991699
	D9EQ1	1397.5407715	1871.8996582	2117.3803711
8	1	0.0000002	47.9561615	−254.7633972
	2	0.0	0.0	0.0
	D1EQ1	−555.4240723	58.0848999	−580.5693359
	D2EQ1	0.0000006	−42.1226044	682.8288574
	D3EQ1	−4165.6796875	290.4240723	−2902.8442383
	D4EQ1	2303.7846680	−0.0000011	−0.0000833
	D5EQ1	−0.0000011	109124.1875000	−1365.6577148
	D6EQ1	−0.0000833	−1365.5577148	118828.5625000

Substructure 2 Reactions (Continued) Are: (Units Are Feet and Kips)

JOINT	LOADING	X FORCE	Y FORCE	Z MOMENT
	D7EQ1	−596.4680176	0.0	4473.5078125
	D8EQ1	0.0	−54519.9765625	0.0
	D9EQ1	−4473.5078125	0.0	22367.5273437
9	1	−8.2584047	67.4999237	−184.0887604
	2	−0.8442421	−3.4904928	14.1320362
	D1EQ1	−16274.9375000	−0.0000044	1397.5410156
	D2EQ1	−0.0000044	−124.7932892	1871.8991699
	D3EQ1	1397.5407715	−1871.8996582	2117.3803711
	D4EQ1	−596.4680176	0.0	−4473.5078125
	D5EQ1	0.0	−54519.9765625	0.0
	D6EQ1	4473.5078125	0.0	22367.5273437
	D7EQ1	16871.4062500	0.0000044	3075.9680176
	D8EQ1	0.0000044	54644.7773437	−1871.8991699
	D9EQ1	3075.9697266	−1871.8996582	98774.6875000

Substructure 3 Reactions Are: (Units Are Feet and Kips)

JOINT	LOADING	X FORCE	Y FORCE	Z MOMENT
8	1	−7.7211161	48.0699005	256.3017578
	2	−2.7662783	0.0808525	0.7694675
	D4EQ1	455.3061523	12.6211119	125.2830353
	D5EQ1	12.6211119	80.5044250	1327.8186035
	D6EQ1	125.2830505	1327.8188477	28973.1289062
	D7EQ1	−292.1584473	−46.9837952	−468.6643066
	D8EQ1	84.0041046	6.5991402	65.2630310
	D9EQ1	838.3061523	65.0914459	638.6240234
9	1	11.9921865	24.1778564	130.0813599
	2	−5.0819130	−5.0140810	−27.7182159
	D4EQ1	−292.1584473	84.0041046	838.3059082
	D5EQ1	−46.9837952	6.5991402	65.0914307
	D6EQ1	−468.6645508	65.2630463	638.6240234
	D7EQ1	227.6214142	−75.1759186	−749.8635254
	D8EQ1	−75.1759186	69.2361145	1214.5786133
	D9EQ1	−749.8637695	1214.5788574	27830.1523437

The matrix equation of equilibrium, $[\hat{S}]\{\hat{D}\} = \{\hat{P}^e\}$, must be formulated and solved for the DOF shown in Figure 14.5 for loading 1 and loading 2. $\{\hat{P}^a\} = \{0\}$ for both loadings; $\{\hat{P}^f\}$ for loading 1 and loading 2 are obtained by algebraically summing the individual reactions at joint 5 from substructures 1 and 2 as well as the individual reactions at joints 8 and 9 from substructures 2 and 3. Then, $\{\hat{P}^e\} = \{\hat{P}^a\} - \{\hat{P}^f\}$ is computed. The individual reactions at joint 5 from substructures 1 and 2 for loadings D1EQ1 through D3EQ1 are algebraically summed to obtain the entries in columns 1 through 3 of $[\hat{S}]$ for matrix row numbers 1 through 3. Only substructure 2 contributes to the entries in matrix row numbers 4 through 9 of columns 1 through 3 of $[\hat{S}]$; the contributions are the reactions at joints 8 and 9 of substructure 2. Similarly, the entries in columns 4 through 9 of $[\hat{S}]$ are obtained from the reactions at joints 5, 8, and 9 of substructures 2 and 3.

$[\hat{S}]\{\hat{D}\} = \{\hat{P}^e\}$ is

$$
[S] = \begin{bmatrix}
16697.4 & -10.8651 & 1912.75 & -555.424 & 58.0849 & -580.570 & -16274.9 & 0 & 1397.54 \\
& 27459.2 & 670.15 & 0 & -42.1226 & 682.829 & 0 & -124.793 & 1871.90 \\
& & 119734. & -4165.68 & 290.424 & -2902.84 & 1397.54 & -1871.90 & 2117.38 \\
& & & 2759.09 & 12.6211 & 125.283 & -888.626 & 84.0041 & -3635.20 \\
& & & & 109205. & -37.84 & -46.9838 & -54513.4 & 65.0914 \\
& & & & & 147802. & 4004.85 & 65.2630 & 23006.1 \\
& \text{symmetric} & & & & & 17099.0 & -75.1759 & 2326.11 \\
& & & & & & & 54714.0 & -657.32 \\
& & & & & & & & 126605.
\end{bmatrix}
$$

Note: six-digit precision was used in obtaining the entries in $[\hat{S}]$ and $\{\hat{P}^e\}$. Units are feet and kips.

$$
\{\hat{P}^e\} = \begin{Bmatrix}
1.33731 & 9.89287 \\
-114.727 & 10.7438 \\
7.344 & -60.498 \\
7.72112 & 2.76628 \\
-96.0261 & -0.089853 \\
-1.539 & -0.769467 \\
-3.73379 & 5.92615 \\
-91.6778 & 8.50457 \\
54.008 & 13.5862
\end{Bmatrix} ;
$$

loading 1   loading 2

loading 1   loading 2

$$
\{\hat{D}\} = \begin{Bmatrix}
0.00154603 & 0.0466952 \\
-0.00424523 & 0.000488241 \\
0.000122680 & -0.000997221 \\
0.000467940 & 0.0233684 \\
-0.00342334 & 0.000134739 \\
-0.000112355 & -0.00106043 \\
0.00140268 & 0.0463897 \\
-0.00509008 & 0.000281166 \\
0.000574574 & -0.000385932
\end{Bmatrix}
$$

Substructure 1 Final Results Are: (Units Are Feet and Kips)

JOINT DISPLACEMENTS — SUPPORTS

JOINT	LOADING	/ ─────────── DISPLACEMENTS ───────────/		
		X DISP	Y DISP	Z ROT
1	GLOBAL			
	1	0.0	0.0	0.0011729
	2	0.0	0.0	−0.0024780
	3	0.0	0.0	−0.0009789
5	GLOBAL			
	1	0.0015460	−0.0042452	0.0001227
	2	0.0466952	0.0004882	−0.0009972
	3	0.0361809	−0.0028177	−0.0006559

JOINT DISPLACEMENTS — FREE JOINTS

JOINT	LOADING	/ ─────────── DISPLACEMENTS ───────────/		
		X DISP	Y DISP	Z ROT
2	GLOBAL			
	1	0.0016310	−0.0007015	−0.0025088
	2	0.0470037	0.0002089	0.0002556
	3	0.0364761	−0.0003695	−0.0016899

JOINT LOADS — SUPPORTS

JOINT	LOADING	/ ─────────── FORCES ───────────/		
		X FORCE	Y FORCE	Z MOMENT
1	GLOBAL			
	1	1.373	19.123	−0.000
	2	−1.019	−5.694	−0.000
	3	0.265	10.072	−0.000
5	GLOBAL			
	1	−1.373	25.877	−142.488
	2	−4.981	−3.306	−5.253
	3	−4.765	16.928	−110.805

MEMBER FORCES

MEMBER	LOADING	JOINT	/ ─────────── FORCES ───────────/		
			AXIAL	SHEAR Y	MOMENT Z
1					
	1	1	19.123	−1.373	−0.000
		2	−19.123	1.373	−41.175
	2	1	−5.694	1.019	−0.000
		2	5.694	−1.019	30.572
	3	1	10.072	−0.265	−0.000
		2	−10.072	0.265	−7.952
10					
	1	2	1.373	19.123	41.175
		5	−1.373	25.877	−142.488

Substructure 1 Final Results (Continued) Are: (Units Are Feet and Kips)

      MEMBER FORCES

MEMBER	LOADING	JOINT	/ ------------- FORCES --------------- /		
			AXIAL	SHEAR Y	MOMENT Z
	2	2	4.981	−5.694	−30.572
		5	−4.981	−3.306	−5.253
	3	2	4.765	10.072	7.952
		5	−4.765	16.928	−110.805

Substructure 2 Final Results Are: (Units Are Feet and Kips)

      JOINT DISPLACEMENTS − SUPPORTS

JOINT	LOADING		/ ------------ DISPLACEMENTS ------------- /		
			X DISP	Y DISP	Z ROT
3	GLOBAL				
		1	0.0	0.0	0.0
		2	0.0	0.0	0.0
		3	0.0	0.0	0.0
5	GLOBAL				
		1	0.0015460	−0.0042452	0.0001227
		2	0.0466951	0.0004882	−0.0009972
		3	0.0361808	−0.0028177	−0.0006559
7	GLOBAL				
		1	0.0	0.0	0.0
		2	0.0	0.0	0.0
		3	0.0	0.0	0.0
8	GLOBAL				
		1	0.0046794	−0.0034233	−0.0001124
		2	0.0233684	0.0001347	−0.0010604
		3	0.0210358	−0.0024664	−0.0008796
9	GLOBAL				
		1	0.0014027	−0.0050901	0.0005746
		2	0.0463897	0.0002812	−0.0003859
		3	0.0358443	−0.0036067	0.0001415

      JOINT DISPLACEMENTS − FREE JOINTS

JOINT	LOADING		/ ------------ DISPLACEMENTS ------------- /		
			X DISP	Y DISP	Z ROT
4	GLOBAL				
		1	0.0044422	−0.0025066	−0.0019385
		2	0.0231096	0.0002795	−0.0014076
		3	0.0206639	−0.0016703	−0.0025095
6	GLOBAL				
		1	−0.0011968	−0.0046621	−0.0017484
		2	0.0593714	0.0005969	0.0000361
		3	0.0436309	−0.0030489	−0.0012843

Substructure 2 Final Results (Continued) Are: (Units Are Feet and Kips)

JOINT DISPLACEMENTS – FREE JOINTS

JOINT	LOADING	/ ------------ DISPLACEMENTS ------------/		
		X DISP	Y DISP	Z ROT
10	GLOBAL			
	1	−0.0017487	−0.0054986	0.0018371
	2	0.0593876	0.0003376	−0.0009584
	3	0.0432291	−0.0038707	0.0006590

JOINT LOADS – SUPPORTS

JOINT	LOADING	/ ------------- FORCES ----------------/		
		X FORCE	Y FORCE	Z MOMENT
3	GLOBAL			
	1	6.022	136.662	−23.486
	2	−7.487	−15.240	71.898
	3	−1.099	91.066	36.308
5	GLOBAL			
	1	1.372	−25.877	142.490
	2	4.979	3.306	5.252
	3	4.764	−16.928	110.807
7	GLOBAL			
	1	−2.288	186.640	18.420
	2	−9.195	−7.346	80.820
	3	−8.612	134.471	74.430
8	GLOBAL			
	1	6.004	−47.645	−248.465
	2	6.111	3.224	48.817
	3	9.086	−33.315	−149.736
9	GLOBAL			
	1	−11.110	−24.781	−142.465
	2	0.592	7.056	53.981
	3	−7.888	−13.294	−66.363

MEMBER FORCES

MEMBER	LOADING	JOINT	/ ------------ FORCES ------------/		
			AXIAL	SHEAR Y	MOMENT Z
2					
	1	3	136.662	−6.022	−23.486
		4	−136.662	6.022	−66.845
	2	3	−15.240	7.487	71.898
		4	15.240	−7.487	40.414
	3	3	91.066	1.099	36.308
		4	−91.066	−1.099	−19.823
3					
	1	4	94.788	−9.850	−96.930
		5	−94.788	9.850	−50.827

Substructure 2 Final Results (Continued) Are: (Units Are Feet and Kips)

MEMBER FORCES

MEMBER	LOADING	JOINT	/ -------------FORCES -------------/		
			AXIAL	SHEAR Y	MOMENT Z
	2	4	−11.379	3.310	20.237
		5	11.379	−3.310	29.415
	3	4	62.557	−4.905	−57.520
		5	−62.557	4.905	−16.059
4					
	1	5	22.727	−8.909	−45.890
		6	−22.727	8.909	−87.742
	2	5	−5.922	3.261	12.903
		6	5.922	−3.261	36.015
	3	5	12.604	−4.236	−24.740
		6	−12.604	4.236	−38.795
5					
	1	7	186.640	2.288	18.420
		8	−186.640	−2.288	15.907
	2	7	−7.346	9.195	80.820
		8	7.346	−9.195	57.100
	3	7	134.471	8.612	74.430
		8	−134.471	−8.612	54.756
6					
	1	8	90.870	0.113	−6.833
		9	−90.870	−0.113	8.532
	2	8	−7.983	7.261	46.915
		9	7.983	−7.261	62.002
	3	8	62.165	5.531	30.062
		9	−62.165	−5.531	52.901
7					
	1	9	22.273	8.909	52.697
		10	−22.273	−8.909	80.936
	2	9	−3.078	1.739	19.443
		10	3.078	−1.739	6.638
	3	9	14.396	7.986	54.105
		10	−14.396	−7.986	65.680
11					
	1	4	−3.828	41.874	163.775
		8	3.828	48.125	−257.539
	2	4	−4.177	−3.862	−60.651
		8	4.177	3.862	−55.198
	3	4	−6.004	28.510	77.343
		8	6.004	38.990	−234.553
12					
	1	5	2.314	46.184	239.207
		9	−2.314	43.816	−203.694

Substructure 2 Final Results (Continued) Are: (Units Are Feet and Kips)

MEMBER FORCES

MEMBER	LOADING	JOINT	/ --------FORCES--------/		
			AXIAL	SHEAR Y	MOMENT Z
	2	5	4.930	−2.151	−37.066
		9	−4.930	2.151	−27.464
	3	5	5.433	33.025	151.605
		9	−5.433	34.475	−173.369
13					
	1	6	8.909	22.727	87.742
		10	−8.909	22.273	−80.936
	2	6	−0.261	−5.922	−36.015
		10	0.261	−3.078	−6.638
	3	6	6.486	12.604	38.795
		10	−6.486	14.396	−65.680

Substructure 3 Final Results Are: (Units Are Feet and Kips)

JOINT DISPLACEMENTS – SUPPORTS

JOINT		LOADING	/ --------DISPLACEMENTS--------/		
			X DISP	Y DISP	Z ROT
8	GLOBAL				
		1	0.0046794	−0.0034233	−0.0001124
		2	0.0233684	0.0001347	−0.0010604
		3	0.0210358	−0.0024664	−0.0008796
9	GLOBAL				
		1	0.0014027	−0.0050901	0.0005746
		2	0.0463897	0.0002812	−0.0003859
		3	0.0358443	−0.0036067	0.0001415
11	GLOBAL				
		1	0.0	0.0	−0.0014782
		2	0.0	0.0	−0.0018733
		3	0.0	0.0	−0.0025136

JOINT DISPLACEMENTS – FREE JOINTS

JOINT		LOADING	/ --------DISPLACEMENTS--------/		
			X DISP	Y DISP	Z ROT
12	GLOBAL				
		1	0.0050513	−0.0011478	0.0019460
		2	0.0237469	−0.0000235	−0.0010027
		3	0.0215986	−0.0008784	0.0007075
13	GLOBAL				
		1	0.0007145	−0.0015186	0.0011156
		2	0.0464263	0.0000122	−0.0012592
		3	0.0353556	−0.0011298	−0.0001077

Substructure 3 Final Results (Continued) Are: (Units Are Feet and Kips)

JOINT LOADS – SUPPORTS

JOINT	LOADING		/ ————————— FORCES ————————— /		
			X FORCE	Y FORCE	Z MOMENT
8	GLOBAL				
		1	−6.004	47.642	248.465
		2	−6.111	−3.224	−48.817
		3	−9.086	33.313	149.735
9	GLOBAL				
		1	11.110	24.781	142.466
		2	−0.591	−7.056	−53.982
		3	7.889	13.294	66.363
11	GLOBAL				
		1	−5.106	62.577	0.000
		2	−1.298	1.280	0.000
		3	−4.803	47.893	0.000

MEMBER FORCES

MEMBER	LOADING	JOINT	/ ————————— FORCES ————————— /		
			AXIAL	SHEAR Y	MOMENT Z
8					
	1	11	62.577	5.106	0.000
		12	−62.577	−5.106	76.591
	2	11	1.280	1.298	0.000
		12	−1.280	−1.298	19.474
	3	11	47.893	4.803	0.000
		12	−47.893	−4.803	72.049
9					
	1	12	20.219	11.110	92.610
		13	−20.219	−11.110	74.035
	2	12	−1.944	3.409	28.436
		13	1.944	−3.409	22.699
	3	12	13.706	10.889	90.784
		13	−13.706	−10.889	72.551
14					
	1	8	−6.004	47.642	248.465
		12	6.004	42.358	−169.201
	2	8	−6.111	−3.224	−48.817
		12	6.111	3.224	−47.910
	3	8	−9.086	33.313	149.735
		12	9.086	34.187	−162.833
15					
	1	9	11.110	24.781	142.466
		13	−11.110	20.219	−74.035
	2	9	−0.591	−7.056	−53.982
		13	0.591	−1.944	−22.699

Substructure 3 Final Results (Continued) Are: (Units Are Feet and Kips)

MEMBER FORCES

MEMBER	LOADING	JOINT	/------------ FORCES ------------/		
			AXIAL	SHEAR Y	MOMENT Z
	3	9	7.889	13.294	66.363
		13	−7.889	13.706	−72.551

Results Obtained by Routinely Solving the Structure in Figure 14.4 Are: (Units Are Feet and Kips)

JOINT DISPLACEMENTS – SUPPORTS

JOINT	LOADING		/------------ DISPLACEMENTS ------------/		
			X DISP	Y DISP	Z ROT
1	GLOBAL				
		1	0.0000000	0.0000000	0.0011729
		2	0.0000000	0.0000000	−0.0024781
		3	0.0000000	0.0000000	−0.0009789
3	GLOBAL				
		1	0.0000000	0.0000000	0.0000000
		2	0.0000000	0.0000000	0.0000000
		3	0.0000000	0.0000000	0.0000000
7	GLOBAL				
		1	0.0000000	0.0000000	0.0000000
		2	0.0000000	0.0000000	0.0000000
		3	0.0000000	0.0000000	0.0000000
11	GLOBAL				
		1	0.0000000	0.0000000	−0.0014782
		2	0.0000000	0.0000000	−0.0018734
		3	0.0000000	0.0000000	−0.0025137

JOINT DISPLACEMENTS – FREE JOINTS

JOINT	LOADING		/------------ DISPLACEMENTS ------------/		
			X DISP	Y DISP	Z ROT
2	GLOBAL				
		1	0.0016312	−0.0007015	−0.0025088
		2	0.0470051	0.0002089	0.0002556
		3	0.0364773	−0.0003695	−0.0016899
4	GLOBAL				
		1	0.0044423	−0.0025066	−0.0019385
		2	0.0231103	0.0002795	−0.0014076
		3	0.0206645	−0.0016703	−0.0025096
5	GLOBAL				
		1	0.0015462	−0.0042452	0.0001227
		2	0.0466966	0.0004882	−0.0009972
		3	0.0361821	−0.0028177	−0.0006559
6	GLOBAL				
		1	−0.0011966	−0.0046621	−0.0017484
		2	0.0593730	0.0005969	0.0000361
		3	0.0436323	−0.0030489	−0.0012843

Results Obtained by Routinely Solving the Structure in Figure 14.4 (Continued) Are: (Units Are Feet and Kips)

JOINT DISPLACEMENTS – FREE JOINTS

JOINT	LOADING	/ ----------- DISPLACEMENTS ------------/		
		X DISP	Y DISP	Z ROT
8	GLOBAL			
	1	0.0046795	−0.0034233	−0.0001124
	2	0.0233691	0.0001347	−0.0010605
	3	0.0210364	−0.0024664	−0.0008796
9	GLOBAL			
	1	0.0014028	−0.0050900	0.0005746
	2	0.0463911	0.0002812	−0.0003859
	3	0.0358455	−0.0036066	0.0001415
10	GLOBAL			
	1	−0.0017485	−0.0054986	0.0018371
	2	0.0593892	0.0003376	−0.0009585
	3	0.0432305	−0.0038707	0.0006590
12	GLOBAL			
	1	0.0050514	−0.0011478	0.0019461
	2	0.0237476	−0.0000235	−0.0010027
	3	0.0215992	−0.0008784	0.0007075
13	GLOBAL			
	1	0.0007147	−0.0015186	0.0011156
	2	0.0464277	0.0000122	−0.0012592
	3	0.0353568	−0.0011298	−0.0001077

JOINT LOADS – SUPPORTS (Units are feet and kips)

JOINT	LOADING	/ ----------- FORCES -----------/		
		X FORCE	Y FORCE	Z MOMENT
1	GLOBAL			
	1	1.373	19.123	0.000
	2	−1.019	−5.694	0.000
	3	0.265	10.071	0.000
3	GLOBAL			
	1	6.022	136.662	−23.486
	2	−7.488	−15.240	71.900
	3	−1.099	91.066	36.310
7	GLOBAL			
	1	−2.289	186.638	18.421
	2	−9.195	−7.346	80.822
	3	−8.613	134.469	74.432
11	GLOBAL			
	1	−5.106	62.577	0.000
	2	−1.298	1.280	0.000
	3	−4.803	47.893	0.000

MEMBER FORCES (Units are feet and kips)

MEMBER	LOADING	JOINT	/ AXIAL	FORCES SHEAR Y	/ MOMENT Z
1					
	1	1	19.123	−1.373	0.000
		2	−19.123	1.373	−41.175
	2	1	−5.694	1.019	0.000
		2	5.694	−1.019	30.573
	3	1	10.071	−0.265	0.000
		2	−10.071	0.265	−7.952
2					
	1	3	136.662	−6.022	−23.486
		4	−136.662	6.022	−66.845
	2	3	−15.240	7.488	71.900
		4	15.240	−7.488	40.415
	3	3	91.066	1.099	36.310
		4	−91.066	−1.099	−19.823
3					
	1	4	94.788	−9.851	−96.930
		5	−94.788	9.851	−50.827
	2	4	−11.379	3.310	20.238
		5	11.379	−3.310	29.417
	3	4	62.557	−4.905	−57.519
		5	−62.557	4.905	−16.058
4					
	1	5	22.727	−8.909	−45.891
		6	−22.727	8.909	−87.742
	2	5	−5.922	3.261	12.903
		6	5.922	−3.261	36.016
	3	5	12.604	−4.236	−24.741
		6	−12.604	4.236	−38.795
5					
	1	7	186.638	2.289	18.421
		8	−186.638	−2.289	15.907
	2	7	−7.346	9.195	80.822
		8	7.346	−9.195	57.102
	3	7	134.469	8.613	74.432
		8	−134.469	−8.613	54.757
6					
	1	8	90.870	0.113	−6.833
		9	−90.870	−0.113	8.532
	2	8	−7.983	7.261	46.917
		9	7.983	−7.261	62.004
	3	8	62.165	5.531	30.063
		9	−62.165	−5.531	52.902

MEMBER FORCES (Units are feet and kips)

MEMBER	LOADING	JOINT	AXIAL	SHEAR Y	MOMENT Z
7					
	1	9	22.273	8.909	52.697
		10	−22.273	−8.909	80.936
	2	9	−3.078	1.739	19.443
		10	3.078	−1.739	6.638
	3	9	14.396	7.986	54.105
		10	−14.396	−7.986	65.680
8					
	1	11	62.577	5.106	0.000
		12	−62.577	−5.106	76.591
	2	11	1.280	1.298	0.000
		12	−1.280	−1.298	19.475
	3	11	47.893	4.803	0.000
		12	−47.893	−4.803	72.050
9					
	1	12	20.219	11.110	92.610
		13	−20.219	−11.110	74.036
	2	12	−1.944	3.409	28.437
		13	1.944	−3.409	22.700
	3	12	13.706	10.889	90.785
		13	−13.706	−10.889	72.552
10					
	1	2	1.373	19.123	41.175
		5	−1.373	25.877	−142.489
	2	2	4.981	−5.694	−30.573
		5	−4.981	−3.306	−5.253
	3	2	4.765	10.071	7.952
		5	−4.765	16.928	−110.807
11					
	1	4	−3.828	41.874	163.776
		8	3.828	48.125	−257.541
	2	4	−4.177	−3.862	−60.653
		8	4.177	3.862	−55.200
	3	4	−6.004	28.510	77.342
		8	6.004	38.990	−234.556
12					
	1	5	2.314	46.184	239.207
		9	−2.314	43.816	−203.696
	2	5	4.932	−2.151	−37.067
		9	−4.932	2.151	−27.465
	3	5	5.434	33.024	151.605
		9	−5.434	34.475	−173.370

MEMBER FORCES (Units are feet and kips)

MEMBER	LOADING	JOINT	AXIAL	SHEAR Y	MOMENT Z
13					
	1	6	8.909	22.727	87.742
		10	−8.909	22.273	−80.936
	2	6	−0.261	−5.922	−36.016
		10	0.261	−3.078	−6.638
	3	6	6.486	12.604	38.795
		10	−6.486	14.396	−65.680
14					
	1	8	−6.004	47.642	248.466
		12	6.004	42.358	−169.202
	2	8	−6.111	−3.224	−48.819
		12	6.111	3.224	−47.912
	3	8	−9.086	33.313	149.735
		12	9.086	34.187	−162.835
15					
	1	9	11.110	24.781	142.466
		13	−11.110	20.219	−74.036
	2	9	−0.591	−7.056	−53.983
		13	0.591	−1.944	−22.700
	3	9	7.889	13.294	66.363
		13	−7.889	13.706	−72.552

In the substructuring approach described earlier, the user had to make two computer runs for each substructure. In each first computer run, independent loadings had to be defined for all of the real loadings and a unit prescribed displacement for each interface DOF. The user had to formulate and solve the matrix equation of system equilibrium for the interface DOF. In each second computer run, independent loadings had to be defined for all of the real loadings and the actual prescribed displacements for the interface DOF; a dependent loading was defined to have the results combined. Some available computer programs have the capability of performing the substructuring analysis of a structure in a single computer run. That is, the user does not have to formulate and solve the matrix equation of system equilibrium for the interface displacements. Those programs that perform the substructuring analysis in a single computer run use the subdivided matrix approach described in Section 14.1 to compute the reactions. The subdivided matrix equations defined in Section 14.1 are applicable for each substructure by attaching a subscript M to denote the substructure number as shown below.

$$[S_{ff}]_M\{D_f\}_M = \{P_f^e\}_M - [S_{fs}]_M\{D_s\}_M$$
$$\{P_s^e\}_M = [S_{sf}]_M\{D_f\}_M + [S_{ss}]_M\{D_s\}_M$$

In order to obtain the substructure stiffness matrix, $[\hat{S}_{ss}]_M$, the first matrix equation is solved by matrix inversion to obtain

$$\{D_f\}_M = [S_{ff}]_M^{-1}(\{P_f^e\}_M - [S_{fs}]_M\{D_s\}_M)$$

which is substituted into the second equation to obtain

$$\{P_s^e\}_M = [S_{sf}]_M[S_{ff}]_M^{-1}(\{P_f^e\}_M - [S_{fs}]_M\{D_s\}_M) + [S_{ss}]_M\{D_s\}_M$$

which can be rearranged to obtain

$$([S_{ss}]_M - [S_{sf}]_M[S_{ff}]_M^{-1}[S_{fs}]_M)\{D_s\}_M = (\{P_s^e\}_M - [S_{sf}]_M[S_{ff}]_M^{-1}\{P_f^e\}_M)$$

The terms enclosed in parentheses in the preceding equation can be renamed as follows:

$$[\hat{S}_{ss}]_M\{D_s\}_M = \{\hat{P}_s^e\}_M$$

which is the substructure force-deformation relation and $[\hat{S}_{ss}]_M$ is the substructure stiffness matrix.

14.4 SYMMETRIC AND ANTISYMMETRIC COMPONENTS OF LOADING

Any unsymmetrical loading can be decomposed into symmetric and antisymmetric components. However, it is only beneficial to do so when the structure is symmetric (structural properties, geometry, and boundary conditions define a structure). If the structure is symmetric, the total number of DOF can be reduced by dealing with the symmetric part of the structure subjected to symmetric and antisymmetric loadings. In computerized solutions of symmetric structures having lots of joints, sometimes the bandwidth of the system stiffness matrix can be reduced drastically by analyzing the symmetric part of the structure. If the bandwidth of the entire structure is too large for the computer memory and can be reduced by analyzing the symmetric part, this approach is more economical than using the method of substructuring. In non-computerized solutions, reducing the total number of DOF is time saving even if the bandwidth of the system stiffness matrix is not reduced.

The reader should review Section 10.8 before proceeding to study the following material. Section 10.8 introduced the concepts of structural symmetry, loading symmetry, and loading antisymmetry as applicable to moment distribution. Those concepts need to be modified and extended for structures in which axial deformations of the members are not ignored. Only plane frames are shown in the following figures, but the concepts can be extended as described later to three-dimensional structures. In the following figures, it is assumed that each structure is symmetric with respect to the vertical line located halfway across the structure. It should be noted that there are other types of structural symmetry [14.1] besides the type discussed in this text.

First, see Figure 14.7 which shows the behavior of a simple beam subjected to unsymmetrical loads that are decomposed into symmetric ($) and antisymmetric (a/s) components. Since superposition is involved, the following concepts are only applicable to structures that behave in a linearly elastic manner.

In Figure 14.7, the deflected shapes are shown for the symmetric and antisymmetric components of loading. At the axis of symmetry, observe that:

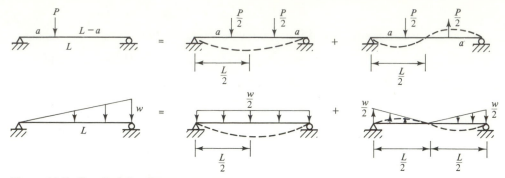

Figure 14.7 $\mathcal{S}$ and a/s loading components for a simple beam.

1. For a symmetric loading:
   a. The slope of the deflection curve is zero.
   b. The deflection is not zero.
   c. The shear is zero.
   d. The moment is not zero.
2. For an antisymmetric loading:
   a. The slope of the deflection curve is not zero.
   b. The deflection is zero.
   c. The shear is not zero.
   d. The moment is zero.

Items 2c and 2d are easily verified by choosing $P = 60^k$, $a = 10$ ft, $L = 30$ ft, and computing the shear and moment at midspan of the beam.

Next, as shown in Figure 14.8a, consider the symmetric plane frame subjected to a general type of unsymmetric loading. Also shown in Figure 14.8 are the deflected shapes of the entire structure for the symmetric and antisymmetric loading components. These deflected shapes are drawn to aid in the visualization of the boundary conditions at the plane of symmetry for each loading. Finally, the geometric boundary conditions at the plane of symmetry are shown in Figure 14.8 for the left half of the structure subjected to the symmetric and antisymmetric loadings. Analyses are performed for the loadings and boundary conditions shown on the half structures in Figures 14.8 (c and e). Each analysis only gives the final member-end forces and reactions for the left half of the entire structure for each loading. To obtain the final member-end forces and reactions of the entire structure for the loading shown in Figure 14.8a from the results of the symmetric and antisymmetric solutions for only the left half of the structure: (1) for the left half of the entire structure, algebraically sum the results of the symmetric and antisymmetric solutions; and (2) for the right half of the entire structure, subtract the results of the antisymmetric solution from the results of the symmetric solution.

As shown in Figure 14.9, we need to consider a plane frame having a member and a concentrated load coincident with the axis of symmetry. Only one of the structural halves is analyzed for the symmetric loading component and boundary conditions as well as for the antisymmetric loading component and boundary conditions. When

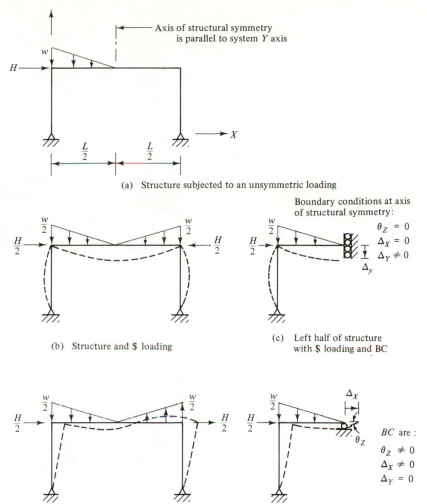

(a) Structure subjected to an unsymmetric loading

(b) Structure and S loading

(c) Left half of structure with S loading and BC

(d) Structure and a/s loading

(e) Left half of structure with a/s loading and BC

Figure 14.8 S and a/s loading components for a plane frame.

the structure is halved, any concentrated load directed along the axis of symmetry is halved and the properties (area and moment of inertia) of the member on the axis of symmetry are halved.

Now, assuming that the reader fully understands the preceding information, the following general statements concerning symmetry and antisymmetry of loading should be comprehensible. (After reading the following general statements, the reader should apply them to each complete structure shown in Figures 14.7 through 14.9.) For a plane frame lying in the plane of this page, subjected to in-plane loads, and having an axis of symmetry at midspan of the full structure (axis of symmetry is perpendicular to the span direction):

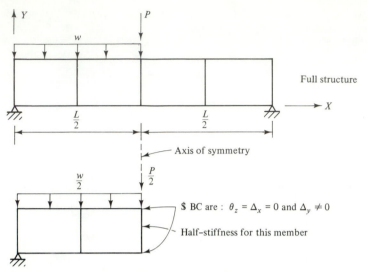

Full structure

Axis of symmetry

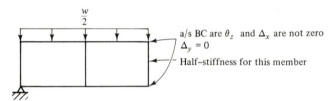

$ BC are : $\theta_z = \Delta_x = 0$ and $\Delta_y \neq 0$

Half–stiffness for this member

Half of symmetric structure subjected to symmetric loading component

a/s BC are θ_z and Δ_x are not zero
$\Delta_y = 0$

Half–stiffness for this member

Figure 14.9   Plane frame with a member and concentrated load on the axis of symmetry.

Half of symmetric structure subjected to antisymmetric loading component

1. A loading is symmetric if the left half of the structure and its loading are rotated 180 degrees about the axis of symmetry and the right half of the structure and loading are exactly duplicated. That is, the described rotation generates the other half of the structure (including boundary conditions) and its loading. Alternatively, one can place a mirror at the axis of symmetry and perpendicular to the left half of the structure, for example. The mirrored image or reflection is what the right half of the structure and loading must be if the structure and loading are symmetric.

2. A loading is antisymmetric if the left half of the structure and its loading are rotated 180 degrees about the axis of symmetry, the loading is then reversed in direction, and the result is identical to the original right half of the structure and loading. Alternatively, using the mirrored image or reflection method—if the loading on the left half of the structure is reversed in direction and the reflection in the mirror is identical to the right half of the structure and loading, then the structure is symmetric and the loading is antisymmetric.

The mirrored image or reflection concept is most useful in three-dimensional structures in which a plane of structural symmetry exists. However, for a special class of three-dimensional structures, there is another concept that may be more useful. Consider a domed roof structure of revolution—for example, a hemisphere that has a structural plan view consisting of a set of concentric circles intersected by several radial lines—having structural symmetry with respect to three or more vertical planes intersecting at the center of the structural plan view:

1. There are three or more pie-shaped sectors in a plan view sense.
2. If the structure is symmetric, one of the pie-shaped sectors can be removed as a piece of pie, rotated around the vertical axis of the dome in a translational sense (translated along a circumferential line), and overlayed in the position of another pie-shaped sector to generate the structure and the symmetric loading in that sector. As in the cutting of a piece of pie, members and concentrated loads on a cutting plane are equally split between adjoining sectors.
3. For antisymmetric loading, one proceeds as in item 2 and then reverses the direction of the loads. If the resulting loading is identical to what existed in the sector being overlayed, the loading is antisymmetric with respect to the cutting planes (meridianal lines) of structural symmetry.

The following statements should be helpful in decomposing an unsymmetric loading into symmetric and antisymmetric components and in arriving at the displacement DOF (boundary conditions) on a plane of symmetry of a structural part:

1. For symmetric loading there can not be any:
   a. Translational displacement normal to the plane of symmetry.
   b. Rotational displacements about two orthogonal axes in the plane of symmetry.
2. For antisymmetric loading there can not be any:
   a. Translational displacement in the plane of symmetry.
   b. Rotational displacement about an axis normal to the plane of symmetry.

The preceding information and examples in Figures 14.7 through 14.9 should be sufficient to assist the reader in learning how to cope with symmetric and antisymmetric loadings and boundary conditions. Three-dimensional structures with nonorthogonally intersecting members can be quite complex in regard to symmetry and antisymmetry of loadings and boundary conditions. Before a structural engineer attempts symmetric and antisymmetric solutions of such structures involving hundreds of DOF, the author recommends that the following should have been done. Create a smaller version of the real structure which must be solved via symmetry and antisymmetry of loading. The smaller version must involve all of the essential details of the real structure in regard to geometry, boundary conditions, structural properties, number of axes of symmetry, and so on. The smaller version should be treated as follows:

1. Solve the entire structure for:
   a. A general loading.
   b. The symmetric component of the loading in a.
   c. The antisymmetric component of the loading in a.
   d. A dependent loading to sum b and c loadings.
   e. Compare solution d to solution a and look at the displacements on the axis of symmetry for each of the solutions b and c to determine which

displacements are zero and which ones are not (this information is valuable in setting up the boundary conditions for a and b in 2 below).

2. Solve the symmetric part of the structure for:

   a. The symmetric loading component and symmetric boundary conditions.

   b. The antisymmetric loading component and antisymmetric boundary conditions.

   c. Algebraically add solutions a and b and also subtract solution b from solution a to obtain the final member-end forces and reactions in each half of the entire structure.

   d. Compare solution c to solution 1a.

Item 1 is done to ensure that the computer program works correctly and that the engineer knows how to correctly input the symmetric and antisymmetric loading components. Note that solutions b and c of item 1 will give zero (or practically zero) system displacement magnitudes in the complete structure where there are no DOF in solutions a and b of item 2.

The concepts described above are demonstrated in the following example of a bridge grid. A grid is a planar structure subjected to out-of-plane loads only. A bridge grid simply supported on the roadway abutments is chosen for illustration purposes. To simplify the discussion, suppose there is no concrete deck on the bridge and that the steel members composing the grid are orthogonally intersecting members. Since the majority of the gravity direction displacements of a bridge are downward, the right-hand coordinate system axes were chosen in Figure 14.10 to have the Z axis be positive in the downward direction. At each joint, let u,v,w be the translational displacements in the X,Y,Z directions, respectively, and let θ_X, θ_Y, θ_Z be the rotations about axes parallel to the X,Y,Z axes, respectively.

Figure 14.10 shows the plan view of a bridge grid which lies in the XY plane and the positive Z direction (gravity direction) is downward. Two concentrated, gravity direction loads are acting on the grid at the locations shown producing an unsymmetric loading. Downward loads are denoted by a circle with an X in it (tail of an arrow as viewed from above) and upward loads are denoted by a circle with a dot in it (head of an arrow as viewed from above). The grid is symmetric with respect to both centerlines. Therefore, only one-quarter of the grid needs to be involved in modeling the symmetric part of the grid subjected to $ and a/s loading components. An analyst can choose to model any one of the four symmetric parts in an analysis and the author chooses to model the quarter of the structure having a corner located at the origin of the XY axes (Northwest quadrant). There are not any in-plane loads. Therefore, the DOFs at a typical interior joint are the downward displacement, w, and the rotations, θ_X and θ_Y, about axes parallel to the X and Y axes. At the abutment (support) joints, only θ_X is nonzero for this simply supported bridge grid.

Figure 14.11 demonstrates items 1b and 1c of the author's recommended procedure. Smaller sketches of the entire grid are drawn such that all four sketches fit on a single page and the abutment lines are omitted for clarity. Since only the northwest quadrant of the structure is to be modeled in item 2, $ and a/s loading components and boundary conditions must be treated with respect to each line of symmetry for that quadrant ($ and a/s per line times 2 lines = 4 conditions). Any concentrated load

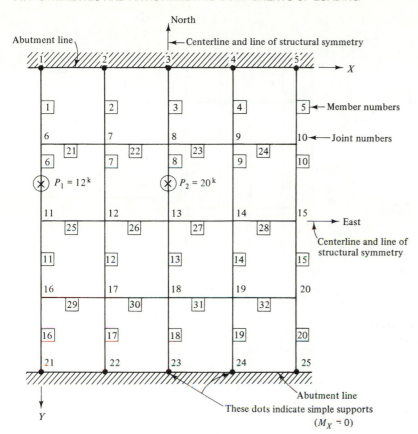

Figure 14.10 Plan view of a bridge grid.

that is not on an axis of symmetry in Figure 14.10 is divided by 4 (there are four symmetric parts) in dealing with $ and a/s loading components. Therefore, the 12^k load on member 6 in Figure 14.10 is divided by 4 and allocated to each quadrant in each of the Figures 14.11 (a through d) in conformance with the indicated $ and a/s boundary conditions on the axes of symmetry. Any concentrated load on an axis of symmetry is halved (it is a symmetric load and has no a/s component with respect to the axis it is located on) unless the load is located on both axes of symmetry in which case the load is halved twice. Therefore, the 20^k load on member 8 in Figure 14.10 is halved and allocated as shown in Figures 14.11 (a and c). Note that the algebraic sum (down loads are positive) of all the loadings in Figure 14.11 gives the original loading shown in Figure 14.10 (this is a check on the $ and a/s loading components).

Figure 14.12 demonstrates items 2a and 2b in the author's recommended procedure. The author chooses to model the northwest quadrant and each of Figures 14.11 (a through d) is cut along the lines of structural symmetry to arrive at the structure, boundary conditions, and four loadings defined in Figure 14.12. The stiffness of each member and each concentrated load on a cutting line is halved. To emphasize that the northwest quadrant is being modeled, the joint and member numbers of the

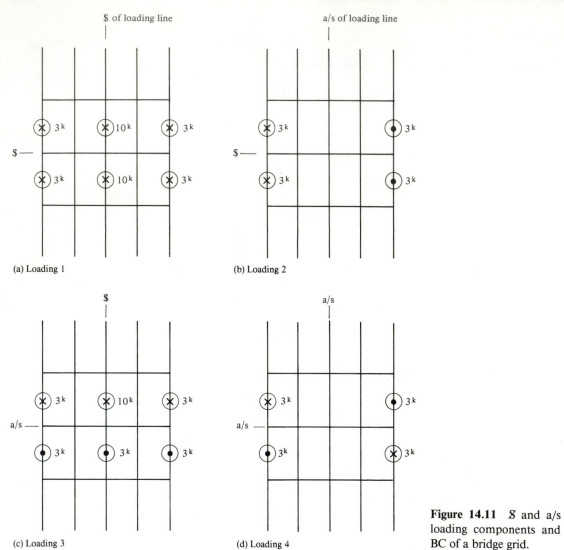

Figure 14.11 $\mathcal{S}$ and a/s loading components and BC of a bridge grid.

original structure (Figure 14.10) are shown in Figure 14.12 (some available programs allow this and it greatly facilitates the execution of item 2d). The computer output for each loading is given for the member numbers and joint numbers in the quadrant being modeled. To obtain the final member-end forces and reactions:

1. For the quadrant being modeled, northwest quadrant, the dependent loading is

$$\text{loading } 1 + \text{loading } 2 + \text{loading } 3 + \text{loading } 4$$

2. For the northeast quadrant, the dependent loading is

$$\text{loading } 1 - \text{loading } 2 + \text{loading } 3 - \text{loading } 4$$

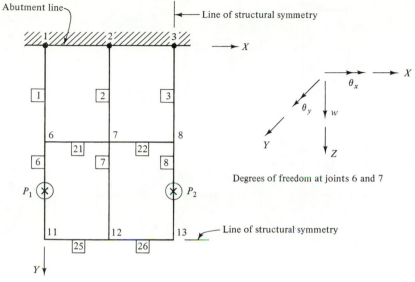

Plan view of the first quadrant of a doubly symmetric, simply supported, bridge grid

Loading number	Loads		Boundary conditions (only the nonzero degrees of freedom are listed)		
	P_1	P_2			
1	3	5	θ_x @ 1,2,3,8;	θ_y @ 11,12;	w @ 8,11,12,13;
2	3	0	θ_x @ 1,2;	θ_y @ 8,11,12,13;	w @ 11,12;
3	3	5	θ_x @ 1,2,3,8,11,12,13;		w @ 8;
4	3	0	θ_x @ 1,2,11,12;	θ_y @ 8;	

Loading numbers correspond to Figure 14.11 (a through d), respectively. Units of load are kips.

Figure 14.12 Modeling details for northwest quadrant portion of Figure 14.11.

3. For the southwest quadrant, the dependent loading is

$$\text{loading } 1 + \text{loading } 2 - \text{loading } 3 - \text{loading } 4$$

4. For the southeast quadrant, the dependent loading is

$$\text{loading } 1 - \text{loading } 2 - \text{loading } 3 + \text{loading } 4$$

The signs in each dependent loading were obtained by looking at the direction of the 3^k force (up is negative) in the quadrant for which the dependent loading was being defined.

Figure 14.13 shows the plan view of a domed roof structure. The radial lines are lines of structural symmetry. This structure might be a space truss or a space frame with lots more members than the number of lines shown (that will be the usual case

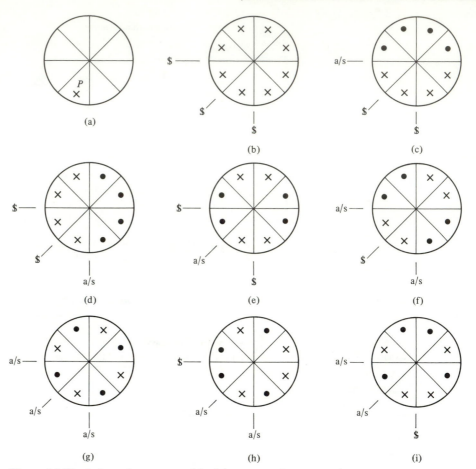

Figure 14.13 A domed structure with eight symmetric structural parts.

for a sports arena, for example). Alternatively, the dome might be a shell of revolution and a finite element analysis would be appropriate for such a structure. Our objective here is to discuss the $ and a/s loading components and boundary conditions for modeling only one-eighth (the smallest symmetric part) of the entire structure. Figure 14.13a shows a single concentrated load which is not on an axis of symmetry. This load must be decomposed into a total of eight combinations of $ and a/s components. The $ and a/s notations on the lines of structural symmetry in each of Figures 14.13 (b through i) denote the boundary conditions associated with the loading in each figure. Downward loads are denoted by an × and upward loads are denoted by a dot. The solution for Figure 14.13a is the algebraic sum of Figures 14.13 (b through i). Any one of the eight sectors can be modeled with the loadings and boundary conditions shown for that sector. In each of the Figures 14.13 (b through i), the value of the load is $P/8$.

14.5 EFFECTS DUE TO TEMPERATURE CHANGES

Consider a steel member, for example. The temperature at the time of fabrication and at the time of erection are not necessarily the same. Also temperature varies with time during the life of the structure. Temperature extremes must be estimated for the life of the member. For purposes of discussion, let

T_1 = the temperature at some initial time (at fabrication or erection)
T_2 = the temperature at some time after erection
C_{te} = the coefficient of thermal expansion (C_{te} = 0.0000065 in./in. per °F for steel)

The effects of temperature changes for a structure analyzed by the Displacement Method can be treated as follows. Temporarily, to simplify the discussion, consider an isolated member of constant EA and EI. First, the isolated member that is perfectly straight and has zero weight is treated as a simply supported beam subjected to a temperature change in order to determine the deformed geometry for these conditions. Then, the member-end forces required to fully remove the member-end displacements computed in the first step are determined. These fixed ended forces are the computer input data for a "temperature loading."

EXAMPLE 14.1

A prismatic member has a constant temperature at any time over the entire cross section and along the member length, but temperature changes with time. Due to the described temperature change, find the fixed ended forces needed as input data for a Displacement Method analysis.

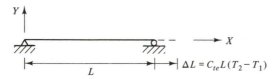

(a) Simply supported beam subjected to temperature

$$F_1^f = \frac{EA}{L} \Delta L = EA \star C_{te} \star (T_2 - T_1)$$

(b) Fixed ended forces due to temperature

Figure 14.14 Member subjected to a constant temperature distribution.

Solution. The member as a simply supported beam subjected to the described temperature change only elongates the amount shown in Figure 14.14a. The member must be compressed by the member-end forces shown in Figure 14.14b to remove this elongation.

EXAMPLE 14.2

A prismatic member has a constant temperature at any time along the length, but the temperature at the top and bottom surfaces are not identical. The variation in temperature between the top and bottom surfaces is linear. Also, the described temperature distribution changes with time.

Let:

T_u = the temperature change for the upper surface
T_b = the temperature change for the bottom surface [for example, $T_b = (T_2 - T_1)_b$]

Due to the described temperature change, find the fixed ended forces needed as input for a Displacement Method analysis.

Solution. For graphical convenience in drawing the deformed shape of the simply supported beam shown in Figure 14.15, it is assumed that the temperature changes at both surfaces are positive and that $T_b > T_u$.

For graphical convenience, consider a differential element from Figure 14.15a such that the left side of the element coincides with the zero slope point of the deflected shape (Fig. 14.16).

$$e_u = C_{te} T_u \, dx; \qquad e_b = C_{te} T_b \, dx; \qquad d\theta = \frac{e_b - e_u}{h}; \qquad e_x = e_u + (h - d_b) \, d\theta$$

$$\Delta L = \int_0^L e_x = C_{te} \left[T_u + \frac{h - d_b}{h} (T_b - T_u) \right] L$$

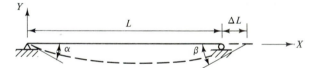

(a) Simply supported beam subjected to temperature change

(b) Fixed ended forces

Figure 14.15 Member subjected to linear variation in temperature along depth.

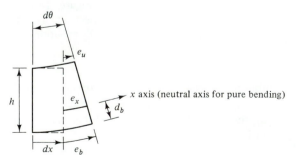

x axis (neutral axis for pure bending)

Figure 14.16 Deformed differential element due to temperature.

At any point x on the deflection curve of Figure 14.15a, v is the deflection and the slope is $\theta = dv/dx$ and the curvature is $\phi = d\theta/dx = d^2v/dx^2 \approx M_z/EI_z$. Consequently, the end slopes of the unrestrained member in Figure 14.15a can be obtained from the conjugate beam loaded with the curvature diagram (Fig. 14.17).

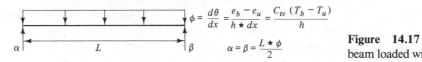

$$\phi = \frac{d\theta}{dx} = \frac{e_b - e_u}{h * dx} = \frac{C_{te}\,(T_b - T_u)}{h}$$

$$\alpha = \beta = \frac{L * \phi}{2}$$

Figure 14.17 Conjugate beam loaded with ϕ diagram.

The fixed ended forces of Figure 14.15b can be computed as follows:

$$F^f_1 = \frac{EA}{L}(\Delta L); \qquad F^f_4 = -F^f_1; \qquad F^f_3 = \frac{EI_z}{L}(4\alpha - 2\beta);$$

$$F^f_6 = \frac{EI_z}{L}(2\alpha - 4\beta); \qquad F^f_2 = \frac{F^f_3 + F^f_6}{L}$$

and $F^f_5 = -F^f_2$. Substitution of the values for α and β gives

$$F^f_3 = EI_z C_{te} \frac{(T_b - T_u)}{h} \quad \text{and} \quad F^f_6 = -F^f_3 \quad \text{and} \quad F^f_2 = F^f_5 = 0$$

Substitution of ΔL gives

$$F^f_1 = EA * C_{te}\left(T_u + \frac{h - d_b}{h}(T_b - T_u)\right).$$

Conceptually, it is easy to extend Example 14.2 to include:

1. Variable member properties along the member length.
2. Any temperature distribution provided the curvature can be computed at a sufficient number of points along the x axis to enable an analyst to perform a numerical integration of the curvature diagram loaded on the conjugate beam.

Item 1 involves using the general definition of the member stiffness matrix given in Figure 13.6 in order to determine the fixed ended forces. In item 2, if the curvature is computed at sufficiently close intervals of x, the curvature can be assumed to vary linearly between the computed points. Thus, the trapezoidal shaped areas of the curvature diagram between the computed points of curvature can be decomposed into a rectangle and a triangle for which the area and centroids are easily computed. Summation of moments at the right end of the conjugate beam loaded with the resultant concentrated angle changes (the rectangular and triangular areas of curvature) gives the left end slope of the conjugate beam. Summation of vertical angle changes gives the right end slope of the conjugate beam.

14.6 EFFECTS DUE TO PRESTRAIN

Prestrain means that a member was accidentally or deliberately fabricated too short or was initially crooked. Precambering of trusses to remove the dead load deflection, for example, may be desirable and is accomplished by selectively fabricating certain members shorter than their idealized, unloaded length. When the truss is assembled, external construction loads have to be applied at appropriate locations to make the truss deform enough to allow the connections to be made. As the construction loads are removed, the truss geometry changes and forces in the members occur due to the changes in the geometry—that is, the truss members are prestrained. Precambering of a steel member in a plane frame may be desirable to remove the dead load deflection. Such a member may be precambered by deliberately bending the member into the strain hardening range and removing the loads to create a member with a desired crookedness.

The effects of prestrain are treated similar to the manner that temperature changes were treated in Section 13.12. That is, fixed ended forces due to the effects of prestrain are computed and input to the Displacement Method analysis. A couple of examples should be sufficient to illustrate the general procedure for the treatment of prestrain effects.

EXAMPLE 14.3 ───

The length of a prismatic, plane frame member is found to be ΔL longer than it should have been. Find the fixed ended forces due to this prestrained condition.

Solution. The isolated, weightless, deformed member only needs to be axially compressed (shortened) to remove the excess length. The only nonzero fixed ended forces needed to accomplish this are

$$F_1^f = \frac{EA}{L} * (\Delta L) \quad \text{and} \quad F_4^f = -F_1^f.$$

EXAMPLE 14.4 ───

A prismatic, plane frame member is found to be permanently bent as shown in Figure 14.18a. However, the member-end connection details were prepared assuming the member was not bent as shown. Assume that axial deformations due to bending are negligible. Find the fixed ended forces due to this prestrained condition.

Solution. The concentrated angle change, θ, in Figure 14.18a is computed and placed on the conjugate beam as shown in Figure 14.18b to obtain the member-end slopes. Figure 14.18c shows the fixed ended forces required to remove these member-end slopes.

If a member is deliberately bent for precambering purposes, usually, it is symmetrically bent at two or three locations as shown in Figure 14.19. Concentrated angle changes at each of the bend points are computed and placed on the conjugate beam to find the member-end slopes needed in the fixed ended force relations shown in Figure 14.18c.

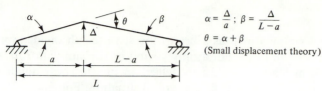

$$\alpha = \frac{\Delta}{a} \; ; \; \beta = \frac{\Delta}{L-a}$$

$$\theta = \alpha + \beta$$

(Small displacement theory)

(a) Bent, weightless member

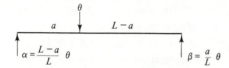

$$\alpha = \frac{L-a}{L}\theta \qquad\qquad \beta = \frac{a}{L}\theta$$

(b) Conjugate beam loaded with
concentrated angle change(s)
(reminder: theory is only valid
for small slopes)

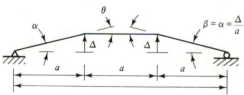

$$F_3^f = \frac{-EI}{L}(4\alpha - 2\beta) \qquad\qquad F_6^f = \frac{-EI}{L}(2\alpha - 4\beta)$$

$$F_2^f = \frac{F_3^f + F_6^f}{L} \qquad\qquad F_5^f = -F_2^f$$

Note: $F_1^f = F_4^f = 0$ (no change in length is assumed to occur)

(c) Fixed ended forces for bent member in Figure (a)

Figure 14.18 A bent member and fixed ended forces.

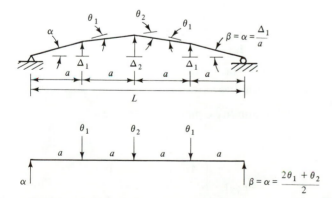

$$\beta = \alpha = \frac{\Delta}{a}$$

(a) Deliberately bent member for precambering purposes

$$\beta = \alpha = \frac{\Delta_1}{a}$$

$$\beta = \alpha = \frac{2\theta_1 + \theta_2}{2}$$

(b) Symmetrically bent member and its conjugate beam

Figure 14.19 Symmetrically bent members.

601

14.7 SHEAR DEFORMATIONS

Typically, for rolled steel sections used as a beam, the span length divided by the depth, L/d, is in the range of 10 to 20. For a simply supported beam with $L/d = 10$ and subjected to a uniform load, the midspan deflection due to shear divided by the deflection due to bending is about 0.18 in a typical situation. For $L/d < 10$, the shear deflection is much more significant. For a cantilever beam with $L/d = 1$, the bending deflection may be negligible compared to the shear deflection. Consequently, there are cases for which shear deformations cannot be ignored.

The general member stiffness matrix definition in Figure 13.6 can be revised to include shear deformations. For the approach described in Section 9.6 and Figure 9.13, we need the solution for a member composed of a set of prismatic segments. For simplicity, the author chooses only to demonstrate the inclusion of shear deformations for a prismatic member.

The following definitions are needed in the derivation of the member stiffness matrix which includes shear deformations for a prismatic member:

$$\text{modulus of shear, } G = \frac{E}{2(1 + \nu)}$$

where ν is Poisson's ratio ($\nu = 0.25$ for steel and concrete) and E is the modulus of elasticity, A_v is the shear area of a cross section (A_v = web area of a steel W section bending about the major axis and the flange areas for bending about the minor axis; for a rectangular section of wood, for example, $A_v = \frac{5}{6}bh$), and $r = 3EI/(GA_vL^2)$ is a convenience ratio.

Since the objective is to include shear deformations in the member stiffness matrix for a prismatic member, we need only to find the member-end moments in Figure 14.20a by the Force Method, for example. All member-end rotations in Figure 14.20 are due to bending and shear.

Virtual work is used to obtain the following member-end rotations due to the unit redundants in Figures 14.20b and 14.20d.

$$\theta_{3,3} = \frac{(1)L}{3EI} + \int_0^L \bar{V}_3 \left(\frac{V\,dz}{GA_v} \right) = \frac{L}{3EI} + \frac{1}{L} \left(\frac{\left(\frac{1}{L} \right) L}{GA_v} \right) = \frac{L}{3EI} + \frac{1}{GA_vL} = \frac{L}{3EI}(1 + r)$$

$$\theta_{6,3} = -\frac{L}{6EI} + \int_0^L \bar{V}_6 \left(\frac{V\,dz}{GA_v} \right) = \frac{-L}{6EI} + \frac{1}{GA_vL} = \frac{-L}{6EI}(1 - 2r)$$

$\theta_{6,6} = \theta_{3,3}$ due to symmetry of member properties

$\theta_{3,6} = \theta_{6,3}$ via Maxwell-Betti reciprocal theorem

$$\begin{bmatrix} \theta_{3,3} & \theta_{3,6} \\ \theta_{6,3} & \theta_{6,6} \end{bmatrix} \begin{Bmatrix} S_{3,3}^M \\ S_{6,3}^M \end{Bmatrix} = \begin{Bmatrix} 1 \\ 0 \end{Bmatrix} \qquad \text{(compatibility equation for Figure 14.20)}$$

Solving via matrix inversion, we get

$$\begin{Bmatrix} S_{3,3}^M \\ S_{6,3}^M \end{Bmatrix} = \frac{1}{\theta_{3,3}^2 - \theta_{6,3}^2} \begin{bmatrix} \theta_{3,3} & -\theta_{6,3} \\ -\theta_{6,3} & \theta_{3,3} \end{bmatrix} = \begin{Bmatrix} \dfrac{4EI}{L} \left(\dfrac{1 + r}{1 + 4r} \right) \\[2mm] \dfrac{2EI}{L} \left(\dfrac{1 - 2r}{1 + 4r} \right) \end{Bmatrix}$$

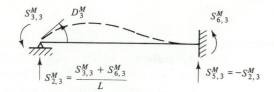

$$S_{2,3}^M = \frac{S_{3,3}^M + S_{6,3}^M}{L}$$

(a) Prismatic member stiffness coefficients with shear deformations included

(b) Effects due to a unit redundant

(c) Real and virtual shear diagram

(d) Effects due to other unit redundant

Figure 14.20 Inclusion of shear deformations in member stiffness coefficients.

where

$$\theta_{3,3}^2 - \theta_{6,3}^2 = \left[\frac{L(1+r)}{3EI}\right]^2 - \left[\frac{-L(1-2r)}{6EI}\right]^2 = \left(\frac{L}{3EI}\right)^2 \left[(1+r)^2 - \left(\frac{1-2r}{2}\right)^2\right]$$

$$= \left(\frac{L}{3EI}\right)^2 \left(\frac{3}{4}+3r\right) = \frac{3}{4}\left(\frac{L}{3EI}\right)^2 (1+4r)$$

Note that the dimensionless ratio r contains the shear area, A_v. If shear deformations are ignored, then A_v = infinity and $r = 0$. Substitution of $r = 0$ into the last matrix equation shown above gives the usual bending stiffness coefficients when shear deformations are not included.

The stiffness coefficients for a unit translational, member-end displacement which includes shear deformations can be obtained as shown in Figure 14.21.

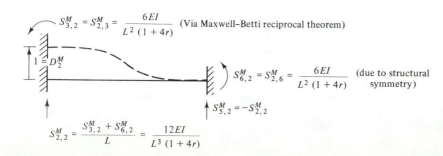

$$S_{3,2}^M = S_{2,3}^M = \frac{6EI}{L^2(1+4r)} \quad \text{(Via Maxwell–Betti reciprocal theorem)}$$

$$1 = D_2^M$$

$$S_{6,2}^M = S_{2,6}^M = \frac{6EI}{L^2(1+4r)} \quad \begin{array}{l}\text{(due to structural}\\ \text{symmetry)}\end{array}$$

$$S_{5,2}^M = -S_{2,2}^M$$

$$S_{2,2}^M = \frac{S_{3,2}^M + S_{6,2}^M}{L} = \frac{12EI}{L^3(1+4r)}$$

Figure 14.21 Member stiffness coefficients including shear deformations.

14.8 ELASTIC CONNECTIONS

In all of the previous examples, the structural joints have been treated as perfectly rigid and infinitesimal in size except for trusses in which case the joints were treated as perfect pins. Actually the joints are neither perfectly pinned nor perfectly rigid and sometimes the joint size is not negligible. Joints that are neither pins nor rigid but are capable of resisting moment are sometimes called semirigid joints. Consequently, it may be necessary to account for deformation of the joint and the joint size in an analysis. In fact, semirigid joints may be preferred for an earthquake resistant design since they allow the structure to be more ductile in behavior (more earthquake motion and energy absorbent).

For multistory buildings, for example, the rotational deformation of beam-to-column connections can be accounted for most economically in a computerized solution as shown below. For simplicity in the following derivation, it is assumed that the member in Figure 14.22 is prismatic and has different elastic connections on each

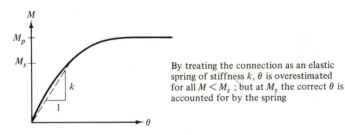

By treating the connection as an elastic spring of stiffness k, θ is overestimated for all $M < M_s$; but at M_s the correct θ is accounted for by the spring

(a) Moment versus rotation behavior of a connection

(b) Prismatic member with elastic end connections — numbers are the subscripts of member-end forces and displacements

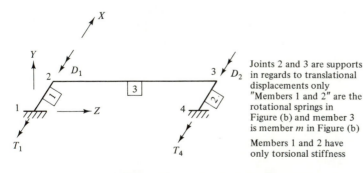

Joints 2 and 3 are supports in regards to translational displacements only "Members 1 and 2" are the rotational springs in Figure (b) and member 3 is member m in Figure (b)

Members 1 and 2 have only torsional stiffness

(c) Plane grid representation of Figure (b)

Figure 14.22 A member attached to elastic connections.

end of the member. Figure 14.22a shows the moment-rotation behavior of a typical semirigid connection; M_s is the maximum expected, service conditions moment (the allowable moment of a member to which the connection is attached). Note that the behavior is nonlinear, but for service condition loads the behavior can be modeled by treating the behavior as a linearly elastic, rotational spring of stiffness k. Subsequently the reader will be shown how the spring stiffness, k, can be computed. In the derivation shown below for Figure 14.22b, the connections are shown as rotational springs of stiffness k_1 and k_2. As shown in Figure 14.22c, for visualization convenience purposes these springs can be treated for as pseudomembers having only torsional stiffness. To find columns 3 and 6 of $[S^M]_m$, unit rotational support movements about the X axis are separately induced at joints 1 and 4 of Figure 14.22c. Therefore, there are only 2 DOF (rotations about the X axis at joints 2 and 3 in Figure 14.22c). The DOFs are depicted as vectors having double headed arrows and conforming to the right hand screw rule. In Figure 14.22c, joints 2 and 3 are supports that prevent translational movements but allow rotational movements—such support conditions are difficult to show in graphical form, but the DOFs shown in Figure 14.22c implicitly account for these conditions.

For independently prescribed unit rotations at joints 1 and 4 of Figure 14.22c, the following fixed ended torques are needed for "members 1 and 2."

$$\left\{\begin{matrix} P_1^f \\ P_2^f \end{matrix}\right\} = \left\{\begin{matrix} -k_1 \\ 0 \end{matrix}\right\} \qquad \left\{\begin{matrix} P_1^f \\ P_2^f \end{matrix}\right\} = \left\{\begin{matrix} 0 \\ -k_2 \end{matrix}\right\}$$

$$\{P^e\} = \{P^a\} - \{P^f\}$$

$$\{P^e\} = \left\{\begin{matrix} 0 \\ 0 \end{matrix}\right\} - \left\{\begin{matrix} -k_1 \\ 0 \end{matrix}\right\} = \left\{\begin{matrix} k_1 \\ 0 \end{matrix}\right\} \qquad \{P^e\} = \left\{\begin{matrix} 0 \\ 0 \end{matrix}\right\} - \left\{\begin{matrix} 0 \\ -k_2 \end{matrix}\right\} = \left\{\begin{matrix} 0 \\ k_2 \end{matrix}\right\}$$

The matrix equation of system equilibrium, $[S]\{D\} = \{P^e\}$, for two independent loadings is

$$\begin{bmatrix} \dfrac{4EI}{L} + k_1 & \dfrac{2EI}{L} \\[2ex] \dfrac{2EI}{L} & \dfrac{4EI}{L} + k_2 \end{bmatrix} \left\{\begin{matrix} D_1 \\ D_2 \end{matrix}\right\} = \left\{\begin{matrix} k_1 & 0 \\ 0 & k_2 \end{matrix}\right\}$$

To simplify the algebra, let $k_1 = r_1 EI/L$ and $k_2 = r_2 EI/L$ where EI and L are for member 3 of the system which is the original member m. This enables us to factor out EI/L on both sides of the matrix equation to give

$$\frac{EI}{L}\begin{bmatrix} 4+r_1 & 2 \\ 2 & 4+r_2 \end{bmatrix}\begin{bmatrix} D_1 \\ D_2 \end{bmatrix} = \frac{EI}{L}\begin{Bmatrix} r_1 & 0 \\ 0 & r_2 \end{Bmatrix}$$

$$\left\{\begin{matrix} D_1 \\ D_2 \end{matrix}\right\} = \frac{1}{(4+r_1)(4+r_2) - 4}\begin{bmatrix} 4+r_2 & -2 \\ -2 & 4+r_1 \end{bmatrix}\begin{Bmatrix} r_1 & 0 \\ 0 & r_2 \end{Bmatrix}$$

Now the final torques required at joints 1 and 4 to produce the independently prescribed unit rotations can be computed as follows:

For a unit rotation at joint 1,

$$\left(T_1 = k_1 - k_1 D_{1,1} = \frac{4EI[r_1(3 + r_2)]}{L[(4 + r_1)(4 + r_2) - 4]}\right) = S^M_{3,3,m}$$

$$\left(T_4 = -k_2 D_{2,1} = \frac{2EI(r_1 r_2)}{L[(4 + r_1)(4 + r_2) - 4]}\right) = S^M_{6,3,m}$$

For a unit rotation at joint 4,

$$\left(T_1 = -k_1 D_{1,2} = \frac{2EI(r_1 r_2)}{L[(4 + r_1)(4 + r_2) - 4]}\right) = S^M_{3,6,m}$$

$$\left(T_4 = k_2 - k_2 D_{2,2} = \frac{4EI[r_2(3 + r_1)]}{L[(4 + r_1)(4 + r_2) - 4]}\right) = S^M_{6,6,m}$$

Recall that our objective was to find columns 3 and 6 of the member stiffness matrix for a prismatic member with different elastic end connections. Four of the desired definitions were obtained in the preceding derivation. By inspection, the first and fourth entries in columns 3 and 6 are zero (no interaction of axial forces and bending moments is permitted in a first order analysis). Since the member-end moments are already available, the member-end shears in Figure 14.22b are obtained by statics and they are

$$\text{for } j = 3 \text{ and } 6, \qquad S^M_{2,j} = \frac{S^M_{3,j} + S^M_{6,j}}{L} \quad \text{and} \quad S^M_{5,j} = -S^M_{2,j}$$

If member m in Figure 14.22b is nonprismatic, the following relations are applicable:

$$S^M_{3,3} = k_1[C_2(C_4 + r_2) - C_3^2]/\bar{D}; \qquad S^M_{6,6} = k_2[C_4(C_2 + r_1) - C_3^2]/\bar{D}$$
$$S^M_{6,3} = k_2 C_3 r_1/\bar{D}; \qquad\qquad\qquad S^M_{3,6} = S^M_{6,3}$$

where $\bar{D} = (C_2 + r_1)(C_4 + r_2) - C_3^2$

C_2, C_3, and C_4 are as defined in Figure 13.6

Consider Figure 14.23 which shows a structure having only one connection of two members. Detail 1 is an example of a welded connection which usually is stiff enough to be treated as a rigid joint, but an analyst may desire to treat the joint size as finite. Details 2 and 3 are examples of elastic (semirigid) connections.

As shown in Figure 14.23, in a structural analysis for the case of the rigid joint shown in detail 1, the structure is represented as a line diagram of the neutral axes of the members. If an analyst wishes to investigate the effect of the joint size and joint stiffness, the author recommends the following approach. A plausible estimate of the joint stiffness, k, can be made as follows:

1. Treat the joint as a truss having dimensions equal to the center-to-center distances between the beam and column flanges.

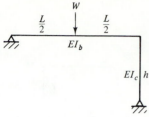

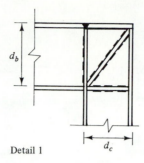

Structure with rigid connection

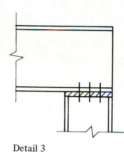

Detail 1

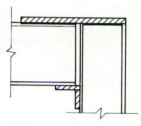

Detail 2

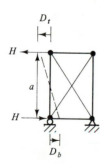

Detail 3

Figure 14.23 Structure and connection details.

2. Areas of the vertical and horizontal truss members should be slightly more than the column and beam flange areas, respectively.

3. Area of the diagonal truss member which is in compression should be equal to the diagonal stiffener area.

4. Area of the tension diagonal, based on plate girder tension field theory, should be $t_{wc}(0.9d_b - 0.4d_c)$.

5. An estimate of k is found from a truss analysis as shown below.

$$a = d_b - t_{fb}$$

$$H = \frac{\text{allowable } M \text{ of beam or column}}{a}$$

$$\theta = \frac{D_t + D_b}{a}; \qquad M = aH$$

$$\text{spring stiffness } k = \frac{M}{\theta}$$

Based on the preceding information, the joint shown in detail 1 of Figure 14.23 could be modeled as shown in joint 1 of Figure 14.24. Similarly, appropriate analyses can be made for details 2 and 3 of Figure 14.23 and modeled as joints 2 and 3 shown in Figure 14.24.

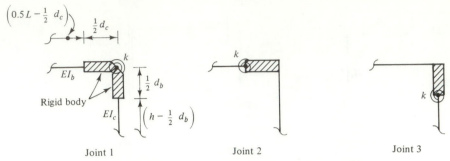

Figure 14.24 Analytical joints for connection details of Figure 14.23.

In case the joint size can be ignored and the joint deformation cannot be ignored, the structure is treated as shown below.

$$W = 32^{\text{k}}; \qquad L = 32\,\text{ft}; \qquad h = 16\,\text{ft}$$

$$\left(\frac{EI}{L}\right)_b = 8370\,\text{ftk}$$

$$\left(\frac{EI}{L}\right)_c = 7513\,\text{ftk}$$

$$k = 5000\,\text{ftk/rad}$$

For a noncomputerized solution, the following approach is recommended. Until a sufficient familiarity is gained in dealing with rotational springs, treat the spring as an equivalent member having only torsional stiffness. This enables the analyst to quickly determine the number of DOF and the proper directions of the member-end forces in formulating the matrix equation of system equilibrium. The matrix equation of system equilibrium is

$$\begin{bmatrix} \dfrac{3EI_b}{L} + k & -k \\[2ex] -k & \dfrac{3EI_c}{h} + k \end{bmatrix} \begin{Bmatrix} D_1 \\ D_2 \end{Bmatrix} = \begin{Bmatrix} \dfrac{3WL}{16} \\[2ex] 0 \end{Bmatrix}$$

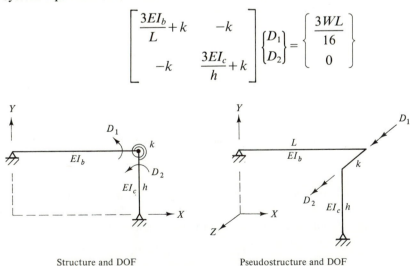

Structure and DOF   Pseudostructure and DOF

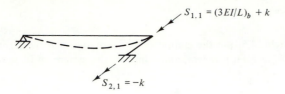

$$S_{1.1} = (3EI/L)_b + k$$

$$S_{2,1} = -k$$

Solution for column 1 of $[S]$

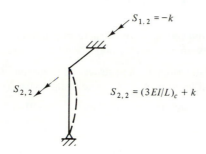

$$S_{1,2} = -k$$

$$S_{2,2}$$

$$S_{2,2} = (3EI/L)_c + k$$

Solution for column 2 of $[S]$

Substitution of $W = 32^k$, $L = 32$ ft, $h = 16$ ft, $(EI/L)_b = 8370$ ftk, $(EI/h)_c = 7513$ ftk, and $k = 5000$ ftk/rad into the matrix equation of system equilibrium gives

$$\begin{bmatrix} 30110 & -5000 \\ -5000 & 27539 \end{bmatrix} \begin{Bmatrix} D_1 \\ D_2 \end{Bmatrix} = \begin{Bmatrix} 192 \\ 0 \end{Bmatrix}$$

$$\begin{Bmatrix} D_1 \\ D_2 \end{Bmatrix} = \begin{Bmatrix} 0.0065748 \\ 0.0011937 \end{Bmatrix}$$

The final member-end moments at the connection are

$$\text{beam moment} = -192 + 3*8370*0.0065748 = -26.9 \text{ ftk}$$
$$\text{column moment} = 0 + 3*7513*0.0011937 = 26.9 \text{ ftk}$$

If the joint deformation is ignored,

$$D_1 = 0.0040295, \quad \text{beam moment} = -90.8 \text{ ftk}, \quad \text{and} \quad \text{column moment} = 90.8 \text{ ftk}$$

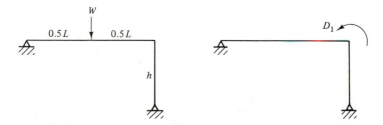

It should be obvious to the reader from the comparison of results that the elastic connection value of $k = 5000$ ftk/rad corresponds to connection detail 3 in Figure 14.23.

PROBLEMS

For Problems 14.1 through 14.5, use the computer program specified by the instructor, the indicated condensed DOF in each problem, and compute the system stiffness matrix for the condensed DOF.

14.1. See Example 9.10 for the geometry and structural properties. Choose $\hat{D}_1$ downward at joint 2. Also, assume $A_1 = A_2 = 20$ in.^{2}

14.2. See Problem 9.32 for the geometry and structural properties. Choose $\hat{D}_1$ downward at the 40^k load point.

14.3. See Problem 9.48 for the geometry and structural properties. Choose $\hat{D}_1$ acting to the right at joint 3. Also, assume $EA = 300,000^k$ for all members.

14.4. See Problem 13.18 for the geometry and structural properties. Choose $\hat{D}_1$ and $\hat{D}_2$ acting to the right at the beam levels on the right side of the structure.

14.5. For all members, $E = 29,000$ ksi, $A = 16.2$ in.^{2}, $I = 1350$ in.^{4}

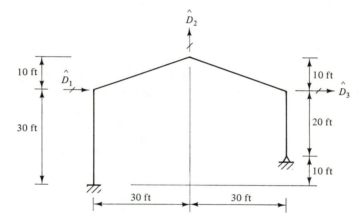

Figure P14.5

Problems 14.6 through 14.12 are to be solved by the noncomputerized approach. For the substructuring problems, do NOT use the subdivided matrix approach.

14.6. Finish numerically formulating the substructure stiffness matrix, $[\hat{S}]$. Both members are axially rigid. Member 1: $EI/L = 6000$ ftk. member 2: $EI/L = 4000$ ftk.

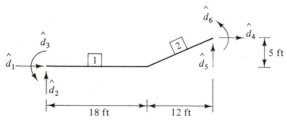

Substructure DOF

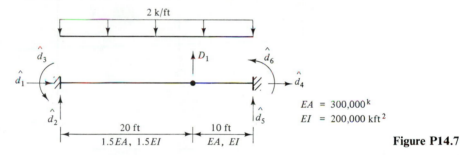

$$[\hat{S}] =$$

	1	2	3	4	5	6	
							1
	$\dfrac{1600}{3}$						2
	4800						3
							4
							5
	-4800		-2400			14400	6

Figure P14.6

14.7. Numerically formulate the substructure force-deformation relation. There is an internal hinge ($M = 0$ condition) located at 20 ft from the left support. Do NOT use a subdivided matrix approach. Reminder: After $\hat{S}_{1,1}$, $\hat{S}_{3,3}$, $\hat{S}_{6,3}$, $\hat{S}_{6,6}$ have been obtained, the other entries in $[\hat{S}]$ can be obtained by statics, MBRT, and logic.

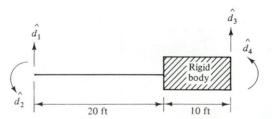

Figure P14.7

14.8. A substructure consists of a rigid body and one bending member as shown. Numerically formulate the substructure stiffness matrix for the DOF shown. $EI = 80,000$ kft^2 for the bending member.

Figure P14.8

14.9. For all members $EI = 235,625$ kft^2 and $EA =$ infinity.

  (a) Numerically formulate the substructure stiffness matrix for the interface DOF shown in Figure P14.9a.

  (b) Using the substructure stiffness matrix obtained in part a, numerically formulate the system stiffness matrix for the structure shown in Figure P14.9b.

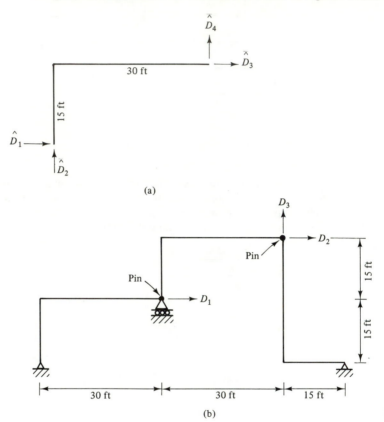

(a)

(b)   **Figure P14.9**

14.10. For all members $EI = 225,000$ kft^2, $EA =$ infinity.

  (a) Numerically formulate the substructure stiffness matrix for the interface DOF shown in Figure P14.10a.

  (b) Using logic and the results obtained in part a, numerically obtain the system stiffness matrix for the structure shown in Figure P14.10b.

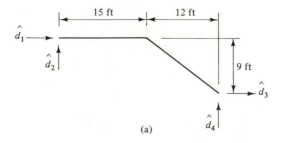

(a)

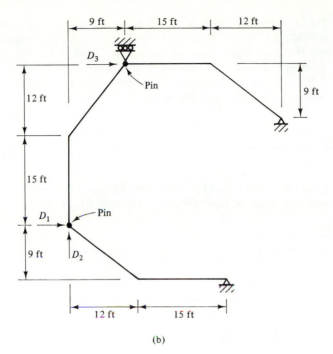

(b)

Figure P14.10

14.11. D_i are the system DOFs and $\hat{d}_i$ are the substructure DOFs. Numerically formulate the substructure force-deformation relation for the loading shown.

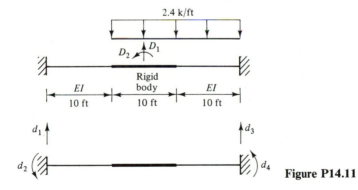

Figure P14.11

Structural part number	EA (kips)	EI (kft^2)	Length (ft)
1	infinity	300,000	10
2	infinity	infinity	10
3	infinity	150,000	10
4	2000	0	10

Figure P14.12a

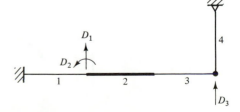

14.12. For all deformable parts $E = 29,000$ ksi and the thermal coefficient of expansion is 0.0000065 in./in.-°F. Numerically find (a) column 2 of the system stiffness matrix; (b) effective joint load vectors for loadings 1, 2, and 3. Loading 1: 3 k/ft downward on member 1. Loading 2: Temperature in member 4 uniformly decreases 100°F. Loading 3: For member 3 treated as a simply supported beam, a temperature change causes the curvature diagram and deflected shape in Figure P14.12b.

Figure P14.12b

14.13. For all members $L = 30$ ft, $EI/L = 7000$ ftk. Using symmetry and antisymmetry loading components for the loading shown in Figure P14.13, find the reactions for Figure P14.13.

Figure P14.13

14.14. For all members $EA = $ infinity and $EI = 100,000$ kft². For the left half of the symmetric structure, numerically formulate the matrix equation of system equilibrium due to (a) symmetric loading component of the loading shown, (b) antisymmetric component of the loading shown.

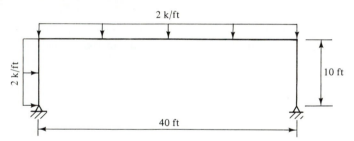

Figure P14.14

14.15. For all members $L = 20$ ft, $EI/L = 6000$ ftk, $EA/L = $ infinity. Numerically formulate the matrix equation of system equilibrium for the left half of the symmetric structure subjected to (a) the symmetric loading component of the loading shown; (b) the antisymmetric loading component of the loading shown.

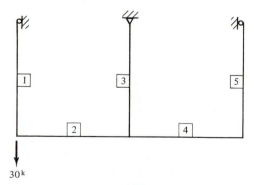

Figure P14.15

14.16. All members are axially rigid, prismatic, and have the same *EI*. For the left symmetric half of the structure, numerically formulate only the effective joint load vector due to (a) symmetric loading component of the loading shown; (b) antisymmetric loading component of the loading shown.

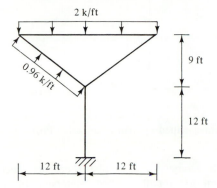

Figure P14.16

14.17. Plan view of a plane grid is shown in Figure P14.17. Grid lies in *XY* plane; *Z* is downward. All members: $L = 100$ in.; $EI_y = 1,000,000$ k in.^{2}; $GI_x = 400,000$ k in.^{2}; *y* axis is local bending axis; *x*-axis is local torsional axis (length direction axis). The only loading is 0.6 k/in. acting downward on member 2. For the symmetric half of the structure which includes member 2, numerically formulate (but do not solve) the matrix equation of system equilibrium for the following loadings and DOF at joint 2: (a) the symmetric loading component of the loading; D_1 (downward); D_2 $(-\theta_Y)$. (b) the antisymmetric loading component of the loading; D_1 $(-\theta_X)$.

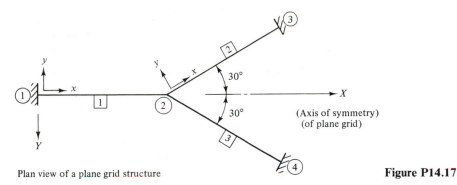

Plan view of a plane grid structure

Figure P14.17

14.18. All members are W24 × 55 and the thermal coefficient of expansion is 0.0000065 (strain)/ °F. At the time of construction, all members had a uniform temperature of 85°F. On a cold winter day after construction, the inside temperature is 70°F and the outside temperature is 0°F. Assume the temperature variation is linear across the depth of each member. Numerically formulate the effective joint load vector due to the described "temperature loading" assuming (a) axial deformations are not negligible (6 DOF for noncomputerized solution); (b) axial deformations are negligible (3 DOF for noncomputerized solution).

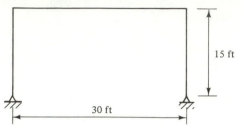

Figure P14.18

14.19. Using computer program specified by instructor, solve Problem 14.18.

14.20. See Problem 7.66 for the geometry and member properties. Thermal coefficient of expansion = 0.000012 (strain)/°C for all members. Members 1 through 4 and 6 experience a 40°C decrease in temperature relative to their construction time temperature. The other members experience a 10°C decrease in temperature. Numerically formulate the effective joint load vector due to the described "temperature loading."

14.21. Using computer program specified by instructor, solve Problem 14.20.

14.22. Solve Problem 10.24 using symmetry and antisymmetry concepts.

14.23. Solve Problem 10.25 using symmetry and antisymmetry concepts.

14.24. Solve Problem 10.26 using symmetry and antisymmetry concepts.

14.25. Solve Problem 10.27 using symmetry and antisymmetry concepts.

14.26. Solve Problem 10.28 using symmetry and antisymmetry concepts.

14.27. For the DOF shown, numerically formulate the matrix equation of system equilibrium. $EI = 100,000$ kft^2; $k = 30,000$ ftk/rad.

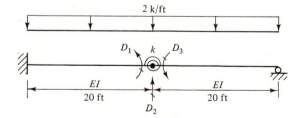

Figure P14.27

14.28. Members 2 through 4: EA = infinity; EI = constant. Structural part 5 is a rigid body. Structural part 1 is a linearly elastic translational spring. Structural part 2 is a linearly elastic rotational spring. Numerically formulate the effective joint load vector for the loading shown in Figure P14.28.

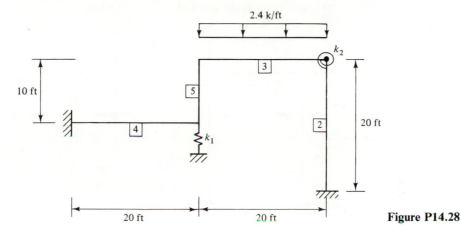

Figure P14.28

For the following problems, use the computer program specified by the instructor.

14.29. See Figure 14.1a for the geometry, loading, and member properties. Solve this problem as a substructuring problem using the indicated interface DOF.

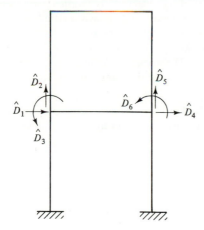

Figure P14.29

14.30. See Example 10.8 for the geometry, loading, and member properties. Choose E = 29,000 ksi; I = 1830 in.^{4}; A = 20 in.^{2}; $1.5I$ = 2745 in.^{4}; A = 28 in.^{2}; $2I$ = 3660 in.^{4}; A = 30 in.^{2} Choose joint D as the interface joint between two substructures. Solve by the method of substructuring.

14.31. See Figure 14.4 for the geometry, loading, and member properties. Solve by the method of substructuring. Choose only joints 8 and 9 as the interface joints.

14.32. Solve Problem 14.31 with members 1 and 10 deleted.

14.33. Loading 1 (dead load): 1.68 k/ft on members 15 through 26 (includes member weight), 0.061 k/ft on members 1 through 12 for member weight. Loading 2 (a live load configuration) is shown in Figure P14.33. On members 16, 18, 21, and 23: 1.2 k/ft ↓. No loading on any of the other members. Loading 3 = 0.75 ∗ (loading 1 + loading 2). Using the computer program specified by the instructor:

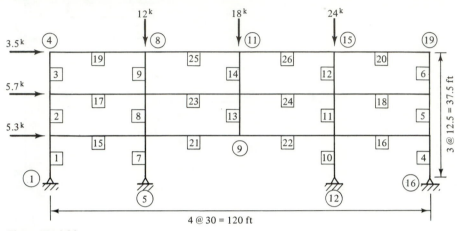

Figure P14.33

(a) Routinely solve the entire structure for the three loadings described above.

(b) Using only the left half of the structural properties, joints, etc., solve the problem defined in part a using symmetric and antisymmetric loading components for each of the independent loadings. Find all member-end forces and reactions in both halves of the structure. Members 1 through 14: W14 × 61. Members 15 through 20: W24 × 62. Members 21 through 26: W36 × 135. Note: For ease of comparison of results for parts a and b, keep the joint numbers and member numbers in both parts the same (if possible).

The following problems can be solved by symmetry and antisymmetry concepts using the computer program specified by the instructor. Wherever numerical values of A and I are not given, choose appropriate values of A and I.

Problems 10.23 through 10.32.

Problems 14.14 through 14.18.

Problems 7.55, 7.57, 7.60, 7.62, 7.65, 7.67, 7.79.

Displacement Method— Second Order Analysis

15.1 INTRODUCTION

In Chapters 8 through 14, for linearly elastic structures, the equilibrium equations were written in terms of the undeformed structural geometry and the axial forces in the members were assumed to be independent of the member-end moments. Such analyses are defined as first order analyses (FOA).

In a second order analysis (SOA) of a linearly elastic structure, the equilibrium equations are written in terms of the deformed structure and the interactive effect of axial forces and member-end moments is taken into account. Although the structure is linearly elastic, the load versus deflection curve for a particular DOF is nonlinear as shown subsequently. For indeterminate structures, the deformed structural geometry and final member-end forces are unknown. Writing the equilibrium equations in terms of the unknown deformed geometry does not pose an insurmountable problem, but it is subsequently shown that the axial forces must be known in order to compute the interactive effect of axial force and member-end moments. Consequently, in a computerized second order analysis, the first step is to obtain a first order analysis solution which gives estimated values of (1) the axial forces in the members; and (2) the joint displacements (the deformed geometry). With these estimated axial force values, a second order analysis can be made to obtain a better approximation of the axial forces and the deformed geometry. The second order analysis is repeatedly performed to obtain better and better approximate values of the axial forces until the deformed geometry solution converges. Convergence is obtained when the difference between the current solution and the previous solution does not disagree by more than a specified percentage (3%, for example) for all displacements. Therefore, a second order analysis is iterative and the converged solution gives the final member-end forces and joint displacements. It should be mentioned that the iterative solution may not converge for improperly designed structures; in that case, the structure continues to deform until collapse occurs. Indeed, in some structural investigations the main purpose for performing a second order analysis may be to determine at what level of overload does the second order analysis fail to converge. For an overload analysis, it is likely that the elastic limit of one or more members is exceeded and an inelastic SOA is needed for such cases. This text does not discuss how to perform an inelastic SOA.

15.2 SOME BASIC CONCEPTS

For reasons that will become apparent later, the discussion is limited to plane frames composed of prismatic members that are linearly elastic. To set the stage for the needed theoretical derivations to be performed later, consider Figure 15.1 which shows that in a second order analysis:

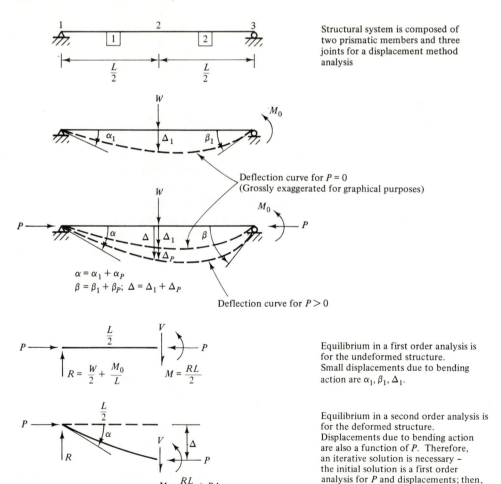

Structural system is composed of two prismatic members and three joints for a displacement method analysis

Deflection curve for $P = 0$
(Grossly exaggerated for graphical purposes)

$$\alpha = \alpha_1 + \alpha_P$$
$$\beta = \beta_1 + \beta_P; \ \Delta = \Delta_1 + \Delta_P$$

Deflection curve for $P > 0$

$$R = \frac{W}{2} + \frac{M_0}{L} \qquad M = \frac{RL}{2}$$

Equilibrium in a first order analysis is for the undeformed structure. Small displacements due to bending action are $\alpha_1, \beta_1, \Delta_1$.

$$M = \frac{RL}{2} + P\Delta$$

Equilibrium in a second order analysis is for the deformed structure. Displacements due to bending action are also a function of P. Therefore, an iterative solution is necessary – the initial solution is a first order analysis for P and displacements; then, iteration is required until convergence within acceptable limits is achieved on the system displacements.

Figure 15.1 Some basic concepts of second order analysis.

1. The system displacements are a function of the axial forces in the members.

2. The member-end forces are a function of the system displacements; this is true for the fixed ended forces as well as the final member-end forces.

Consequently, to perform a second order Displacement Method analysis, the interdependence of the member-end forces and displacements must be accounted for in the member force-deformation relation, $\{F\} = [S^M]\{D^M\} + \{F^f\}$.

15.3 MEMBER FORCE-DEFORMATION RELATION

Derivations of the member stiffness matrix and fixed ended forces are confined to the case for which the member is prismatic and linearly elastic. A nonprismatic member can be subdivided into a set of prismatic segments and each segment is treated as a member. Since the Force Method is involved in deriving the fixed ended forces and the member stiffness matrix, the fundamental solution for a simple beam subjected to axial compression and a moment at one end is needed. Eventually a unit valued end moment is used, but in the following fundamental solution the end moment is arbitrary and assumed to be known. Also, the axial compression is assumed to be known.

$$M_z = \frac{M_2 x}{L} + Py$$

$$\left(\frac{d^2y}{dx^2} \cong -\frac{M_z}{EI_z}\right) = \frac{-(M_2 x/L + Py)}{EI_z}$$

$$\frac{d^2y}{dx^2} + \frac{P}{EI_z}y = -\frac{M_2}{EI_z}\frac{x}{L} \quad \text{and for convenience, let} \quad k = \sqrt{\frac{P}{EI_z}}; \quad EI_z = \frac{P}{k^2}$$

$$\frac{d^2y}{dx^2} + k^2y = -\frac{M_2 k^2}{P}\frac{x}{L}$$

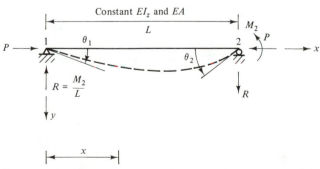

Figure 15.2

$y = A \cos kx + B \sin kx - (M_2/P)(x/L)$ is the solution to the preceding differential equation. At $x = 0$, $y = 0$ requires $A = 0$. At $x = L$, $y = 0$ requires $B = M_2/(P \sin kL)$. Therefore,

$$y = \frac{M_2}{P}\left(\frac{\sin kx}{\sin kL} - \frac{x}{L}\right)$$

For convenience, let $\phi = kL$ which also means that

$$\phi = kL = \sqrt{\frac{P}{EI_z}}\,L = \sqrt{\frac{PL^2}{EI_z}\frac{\pi^2}{\pi^2}} = \pi\sqrt{\frac{P}{(\pi^2 EI_z/L^2)}} = \pi\sqrt{\frac{P}{P_E}} = \pi\sqrt{\rho}$$

where $\rho = P/P_E$ is the ratio of the axial force to the Euler force, $P_E = \pi^2 EI_z/L^2$. For this notation,

$$y = \frac{M_2}{P}\left(\frac{\sin kx}{\sin \phi} - \frac{x}{L}\right)$$

$$\frac{dy}{dx} = \frac{M_2}{P}\left(\frac{k \cos kx}{\sin \phi} - \frac{1}{L}\right)$$

$$\theta_1 = \left.\frac{dy}{dx}\right|_{x=0} = \frac{M_2}{P}\left(\frac{k}{\sin \phi} - \frac{1}{L}\right) = \frac{M_2}{PL}\left(\frac{\phi - \sin \phi}{\sin \phi}\right)$$

$$\theta_2 = \left.\frac{dy}{dx}\right|_{x=L} = \frac{M_2}{P}\left(\frac{k}{\tan \phi} - \frac{1}{L}\right) = \frac{-M_2}{PL}\left(\frac{\phi - \tan \phi}{\tan \phi}\right)$$

Let

$$\alpha = \frac{\theta_2}{M_2} = \frac{-1}{PL}\left(\frac{\phi - \tan \phi}{\tan \phi}\right) \quad \text{and} \quad \beta = \frac{\theta_1}{M_2} = \frac{1}{PL}\left(\frac{\phi - \sin \phi}{\sin \phi}\right)$$

$$\text{where} \quad P = \frac{EI_z\phi^2}{L^2} \equiv \rho P_E$$

which are the member-end flexibility coefficients needed in the next two figures.

The derivation of $[S^M]$ begins here using the Force Method and the fundamental solution shown above to obtain the member-end stiffness coefficients which account for a known axial compressive force, P (Fig. 15.3).

The equation of system compatibility for Figure 15.3 is

$$\begin{bmatrix} \alpha & -\beta \\ -\beta & \alpha \end{bmatrix}\begin{Bmatrix} S^M_{3,3} \\ S^M_{6,3} \end{Bmatrix} = \begin{Bmatrix} 1 \\ 0 \end{Bmatrix}$$

$$\begin{Bmatrix} S^M_{3,3} \\ S^M_{6,3} \end{Bmatrix} = \frac{1}{\alpha^2 - \beta^2}\begin{bmatrix} \alpha & \beta \\ \beta & \alpha \end{bmatrix}\begin{Bmatrix} 1 \\ 0 \end{Bmatrix} = \frac{1}{\alpha^2 - \beta^2}\begin{Bmatrix} \alpha \\ \beta \end{Bmatrix} \equiv \begin{Bmatrix} \dfrac{C_2 EI}{L} \\ \dfrac{C_3 EI}{L} \end{Bmatrix}$$

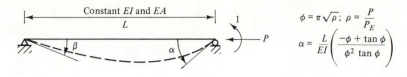

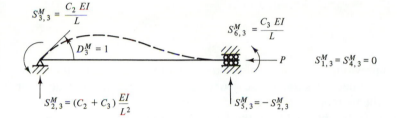

Figure 15.3

Solving the identity relation shown above gives

$$C_2 = \frac{\phi \sin \phi - \phi^2 \cos \phi}{2 - 2 \cos \phi - \phi \sin \phi} \quad \text{and} \quad C_3 = \frac{\phi^2 - \phi \sin \phi}{2 - 2 \cos \phi - \phi \sin \phi}$$

Recall that

$$\phi = \pi \sqrt{\rho} = \pi \sqrt{\frac{P}{P_E}} \quad \text{and} \quad P_E = \frac{\pi^2 EI}{L^2}$$

For $P = 0$ ($\phi = 0$), the formulas for C_2 and C_3 become indeterminate forms in mathematical terminology. However, as ϕ approaches zero (for example, if $\phi = 0.00001$ then $\rho = 1.0132118 \times 10^{-11}$) C_2 approaches 4 and C_3 approaches 2 which are the correct values for a prismatic member when $P = 0$ (first order analysis case). See Appendix C for a table of C_2 and C_3 values versus ϕ; that table is useful for noncomputerized solutions.

Since the member is prismatic, the solution for $D_6^M = 1$ can be obtained from the solution for $D_3^M = 1$ as shown in Figure 15.4. Note that $C_4 = C_2$, since the member is prismatic.

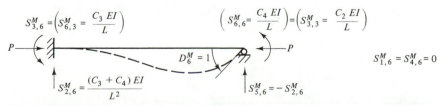

Figure 15.4 Note that $C_4 = C_2$, since the member is prismatic.

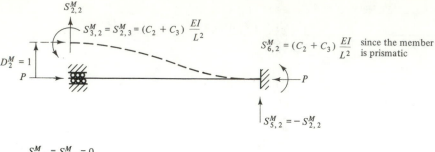

$$S_{1,2}^M = S_{4,2}^M = 0$$

$$S_{2,2}^M = \frac{2(C_2 + C_3)\dfrac{EI}{L^2} - P \star 1}{L} = [2(C_2 + C_3) - \pi^2\rho]\frac{EI}{L^3}$$

Figure 15.5

As mentioned previously, the force-deformation relation is needed for the member in the deformed geometry position of the structure. Figures 15.5 and 15.6 account for the $P\Delta$ effect in the member stiffness coefficients. Since the member is prismatic, the $D_5^M = 1$ solution can be obtained from the $D_2^M = 1$ solution (Fig. 15.6). Note that in the preceding derivations the axial load, P, is known and applied before the member-end displacements are invoked. No additional axial deformation is assumed to occur due to bending. Consequently, the following axial stiffness relations are taken to be independent of the bending stiffnesses (Fig. 15.7).

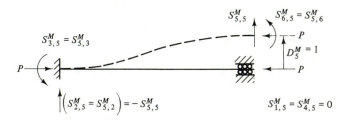

$S_{5,5}^M = S_{2,2}^M$ since the member is prismatic   **Figure 15.6**

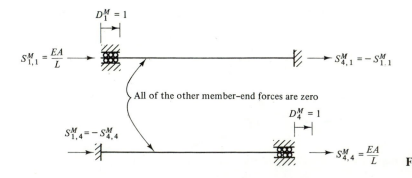

Figure 15.7

In the preceding member stiffness matrix derivation, it was implicitly assumed that the axial compressive force, P, was a positive value. For the case when the axial force is tensile, P is a negative value and

$$\rho = \frac{P}{P_E}; \qquad \phi = \pi \sqrt{-\rho}$$

$$C_4 = C_2 = \frac{-\phi \sinh \phi + \phi^2 \cosh \phi}{2 - 2 \cosh \phi + \phi \sinh \phi}$$

$$C_3 = \frac{-\phi^2 + \phi \sinh \phi}{2 - 2 \cosh \phi + \phi \sinh \phi}$$

where sinh and cosh are hyperbolic sine and cosine functions, respectively. Note that an axial tensile force has a strengthening effect on the member with regard to bending stiffness—see Appendix C for any negative value of ρ; that is, $C_2 > 4$ and $C_3 < 2$ for a member subjected to an axial tensile force.

For a noncomputerized solution, it is helpful to have the relations of a member hinged at one end as shown in Figures 15.8 and 15.9 in order to minimize the number of DOF of the system. By superposition,

$$C_h \frac{EI}{L} \theta = C_2 \frac{EI}{L} \theta - \frac{C_3}{C_2}\left(C_3 \frac{EI}{L} \theta\right); \qquad \text{therefore} \qquad C_h = \frac{C_2^2 - C_3^2}{C_2}$$

In order to complete the force-deformation relation for a member in its deformed position, member fixed ended forces are also needed as a function of the axial force. The Force Method can be used as shown in Figure 15.10 to find the fixed ended moments and then the fixed ended shears can be found from statics. Fixed ended moments are found for only two cases: (1) a uniformly distributed load; and (2) a single concentrated load located at an arbitrary position along the member length. Superposition is valid and can be used to obtain other loading combinations. The

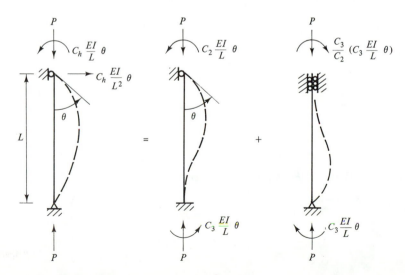

Figure 15.8

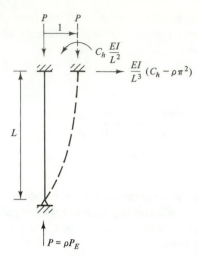

Figure 15.9

flexibility matrix and its inverse are the same as in the derivation for $S_{3,3}^M$ and $S_{6,3}^M$. To avoid difficulties with signs, P is a magnitude in the following relations and has no sign; likewise, θ_1 and θ_2 are magnitudes.

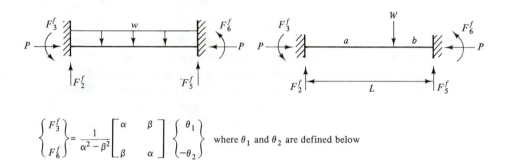

$$\begin{Bmatrix} F_3^f \\ F_6^f \end{Bmatrix} = \frac{1}{\alpha^2 - \beta^2} \begin{bmatrix} \alpha & \beta \\ \beta & \alpha \end{bmatrix} \begin{Bmatrix} \theta_1 \\ -\theta_2 \end{Bmatrix} \quad \text{where } \theta_1 \text{ and } \theta_2 \text{ are defined below}$$

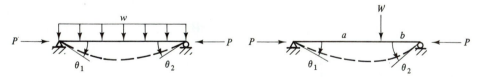

Figure 15.10

For a compressive P, $\rho = P/P_E$ is positive and $\phi = \pi\sqrt{\rho}$:

$$\theta_1 = \theta_2 = \frac{wL^3}{EI}\left(\frac{[\tan(\phi/2) - \phi/2]}{\phi^3}\right)$$

$$\theta_1 = \frac{WL^2}{EI\phi^2}\left(\frac{\sin(b\phi/L)}{\sin\phi} - \frac{b}{L}\right)$$

$$\theta_2 = \frac{WL^2}{EI\phi^2}\left(\frac{\sin(a\phi/L)}{\sin\phi} - \frac{a}{L}\right)$$

$$F_3^f = \frac{wL^2}{12}\left(\frac{12[\tan{(\phi/2)} - (\phi/2)]}{\phi^2 \tan{(\phi/2)}}\right) = \frac{wL^2}{12} C_u$$

$$F_6^f = -F_3^f$$

For the concentrated load case, solve for F_3^f and F_6^f numerically from the matrix equation shown above.

For a tensile P, $\rho = -P/P_E$ is negative and $\phi = \pi\sqrt{-\rho}$:

$$F_3^f = \frac{wL^2}{12}\left(\frac{12[(\phi/2) - \tanh{(\phi/2)}]}{\phi^2 \tanh{(\phi/2)}}\right) = \frac{wL^2}{12} C_u \qquad \theta_1 = \frac{WL^2}{EI\phi^2}\left(\frac{b}{L} - \frac{\sinh{(b\phi/L)}}{\sinh{\phi}}\right)$$

$$F_6^f = -F_3^f \qquad\qquad\qquad\qquad\qquad\qquad \theta_2 = \frac{WL^2}{EI\phi^2}\left(\frac{a}{L} - \frac{\sinh{(a\phi/L)}}{\sinh{\phi}}\right)$$

See Appendix C for a table of C_2, C_3, C_h, and C_u. As shown by Goldberg [15.1, 15.2], and Livesley and Chandler [15.3], such a table is useful for noncomputerized solutions.

In Example 15.2, an alternate procedure for finding the fixed ended moments due to a single concentrated load is demonstrated.

15.4 NONCOMPUTERIZED SOLUTIONS

A few noncomputerized solutions are shown in order to demonstrate the method of SOA. Residual stresses exist in steel rolled members due to the thermal cooling process. Residual compressive stresses are greatest at the flange tips of the cross section of a rolled member and residual tensile stresses are greatest in the section at the junction of the flanges and the web. When a steel rolled member is subjected to only an axial compressive force, the flange tips begin to yield when the axial stress, P/A, reaches a value of about $0.5F_y$ where F_y is the yield stress of the steel and A is the gross area of the cross section. If the axial stress exceeds $0.5F_y$, the tangent modulus of elasticity, E_t, should be used in computing the Euler load, P_E, of the member since ρ will be larger in this case. Examples involving E_t are shown in Chapter 16, but in the following examples it is assumed that the steel yield stress is large enough such that the members remain elastic.

In a computerized solution, a FOA would be performed to obtain the estimated axial force values needed in the first cycle of the SOA. For noncomputerized solutions, the author recommends that the axial forces in the members be estimated by an approximate analysis and used in the first cycle of the iterative SOA. For example, treat all members in a braced frame as being pinned ended in order to obtain estimated axial force values. For unbraced frames, use the reactions if they are obtainable from statics or the beam fixed ended shears to find the axial forces in the columns. Axial forces in the beams can be assumed to be zero in the first cycle of a SOA.

EXAMPLE 15.1 ───

Perform an elastic SOA for the structure shown in Figure 15.11a. The moment of inertia for the column is deliberately chosen to be smaller than is customary in such a structure to accentuate the SOA results. Assume that the members are continuously laterally braced to prevent buckling perpendicular to the plane of the frame.

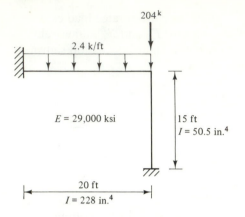

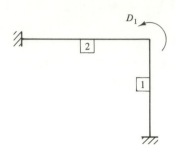

Figure 15.11a

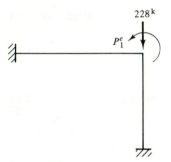

Figure 15.11b

Solution. Member 2—fixed ended forces:

$$M = \frac{2.4(20)^2}{12} = 80 \text{ ftk} = 960 \text{ in.-k}$$

$$V = \frac{2.4(20)}{2} = 24^k$$

$$P_1^e = P_1^a - P_1^f = 0 - (-960) = 960 \text{ in.-k}$$

Second order analysis—cycle 1:

Member	P (kips)	P_E (kips)	ρ	C_2	C_3	C_u
1	228	446	0.511	3.278	2.199	N.A.
2	0	1133	0	4	2	1.00000

$$S_{1,1} = \left(\frac{C_2 EI}{L}\right)_1 + \left(\frac{C_2 EI}{L}\right)_2 = 29,000\left(\frac{3.278 * 50.5}{180} + \frac{4 * 228}{240}\right) = 136,870 \text{ in.-k/rad}$$

$$[S]\{D\} = \{P^e\} \quad \text{is} \quad 136,870 D_1 = 960$$

$$D_1 = 0.007014 \text{ rad}$$

Member-End Forces
Member 1—

$$\frac{C_2 EI}{L} * D_1 = \frac{3.278 * 29,000 * 50.5 * 0.007014}{180} = 187.1 \text{ in.-k} = 15.6 \text{ ftk}$$

$$\frac{C_3 EI}{L} * D_1 = \frac{2.199 * 29,000 * 50.5 * 0.007014}{180} = 124.5 \text{ in.-k} = 10.5 \text{ ftk}$$

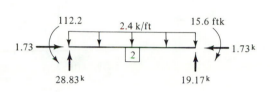

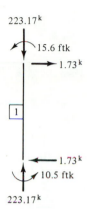

Figure 15.11c

Second order analysis—cycle 2

Member	P (kips)	P_E (kips)	ρ	C_2	C_3	C_u
1	223.2	446	0.500	3.29445	2.19365	N.A.
2	1.73	1133	0.00153	3.998	2.0005	1.00025

$$[S]\{D\} = \{P^e\} \quad \text{is} \quad 136,949 D_1 = 1.00025 * 960; \qquad D_1 = 0.007012 \text{ rad}$$

Since $0.007012/0.007014 = 0.9997$ (only 0.03% change), convergence is reached and the member-end forces are very nearly the same as the ones for cycle 1.

The preceding example was run on the computer using the author's SOA program and accounting for axial and shear deformations. Member input data and output were as shown below: member 2—$12 \times 6 \times 0.375$ structural tube; member 1—$6 \times 6 \times 0.5$ structural tube.

Member	A (in.²)	Shear area (in.²)	I (in.⁴)
1	10.4	6.00	50.5
2	12.6	9.00	228.0

Since $0.006679/0.006148 = 1.086$, the SOA rotational displacement is 8.6% larger than the FOA results. Also, since $0.007012/0.006679 = 1.05$, inclusion of axial and shear deformations amounted to 5% in the displacement solution.

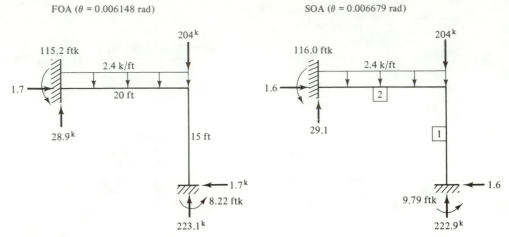

Figure 15.11d

However, $116.0/112.2 = 1.034$ which means that inclusion of axial and shear deformations affected the maximum moment by 3.4%.

EXAMPLE 15.2 ──

Compute the fixed ended moments of a member subjected to a single concentrated load and an axial compressive force by performing a SOA. To perform the SOA, choose an internal joint at the concentrated load location; this defines a plane frame composed of two members and the concentrated load becomes an applied joint load. The axial compressive load is constant; that is, the fixed ended moments are desired for a specific value of axial force. Consequently, this solution is not iterative since the axial force in the members does not change. See Figure 15.12a.

$$E = 29,000 \text{ ksi}$$
$$I = 300 \text{ in.}^4$$

$$P_E = \frac{\pi^2 EI}{L^2} = 954^k; \qquad \rho = 0.5$$

$$P_{E(1)} = 2146.6^k; \qquad \rho_1 = 0.222$$
$$P_{E(2)} = 8586.6^k; \qquad \rho_2 = 0.0556$$

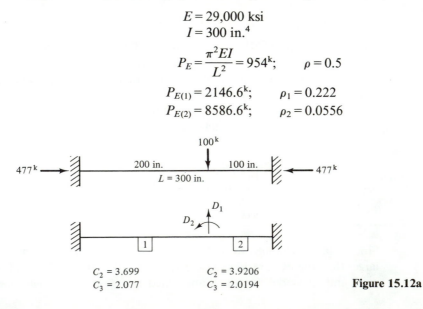

Figure 15.12a

Solution

$$[S]\{D\} = \{P^e\}$$

$$\begin{bmatrix} 108.762 & 3911.52 \\ 3911.52 & 501998.7 \end{bmatrix} \begin{Bmatrix} D_1 \\ D_2 \end{Bmatrix} = \begin{Bmatrix} -100 \\ 0 \end{Bmatrix}$$

$$\begin{Bmatrix} D_1 \\ D_2 \end{Bmatrix} = \begin{Bmatrix} -1.2774 \text{ in.} \\ 0.0099534 \text{ rad} \end{Bmatrix}$$

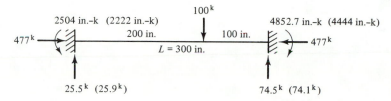

Figure 15.12b

The values shown in parentheses are the fixed ended forces for an axial compression force of $P = 0$. For $P = 477^k$ versus $P = 0$, the left end moment ratio is $2504/2222 = 1.127$ (12.7% increase) and the right end moment ratio is $4852.7/4444 = 1.09$ (9% increase) (Fig. 15.12b). Note that this solution is not iterative since the axial force is constant and the deformed geometry position was used in all calculations.

EXAMPLE 15.3 _____

All members are steel W sections (W6 × 16 $F_y = 36$ ksi): $A = 4.74$ in.; $I = 32.1$ in.; radii of gyrations are $r = 2.60$ in. (strong axis) and $r = 0.966$ in. (weak axis). All members are bending about the strong axis of the section and are laterally braced such that the allowable bending stress is $1.33 * 24$ ksi $= 32$ ksi (horizontal force of 2^k is due to wind and 33.3% increase in allowable stress is permitted). The loads were chosen such that the stability interaction equation of the AISC Specifications [15.4] sums to 1.037 based on FOA results. For a noncomputerized solution ignoring axial and shear deformations, only the right half of the structure can be analyzed since the structure is symmetric (Fig. 15.13a). Note that the axial compression in the right side column is greater than in the left side column; therefore, modeling the right half of the structure is conservative. This approach requires that the horizontal loading be

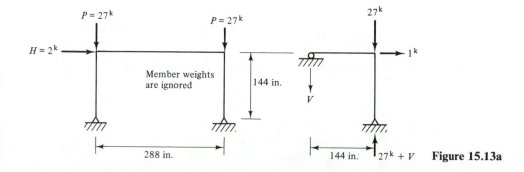

Figure 15.13a

decomposed into $ and a/s loading components. Only the a/s loading component causes any joint displacements.

The DOF are shown in Figure 15.13b. Note that both members of the half structure have the same properties: $E = 29,000$ ksi, $I = 32.1$ in.^{4}, and $L = 144$ in. At the right end of member 2, the moment $= 144$ in. $* V$ where

$$V = \frac{144 * 1 + 27 * D_1}{144}$$

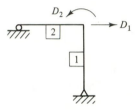

Figure 15.13b

Also note that the axial force in member 2 is always zero. The matrix equation of system equilibrium, $[S]\{D\} = \{P^e\}$, is

$$\begin{bmatrix} (C_h - \pi^2\rho)_{(1)}\dfrac{EI}{L^3} & C_{h(1)}\dfrac{EI}{L^2} \\ C_{h(1)}\dfrac{EI}{L^2} & (C_{h(1)} + C_{h(2)})\dfrac{EI}{L} \end{bmatrix} \begin{Bmatrix} D_1 \\ D_2 \end{Bmatrix} = \begin{Bmatrix} 1 \\ 0 \end{Bmatrix}$$

where C_h and ρ are as shown in the following table:

SOA cycle number	Member number	P (kips)	P_E (kips)	$\rho = \dfrac{P}{P_E}$	C_h	$C_h - \pi^2$	D_1 (in.)	D_2 (rad)	V (kips)
All	2	0	NA	0	3	3	NA	NA	NA
1	1	28.0	443.0	0.0632	2.87286	2.24910	3.8016	−0.0129141	1.713
2	1	28.75	443.0	0.0650	2.86918	2.22766	3.8878	−0.0131985	1.729

Each displacement only increased by about 2% from cycle 1 to cycle 2. For the converged SOA results, the maximum moment $= 144 * V = 249.0$ in. kips.

For a FOA, $S_{1,1} = C_{h(1)}(EI/L^3)$ and $C_{h(1)} = C_{h(2)} = 3$,

$$\begin{Bmatrix} D_1 \\ D_2 \end{Bmatrix} = \begin{Bmatrix} 2.138 \text{ in.} \\ -0.007425 \text{ rad} \end{Bmatrix}; \qquad V = 1^k \text{ and maximum } M = 144 \text{ in.-k}$$

Using the author's SOA computer program and accounting for axial and shear deformations

$$D_1 = 3.755 \text{ in.}$$
$$D_2 = -0.012821 \text{ rad}$$
$$\text{max } M = 245.9 \text{ in.-k}$$

Therefore,

$$\frac{\text{SOA } M}{\text{FOA } M} = \frac{245.9}{144.1} = 1.7065 \quad \text{and} \quad \frac{\text{SOA } D_1}{\text{FOA } D_1} = \frac{3.755}{2.157} = 1.741$$

Ratio of noncomputerized results (they do not include axial and shear deformations) and the computerized results for moment and displacements are as follows:

$$\text{moment:} \quad \frac{249.0}{245.9} = 1.0126$$

$$D_1: \quad \frac{3.8878}{3.755} = 1.035$$

$$D_2: \quad \frac{-0.0131985}{-0.012821} = 1.0294$$

That is, the noncomputerized results are greater than the computerized results by 1.3% for moment and 3.5% for displacements.

According to the AISC Specifications, when a FOA is used in checking the stability interaction equation, the FOA maximum moment should be modified by a factor of 1.420 for the columns in this example problem. However, the ratio of SOA moment to FOA moment is 1.71 and the interaction equation sums to 1.174 using SOA moment and axial force (using FOA moment and axial force, the stability interaction equation sums to 1.037). When SOA moment and axial force are used in checking the interaction equation, the SOA moment is not modified as it must be when using the FOA moment.

It should be noted that this example was chosen to emphasize the SOA effects. Example 15.5 shows the results for a more commonly encountered structure and loading.

EXAMPLE 15.4

The purpose of this example is to show that if a SOA is needed, the effects due to out-of-plumbness of the columns probably should not be ignored. The AISC Code of Standard Practice stipulates that the maximum out-of-plumbness of the columns in Example 15.3 is $0.002 * 144$ in. $= 0.288$ in. If both columns are out-of-plumb by 0.288 in. in the same direction and coincident with the wind direction, this is the worst condition for the structure in Example 15.3. Therefore, solve Example 15.3 assuming the initial positions of the column tops are 0.288 in. to the right of the perfectly plumb positions. Compare the results of Example 15.4 to the results of Example 15.3 to illustrate the effects of out-of-plumbness when a SOA is needed.

Solution

Left side column	Right side column
Horizontal displacement = 3.961 in.	
Rotational displacement = −0.0135 rad	
Axial force = 25.2^k	Axial force = 28.8^k
Moment at top = 259.2 in.-k	Moment at top = 258.8 in.-k

Comparison of results for maximum out-of-plumbness to zero out-of-plumbness:

$$\text{deflection: } \frac{3.961}{3.755} = 1.055 \quad \text{(5.5\% increase due to out-of-plumbness)}$$

$$\text{moment: } \frac{258.8}{245.9} = 1.052 \quad \text{(5.2\% increase due to out-of-plumbness)}$$

EXAMPLE 15.5

The purpose of this example is to show that an SOA is not necessary for a commonly encountered design. An extra joint is chosen at midspan of the beam to have the moment and deflection computed at this point for ease in comparing the SOA and FOA results. All members are $F_y = 36$ ksi steel. Members 1 and 2 are $A = 11.8$ in.^{2}; $I = 310$ in.^{4}; W12 × 40. Members 3 and 4 are $A = 9.12$ in.^{2}; $I = 375$ in.^{4}; W16 × 31. All members are bending about the strong axis of the section and are laterally braced such that the allowable moments are 1248 in.-k for members 1 and 2 and 1128 in.-k for members 3 and 4 in loading 1; the allowable moment values given for loading 1 are multiplied by 1.33 to get the values for loading 2 since the 6^k load in loading 2 is a wind load. The member sizes are optimum (lightest members that satisfy AISC Specifications are used) and they are dictated by loading 1 (Fig. 15.14).

Comparison of Results. The following results were obtained by using the author's SOA program. Ratios of SOA results divided by FOA results are:

$$\text{loading 1: } \quad \text{moments: at joints 2 and 4:} \frac{923.2 \text{ in.-k}}{913.9 \text{ in.-k}} = 1.010$$

$$\text{at joint 3:} \frac{1025.5}{1030.1} = 0.996$$

$$\text{deflection at joint 3:} \frac{0.7407 \text{ in.}}{0.7229 \text{ in.}} = 1.025$$

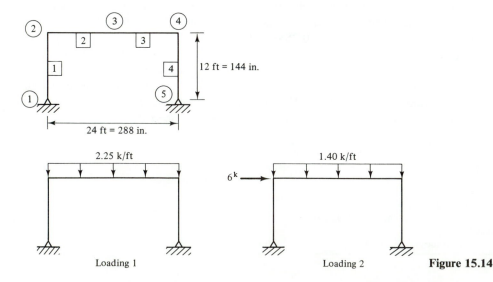

Loading 1   2.25 k/ft

Loading 2   1.40 k/ft 6^k

Figure 15.14

$$\text{loading 2:} \quad \text{moments: at joint 4:} \frac{1015.8}{1000.1} = 1.016$$

$$\text{at joint 3:} \frac{640.1}{641.5} = 0.998$$

$$\text{deflections: at joint 3:} \frac{0.4636}{0.4504} = 1.029 \text{ (downward deflection)}$$

$$\text{at joint 2:} \frac{0.6598}{0.6309} = 1.046 \text{ (horizontal deflection)}$$

The SOA moments are only 1.6% larger than the FOA moments at worst and the SOA deflections are only 4.6% larger than the FOA deflections at worst.

15.5 CONCLUDING REMARKS

Examples 15.3 and 15.4 were devised to accentuate that if a SOA is necessary, the SOA of a plane frame or space frame should account for:

1. Axial and shear deformations.
2. Out-of-plumbness of the columns.

If the SOA results are used in checking a member to determine if it satisfies the AISC Specifications for combined axial compression and bending, the member can be treated as a member from a braced frame since:

1. The SOA results converged; therefore, overall frame stability is assured; the allowable axial compressive stress is computed for the maximum KL/r where K is the effective length factor for a braced frame; and the load modification factor, C_m, is for a braced frame.
2. The SOA directly gives the amplified member-end moments (that is, the amplification factor needed for FOA results is ignored if SOA results are used).

In a steel, highrise (multistory) building, the axial compression force in the columns that are attached to and close to the foundations generally exceeds $0.5AF_y$. It is shown in Chapter 16 that the behavior of such steel columns should be treated as being inelastic (the tangent modulus of elasticity, E_t, should be used for these columns in an SOA).

Arches and domed-type space trusses or frames having a large radius of curvature are susceptible to snap-through buckling as shown in Figure 15.15, particularly if the structure is subjected to concentrated loads located near the crown. If the radius of curvature is too large, a modest overload produces a significant, nonlinear change in geometry and in the axial compressive forces of the members near the crown. Therefore, an overload SOA of such structures should be performed to determine the level of overload at which snap-through buckling occurs. If the members of such structures

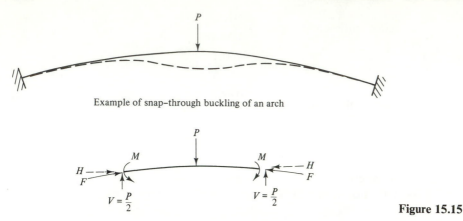

Example of snap-through buckling of an arch

Figure 15.15

are made of steel, the tangent modulus of elasticity may be applicable for some of the members.

It should be noted that if the tangent modulus of elasticity is used for one or more members in an SOA, the structural behavior is inelastic. For inelastic SOA, the structural behavior is load path dependent. That is, only one loading at a time can be performed since the tangent moduli of elasticity are determined during the incremental application of a particular loading and are applicable only to that loading. Such analyses generally are treated as piecewise linearly elastic. That is, for an increment of loading, the tangent moduli are temporary constants computed on the basis of the results obtained in the previous SOA cycle and used in the next incremental loading cycle. After each cycle, new temporary constants for the next SOA cycle are obtained until the SOA either converges or fails to converge in a specified number of cycles.

Most steel structures for which an SOA is necessary behave inelastically. However, the author's SOA program currently can solve only plane frames that remain elastic at all levels of incremental loading (program does not check to determine if the elastic axial compression force limit of a member is exceeded or if plastic bending moments are developed). Unless a computer program has been written to automatically compute the tangent moduli of elasticity, the solution requires an excessive amount of time from the analyst. Therefore, the author chooses not to show any examples of inelastic SOA.

PROBLEMS

Perform a second order displacement analysis of the following problems. If the instructor specifies that a noncomputerized solution is to be obtained, ignore axial deformations of all members except in truss analyses. For a computerized solution, use the computer program specified by the instructor.

15.1. All members are W8 × 67 ($E = 29,000$ ksi; $I = 272$ in.^{4}; $A = 19.7$ in.^{2}).

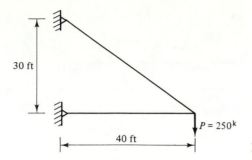

$P = 250^k$

30 ft

40 ft

Figure P15.1

15.2. Treat Problem 15.1 as a truss.

15.3. E = 3000 ksi; I = 3000 in.^{4}; A = 153 in.^{2} for all members. P = 220^k.

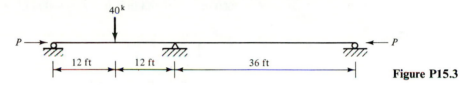

40^k

P →   ← P

12 ft · 12 ft · 36 ft

Figure P15.3

15.4. All members are W8 × 24 (E = 29,000 ksi; I = 82.8 in.^{4}; A = 7.08 in.^{2}). P = 150^k.

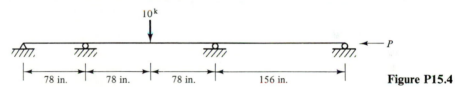

10^k

← P

78 in. · 78 in. · 78 in. · 156 in.

Figure P15.4

15.5. E = 29,000 ksi. Member 1: I = 239 in.^{4}; A = 9.13 in.^{2}; L = 16 ft. Member 2: I = 584 in.^{4}; A = 13.3 in.^{2}; L = 35 ft.

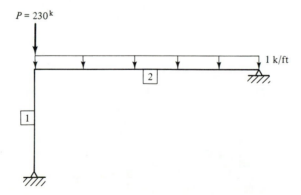

$P = 230^k$

1 k/ft

2

1

Figure P15.5

15.6. In Problem 15.5, change right support to a roller.

15.7. All members are W8 × 31 ($E = 29{,}000$ ksi; $I = 110$ in.^{4}; $A = 9.13$ in.^{2}). $P = 100^k$.

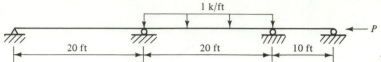

Figure P15.7

15.8. $E = 3000$ ksi; $I_1 = 3000$ in.^{4}; $A_1 = 153$ in.^{2}; $I_2 = I_3 = 6000$ in.^{4}; $A_2 = A_3 = 200$ in.^{2}

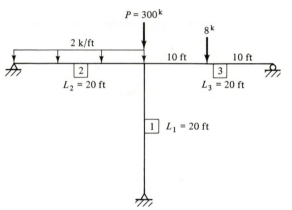

Figure P15.8

15.9. $E = 3000$ ksi; Members 1 and 2: $I = 3000$ in.^{4}; $A = 153$ in.^{2} Members 3 and 4: $I = 6000$ in.^{4}; $A = 200$ in.^{2} Member 5: $I = 9000$ in.^{4}; $A = 230$ in.^{2}

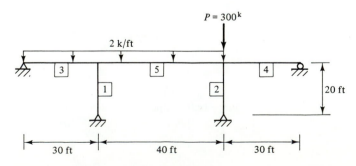

Figure P15.9

15.10. $E = 3000$ ksi. Members 1 and 2: $I = 3000$ in.^{4}; $A = 153$ in.^{2}; $L = 20$ ft. Member 3: $I = 6000$ in.^{4}; $A = 200$ in.^{2}; $L = 40$ ft.

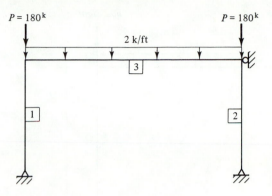

Figure P15.10

15.11. $E = 20,000$ kN/cm^2. Members 1 and 2: $I = 2000$ cm^4; $A = 20$ cm^2; $L = 5$ m. Member 3: $I = 10,000$ cm^4; $A = 45$ cm^2; $L = 10$ m.

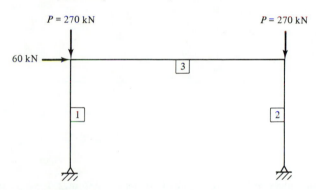

Figure P15.11

15.12. All members: $E = 20,000$ kN/cm^2; $I = 10,000$ cm^4; $A = 50$ cm^2; $L = 5$ m. Load case 1: $w_1 = 100$ kN/m and $w_2 = 0$. Load case 2: $w_1 = 0$; $w_2 = 120$ kN/m.

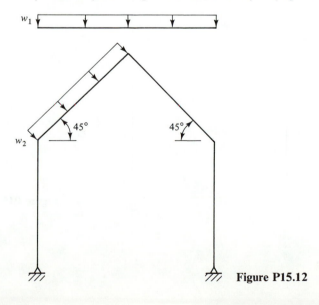

Figure P15.12

15.13. $E = 3000$ ksi. Member 1: $I = 3000$ in.^{4}; $A = 153$ in.^{2} Members 2 and 3: $I = 6000$ in.^{4}; $A = 200$ in.^{2}

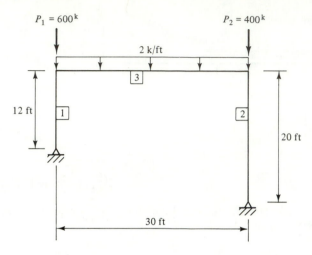

Figure P15.13

15.14. All members: W16 $\times$ 31 ($E = 29{,}000$ ksi; $I = 375$ in.^{4}; $A = 9.12$ in.^{2}).

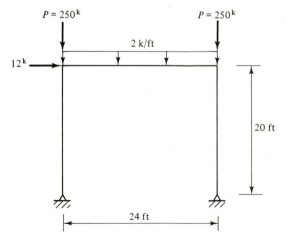

Figure P15.14

15.15. $E = 3000$ ksi. Members 1 through 4: $I = 3000$ in.^{4}; $A = 153$ in.^{2} Members 5 and 6: $I = 6000$ in.^{4}; $A = 200$ in.^{2}

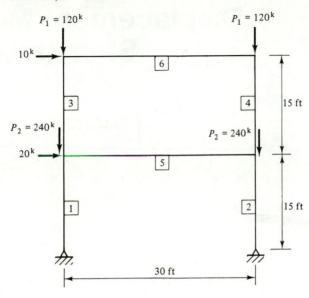

Figure P15.15

Chapter 16

Displacement Method— Stability Analysis

16.1 EULER LOAD

Consider the elastic behavior of a perfectly straight member and an initially crooked member subjected to a gradually increasing axial compressive force, P (see Figure 16.1).

As P is slowly applied, the perfectly straight member remains straight until P attains the critical load value, P_{cr}, and then the member bends into a deformed equilibrium position (buckles). Therefore, P_{cr} is the buckling load and the deformed equilibrium position is referred to as the buckled mode shape. For a pinned ended member, the buckled mode shape is a half-sine wave; the critical (buckling) load, P_{cr}, for this special case is named the Euler load, P_E, in honor of Leonhard Euler, a Swiss mathematician and engineer, who in 1759 [16.1] was the first person to derive the buckling load and mode shape for a linearly elastic, pinned ended member.

All members are initially crooked to some degree. If a member is more crooked than is acceptable, that member is not permanently usable in an engineered construction. The following discussion pertains to an acceptably crooked member. At the midheight of the initially crooked member, let Δ_0 be the deflection for $P = 0$ and let Δ be the final deflection for any $P > 0$. For an initially crooked member, the midheight deflection increases as soon as a small finite value of P is applied, and continues to increase as P is slowly increased. As P approaches P_{cr}, the midheight deflection increases at a faster rate until the member becomes unstable (a SOA equilibrium position can no longer be obtained for the initially crooked member treated as a system of several members, for example). If the initial crooked shape is not a half-sine wave, when the member becomes unstable (buckles), the deflected shape is a half-sine wave with the initial crooked shape superimposed on the half-sine wave shape. An initially crooked, linearly elastic, pinned ended member becomes unstable slightly before the Euler load

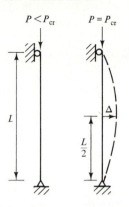

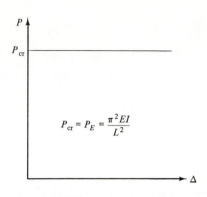

(a) Behavior of a perfectly straight, weightless column

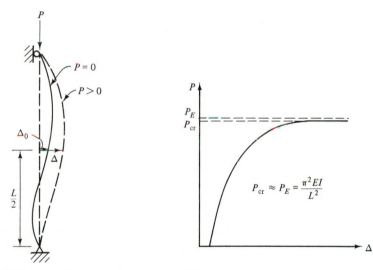

(b) Behavior of an acceptably crooked, weightless column

Figure 16.1 Pinned ended, elastic column behavior.

is attained, but for all practical purposes the critical or buckling load is equal to the Euler load (that is, $P_{cr} = P_E$ for an acceptably crooked, linearly elastic, pinned ended member).

The Euler load for a weightless, linearly elastic, perfectly straight member is obtainable from the derivation in Chapter 15, Section 15.3 by setting $M_2 = 0$. For this case, $B \neq 0$ and $kL = n\pi$ (where $n = 1, 2, 3, \ldots$, infinity) is the only logical condition that satisfies the boundary condition of $y = 0$ at $x = L$. Therefore, $k = n/(\pi L)$ and since $k = \sqrt{P/EI_z}$, then $P = n^2\pi^2 EI_z/L^2$. The fundamental buckling mode, $y = B \sin(\pi x/L)$, occurs for $n = 1$ and the buckling or critical load is $P_{cr} = P_E = \pi^2 EI_z/L^2$.

As shown in Figure 16.2, critical loads can be obtained for linearly elastic members having boundary conditions other than pinned ends provided the effective length, KL, is known or can be determined. KL is the distance between points of inflection

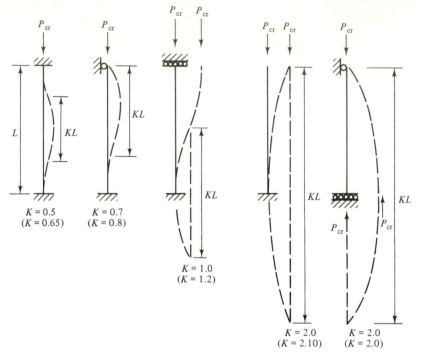

Note: Values not enclosed in parentheses are theoretical K values.
Values enclosed in parentheses are recommended values for usage.
in structural design to account for imperfect boundary conditions.

For a linearly elastic column: $P_{cr} = \dfrac{\pi^2 EI}{(KL)^2}$

Figure 16.2 Effective length factors for isolated columns.

on the buckled mode shape. Points of inflection occur where the curvature and the bending moment are zero. Note, K in KL is not the same as k in the differential equation solution of Chapter 15, Section 15.3.

16.2 INELASTIC BUCKLING

Residual stresses due to cooling exist in steel rolled sections. For W sections that are commonly used as columns, the flange tips and center portion of the web cool faster than the junction of the flanges and web which cools last. Due to the conditions of cooling from a high temperature to room temperature after leaving the rolling machine, there is a residual compressive stress at the flange tips and at the center of the web. Also, there is a residual tensile stress at the junction of the flanges and web. The residual stresses are on the order of 18 to 20 ksi for rolled W sections. For welded W sections and box sections, the residual tensile stress may be on the order of 35 to 40 ksi at the welds and the residual compressive stress may be on the order of 12 to 20 ksi. If a W section is subjected to axial compression and the maximum KL/r is small enough, the flange tips and the center of the web yield first when $P/A + f_{rc} = F_y$ where f_{rc} is the maximum residual compressive stress. As P is increased, more of the cross section yields until inelastic buckling occurs. The AISC Specification writers adopted a conservative assumption that inelastic buckling occurs when $P/A = 0.5F_y$ and assumed $0.5F_y$ occurs at $KL/r = [C_c = \sqrt{\pi^2 E/(0.5F_y)}]$ as shown in Figure 16.3c.

If the 1978 AISC Specification definitions are accepted as being exact, the following definition for the tangent modulus of elasticity, E_t, shown in Figure 16.3 can be derived [16.2] and is exact:

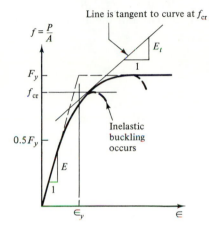

(a) Stub column stress versus strain curve

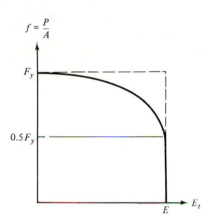

(b) Applied stress versus E_t curve

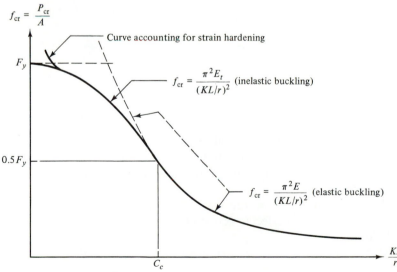

(c) Buckling stress versus $\dfrac{KL}{r}$ Curve

Figure 16.3 Inelastic steel column behavior.

$$E_t = \alpha^2(2 - \alpha^2)E$$

where $E = 29{,}000$ ksi

$$\alpha = \frac{KL/r}{C_c}$$

$\dfrac{KL}{r}$ = the larger ratio for the two principal axes

As shown later in the example problems, for an indeterminate plane frame K is a function of E_t and the procedure to obtain E_t is iterative. After the converged value of E_t is obtained, the inelastic buckling load is computed from $P_{cr} = \pi^2 E_t/(KL)^2$. Some analysts prefer to call this the tangent modulus buckling load, but the author calls it the inelastic buckling load.

16.3 SOME BASIC CONCEPTS

The plane frame shown in Figure 16.4 is symmetric with respect to midspan of the beam and the member weights are assumed to be negligible.

Some or all of the following imperfections exist in an actual steel structure represented by Figure 16.4:

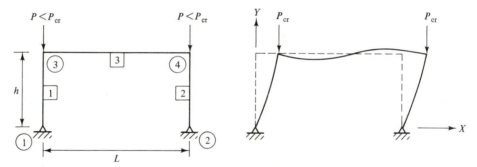

Figure 16.4 Plane frame behavior.

1. Hinged bases are not frictionless.
2. Members are crooked (bent or twisted).
3. Fabrication is less than perfect (fabrication is done only within tolerable limits).
   a. Members are slightly longer or shorter than the lengths shown on the fabrication drawings.
   b. Slight mismatch of connections is unavoidable.
4. Construction is less than perfect (erection of the structure is done only within tolerable limits).
   a. Construction grade elevations do not perfectly agree with the engineering drawings.
   b. Columns are not perfectly plumb.

5. Supposedly symmetrical cross sections are not quite symmetric, but the section does not violate the maximum allowable tolerances.
6. The steel and section properties are not perfect.

If the plane frame in Figure 16.4 is assumed to be weightless and perfect in every way, and the loads are slowly applied, the behavior of this structure is:

1. For $P < P_{cr}$. The only joint displacements occur at joints 3 and 4; these joint displacements are due to the axial deformations of the columns.
2. At $P = P_{cr}$. Joints 3 and 4 translate in the positive or negative X direction causing all members to bend and all joints to rotate about an axis parallel to the Z axis. The value of P at which joints 3 and 4 translate is called the critical load, P_{cr}, and the deflected shape of the structure is called the buckling mode shape.
3. For $P > P_{cr}$. Translational and rotational displacements quickly become excessively large. In terms of the SOA discussion presented in Chapter 15, if a horizontal force of $0.001P$ is applied at joint 3, for P slightly greater than P_{cr} the SOA does not converge. This means that the structure is unstable due to the loading and deformed geometry configuration.

Suppose the columns in Figure 16.4 are not perfectly plumb. How does the behavior of this imperfect structure differ from the behavior described above for a perfect structure? For $P < P_{cr}$, all joint displacements exist as soon as a small value of P is applied and they increase when P is increased. As P approaches the P_{cr} obtained for the perfect structure, a SOA eventually does not converge. That is, an unstable or buckled configuration of the structural system is reached for P slightly larger than the last P for which the SOA converged. Consequently, the P_{cr} obtained for the perfect structure is an upper bound on structural stability for a small displacement theory formulation.

If a SOA is performed for a particular loading to determine the overload factor at which the structure becomes unstable, the author recommends that the worst, maximum out-of-plumbness condition of the columns be assigned in the joint coordinate definitions prior to the application of any amount of loading. The worst, maximum out-of-plumbness condition of the columns is not always obvious, but generally alternating out-of-plumb columns is worst. An envelope of out-of-plumbness of steel columns is available in the Code of Standard Practice part of the AISC Manual.

16.4 EXAMPLE PROBLEMS

In the following noncomputerized solutions, the structures are assumed to be perfect, weightless, and possibly inelastic in obtaining the critical load in each example problem. Also, only joint loads are used in these solutions and actual joint displacements are not found; instead, after P_{cr} is obtained, the buckling mode shape is found to an arbitrary scale—that is, the maximum probable displacement is chosen as unity and the other displacements are computed based on the chosen displacement (if a computed displacement is greater than unity, the set of displacements can be rescaled such that the maximum displacement is unity). This approach is called a first order stability analysis and a critical load value obtained by this approach is an upper bound on the critical load for the imperfect structure.

EXAMPLE 16.1

Verify the effective length factors, K, given in Figures 16.1 and 16.2 by performing a stability analysis which uses Appendix C. Assume that the member buckles elastically in each case.

Solution

$$P_{cr} = \rho * P_E = \rho * \frac{\pi^2 EI}{L^2} \quad \text{and} \quad P_{cr} = \frac{\pi^2 EI}{(KL)^2}$$

Therefore, $K = 1/\sqrt{\rho}$ and if ρ can be determined, we can find K.
1. For the pinned ended case of Figure 16.1:

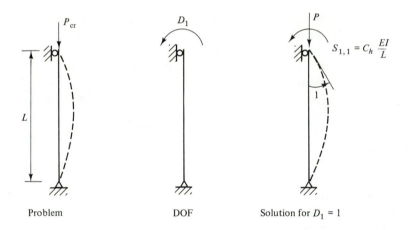

Problem   DOF   Solution for $D_1 = 1$

The matrix equation of system equilibrium, $[S]\{D\} = \{P^e\}$, is

$$\left[C_h \frac{EI}{L} \right] \{D_1\} = \{0\}$$

Obviously $D_1 \neq 0$ when buckling occurs; therefore, $C_h = 0$ since $EI/L \neq 0$. From Appendix B, $C_h = 0$ at $\rho = 1$; therefore, $K = 1/\sqrt{1} = 1$. The physical meaning is that when $P = P_{cr}$, the axial load has used up all of the column's bending stiffness and the column has no rotational member-end stiffness remaining to resist the induced unit displacement.

2. For the second case in Figure 16.2:

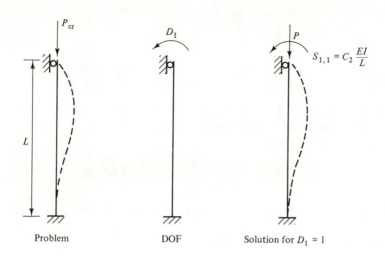

Problem   DOF   Solution for $D_1 = 1$

The equation of equilibrium is

$$C_2 \frac{EI}{L} * D_1 = 0$$

which requires that $C_2 = 0$. From Appendix C, $C_2 = 0$ occurs at $\rho = 2.04$ by interpolation. Therefore, $K = 1/\sqrt{2.04} = 0.70014$.

3. For the first case in Figure 16.2, the column must be subdivided into a system of two members. Symmetric buckling occurs; therefore, there is only 1 DOF.

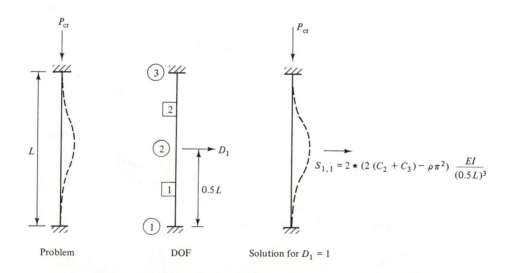

Problem   DOF   Solution for $D_1 = 1$

The equation of system equilibrium is $S_{1,1}*D_1 = 0$ which requires that

$$(C_2 + C_3) - \frac{\rho\pi^2}{2} = 0$$

Since

$$\left(P_{cr} = \rho * P_{E(1)} = \rho\frac{\pi^2 EI}{(0.5L)^2}\right) = \frac{\pi^2 EL}{(KL)^2}$$

then $\rho = (0.5/K)^2$. Fortunately we are verifying K (supposedly, $K = 0.5$ for this case) and we know where to look in Appendix C. Try $K = 0.5$: $\rho = (0.5/0.5)^2 = 1$; $(C_2 + C_3) = 4.93480$ at $\rho = 1$;

$$\frac{\rho\pi^2}{2} = \frac{1*(3.14159)^2}{2} = 4.93480$$

Since $(C_2 + C_3) - \rho\pi^2/2 = 4.93480 - 4.93480 = 0$ at $\rho = 1$, then

$$P_{cr} = \rho * P_{E(1)} = \frac{\pi^2 EI}{(0.5L)^2}$$

and Figure 16.2 gives

$$P_{cr} = \frac{\pi^2 EI}{(KL)^2}$$

Equating P_{cr} expressions and solving for K gives $K = 0.5$.
   4. For the fourth case in Figure 16.2:

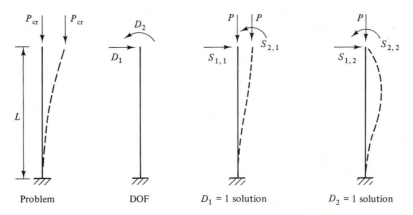

Problem   DOF $D_1 = 1$ solution $D_2 = 1$ solution

$$S_{1,1} = [2(C_2 + C_3) - \rho\pi^2]EI/L^3$$
$$S_{2,1} = S_{1,2} = (C_2 + C_3)EI/L^2$$
$$S_{2,2} = C_2 EI/L$$

The matrix equation of system equilibrium is

$$\begin{bmatrix} S_{1,1} & S_{1,2} \\ S_{2,1} & S_{2,2} \end{bmatrix}\begin{Bmatrix} D_1 \\ D_2 \end{Bmatrix} = \begin{Bmatrix} 0 \\ 0 \end{Bmatrix}$$

Obviously D_1 and D_2 are not zero when buckling occurs. Therefore, the determinant of $[S]$ must be zero; that is, $S_{1,1} * S_{2,2} - S_{1,2} * S_{2,1} = 0$ which gives the following criterion for buckling:

$$[2(C_2 + C_3) - \rho \pi^2] * C_2 - (C_2 + C_3)^2 = 0$$

Try $K = 2$;

$$\rho = \frac{1}{K^2} = 0.25$$

$$C_2 = 3.65978$$

$$(C_2 + C_3) = 5.74878$$

$$(2 * 5.74878 - 0.25 * \pi^2) * 3.65978 - (5.74878)^2$$
$$= 33.048395 - 33.048471 = -0.0000765$$

which is approximately zero. Therefore, $\rho = 0.25$ produces buckling and $K = 1/\sqrt{\rho} = 2.00$ is the effective length factor for this case. Note that this problem could have been solved using only 1 DOF as shown below.

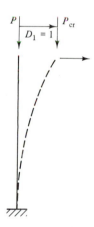

$$S_{1,1} = (C_h - \rho \pi^2) \frac{EI}{L^3}$$

$S_{1,1} = 0$ requires $C_h = \rho \pi^2$ for $\rho = 0.25$: $C_h = 2.46738$ and $\rho \pi^2 = 0.25 \pi^2 = 2.4674011$. Therefore, $K = 1/\sqrt{0.25} = 2.00$.

EXAMPLE 16.2 ───

In Figure 16.5, both members are W6 × 16 $F_y = 36$ ksi steel with the weak axis of the section braced as shown. For the strong axis of the section, bracing is provided only at $X = 0$ and $X = 30$ ft in the form of hinged ends. Find P_{cr}.

For a noncomputerized stability analysis, the local x axes of the members are coincident with the X axis of the system. The section weak axis is the local y axis and the section strong axis is the local z axis. From the AISC Manual

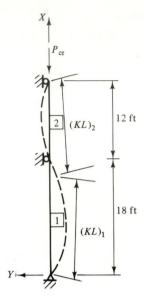

Figure 16.5

$$A = 4.74 \text{ in.}^2$$
$$I_y = 4.43 \text{ in.}^4 \text{ (weak axis inertia)}$$
$$I_z = 32.1 \text{ in.}^4 \text{ (strong axis inertia)}$$
$$r_y = 0.966 \text{ in.}$$
$$r_z = 2.60 \text{ in.}$$

Note that $(KL)_z = 30 \text{ ft} = 360 \text{ in.}$ where y and z are coincident with Y and Z.

Solution. Assume elastic buckling occurs for strong axis:

$$P_{cr(z)} = \frac{\pi^2 E I_z}{(KL)_z^2}$$

$$\left(P_{cr(z)} = \frac{\pi^2 * 29000 * 32.1}{(30 * 12)^2} = 70.89^k \right) < (0.5 F_y A = 85.32^k)$$

Therefore, strong axis buckling load is elastic behavior as assumed. However, the weak axis behavior may give a smaller buckling load.

For weak axis buckling, assume elastic buckling occurs:

$$P_{cr(y)} = \rho_1 \frac{\pi^2 E I_y}{(18 \text{ ft})^2} = \rho_2 \frac{\pi^2 E I_y}{(12 \text{ ft})^2}; \qquad \rho_2 = \frac{(12)^2}{(18)^2} \rho_1 = \frac{4}{9} \rho_1$$

$$[S]\{D\} = \{0\}$$

$$\text{determinant of } [S] = \frac{C_{h(1)} E I_y}{18} + \frac{C_{h(2)} E I_y}{12} = 0$$

which occurs when $C_{h(1)} + 1.5C_{h(2)} = 0$. Assume

$$(KL)_1 = 16 \text{ ft}; \qquad K_1 = \frac{16}{18} = 0.888; \qquad \rho_1 = \frac{1}{K_1^2} = 1.266$$

$$\rho_2 = \frac{4}{9}\rho_1 = 0.5625$$

$$C_{h(1)} = -1.666 \quad \text{and} \quad C_{h(2)} = 1.654$$

check assumption: $\qquad C_{h(1)} + 1.5C_{h(2)} = 0.815 \neq 0 \text{ \underline{assumption N.G.}}$

Assume

$$(KL)_1 = 15.537 \text{ ft}; \qquad \rho_1 = 1.342125; \qquad \rho_2 = 0.5965$$
$$C_{h(1)} = -2.329803; \qquad C_{h(2)} = 1.552009$$

check assumption: $\qquad -2.329803 + 1.5 * 1.552009 = -0.0017895 \approx 0 \text{ \underline{okay}}$

$$(\text{elastic } P_{cr})_y = \frac{\pi^2 E I_y}{(15.537 * 12)^2} = 36.47^k < [(\text{elastic } P_{cr})_z = 85.32^k]$$

Therefore, elastic buckling occurs at $P = 36.47^k$ and the effective lengths for y-axis buckling are $(KL)_1 = 15.537$ ft and $(KL)_2 = 30 - 15.537 = 14.463$ ft. Suppose that weak axis braces were provided at $X = 8$, $X = 18$, and $X = 30$ ft, and P_{cr} were located at $X = 30$ ft. There would be three members and $L_1 = 8$, $L_2 = 10$, and $L_3 = 12$ ft, respectively. For this case,

$$\frac{\rho_1}{(8)^2} = \frac{\rho_2}{(10)^2} = \frac{\rho_3}{(12)^2}$$

which gives $\rho_3 = 2.25\rho_1$ and $\rho_2 = 1.5625\rho_1$ as the inter-related ρ values to be used in searching for the zero value for the determinant of $[S]$ to find P_{cr} for the 2 DOF system of buckling for the weak axis of the section.

EXAMPLE 16.3 ─────────────────────────────────

Change the members in Example 16.2 to W8 $\times$ 31 $F_y = 36$ ksi steel: $I_y = 37.1$ in.^{4}; $r_y = 2.02$ in.; $I_z = 110$ in.^{4}; $r_z = 3.47$ in.; $A = 9.13$ in.^{2} Find P_{cr}.

Solution. *Strong axis buckling:* If the behavior is elastic,

$$P_{cr(z)} = \frac{\pi^2 E I_z}{(KL)_z^2} = 242.9^k$$

However, $242.9^k > (0.5F_y A = 164.34^k)$—inelastic buckling occurs.

$$\alpha = \frac{(KL/r)_z}{C_c} = \frac{(360/3.47)}{126.1} = 0.822731$$

$$\frac{E_t}{E} = \alpha^2(2 - \alpha^2) = 0.8956$$

$$(\text{inelastic } P_{cr})_z = \frac{\pi^2 E_t I_z}{(KL)_z^2} = \frac{E_t}{E} * (\text{elastic } P_{cr})_z = 0.8956 * 242.9 = 217.5^k$$

Weak axis buckling: If the behavior is elastic, from Example 16.2,

$$(KL)_1 = 15.537 \text{ ft} = 186.444 \text{ in}; \qquad \rho_1 = 1.342125; \qquad \rho_2 = \frac{4}{9}\rho_1 = 0.5965$$

$$(\text{elastic } P_{cr})_y = \frac{\pi^2 * 29000 * 37.1}{(186.444)^2} = 305.5^k > (0.5F_y A = 164.34^k)$$

Therefore, inelastic buckling occurs, and

$$\alpha = (KL/r)_y/C_c = \frac{186.444/2.02}{126.1} = 0.731951$$

$$\frac{E_t}{E} = \alpha^2(2 - \alpha^2) = 0.784474; \qquad (\text{inelastic } P_{cr})_y = 0.784474 * 305.5 = 239.7^k$$

Therefore, buckling is inelastic and $P_{cr} = 217.5^k$ (*z*-axis bracing controls).

EXAMPLE 16.4 ———

Change the loading in Example 16.2 to be as shown in Figure 16.6 and find P_{cr}.

Solution. For *y*-axis buckling:

$$P_{E(1)} = \frac{\pi^2 * 29,000 * 4.43}{(18 * 12)^2} = 27.18^k$$

$$P_{E(2)} = \frac{\pi^2 * 29,000 * 4.43}{(12 * 12)^2} = 61.15^k; \qquad P_{E(2)} = 2.25P_{E(1)}$$

$$P_1 = 3P_{cr}; \qquad P_2 = P_{cr}$$

$$\rho_1 = \frac{3P_{cr}}{P_{E(1)}}; \qquad \rho_2 = \frac{P_{cr}}{P_{E(2)}} = \frac{P_{cr}}{2.25P_{E(1)}}$$

$$\frac{\rho_2}{\rho_1} = 0.148148$$

When $C_{h(1)} + 1.5C_{h(2)} = 0$, then determinant of $[S] = 0$. Assume

$$\rho_1 = 1.47 \qquad (\text{that is, } (KL)_1 = 14.846 \text{ ft})$$
$$\rho_2 = 0.148148 * 1.47 = 0.2178$$
$$C_{h(1)} = -3.797646$$
$$C_{h(2)} = 2.540833$$
$$C_{h(1)} + 1.5C_{h(2)} = 0.0136 \approx 0 \,(\text{assumption was correct})$$

Therefore, $(3P_{cr} = 1.47 * 27.18 = 39.95^k) < (0.5F_y A = 85.32^k)$, elastic buckling occurs, and $P_{cr} = 0.2178 * 61.15 = 13.3^k$ or $P_{cr} = 39.95/3 = 13.3^k$. For member 1

$$\frac{\text{critical axial force from Example 16.4}}{\text{critical axial force from Example 16.2}} = \frac{39.95}{36.47} = 1.095$$

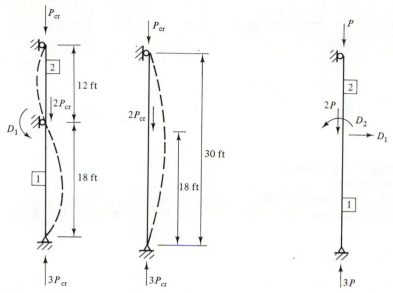

Figure 16.6   **Figure 16.7**

Therefore, if the loads are distributed as shown in Example 16.4 instead of as shown in Example 16.2, member 1 can resist 9.5% more load before buckling.

It should be obvious to the reader that $(3P_{cr})_z$ for Example 16.4 is greater than $P_{cr(z)}$ of Example 16.2 which is greater than 39.95; therefore, y-axis buckling controls for Example 16.4.

If one wishes to find $(3P_{cr})_z$ for Example 16.4, the approach in Figure 16.7 is needed.

EXAMPLE 16.5 _____

Ignore axial deformations and member weight. Find elastic P_{cr} and mode shape of in-plane buckling (Fig. 16.8a).

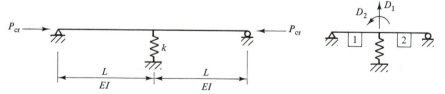

Figure 16.8a

Solution

$$[S] = \begin{bmatrix} k + 2(C_h - \pi^2 \rho)\dfrac{EI}{L^3} & 0 \\ 0 & 2C_h \dfrac{EI}{L^2} \end{bmatrix}; \quad \left(\text{note that } P_{E(1)} = P_{E(2)} = \dfrac{\pi^2 EI}{L^2} \right)$$

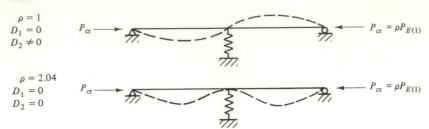

$\rho = 1$
$D_1 = 0$
$D_2 \neq 0$

$\rho = 2.04$
$D_1 = 0$
$D_2 = 0$

Figure 16.8b

The determinant of $[S]$ can become zero in two ways:

1. $C_h = 0$ which occurs for:

    a. $\rho = 1$ ($M = 0$ at both ends of members 1 and 2).

    b. $\rho = 2.04$ (one end pinned and other fixed for members 1 and 2). $KL = 0.7L$.

2. $k = 2(\pi^2\rho - C_h)\dfrac{EI}{L^3}$.

    a. Suppose it is desired to find k such that $P_{cr} = 0.85 P_{E(1)}$. $\rho = 0.85$; $C_h = 0.66707$; $k = 15.444\dfrac{EI}{L^3} = 1.565\dfrac{P_{E(1)}}{L}$.

    b. Suppose it is desired to find k such that $P_{cr} = P_{E(1)}$. $\rho = 1$; $C_h = 0$; $k = 2\dfrac{P_{E(1)}}{L}$. So, spring stiffness need only be greater than or equal to k shown in order to serve as a column brace to prevent sidesway at the spring. That is, an infinite spring stiffness is not needed in order for the spring to serve as a column brace.

EXAMPLE 16.6

Members 1 and 2: $E = 29,000$ ksi; $I = 32.1$ in.[4]; $L = 144$ in. All members are braced out-of-plane such that in-plane buckling occurs. Find P_{cr} for elastic buckling. Note that member 3 is pinned ended. See Figure 16.9a.

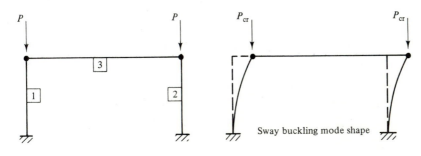

Sway buckling mode shape

Figure 16.9a

Solution

$$P_{cr} = \frac{\pi^2 EI}{(KL)^2} = \frac{\pi^2 * 29,000 * 32.1}{(2.0 * 144)^2} = 110.77^k$$

Total gravity load of 221.54^k equally shared by members 1 and 2 produces elastic buckling. Buckling mode is as shown in Figure 16.9a.

Now, find P_{cr} if member 2 does not have any axial load in it. See Figure 16.9b.

Solution

$$S_{1,1} = \left[(C_h - \rho\pi^2)\frac{EI}{L^3} \right]_1 + \left(\frac{3EI}{L^3} \right)_2 = 0 \text{ is the buckling criterion}$$

Since members 1 and 2 have identical member properties, the buckling criterion reduces to $(C_h - \rho\pi^2)_1 + 3 = 0$. By trial and checking procedure, for $\rho_1 = 0.492$, $C_h = 1.856064$. The buckling criterion evaluates to $0.0002186 \approx 0$ and $P_{cr} = \rho_1 * P_{E(1)}$ where $P_{E(1)} = \pi^2 EI/L^2 = 443.08^k$.

Therefore, if sway buckling occurs, $P_{cr} = 0.492 * 443.08 = 218.0^k$. Note that this buckling load of 218.0^k is very nearly equal to the total gravity load of 221.54^k that produced buckling when both columns equally shared the total gravity load at buckling.

Now find P_{cr} if member 2 does not have any axial load in it and no sway buckling occurs (see Figure 16.9c). Since the sway buckling load of 218.0^k is less than the no sway buckling load of 904.24^k, $P_{cr} = 218.0^k$ when member 2 has no axial load in it.

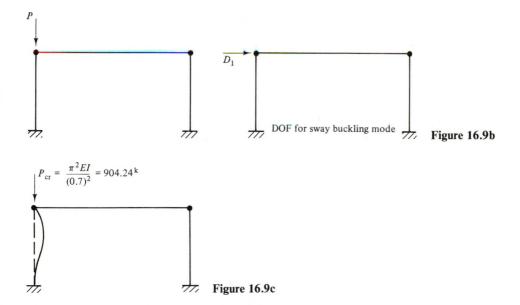

DOF for sway buckling mode   **Figure 16.9b**

$P_{cr} = \dfrac{\pi^2 EI}{(0.7)^2} = 904.24^k$

Figure 16.9c

EXAMPLE 16.7 ——————————————————————————————

Repeat the solution described in Example 16.6 with member 2 changed as follows:

1. $I_2 = 2I_1 = 64.2 \text{ in.}^4$
2. $I_2 = 8I_1 = 256.8 \text{ in.}^4$

Solution for (1). ($I_2 = 2I_1$). Note that $P_{cr(1)}$ for no sway buckling is 904.24^k (same as in Example 16.6). Therefore, $P_{cr(1)} = 320.52^k$ and it is very nearly equal to 332.31^k which is the total gravity load that produces sway buckling when both columns equally share the total buckling load. See Figure 16.10.

Solution for (2). ($I_2 = 8I_1$). In this case note that $P_{cr(1)}$ for no sway buckling is 904.24^k as before and it is less than the sway buckling load of 1314.62^k. Therefore, no sway buckling controls and the buckling load is 904.24^k. See Figure 16.11.

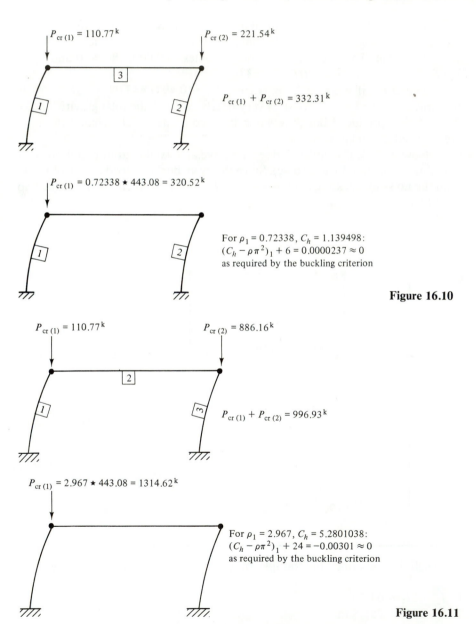

$P_{cr(1)} = 110.77^k$ $P_{cr(2)} = 221.54^k$

$P_{cr(1)} + P_{cr(2)} = 332.31^k$

$P_{cr(1)} = 0.72338 \star 443.08 = 320.52^k$

For $\rho_1 = 0.72338$, $C_h = 1.139498$:
$(C_h - \rho\pi^2)_1 + 6 = 0.0000237 \approx 0$
as required by the buckling criterion

Figure 16.10

$P_{cr(1)} = 110.77^k$ $P_{cr(2)} = 886.16^k$

$P_{cr(1)} + P_{cr(2)} = 996.93^k$

$P_{cr(1)} = 2.967 \star 443.08 = 1314.62^k$

For $\rho_1 = 2.967$, $C_h = 5.2801038$:
$(C_h - \rho\pi^2)_1 + 24 = -0.00301 \approx 0$
as required by the buckling criterion

Figure 16.11

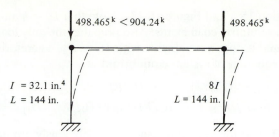

Figure 16.12

Conclusion: If elastic buckling occurs, the total gravity load that produces sway buckling when both columns are ideally loaded can be allocated in any manner desired provided each column is not required to resist more than the no sway buckling load in that column. That is, for Solution (1), the loads shown in Figure 16.12 cause sway buckling (note: $996.93/2 = 498.465$).

EXAMPLE 16.8

Using a noncomputerized approach and ignoring axial deformations, formulate the matrix equation of system equilibrium for a first order stability analysis. Allow for the possibility that inelastic buckling may occur. See Figure 16.13a.

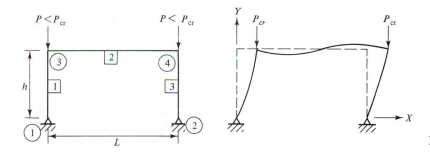

Figure 16.13a

Solution. Since the structure is symmetric and the buckling mode is antisymmetric, only half of the structure need be involved in the stability analysis. See Figure 16.13b. Note that immediately prior to buckling, $D_1 = D_2 = 0$ and the axial force in member 1 is P_{cr} and the axial force in member 2 is zero. Refer to

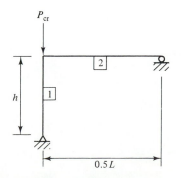

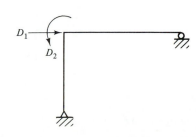

Figure 16.13b

Chapter 15 Section 15.3 and Figure 16.13c. Note that $L_1 = h$ and $L_2 = 0.5L$ in the system stiffness matrix coefficients. Also note that the only load is an applied joint load, P_{cr}, and it does not coincide with D_1 or D_2. Therefore, $P_1^e = P_2^e = 0$ and the matrix equation of system equilibrium is

$$\begin{bmatrix} (C_h - \rho\pi^2)_1(EI/L^3)_1 & (C_h EI/L^2)_1 \\ (C_h EI/L^2)_1 & (C_h EI/L)_1 + (3EI/L)_2 \end{bmatrix} \begin{Bmatrix} D_1 \\ D_2 \end{Bmatrix} = \begin{Bmatrix} 0 \\ 0 \end{Bmatrix}$$

Since $\{P^e\} = \{0\}$ and the system displacements obviously are not zero for the buckled mode shape, the determinant of $[S]$ must be zero. That is, the stiffness of the system has been used up—the effect of the axial force, P_{cr}, in the column has significantly reduced the member stiffness coefficients of the column and nothing is left to combine with the beam rotational stiffness to resist any more growth in the system displacements.

Since $C_{h(1)}$ is a function of ρ_1, the value of ρ_1 which makes the determinant of $[S]$ become zero enables us to determine P_{cr} since $P_{cr} = (P_1 = \rho_1 * P_{E(1)})$ and $P_{E(1)} = (\pi^2 EI/L^2)_1$. Also we know, from Figure 16.2, that $P_{cr(1)} = (\pi^2 EI/(KL)^2)_1$. Therefore, $\rho_1 = 1/K_1^2$ is obtained by equating the two different definitions of $P_{cr(1)}$. It is shown in the next example that we can estimate K_1 by using the AISC

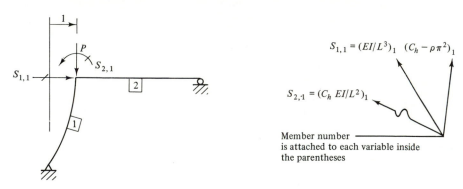

Solution for $D_1 = 1$ and $D_2 = 0$ (to obtain column 1 of $[S]$)

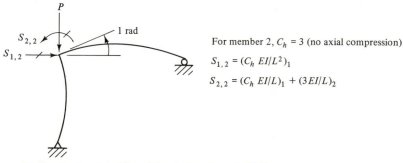

Solution for $D_2 = 1$ and $D_1 = 0$ (to obtain column 2 of $[S]$)

Figure 16.13c

nomograph for an unbraced frame provided the column and beam properties are numerically known. Consequently, if K_1 can be estimated, we have an estimate of ρ_1 and a search can be conducted in the neighborhood of the estimated ρ_1 to find the value of ρ_1 which makes the determinant of $[S]$ become zero. Actually we know that $K > 1$ for each column in an unbraced frame. Therefore, a plot of the determinant of $[S]$ versus ρ_1 can be obtained beginning at $\rho_1 = 0.05$. When the determinant of $[S]$ changes sign, a finer search can be made in that neighborhood to obtain the smallest value of ρ_1 that causes the determinant of $[S]$ to be zero.

Also note that if inelastic buckling occurs, for member 1 the analyst must determine the tangent modulus of inelastic buckling; that is, $E_1 = E_t$ for member 1.

EXAMPLE 16.9 _____

Find P_{cr} for the structure in Example 15.3 by using the results of the half-structure in Example 16.8 for which $E_1 = E_2 = 29,000$ ksi, $I_1 = I_2 = 32.1$ in.^{4}, and $L_1 = L_2 = 12$ ft $= 144$ in. Note: Since P_{cr} is obtained for the case where $\{P^e\} = \{0\}$, it does not matter which half of the structure in Example 15.3 is analyzed; also, the horizontal load in Example 15.3 does not enter into the analysis—that is, P_{cr} is obtained by setting the determinant of $[S] = 0$.

Solution. In this case, the determinant of $[S]$ becomes

$$((C_{h(1)} - \pi^2\rho_1)(C_{h(1)} + 3) - C_{h(1)}^2)\left(\frac{EI}{L^2}\right)^2 = 0$$

An estimate of $(KL)_1$ can be obtained by using the entire structure in Example 15.3 and the nomograph on page 5-125 of the 1980 AISC Manual [16.1] with $G_{bottom} = $ infinity and $G_{top} = 2$ which gives $K = 2.6$ approximately. Therefore, an estimate of ρ_1 is $\rho_1 = 1/K_1^2 = 1/(2.6)^2 = 0.148$ and the results of the search for the determinant of $[S] = 0$ in the neighborhood of $\rho_1 = 0.148$ are shown in the following table:

ρ_1	$C_{h(1)}$	$(C_{h(1)} - \pi^2\rho_1)(C_{h(1)} + 3) - C_{h(1)}^2$
0.1375	2.717355	0.393
0.140	2.71200	0.243
0.145	2.70126	−0.055
0.144	2.703408	$0.0044 \approx 0$

Consequently, $(P_{cr} = 0.144 * P_{E(1)} = 63.8^k) < (0.5F_yA_1 = 85.3^k)$. Therefore, the frame in Example 15.3 buckles elastically for $P = 63.8^k$ if the column bases are perfect hinges.

If one prefers to program the expression for the determinant of $[S]$ and have the computer search for the first value of ρ_1 that causes the determinant to be zero, the following approach is recommended. Start with $\rho_1 = 0.05$ and evaluate the determinant which is a large positive number whenever a sizable structural system (several joints and members) is involved and the values of E, I, and L of the members are included in $[S]$. An incremental size of 0.05 can be used for ρ_1 to repeatedly evaluate the determinant until a negative value of the determinant is computed. At this stage, the incremental size can be divided by 10 and the search is made starting with the last value of ρ_1 which gave a positive value for the determinant. If greater precision is desired for the value of ρ_1 which causes the determinant to be zero, the incremental size can again be divided by 10 and used for the third search.

If one wishes to account for the rotational resistance of the hinged bases as recommended on the AISC nomograph page, a rotational spring can be inserted between the bottom end of the column and the support. Of course this will necessitate another DOF and a longer expression for the determinant.

EXAMPLE 16.10

Change the value of I_2 in Example 16.9 to $I_2 = 3210$ in.[4] which is 100 times the original value of I_2. Do not change any other member properties for members 1 and 2. Find P_{cr} for the revised structure.

Solution. In this case, member 2 is so stiff that it can be considered as rigid in regard to bending and $(KL)_1 = 2*12$ ft $= 24$ ft. $\rho_1 = 1/(2)^2 = 0.25$ and $(P_{cr} = 0.25*P_{E(1)} = 110.8^k) > (0.5F_yA_1 = 85.3^k)$. Consequently, inelastic buckling occurs prior to $P = 110.8^k$ and the following approach to finding the inelastic P_{cr} is recommended:

$$\alpha = \frac{(KL/r)}{C_c} = \frac{2.0*144/2.60}{126.1} = 0.878424$$

$$\frac{E_t}{E} = \alpha^2(2 - \alpha^2) = 0.947846$$

$$(\text{inelastic } P_{cr}) = \frac{E_t}{E}*(\text{elastic } P_{cr}) = 105.0^k$$

For a general case of inelastic buckling of the frame with a flexible member 2 (say, $5I_1 \geq I_2 \geq 2I_1$), the following approach is recommended:

1. Assume elastic buckling occurs; find (elastic P_{cr}). If (elastic P_{cr}) $> 0.5F_yA_1$, proceed to item 2; otherwise quit.
2. Compute E_t/E using the formula shown above. Now, $E_1 = E_t$ and $E_2 = E$ must be used in searching for the value of ρ_1 which causes the determinant of $[S]$ to be zero. The new value of ρ_1 gives a new value of $K_1 = 1/\sqrt{\rho_1}$ which gives a new value of $(KL/r)_1$.
3. Repeat item 2 until sufficient convergence has been obtained. Usually only 1 or 2 cycles are sufficient. Then, (inelastic P_{cr}) $= \rho_1*(E_t/E)*P_{E(1)}$ where $P_{E(1)} = \pi^2EI/L_1^2$. Alternatively, (inelastic P_{cr}) $= [\pi^2E_tI/(KL)^2]_1$.

EXAMPLE 16.11

All members are W6 $\times$ 16 $F_y = 36$ ksi steel and there is only one P load which becomes P_{cr} when buckling occurs. This is the same structure, but different loading, as in Ex-

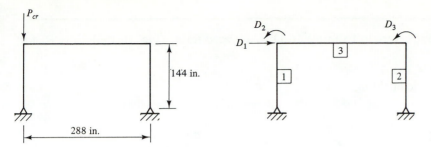

Figure 16.14

ample 15.3 and Example 16.9 half-structure. Find P_{cr}. Since the loading is not symmetric, the entire structure must be analyzed. See Figure 16.14.

Solution. The system stiffness matrix is

$$[S] = \begin{bmatrix} (C_{h(1)} - \pi^2 \rho_1)\dfrac{(EI)_1}{L_1^3} + \dfrac{3(EI)_2}{L_2^3} & C_{h(1)}\dfrac{(EI)_1}{L_1^2} & \dfrac{3(EI)_2}{L_2^2} \\[2ex] C_{h(1)}\dfrac{(EI)_1}{L_1^2} & C_{h(1)}\dfrac{(EI)_1}{L_1} + \dfrac{4(EI)_3}{L_3} & \dfrac{2(EI)_3}{L_3} \\[2ex] \dfrac{3(EI)_2}{L_2^2} & \dfrac{2(EI)_3}{L_3} & \dfrac{3(EI)_2}{L_2} + \dfrac{4(EI)_3}{L_3} \end{bmatrix}$$

$$\text{determinant of } [S] = S_{1,1}*(S_{2,2}*S_{3,3} - S_{2,3}*S_{3,2}) - S_{2,1}*(S_{1,2}*S_{3,3} - S_{1,3}*S_{3,2}) + S_{3,1}*(S_{1,2}*S_{2,3} - S_{1,3}*S_{2,2})$$

A computer program was written to evaluate the determinant of $[S]$ in terms of the variables ρ_1 and $C_{h(1)}$. From Section 15.3, $C_h = (C_2^2 - C_3^2)/C_2$ which required that the expressions for C_2 and C_3 had to be programmed. All of the other symbols in $[S]$ were computed on a calculator and typed in the determinant of $[S]$ expression as constants. The search for the value of ρ_1 to produce a zero determinant gave the following results: For $\rho_1 = 0.3198$, the determinant of $[S] = 24,434$. For $\rho_1 = 0.3199$, the determinant of $[S] = -24,460$. To be conservative, the author chose to say that $\rho_1 = 0.3198$ was the value at which the determinant became zero, and at that value of ρ_1, $C_{h(1)} = 2.302394$; $E_{t(1)} = 24,434$; $E_t/E = 0.84255$; and (inelastic P_{cr}) = 119.4^k.

The buckling mode shape can be determined as follows: Choose $D_1 = 1$ and solve equations 2 and 3 of the matrix equation of system equilibrium for D_2 and D_3 when $\rho_1 = 0.3198$:

$$\begin{bmatrix} S_{2,2} & S_{2,3} \\ S_{3,2} & S_{3,3} \end{bmatrix}\begin{bmatrix} D_2 \\ D_3 \end{bmatrix} = \begin{Bmatrix} -S_{2,1}D_1 \\ -S_{3,1}D_1 \end{Bmatrix}; \qquad \begin{Bmatrix} D_1 \\ D_2 \\ D_3 \end{Bmatrix} = \begin{Bmatrix} 1 \\ -0.002488 \\ -0.003669 \end{Bmatrix}$$

If one uses $E_1 = E = 29,000$ ksi (assumes elastic buckling occurs): At $\rho_1 = 0.286$: the determinant of $[S] = 107,805$; $C_{h(1)} = 2.38331$; $P_{cr} = 126.72^k$. At $\rho_1 = 0.287$: the determinant of $[S] = -2,792,560$.

EXAMPLE 16.12

Change the structure in Example 16.11 to a braced frame (Fig. 16.15a). Find P_{cr} for:

1. $F_y = 36$ ksi.
2. Elastic buckling.

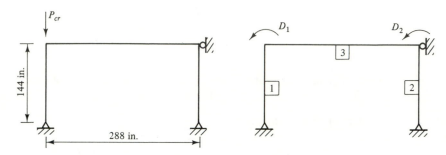

Figure 16.15a

Solution

$$[S] = \begin{bmatrix} \left(C_h \dfrac{EI}{L}\right)_1 + \left(4\dfrac{EI}{L}\right)_3 & \left(2\dfrac{EI}{L}\right)_3 \\[4mm] \left(2\dfrac{EI}{L}\right)_3 & \left(4\dfrac{EI}{L}\right)_3 + \left(3\dfrac{EI}{L}\right)_2 \end{bmatrix}$$

Determinant of $[S] = S_{1,1} * S_{2,2} - S_{1,2} * S_{2,1} = 0$ is the buckling criterion.

1. $F_y = 36$ ksi (inelastic buckling occurs) for $\rho_1 = 1.685$: determinant of $[S] = 1,189,450$; $C_{h(1)} = -8.31225$; $E_1 = 6260$ ksi; $P_{cr} = 1.685 * \dfrac{6260}{29,000} * 443.08 = 161.16^k$ for $\rho_1 = 1.686$: determinant of $[S] = -85,907$

2. Elastic buckling for $\rho_1 = 1.283$: determinant of $[S] = 47,424$; $C_{h(1)} = -1.79977$; $E_1 = 29,000$ ksi; $P_{cr} = 1.283 * 443.08 = 568.47^k$ for $\rho_1 = 1.284$: determinant of $[S] = -1,689,130$

A summary of buckling load(s) obtained in Examples 16.9, 16.11, and 16.12 is shown in Figure 16.15b. For elastic buckling, note that a single gravity load on only one column is equal to the sum of the gravity load when both columns are equally loaded provided the singly loaded column load does not exceed the braced frame load for the singly loaded column. This statement is true for different column sizes and for any beam size provided elastic buckling occurs.

For inelastic buckling, a single gravity load on only one column may be considerably less than the sum of the gravity load for both columns equally loaded provided the single column load does not exceed the braced frame load for the singly loaded column.

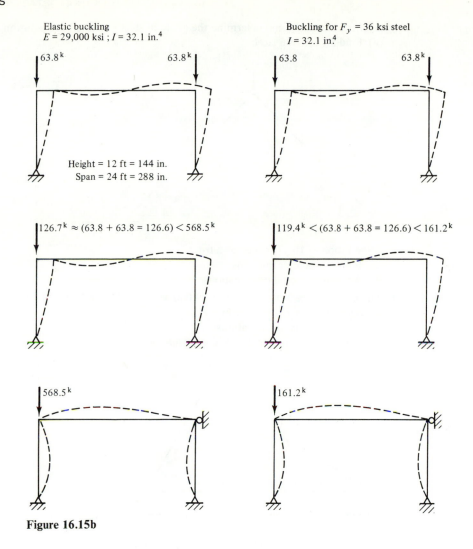

Elastic buckling
$E = 29{,}000$ ksi ; $I = 32.1$ in.^{4}

63.8^k 63.8^k

Height = 12 ft = 144 in.
Span = 24 ft = 288 in.

Buckling for $F_y = 36$ ksi steel
$I = 32.1$ in.^{4}

63.8 63.8^k

$126.7^k \approx (63.8 + 63.8 = 126.6) < 568.5^k$

$119.4^k < (63.8 + 63.8 = 126.6) < 161.2^k$

568.5^k

161.2^k

Figure 16.15b

PROBLEMS

Perform a noncomputerized stability analysis of the following problems. Ignore axial deformations of all members except in truss analyses.

16.1–16.15. See Problems 15.1 through 15.15 for the geometry, member properties, and loading. Find P_{cr} in each of these problems with all loads except P deleted. In Problem 15.12, load case 1, replace the distributed load with P at the eaves and $2P$ at the crown.

16.16. Solve Problem 16.15 with P_2 deleted.

16.17. Treat the structure as a plane frame and find P_{cr}. All members are W8 × 67 ($E = 29{,}000$ ksi; $I = 272$ in.^{4}; $A = 19.7$ in.^{2}; $L = 30$ ft).

(a) Find elastic P_{cr} and determine the grade of steel for which this solution is valid.

(b) Find P_{cr} for A36 steel.

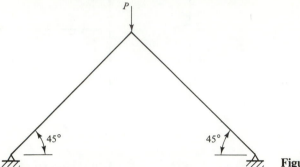

Figure P16.17

16.18. Solve Problem 16.17 treated as a truss.

(a) Ignore axial deformations.

(b) Account for axial deformations.

16.19. All members are W8 × 31 ($E = 29,000$ ksi; $I = 110$ in.^{4}; $A = 9.13$ in.^{2}). Treat the structure as a plane truss and find P_{cr}.

(a) Ignore all axial deformations.

(b) Ignore only compressive axial deformations.

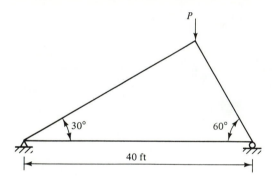

Figure P16.19

16.20. Delete the tension member in Problem 16.19 and change the right support to a hinge. Treat the structure as a plane frame and find P_{cr}.

16.21. Find P_{cr}.

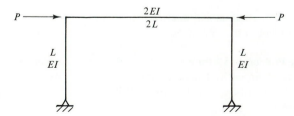

Figure P16.21

16.22. The cross-hatched structural parts are rigid bodies. All members have the same *EI*. Find P_{cr}.

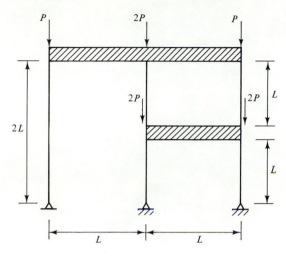

Figure P16.22

Areas and Centroids

The purpose of this appendix is to show the calculus derivations of the area and the centroid of the area beneath a curve, $y = ax^n$ where n is an integer and $n \geq 2$. See Figure A.1.

At $x = b$, $[y = ab^n] = h$; $a = h/b^n$. The shaded area, A_s, is

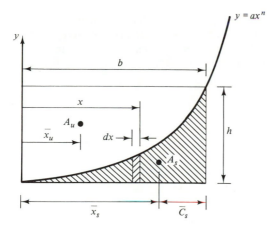

Figure A.1

$$A_s = \int dA = \int_0^b y\,dx = a\int_0^b x^n\,dx = \frac{h}{b^n}\left[\frac{x^{n+1}}{n+1}\right]_0^b = \frac{bh}{n+1}$$

The unshaded area, A_u, is

$$A_u = bh - A_s = \frac{n}{n+1}*(bh)$$

The centroid of the shaded area with respect to the y axis is

$$A_s \bar{x}_s = \int x\, dA = \int_0^b x(y\, dx) = a\int_0^b x^{n+1}\, dx = a\left[\frac{x^{n+2}}{n+2}\right]_0^b$$

$$\frac{bh}{n+1} * \bar{x}_s = \frac{h}{b^n}\left[\frac{b^{n+2}}{n+2}\right]$$

$$\bar{x}_s = \frac{n+1}{n+2} * b$$

It is easier to remember $\bar{C}_s$ than to remember $\bar{X}_s$; therefore, find $\bar{C}_s$:

$$\bar{C}_s = b - \bar{X}_s = \frac{(n+2)-(n+1)}{n+2} * b = \frac{b}{n+2}$$

Find $\bar{X}_u$:

$$A_u \bar{X}_u = (bh) * \frac{b}{2} - A_s \bar{X}_s$$

$$\frac{n*(bh)}{n+1} * \bar{X}_u = b^2 h\left[\frac{1}{2} - \frac{(n+1)}{(n+1)*(n+2)}\right]$$

$$\bar{X}_u = \frac{(n+1)}{(n+2)} * \frac{b}{2} = \frac{1}{2} * \bar{X}_s$$

Thus, the author memorizes

$$A_s = \frac{bh}{n+1}$$

$$\bar{C}_s = \frac{b}{n+2}$$

(actually these are valid
for $n \geq 1$ for a power curve
having a vertex at the origin:
$y = ax^n$)

$$\bar{X}_s = b - \bar{C}_s$$
$$A_u = (bh) - A_s$$

$$\bar{X}_u = \frac{1}{2} * \bar{X}_s$$

n	A_s	$\bar{C}_s$	$\bar{X}_s = b - \bar{C}_s$	$A_u = (bh) - A_s$	$\bar{X}_u = \dfrac{\bar{X}_s}{2}$
1	$\dfrac{bh}{2}$	$\dfrac{b}{3}$	$\dfrac{2b}{3}$	$\dfrac{bh}{2}$	$\dfrac{b}{3}$
2	$\dfrac{bh}{3}$	$\dfrac{b}{4}$	$\dfrac{3b}{4}$	$\dfrac{2bh}{3}$	$\dfrac{3b}{8}$
3	$\dfrac{bh}{4}$	$\dfrac{b}{5}$	$\dfrac{4b}{5}$	$\dfrac{3bh}{4}$	$\dfrac{2b}{5}$
4	$\dfrac{bh}{5}$	$\dfrac{b}{6}$	$\dfrac{5b}{6}$	$\dfrac{4bh}{5}$	$\dfrac{5b}{12}$

Plane Grid and Space Frame Member Stiffness Matrices and Transformation Matrices

B.1 PLANE GRID

A plane grid is a structural system for which:

1. All of the joints lie in a common plane (global XY plane, for example, is assumed in the following discussion).
2. The local axis direction and one of the principal axes of the cross section lie in the XY plane; the other principal axis is perpendicular to the XY plane.
3. Only loadings that are perpendicular to the XY plane are permitted; that is, no in-plane joint translations and joint rotations are permitted.

The member-end displacement and force subscript numbers for a typical, nonprismatic, plane grid member are shown in Figure B.1. The double headed arrowed vectors in Figure B.1 are rotational member-end displacements and forces that conform to the right-hand screw rule.

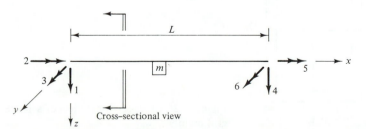

Figure B.1

The member stiffness matrix is

$$[S^M]_m = \begin{bmatrix} \dfrac{(C_2+2C_3+C_4)EI_y}{L^3} & 0 & \dfrac{-(C_2+C_3)EI_y}{L^2} & \dfrac{-(C_2+2C_3+C_4)EI_y}{L^3} & 0 & \dfrac{-(C_3+C_4)EI_y}{L^2} \\[2ex] 0 & \dfrac{C_1 GI_x}{L} & 0 & 0 & \dfrac{-C_1 GI_x}{L} & 0 \\[2ex] \dfrac{-(C_2+C_3)EI_y}{L^2} & 0 & \dfrac{C_2 EI_y}{L} & \dfrac{(C_2+C_3)EI_y}{L^2} & 0 & \dfrac{C_3 EI_y}{L} \\[2ex] \dfrac{-(C_2+2C_3+C_4)EI_y}{L^3} & 0 & \dfrac{(C_2+C_3)EI_y}{L^2} & \dfrac{(C_2+2C_3+C_4)EI_y}{L^3} & 0 & \dfrac{(C_3+C_4)EI_y}{L^2} \\[2ex] 0 & \dfrac{-C_1 GI_x}{L} & 0 & 0 & \dfrac{C_1 GI_x}{L} & 0 \\[2ex] \dfrac{-(C_3+C_4)EI_y}{L^2} & 0 & \dfrac{C_3 EI_y}{L} & \dfrac{(C_3+C_4)EI_y}{L^2} & 0 & \dfrac{C_4 EI_y}{L} \end{bmatrix}_m$$

where I_x = member's nonwarping torsional constant at origin of member (for a solid circular section, I_x is polar moment of inertia)

I_y = member's bending moment of inertia at origin of member C_1, C_2, C_3, C_4 are member section property coefficients to complete stiffness definitions (for a prismatic member: $C_1 = 1$; $C_2 = 4$; $C_3 = 2$; $C_4 = 4$)

$G = \dfrac{E}{2(1+\nu)}$ where ν is Poisson's ratio and E is modulus of elasticity

For a pipe section

$$I_x = 0.5\pi(r_o^4 - r_i^4) \text{ and torsional stress} = \frac{Tr}{I_x}$$

where r_o = outside diameter
r_i = inside diameter
$r_i \le r \le r_o$ and T is the torque

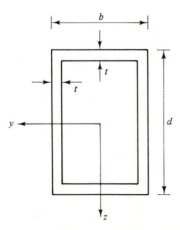

Figure B.2 Cross-sectional view of a plane grid member.

For a structural tube (hollow rectangular cross section)

$$I_x = \frac{2t(d-t)^2(b-t)^2}{d+b-2t}; \qquad \frac{\text{torsional stress}}{\text{(average)}} = \frac{T}{2t(d-t)(b-t)}$$

An example of a plane grid is shown in Figure B.3.

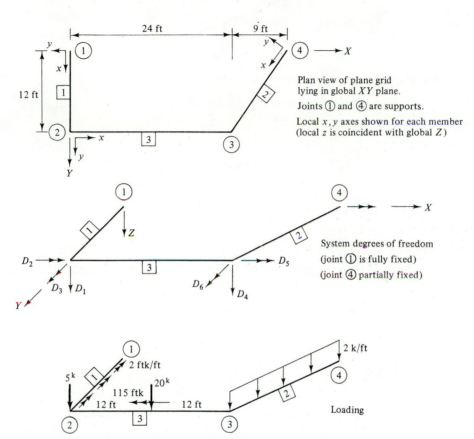

Figure B.3 Members are cold formed structural steel tubes: depth = 20 in.; width = 12 in.; and wall thickness = 0.5 in.; E = 29,000 ksi; Poisson's ratio = 0.25; sections are bending about the strong axis. I_x = 1622 in.^{4}; I_z = 1650 in.^{4}

The transformation matrix for a plane grid member is (member m is lying in XY plane in Fig. B.4):

$$[T]_m = \begin{bmatrix} 1 & 0 & 0 & 0 & 0 & 0 \\ 0 & \cos\theta & \sin\theta & 0 & 0 & 0 \\ 0 & -\sin\theta & \cos\theta & 0 & 0 & 0 \\ 0 & 0 & 0 & 1 & 0 & 0 \\ 0 & 0 & 0 & 0 & \cos\theta & \sin\theta \\ 0 & 0 & 0 & 0 & -\sin\theta & \cos\theta \end{bmatrix}_m$$

$$\{F\}_m = [T]_m \{F^G\}_m \quad \text{and} \quad [S^{MG}]_m = [T]_m^T [S^M]_m [T]_m$$

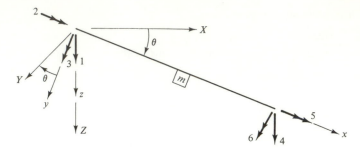

Local member–end displacement and force subscript numbers

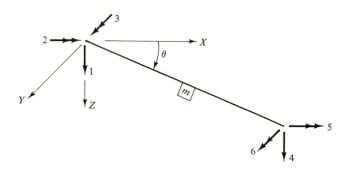

Global member–end displacement and force subscript numbers   **Figure B.4**

B.2 SPACE FRAME

The author chooses to restrict the discussion to a prismatic member for which the shear center and the centroid of the cross section coincide (this condition occurs for cross sections which are doubly symmetric). The local y and z axes are the principal axes of the cross section. The member stiffness matrix definition is made for the local x,y,z axes. To obtain the system stiffness matrix by summing up the contributions from each global member stiffness matrix, a transformation must be made.

For graphical convenience, consider a space frame member arbitrarily skewed in space such that it lies completely in the first quadrant of the global reference axes X_G, Y_G, and Z_G (Fig. B.5). Point i is at the origin of the member's x axis and point j is at the end of the member's x axis. Locations of points i and j are assumed to be known with respect to the global reference axes. For convenience in the notation, let X, Y, Z be axes located at points i and j and parallel to the X_G, Y_G, and Z_G axes, respectively.

The force-deformation relation for a prismatic space frame member with a doubly symmetric cross section is $\{F\} = [S^M]\{D^M\} + \{F^J\}$ and the member stiffness matrix is as shown below.

$$
\begin{bmatrix}
EA/L & 0 & 0 & 0 & 0 & 0 & -EA/L & 0 & 0 & 0 & 0 & 0 \\
0 & 12EI_z/L^3 & 0 & 0 & 0 & 6EI_z/L^2 & 0 & -12EI_z/L^3 & 0 & 0 & 0 & 6EI_z/L^2 \\
0 & 0 & 12EI_y/L^3 & 0 & -6EI_y/L^2 & 0 & 0 & 0 & -12EI_y/L^3 & 0 & -6EI_y/L^2 & 0 \\
0 & 0 & 0 & GI_x/L & 0 & 0 & 0 & 0 & 0 & -GI_x/L & 0 & 0 \\
0 & 0 & -6EI_y/L^2 & 0 & 4EI_y/L & 0 & 0 & 0 & 6EI_y/L^2 & 0 & 2EI_y/L & 0 \\
0 & 6EI_z/L^2 & 0 & 0 & 0 & 4EI_z/L & 0 & -6EI_z/L^2 & 0 & 0 & 0 & 2EI_z/L \\
-EA/L & 0 & 0 & 0 & 0 & 0 & EA/L & 0 & 0 & 0 & 0 & 0 \\
0 & -12EI_z/L^3 & 0 & 0 & 0 & -6EI_z/L^2 & 0 & 12EI_z/L^3 & 0 & 0 & 0 & -6EI_z/L^2 \\
0 & 0 & -12EI_y/L^3 & 0 & 6EI_y/L^2 & 0 & 0 & 0 & 12EI_y/L^3 & 0 & 6EI_y/L^2 & 0 \\
0 & 0 & 0 & -GI_x/L & 0 & 0 & 0 & 0 & 0 & GI_x/L & 0 & 0 \\
0 & 0 & -6EI_y/L^2 & 0 & 2EI_y/L & 0 & 0 & 0 & 6EI_y/L^2 & 0 & 4EI_y/L & 0 \\
0 & 6EI_z/L^2 & 0 & 0 & 0 & 2EI_z/L & 0 & -6EI_z/L^2 & 0 & 0 & 0 & 4EI_z/L
\end{bmatrix}
$$

where I_x is the nonwarping (or simple) torsional constant—for a solid or hollow circular section I_x is the polar moment of inertia.

The global member stiffness matrix is $[S^{MG}]_m = [T]_m^T \times [S^M]_m [T]_m$ where $[T]_m$ is the transformation matrix defined below.

At points i and j in Figure B.5, if $\hat{X}, \hat{Y}, \hat{Z}$ and $\hat{x}, \hat{y}, \hat{z}$ are defined as unit vectors along the X, Y, Z and x, y, z directions, respectively, then for example,

$$\hat{X} \cdot \hat{x} = \cos \alpha_x$$
$$\hat{Y} \cdot \hat{x} = \cos \beta_x$$
$$\hat{Z} \cdot \hat{x} = \cos \gamma_x$$

The preceding are examples of the dot product of unit vectors; similar quantities exist for the y and z axes (Fig. B.6).

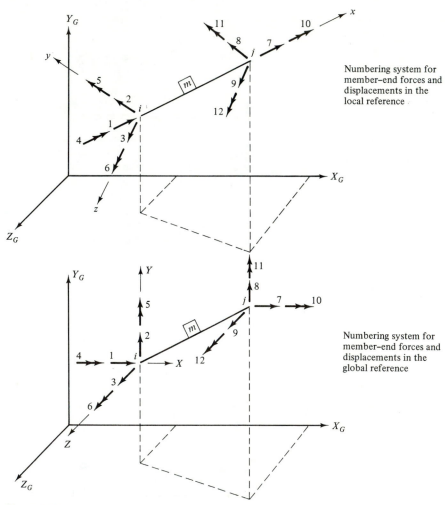

Numbering system for member-end forces and displacements in the local reference

Numbering system for member-end forces and displacements in the global reference

Figure B.5

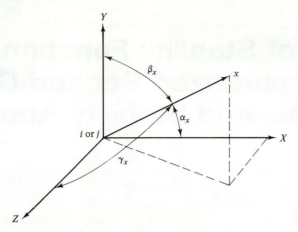

Figure B.6

Resolution of the single headed arrows at point i for the system reference along the local reference axes gives

$$F_1 = F_1^G \hat{X} \cdot \hat{x} + F_2^G \hat{Y} \cdot \hat{x} + F_3^G \hat{Z} \cdot \hat{x}$$
$$F_2 = F_1^G \hat{X} \cdot \hat{y} + F_2^G \hat{Y} \cdot \hat{y} + F_3^G \hat{Z} \cdot \hat{y}$$
$$F_3 = F_1^G \hat{X} \cdot \hat{z} + F_2^G \hat{Y} \cdot \hat{z} + F_3^G \hat{Z} \cdot \hat{z}$$

Let

$$[C_{Gm}] = \begin{bmatrix} \hat{X} \cdot \hat{x} & \hat{Y} \cdot \hat{x} & \hat{Z} \cdot \hat{x} \\ \hat{X} \cdot \hat{y} & \hat{Y} \cdot \hat{y} & \hat{Z} \cdot \hat{y} \\ \hat{X} \cdot \hat{z} & \hat{Y} \cdot \hat{z} & \hat{Z} \cdot \hat{z} \end{bmatrix} \quad \text{and} \quad [T] = \begin{bmatrix} [C_{Gm}] & [0] & [0] & [0] \\ [0] & [C_{Gm}] & [0] & [0] \\ [0] & [0] & [C_{Gm}] & [0] \\ [0] & [0] & [0] & [C_{Gm}] \end{bmatrix}$$

where [0] is a null matrix (all entries are zero).

Then, $\{F\} = [T]\{F^G\}$ and $\{F^G\} = [T]^T\{F\}$ since $[T]^T[T] = [I]$. $[T]$ is the transformation matrix for a space frame member. $[C_{Gm}]$ is a matrix of direction cosines for the angles between the global and member axes.

Table of Stability Functions for Noncomputerized Second Order Analysis and Stability Analysis

Definitions of the column headings for the following table are

$$\rho = \frac{P}{P_E}$$

where P = the axial force in the member
an axial compressive force is positive
an axial tensile force is negative

$$P_E = \frac{\pi^2 EI}{L^2}$$

$\phi = \pi\sqrt{\rho}$ when ρ is positive and $\phi = \pi\sqrt{-\rho}$ when ρ is negative.

C_2 is a term in the rotational member-end stiffness, $C_2(EI/L)$, required to induce a unit rotation at one end of the member when the other end is fixed against rotation.

C_3 is a term in the carry over moment, $C_3(EI/L)$, required at the fixed end when a unit rotation is induced at the other end of the member.

C_h is a term in the rotational member-end stiffness, $C_h(EI/L)$, required to induce a unit rotation at one end of the member when the other end is hinged (free to rotate).

C_u is a term in the fixed ended moments for a uniformly loaded member; that is, the fixed ended moments are $M_f = \pm C_u*(wL^2/12)$.

ρ	ϕ	C_2	C_3	$C_2 + C_3$	C_h	C_u
Axial tension						
−10.000	9.935	11.18638	1.25082	12.43720	11.04651	0.48242
−9.500	9.683	10.94300	1.25875	12.20175	10.79821	0.49173
−9.000	9.425	10.69368	1.26738	11.96106	10.54348	0.50163
−8.500	9.159	10.43798	1.27682	11.71480	10.28180	0.51217
−8.000	8.886	10.17540	1.28718	11.46258	10.01258	0.52344
−7.500	8.604	9.90537	1.29861	11.20398	9.73512	0.53552
−7.000	8.312	9.62724	1.31128	10.93852	9.44864	0.54852
−6.500	8.010	9.34026	1.32542	10.66568	9.15218	0.56255
−6.000	7.695	9.04358	1.34128	10.38486	8.84465	0.57776
−5.500	7.368	8.73619	1.35919	10.09538	8.52472	0.59433
−5.000	7.025	8.41690	1.37958	9.79648	8.19078	0.61247
−4.500	6.664	8.08431	1.40296	9.48727	7.84084	0.63243
−4.000	6.283	7.73672	1.43003	9.16675	7.47240	0.65454
−3.500	5.877	7.37208	1.46167	8.83375	7.08228	0.67921
−3.000	5.441	6.98784	1.49907	8.48691	6.66625	0.70697
−2.500	4.967	6.58078	1.54384	8.12462	6.21860	0.73850
−2.000	4.443	6.14682	1.59818	7.74499	5.73129	0.77469
−1.500	3.848	5.68057	1.66520	7.34577	5.19243	0.81680
−1.000	3.142	5.17479	1.74941	6.92421	4.58338	0.86653
−0.950	3.062	5.12168	1.75901	6.88070	4.51756	0.87200
−0.900	2.980	5.06807	1.76886	6.83693	4.45071	0.87759
−0.850	2.896	5.01395	1.77896	6.79290	4.38277	0.88327
−0.800	2.810	4.95929	1.78932	6.74861	4.31370	0.88907
−0.750	2.721	4.90409	1.79995	6.70404	4.24345	0.89498
−0.700	2.628	4.84833	1.81086	6.65920	4.17197	0.90101
−0.650	2.533	4.79201	1.82207	6.61407	4.09920	0.90716
−0.600	2.433	4.73509	1.83358	6.56867	4.02507	0.91343
−0.550	2.330	4.67758	1.84540	6.52297	3.94953	0.91983
−0.500	2.221	4.61944	1.85754	6.47699	3.87250	0.92636
−0.450	2.107	4.56067	1.87003	6.43070	3.79390	0.93302
−0.400	1.987	4.50125	1.88287	6.38412	3.71365	0.93983
−0.350	1.859	4.44116	1.89607	6.33723	3.63167	0.94679
−0.300	1.721	4.38038	1.90965	6.29003	3.54785	0.95389
−0.250	1.571	4.31888	1.92362	6.24251	3.46210	0.96115
−0.200	1.405	4.25666	1.93801	6.19467	3.37430	0.96857
−0.150	1.217	4.19368	1.95282	6.14650	3.28433	0.97616
−0.100	0.993	4.12993	1.96808	6.09801	3.19206	0.98393
−0.050	0.702	4.06538	1.98380	6.04918	3.09733	0.99187
Axial compression						
0.000	0.010	3.99999	2.00000	5.99999	2.99998	1.00000
0.005	0.222	3.99340	2.00165	5.99505	2.99010	1.00083
0.010	0.314	3.98681	2.00330	5.99011	2.98019	1.00165
0.015	0.385	3.98021	2.00496	5.98517	2.97025	1.00248
0.020	0.444	3.97360	2.00662	5.98022	2.96028	1.00331
0.025	0.497	3.96698	2.00829	5.97527	2.95028	1.00414
0.030	0.544	3.96035	2.00996	5.97032	2.94025	1.00497
0.035	0.588	3.95372	2.01164	5.96536	2.93020	1.00581
0.040	0.628	3.94707	2.01333	5.96040	2.92012	1.00664
0.045	0.667	3.94042	2.01501	5.95544	2.91000	1.00748

ρ	ϕ	C_2	C_3	$C_2 + C_3$	C_h	C_u
Axial compression						
0.050	0.703	3.93376	2.01671	5.95047	2.89986	1.00832
0.055	0.737	3.92709	2.01841	5.94550	2.88969	1.00917
0.060	0.770	3.92041	2.02011	5.94052	2.87949	1.01001
0.065	0.801	3.91372	2.02182	5.93554	2.86925	1.01086
0.070	0.831	3.90702	2.02353	5.93056	2.85899	1.01171
0.075	0.860	3.90032	2.02525	5.92557	2.84870	1.01256
0.080	0.889	3.89360	2.02698	5.92058	2.83837	1.01341
0.085	0.916	3.88688	2.02871	5.91559	2.82802	1.01427
0.090	0.943	3.88015	2.03045	5.91059	2.81763	1.01513
0.095	0.968	3.87341	2.03219	5.90559	2.80721	1.01599
0.100	0.994	3.86665	2.03394	5.90059	2.79676	1.01685
0.105	1.018	3.85989	2.03569	5.89558	2.78628	1.01771
0.110	1.042	3.85313	2.03745	5.89057	2.77577	1.01858
0.115	1.065	3.84635	2.03921	5.88556	2.76522	1.01944
0.120	1.088	3.83956	2.04098	5.88054	2.75465	1.02031
0.125	1.111	3.83276	2.04275	5.87552	2.74403	1.02119
0.130	1.133	3.82596	2.04453	5.87049	2.73339	1.02206
0.135	1.154	3.81914	2.04632	5.86546	2.72271	1.02294
0.140	1.176	3.81232	2.04811	5.86043	2.71200	1.02382
0.145	1.196	3.80549	2.04991	5.85539	2.70126	1.02470
0.150	1.217	3.79864	2.05171	5.85035	2.69048	1.02558
0.155	1.237	3.79179	2.05352	5.84531	2.67967	1.02646
0.160	1.257	3.78493	2.05533	5.84026	2.66882	1.02735
0.165	1.276	3.77806	2.05715	5.83521	2.65794	1.02824
0.170	1.295	3.77118	2.05898	5.83016	2.64702	1.02913
0.175	1.314	3.76429	2.06081	5.82510	2.63607	1.03003
0.180	1.333	3.75739	2.06265	5.82004	2.62508	1.03092
0.185	1.351	3.75048	2.06449	5.81497	2.61406	1.03182
0.190	1.369	3.74356	2.06634	5.80990	2.60300	1.03272
0.195	1.387	3.73663	2.06820	5.80483	2.59190	1.03362
0.200	1.405	3.72969	2.07006	5.79975	2.58077	1.03453
0.205	1.422	3.72275	2.07192	5.79467	2.56960	1.03543
0.210	1.440	3.71579	2.07380	5.78959	2.55840	1.03634
0.215	1.457	3.70882	2.07568	5.78450	2.54715	1.03725
0.220	1.474	3.70185	2.07756	5.77941	2.53587	1.03817
0.225	1.490	3.69486	2.07945	5.77431	2.52455	1.03908
0.230	1.507	3.68786	2.08135	5.76921	2.51319	1.04000
0.235	1.523	3.68086	2.08325	5.76411	2.50180	1.04092
0.240	1.539	3.67384	2.08516	5.75900	2.49036	1.04185
0.245	1.555	3.66681	2.08708	5.75389	2.47889	1.04277
0.250	1.571	3.65978	2.08900	5.74878	2.46738	1.04370
0.255	1.586	3.65273	2.09093	5.74366	2.45583	1.04463
0.260	1.602	3.64568	2.09286	5.73854	2.44423	1.04556
0.265	1.617	3.63861	2.09480	5.73341	2.43260	1.04650
0.270	1.632	3.63153	2.09675	5.72828	2.42093	1.04743
0.275	1.647	3.62445	2.09870	5.72315	2.40921	1.04837
0.280	1.662	3.61735	2.10066	5.71801	2.39746	1.04932
0.285	1.677	3.61024	2.10263	5.71287	2.38566	1.05026
0.290	1.692	3.60313	2.10460	5.70773	2.37382	1.05121
0.295	1.706	3.59600	2.10658	5.70258	2.36194	1.05216
0.300	1.721	3.58886	2.10857	5.69743	2.35001	1.05311

ρ	ϕ	C_2	C_3	$C_2 + C_3$	C_h	C_u
Axial compression						
0.305	1.735	3.58171	2.11056	5.69227	2.33805	1.05406
0.310	1.749	3.57455	2.11256	5.68711	2.32604	1.05502
0.315	1.763	3.56738	2.11456	5.68195	2.31398	1.05598
0.320	1.777	3.56021	2.11657	5.67678	2.30189	1.05694
0.325	1.791	3.55302	2.11859	5.67161	2.28974	1.05790
0.330	1.805	3.54581	2.12062	5.66643	2.27756	1.05887
0.335	1.818	3.53860	2.12265	5.66125	2.26532	1.05984
0.340	1.832	3.53138	2.12469	5.65607	2.25305	1.06081
0.345	1.845	3.52415	2.12673	5.65088	2.24072	1.06178
0.350	1.859	3.51691	2.12878	5.64569	2.22836	1.06276
0.355	1.872	3.50965	2.13084	5.64049	2.21594	1.06374
0.360	1.885	3.50239	2.13291	5.63529	2.20348	1.06472
0.365	1.898	3.49511	2.13498	5.63009	2.19097	1.06570
0.370	1.911	3.48783	2.13706	5.62488	2.17841	1.06669
0.375	1.924	3.48053	2.13914	5.61967	2.16580	1.06768
0.380	1.937	3.47322	2.14124	5.61446	2.15315	1.06867
0.385	1.949	3.46590	2.14334	5.60924	2.14045	1.06966
0.390	1.962	3.45857	2.14544	5.60402	2.12770	1.07066
0.395	1.974	3.45123	2.14756	5.59879	2.11489	1.07166
0.400	1.987	3.44388	2.14968	5.59356	2.10204	1.07266
0.405	1.999	3.43652	2.15181	5.58833	2.08914	1.07367
0.410	2.012	3.42914	2.15394	5.58309	2.07619	1.07467
0.415	2.024	3.42176	2.15609	5.57784	2.06318	1.07568
0.420	2.036	3.41436	2.15824	5.57260	2.05012	1.07670
0.425	2.048	3.40695	2.16040	5.56735	2.03702	1.07771
0.430	2.060	3.39953	2.16256	5.56209	2.02385	1.07873
0.435	2.072	3.39210	2.16473	5.55683	2.01064	1.07975
0.440	2.084	3.38466	2.16691	5.55157	1.99737	1.08078
0.445	2.096	3.37721	2.16910	5.54630	1.98405	1.08180
0.450	2.107	3.36974	2.17129	5.54103	1.97067	1.08283
0.455	2.119	3.36226	2.17349	5.53576	1.95724	1.08386
0.460	2.131	3.35478	2.17570	5.53048	1.94375	1.08490
0.465	2.142	3.34728	2.17792	5.52520	1.93020	1.08593
0.470	2.154	3.33977	2.18014	5.51991	1.91660	1.08697
0.475	2.165	3.33224	2.18238	5.51462	1.90295	1.08802
0.480	2.177	3.32471	2.18462	5.50932	1.88923	1.08906
0.485	2.188	3.31716	2.18686	5.50402	1.87546	1.09011
0.490	2.199	3.30960	2.18912	5.49872	1.86162	1.09116
0.495	2.210	3.30203	2.19138	5.49341	1.84773	1.09222
0.500	2.221	3.29445	2.19365	5.48810	1.83378	1.09327
0.505	2.233	3.28686	2.19593	5.48278	1.81977	1.09433
0.510	2.244	3.27925	2.19821	5.47746	1.80570	1.09540
0.515	2.255	3.27163	2.20051	5.47214	1.79156	1.09646
0.520	2.265	3.26400	2.20281	5.46681	1.77737	1.09753
0.525	2.276	3.25636	2.20512	5.46148	1.76311	1.09860
0.530	2.287	3.24870	2.20744	5.45614	1.74879	1.09968
0.535	2.298	3.24104	2.20976	5.45080	1.73440	1.10076
0.540	2.309	3.23336	2.21210	5.44546	1.71995	1.10184
0.545	2.319	3.22567	2.21444	5.44011	1.70544	1.10292
0.550	2.330	3.21796	2.21679	5.43476	1.69086	1.10401
0.555	2.340	3.21025	2.21915	5.42940	1.67622	1.10509

ρ	ϕ	C_2	C_3	$C_2 + C_3$	C_h	C_u
Axial compression						
0.560	2.351	3.20252	2.22152	5.42404	1.66150	1.10619
0.565	2.361	3.19478	2.22389	5.41867	1.64672	1.10728
0.570	2.372	3.18703	2.22628	5.41330	1.63188	1.10838
0.575	2.382	3.17926	2.22867	5.40793	1.61696	1.10948
0.580	2.393	3.17148	2.23107	5.40255	1.60197	1.11059
0.585	2.403	3.16369	2.23348	5.39716	1.58692	1.11169
0.590	2.413	3.15588	2.23589	5.39178	1.57179	1.11281
0.595	2.423	3.14807	2.23832	5.38639	1.55659	1.11392
0.600	2.433	3.14024	2.24075	5.38099	1.54132	1.11504
0.605	2.444	3.13239	2.24320	5.37559	1.52598	1.11616
0.610	2.454	3.12454	2.24565	5.37019	1.51056	1.11728
0.615	2.464	3.11667	2.24811	5.36478	1.49507	1.11841
0.620	2.474	3.10879	2.25058	5.35936	1.47951	1.11954
0.625	2.484	3.10089	2.25305	5.35395	1.46387	1.12067
0.630	2.494	3.09299	2.25554	5.34853	1.44815	1.12180
0.635	2.503	3.08507	2.25804	5.34310	1.43235	1.12294
0.640	2.513	3.07713	2.26054	5.33767	1.41648	1.12409
0.645	2.523	3.06918	2.26305	5.33224	1.40053	1.12523
0.650	2.533	3.06122	2.26557	5.32680	1.38450	1.12638
0.655	2.543	3.05325	2.26811	5.32135	1.36839	1.12753
0.660	2.552	3.04526	2.27065	5.31591	1.35219	1.12869
0.665	2.562	3.03726	2.27320	5.31046	1.33592	1.12985
0.670	2.572	3.02924	2.27575	5.30500	1.31956	1.13101
0.675	2.581	3.02122	2.27832	5.29954	1.30312	1.13217
0.680	2.591	3.01317	2.28090	5.29407	1.28659	1.13334
0.685	2.600	3.00512	2.28349	5.28860	1.26998	1.13451
0.690	2.610	2.99705	2.28608	5.28313	1.25328	1.13569
0.695	2.619	2.98897	2.28869	5.27765	1.23649	1.13687
0.700	2.628	2.98087	2.29130	5.27217	1.21962	1.13805
0.705	2.638	2.97276	2.29392	5.26668	1.20265	1.13924
0.710	2.647	2.96463	2.29656	5.26119	1.18560	1.14043
0.715	2.656	2.95649	2.29920	5.25569	1.16845	1.14162
0.720	2.666	2.94834	2.30185	5.25019	1.15122	1.14281
0.725	2.675	2.94017	2.30452	5.24469	1.13388	1.14401
0.730	2.684	2.93199	2.30719	5.23918	1.11646	1.14522
0.735	2.693	2.92379	2.30987	5.23367	1.09894	1.14642
0.740	2.703	2.91558	2.31256	5.22815	1.08132	1.14763
0.745	2.712	2.90736	2.31526	5.22262	1.06361	1.14885
0.750	2.721	2.89912	2.31798	5.21710	1.04580	1.15006
0.755	2.730	2.89087	2.32070	5.21157	1.02789	1.15129
0.760	2.739	2.88260	2.32343	5.20603	1.00987	1.15251
0.765	2.748	2.87432	2.32617	5.20049	0.99176	1.15374
0.770	2.757	2.86602	2.32892	5.19494	0.97354	1.15497
0.775	2.766	2.85771	2.33168	5.18939	0.95522	1.15620
0.780	2.775	2.84938	2.33446	5.18384	0.93679	1.15744
0.785	2.783	2.84104	2.33724	5.17828	0.91826	1.15869
0.790	2.792	2.83268	2.34003	5.17271	0.89962	1.15993
0.795	2.801	2.82431	2.34284	5.16715	0.88087	1.16118
0.800	2.810	2.81592	2.34565	5.16157	0.86201	1.16244
0.805	2.819	2.80752	2.34847	5.15600	0.84304	1.16369
0.810	2.827	2.79911	2.35131	5.15041	0.82396	1.16495

ρ	ϕ	C_2	C_3	$C_2 + C_3$	C_h	C_u
Axial compression						
0.815	2.836	2.79067	2.35415	5.14483	0.80476	1.16622
0.820	2.845	2.78223	2.35701	5.13924	0.78545	1.16749
0.825	2.854	2.77376	2.35988	5.13364	0.76602	1.16876
0.830	2.862	2.76529	2.36275	5.12804	0.74647	1.17004
0.835	2.871	2.75679	2.36564	5.12243	0.72680	1.17132
0.840	2.879	2.74828	2.36854	5.11682	0.70702	1.17260
0.845	2.888	2.73976	2.37145	5.11121	0.68711	1.17389
0.850	2.896	2.73122	2.37437	5.10559	0.66707	1.17518
0.855	2.905	2.72266	2.37730	5.09997	0.64691	1.17648
0.860	2.913	2.71409	2.38025	5.09434	0.62662	1.17778
0.865	2.922	2.70550	2.38320	5.08870	0.60621	1.17908
0.870	2.930	2.69690	2.38617	5.08307	0.58566	1.18039
0.875	2.939	2.68828	2.38914	5.07742	0.56499	1.18170
0.880	2.947	2.67964	2.39213	5.07178	0.54418	1.18302
0.885	2.955	2.67099	2.39513	5.06612	0.52323	1.18434
0.890	2.964	2.66232	2.39814	5.06047	0.50215	1.18566
0.895	2.972	2.65364	2.40116	5.05480	0.48093	1.18699
0.900	2.980	2.64494	2.40420	5.04914	0.45957	1.18832
0.905	2.989	2.63622	2.40724	5.04347	0.43807	1.18966
0.910	2.997	2.62749	2.41030	5.03779	0.41642	1.19100
0.915	3.005	2.61874	2.41337	5.03211	0.39463	1.19234
0.920	3.013	2.60997	2.41645	5.02642	0.37269	1.19369
0.925	3.022	2.60119	2.41954	5.02073	0.35060	1.19504
0.930	3.030	2.59239	2.42265	5.01504	0.32836	1.19640
0.935	3.038	2.58357	2.42577	5.00934	0.30597	1.19776
0.940	3.046	2.57474	2.42890	5.00363	0.28342	1.19913
0.945	3.054	2.56589	2.43204	4.99792	0.26072	1.20050
0.950	3.062	2.55702	2.43519	4.99221	0.23785	1.20187
0.955	3.070	2.54813	2.43836	4.98649	0.21483	1.20325
0.960	2.078	2.53923	2.44153	4.98076	0.19164	1.20463
0.965	3.086	2.53031	2.44472	4.97503	0.16828	1.20602
0.970	3.094	2.52137	2.44793	4.96930	0.14476	1.20741
0.975	3.102	2.51242	2.45114	4.96356	0.12106	1.20881
0.980	3.110	2.50345	2.45437	4.95782	0.09719	1.21021
0.985	3.118	2.49446	2.45761	4.95207	0.07315	1.21162
0.990	3.126	2.48545	2.46086	4.94631	0.04893	1.21302
0.995	3.134	2.47643	2.46413	4.94055	0.02453	1.21444
1.000	3.142	.2.46740	2.46740	4.93480	-0.00000	1.21585
1.050	3.219	2.37598	2.50091	4.87689	-0.25644	1.23029
1.100	3.295	2.28268	2.53579	4.81847	-0.53430	1.24521
1.150	3.369	2.18742	2.57212	4.75954	-0.83705	1.26063
1.200	3.441	2.09011	2.60997	4.70008	-1.16902	1.27658
1.250	3.512	1.99064	2.64944	4.64008	-1.53561	1.29308
1.300	3.582	1.88892	2.69062	4.57953	-1.94366	1.31018
1.350	3.650	1.78481	2.73362	4.51843	-2.40199	1.32789
1.400	3.717	1.67821	2.77855	4.45676	-2.92211	1.34627
1.450	3.783	1.56898	2.82553	4.39451	-3.51943	1.36534
1.500	3.848	1.45697	2.87469	4.33166	-4.21497	1.38515
1.550	3.911	1.34203	2.92618	4.26821	-5.03824	1.40574
1.600	3.974	1.22399	2.98014	4.20414	-6.03198	1.42717
1.650	4.035	1.10268	3.03675	4.13943	-7.26049	1.44947

ρ	ϕ	C_2	C_3	$C_2 + C_3$	C_h	C_u
Axial compression						
1.700	4.096	0.97789	3.09620	4.07408	−8.82531	1.47272
1.750	4.156	0.84941	3.15866	4.00807	−10.89656	1.49698
1.800	4.215	0.71701	3.22438	3.94139	−13.78285	1.52230
1.850	4.273	0.58045	3.29358	3.87402	−18.10804	1.54878
1.900	4.330	0.43943	3.36652	3.80595	−25.35205	1.57648
1.950	4.387	0.29365	3.44350	3.73715	−40.08598	1.60550
2.000	4.443	0.14280	3.52482	3.66762	−86.86440	1.63594
2.050	4.498	−0.01351	3.61085	3.59733	964.86257	1.66790
2.100	4.553	−0.17568	3.70196	3.52628	77.83285	1.70151
2.150	4.606	−0.34415	3.79858	3.45443	41.58303	1.73690
2.200	4.660	−0.51942	3.90120	3.38178	28.78126	1.77421
2.250	4.712	−0.70204	4.01035	3.30830	22.20661	1.81362
2.300	4.764	−0.89264	4.12661	3.23398	18.18451	1.85530
2.350	4.816	−1.09189	4.25067	3.15879	15.45582	1.89946
2.400	4.867	−1.30057	4.38328	3.08271	13.47226	1.94634
2.450	4.917	−1.51956	4.52528	3.00572	11.95678	1.99620
2.500	4.967	−1.74986	4.67765	2.92780	10.75427	2.04932
2.550	5.017	−1.99257	4.84149	2.84892	9.77113	2.10606
2.600	5.066	−2.24899	5.01805	2.76906	8.94751	2.16680
2.650	5.144	−2.52059	5.20879	2.68819	8.24333	2.23198
2.700	5.162	−2.80907	5.41537	2.60630	7.63076	2.30212
2.750	5.210	−3.11640	5.63974	2.52334	7.08983	2.37780
2.800	5.257	−3.44486	5.88416	2.43930	6.60587	2.45972
2.850	5.304	−3.79716	6.15130	2.35414	6.16780	2.54870
2.900	5.350	−4.17646	6.44430	2.26784	5.76713	2.64569
2.950	5.396	−4.58656	6.76691	2.18036	5.39721	2.75184
3.000	5.441	−5.03198	7.12364	2.09167	5.05278	2.86853
3.050	5.487	−5.51820	7.51993	2.00173	4.72959	2.99741
3.100	5.531	−6.05195	7.96246	1.91052	4.42415	3.14051
3.150	5.576	−6.64147	8.45946	1.81799	4.13361	3.30036
3.200	5.620	−7.29713	9.02123	1.72410	3.85556	3.48007
3.250	5.664	−8.03200	9.66083	1.62883	3.58797	3.68363
3.300	5.707	−8.86293	10.39504	1.53212	3.32909	3.91615
3.350	5.750	−9.81190	11.24583	1.43393	3.07742	4.18430
3.400	5.793	−10.90824	12.24247	1.33422	2.83164	4.49700
3.450	5.835	−12.19195	13.42490	1.23295	2.59058	4.86639
3.500	5.877	−13.71898	14.84904	1.13006	2.35320	5.30947
3.550	5.919	−15.57018	16.59568	1.02550	2.11854	5.85080
3.600	5.961	−17.86678	18.78601	0.91922	1.88574	6.52724
3.650	6.002	−20.79929	21.61046	0.81117	1.65398	7.39668
3.700	6.043	−24.68519	25.38649	0.70129	1.42251	8.55562
3.750	6.084	−30.09596	30.68548	0.58952	1.19059	10.17774
3.800	6.124	−38.17446	38.65026	0.47580	0.95752	12.61044
3.850	6.164	−51.58742	51.94747	0.36005	0.72262	16.66429
3.900	6.204	−78.33486	78.57708	0.24222	0.48519	24.77099
3.950	6.244	−158.41682	158.53904	0.12223	0.24455	49.08909

References

CHAPTER 5

5.1. Muller-Breslau, Heinrich F.B., *Die Neueren Methoden der Festigkeitslehre und der Statik der Baukonstruktionen,* Baumgartner's Buchhandlung, Leipzig, 1886, pp. 33–40.

CHAPTER 6

6.1. Galambos, Theodore V., and Ellingwood, Bruce, "Serviceability Limit States: Deflection," *Journal of the Structural Division,* ASCE, Vol. 112, No. 1, January 1986, pp. 68–84.

6.2. Borg, S.F., and Gennaro, J.J., *Advanced Structural Analysis,* Princeton, NJ: D. Van Nostrand Company, Inc., 1959, p. 5.

6.3. Borg, S.F., and Gennaro, J.J., *Advanced Structural Analysis,* Princeton, NJ: D. Van Nostrand Company, Inc., 1959, pp. 47–51.

6.4. Betti, E., *Il Nuovo Cimento* (2), Vols. 7 and 8, 1872.

6.5. Maxwell, J.C., "On the Calculations of the Equilibrium and Stiffness of Frames," *Philosophical Magazine* (4), 27(1864), pp. 294–299.

CHAPTER 7

7.1. Maxwell, J.C., "On the Calculations of the Equilibrium and Stiffness of Frames," *Philosophical Magazine* (4), 27(1864), pp. 294–299.

7.2. Mohr, Otto, "Beitrag zur Theorie des Fachwerks," *Zeitschrift des Architekten-Ingenieur-Vereins zu Hannover,* 20(1874), pp. 509–526.

7.3. Beaufait, Fred W., *Basic Concepts of Structural Analysis,* Englewood Cliffs, NJ: Prentice-Hall, 1977, pp. 310–374.

7.4. Meyers, V. James, *Matrix Analysis of Structures,* New York: Harper & Row, 1983, pp. 12–88, 277–340.

7.5. Meek, John L., *Matrix Structural Analysis,* New York: McGraw-Hill, 1971, pp. 224–329.

7.6. Vanderbilt, M. Daniel, *Matrix Structural Analysis,* New York: Quantum, 1974, pp. 69–144.

CHAPTER 8

8.1. Manderla, Heinrich, "Die Berechnung der Sekundarspannungen, welche im einfachen Fachwerke in Folge starrer Knotenverbindungen auftreten," *Forster's Bauzeitung,* 45(1880), p. 34.

8.2. Mohr, Otto, "Beitrag zur Theorie des Fachwerks," *Zeitschrift des Architekten-Ingenieur-Vereins zu Hannover,* 20(1874), pp. 509–526.

8.3. Bendixen, Axel, *Die Methode der Alpha-Gleichungen zur Berechnung von Rahmenkonstruktionen,* Berlin, 1914.

8.4. Maney, George A., *Studies in Engineering,* No. 1, University of Minnesota, March 1915.

8.5. Cross, Hardy, "Analysis of Continuous Frames by Distributing Fixed-End Moments," *Proceedings ASCE,* 56(May 1930), pp. 919–928.

CHAPTER 10

10.1. Cross, Hardy, "Analysis of Continuous Frames by Distributing Fixed-End Moments," *Proceedings ASCE,* 56(May 1930), pp. 919–928.

10.2. Morris, C.T., "Morris on Analysis of Continuous Frames," *Transactions of the American Society of Civil Engineers,* 96(1932), pp. 66–69.

CHAPTER 12

12.1. Sutherland, H., and Bowman, H.L., *Structural Theory,* John Wiley & Sons, New York, 1950, pp. 295–301.

12.2. Norris, C.H., and Wilbur, J.B., *Elementary Structural Analysis,* New York: McGraw-Hill, 1960, Second Edition, pp. 289–317.

12.3. *Manual of Steel Construction,* American Institute of Steel Construction, Inc., Chicago, Eighth Edition, 1980, pp. 2-117 through 2-127.

12.4. *Building Code Requirements for Reinforced Concrete* (ACI 318-83), American Concrete Institute, Detroit, 1983, pp. 28–29.

CHAPTER 13

13.1. Wilson, E.L., *CAL 78 User Information Manual,* Report No. UC SESM 79-1, Department of Civil Engineering, University of California, Berkeley, California, November 1978.

CHAPTER 14

7.6. Vanderbilt, M. Daniel, *Matrix Structural Analysis,* New York: Quantum, 1974, pp. 23–31.

CHAPTER 15

15.1. Goldberg, John E., "Stiffness Charts for Gusseted Members Under Axial Load," *Transactions of the American Society of Civil Engineers,* 119 (1954), pp. 43–59.
15.2. Goldberg, John E., "Buckling of One-Story Frames and Buildings," *Journal of the Structural Division, ASCE,* October 1960, pp. 53–85.
15.3. Livesley, R. K., and Chandler, D. B., *Stability Functions for Structural Frameworks,* Manchester University Press, 1956.
15.4. *Manual of Steel Construction,* American Institute of Steel Construction, Inc., Chicago, Eighth Edition, 1980, pp. 5–26, Eqn(1.6-1a).

CHAPTER 16

16.1. Euler, Leonhard, "Sur la force de colonnes," *Memoires de l'Academie de Berlin,* 1759.
16.2. Smith, Jr., C.V., "On Inelastic Column Buckling," *Engineering Journal,* American Institute of Steel Construction, Vol. 13, No. 3, 1976, pp. 86–88.

Answers to Selected Problems*

2.2. $A_X = 0,\quad A_Y = 88.75,\quad B_Y = 6.25$

2.3. $A_X = -80,\quad A_Y = 92.25,\quad B_Y = 187.75$

2.4. $A_X = 50,\quad A_Y = 200$

2.5. $B_Y = 19$

2.7. $A_Y = 221.25,\quad C_Y = 156.25$

2.8. $B_Y = 9.5,\quad C_Y = 4$

2.9. $A_Y = 11,\quad M_A = 44$ (CCW)

2.10. $A_X = 190,\quad A_Y = 120$

2.11. $A_X = 123.5,\quad A_Y = 164.7$

2.12. $A_X = A_Y = \frac{10}{7}$

2.13. $A_X = 90$

2.14. $B_X = -8.75,\quad B_Y = -49.92$

2.15. $A_X = -19.33,\quad A_Y = 35.5$

2.16. $A_X = -6,\quad A_Y = 21.75,\quad M_A = 235$ (CCW)

2.17. $B_X = 237.47,\quad B_Y = 178.11$

* Partial answers given in some problems.

CHAPTER 3

3.11. $F_1 = 30\sqrt{5}, \quad F_2 = -10\sqrt{5}, \quad F_3 = -40$

3.12. $F_1 = 6.25, \quad F_3 = -45, \quad F_4 = -25$

3.13. $F_1 = \dfrac{-5\sqrt{13}}{6}, \quad F_2 = \dfrac{-35\sqrt{5}}{3}, \quad F_3 = 35$

CHAPTER 4

Values given are for the maximum absolute shear and moment ordinates.

4.7. $V = 76, \quad M = -380$

4.10. $V = 40, \quad M = 500$

4.14. $V = 142.75, \quad M = 1376.25$

CHAPTER 5

5.1. IL for V_2: ordinate at dx to left of point 2 is -0.5
ordinate at dx to right of point 2 is 0.5
ordinate at point 4 is -0.25

IL for $V_{3(\text{left})}$: ordinate at dx to left of point 3 is -1
ordinate at point 4 is -0.25

IL for $V_{3(\text{right})}$: ordinate at dx to right of point 3 is 1
ordinate at point 4 is 1

IL for M_2: ordinate at point 2 is 10
ordinate at point 4 is -5

IL for M_3: ordinate at point 4 is -10

5.4. IL for M_3: ordinate at point 2 is -10
other ordinates are zero

IL for M_4: ordinate at 2 is -5
ordinate at 4 is 7.5

IL for $V_{3(\text{left})}$: ordinate at point 2 is -1
ordinate at dx to left of point 3 is -1
other ordinates are zero

IL for $V_{3(\text{right})}$: ordinate at point 2 is 0.333

5.7. IL for $V_{2(\text{right})}$: ordinate at point 1 is 0.5
ordinate at point 3 is 0.5

IL for M_3: ordinate at point 1 is -10
ordinate at point 3 is 10

5.8. IL for A_Y: ordinate at point 2 is 0.833
ordinates at points 1 and 5 are zero

IL for M_C: ordinate at point 3 is 15

IL for V_{B-C}: ordinate at point 2 is -0.167
ordinate at point 3 is 0.5

5.10. IL for A_Y: ordinate at point 1 is 0.6
ordinate at point 4 is −0.2
IL for M_B: ordinate at point 1 is 6
ordinate at point 4 is −2
IL for V_{A-B}: ordinate at point 1 is 0.6
ordinate at point 4 is −0.2

5.11. IL for A_Y: ordinates at point 3 are 2.67 and 1
ordinate at point 6 is −0.333
IL for B_Y: ordinates at point 3 are −0.67 and 0
ordinate at point 6 is 1.33

5.12. IL for A_Y: ordinate at point 1 is 1.33
ordinates at point 7 are −0.167 and 0
IL for M_A: ordinate at 1 is −6.67
ordinate at 3 is −13.33
ordinates at 7 are 3.33 and 0
IL for V_{F-G}: ordinate at point 1 is −0.667
ordinate at point 3 is 0.667
ordinates at point 7 are −0.167 and 0.5
ordinate at point 9 is −1

CHAPTER 6

6.1. $v_A = \dfrac{-1100.25}{EI}$; $\theta_A = \dfrac{171}{EI}$

6.2. $v_A = \dfrac{-491.4}{EI}$; $\theta_A = \dfrac{74.25}{EI}$

6.3. $v_A = \dfrac{-437.4}{EI}$; $\theta_A = \dfrac{60.75}{EI}$

6.4. $v_A = \dfrac{-1202.85}{EI}$; $\theta_A = \dfrac{182.25}{EI}$

6.5. $\theta_A = \dfrac{-45}{EI}$; $\theta_B = \dfrac{90}{EI}$; $\theta_C = \dfrac{-90}{EI}$; $v_C = \dfrac{-180}{EI}$

6.6. $v_C = \dfrac{-5062.5}{EI}$

6.10. $\theta_C = 0.0013596$ rad; $v_C = -0.403$ in.

6.11. $v_C = -0.726$ in.

6.14. $v_B = -0.590$ in.; $\theta_{BA} = 0.00615$ rad; $\theta_{BC} = 0.00835$ rad; $v_E = -0.108$ in.

6.16. $v = -9$ in.

6.17. $\theta_A = -0.0207$ rad; $u_A = -0.3$ in.; $v_A = -1.32$ in.

6.18. $u_B = 0.484$ in.; $v_C = -0.0583$ in.; $u_E = 0.536$ in.

6.19. $u_D = \dfrac{25,360}{EI}$

6.20. $u_1 = -0.426$ in.; $u_2 = -0.004$ in.; $v_3 = -1.417$ in.

6.26. $v_3 = -0.200$ in.; $u_6 = -0.0500$ in.

6.27. $u_4 = 0.484$ in.

6.30. $v_2 = -0.205$ in.

CHAPTER 7

In obtaining the following answers, positive redundants:

1. Acted in the $+X$ and $+Y$ directions if the redundants were reactions.
2. Produced tension in the top of a beam if the redundant was a bending moment.
3. Produced tension in a truss member, cable, etc. if the redundant was an axial force in a member.

7.1. See Example 7.1

7.2. See Example 7.2

7.3. See Example 7.3

7.4. See Example 7.4

7.5. $Y_2 = 688.6$ kN

7.6. $M_2 = 509.6$ mkN

7.7. $M_2 = 292.0$ ftk

7.8. $M_2 = 38.4$ ftk

7.10. $Y_4 = 26.67$ kN

7.15. $M_2 = 115.7$ mkN; $M_3 = 77.1$ mkN

7.17. $M_1 = 162.8$ ftk

7.19. See Example 7.8

7.21. $M_2 = 152.7$ ftk; $M_3 = 74.3$ ftk

7.26. $M_3 = 123.4$ mkN

7.27. $\begin{bmatrix} 0.01125 & 0.0005625 \\ 0.0005625 & 0.0000500 \end{bmatrix} \begin{Bmatrix} R_1 \\ R_2 \end{Bmatrix} + \begin{Bmatrix} -0.81 \\ -0.0432 \end{Bmatrix} = \begin{Bmatrix} -0.06 \\ -0.0524 \end{Bmatrix}$

7.28. See Example 7.9

7.29. See Example 7.12

7.30. See Example 7.13

7.31. $M_1 = 578.8$ mkN

7.33. $M_2 = 266.7$ mkN

7.35. $M_2 = -147.1$ mkN; $M_1 = 441.2$ mkN

7.36. $X_4 = -8.78$ k

7.38. $M_2 = 346.3$ mkN

7.39. $M_1 = 375$ mkN

7.40. $M_3 = 161.5$ ftk

7.41. $M_2 = 63.0$ ftk

7.42. $M_2 = 201.93$ ftk

7.43. $M_2 = 150$ ftk

7.44. $M_3 = 333.3$ ftk

7.48. $M_2 = 194.5$ ftk; $M_3 = 201.3$ ftk

7.49. $X_1 = -5.3^k$; $Y_1 = -4.2^k$

7.50. $M_{2(right)} = -75.1$ mkN; $M_4 = 37.6$ mkN

7.52. See Example 7.15

7.53. See Example 7.16

7.54. See Example 7.17

7.57. $Y_2 = 162.5^k$

7.58. $F_9 = -44.2^k$

7.60. $F_5 =$ zero

7.61. $Y_2 = 63.9^k$

7.62. $Y_3 = -0.3^k$

7.63. $F_2 = -3.8^k$

7.64. $F_2 = -10.0^k$

7.65. $F_9 = 4.0^k$; $F_{11} = 0.9^k$

7.66. $F_7 = 69.8$ kN; $F_{10} = -56.7$ kN

7.67. $F_{12} = 36.0^k$; $F_{15} = -17.8^k$

7.68. $X_2 = -200$ kN

7.69. See Example 7.22

7.70. See Example 7.23

7.71. See Example 7.24

7.72. $M_{3(right)} = 8.58$ ftk

7.73. $M_1 = 88.8$ mkN

7.74. $F_3 = 6.6$ kN

7.75. $Y_3 = 28.2^k$

7.76. $L_3 = 143.73$ in.

7.79. $F_5 = -9.09^k$

7.80. $M_1 = -360.6$ mkN; $Y_4 = 62.9$ kN

CHAPTER 9

9.1. See Example 7.1

9.2. See Problem 7.5 answer

9.3. See Problem 7.15 answer

9.4. $M_3 = -201.4$ mkN

9.5. See Example 7.12

9.7. $Y_2 = -57.1$ kN; $M_3 = -342.9$ mkN

9.8. See Example 7.2

9.9. See Example 7.3

9.10. See Example 7.4

9.11. See Problem 7.31 answer

9.13. See Problem 7.21 answer

9.16. See Problem 7.50 answer

9.18. See Problem 7.35 answer

9.19. See Example 7.8

9.20. See Example 7.9

9.21. See Example 7.13

9.22. See Problem 7.49 answer

9.24. See Problem 7.61 answer

9.25. See Problem 7.72 answer

9.26. See Problem 7.73 answer

9.27. See Problem 7.75 answer

9.29. See Problem 7.80 answer

9.30. See Example 7.22

9.31. See Example 7.23

9.32. $T_1 = 8.5^k$; $T_2 = 16.5^k$

9.34. $Y_3 = Y_5 = 2.5^k$

9.39. $T_1 = 10.05^k$; $T_2 = 11.39^k$

9.41. $S_{1,1} = 400/3$; $S_{2,2} = 225$; $S_{3,3} = 1100$

9.42. $X = 4.22^k$ at left support

9.43. $\begin{bmatrix} 2790 & -870 \\ -870 & 1830 \end{bmatrix} \begin{Bmatrix} D_1 \\ D_2 \end{Bmatrix} = \begin{Bmatrix} -30 \\ -7.5 \end{Bmatrix}$

9.47. Center support reactions are: $X = -12^k$; $Y = 71.5^k$

9.48. $X_1 = -4.88^k$; $Y_1 = 15^k$

9.49. $X_2 = -3.54^k$; $Y_2 = 25.6^k$; $M_2 = 62.1$ ftk CCW

9.55. $F_1 = 109{,}090.9$ ftk; $F_2 = 9090.9^k$

9.58.

Col. 4	Col. 5	Col. 6	P_e
−3,000	21,000	0	8.4
0	0	0	0
−21,000	98,000	0	19.6
3,216	−15,384	0	37.9
−15,384	598,016	128,000	81.8
0	128,000	256,000	130

9.62. $X_1 = -9.23^k$; $Y_1 = 6.92^k$; $X_5 = 9.23^k$; $Y_5 = 53.08^k$

CHAPTER 10

10.1. See Problem 7.10 answer

10.3. See Problem 7.8 answer

10.5. See Problem 7.17 answer

10.8. See Problem 7.33 answer

10.9. See Problem 7.38 answer

10.10. See Problem 7.41 answer

10.11. See Problem 7.42 answer

10.12. See Problem 7.43 answer

10.13. See Problem 7.27 answer

10.18. See Problem 7.44 answer

10.19. See Problem 7.48 answer

10.20. See Problem 7.36 answer

10.23. $Y_1 = 51.96^k$; $Y_2 = 78.22^k$; $Y_3 = -12.86^k$; $Y_4 = 3.22^k$; $Y_5 = -0.54^k$

10.27. See Problem 7.40 answer

CHAPTER 13

13.1. $M_1 = 119,670\,\theta_1$; $M_2 = 171,024\,\theta_2$

13.2. $\phi = 0.0800$ rad CCW; $D = 0.375$ ft; $P = 1620^k$

13.3. $v_{1(2)} = 0.320$ in; $v_{1(3)} = 0.512$ in.; $v_{2(3)} = 1.792$ in.; $R_{3(4)} = 21.48^k$; $R_{3(5)} = 36^k$. To find $R_{3(5)}$, the area under the deflection curve of system 3 must first be found: $A_{3(3)} = 3.24$ ft^2

13.4. $v_{2(1)} = 9000/EI_1$; $M_{1(3)} = 620.7$ ftk CCW. It is impossible to find $v_{2(2)}$ unless $v_{2(3)}$ is given.

Index

Fixed ended moments for prismatic members

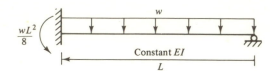

$$\frac{wL^2}{8}$$

w

Constant EI

L

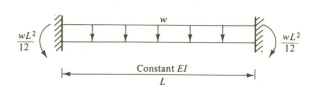

$$\frac{wL^2}{12}$$

w

$$\frac{wL^2}{12}$$

Constant EI

L

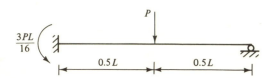

$$\frac{3PL}{16}$$

P

$0.5L$ $0.5L$

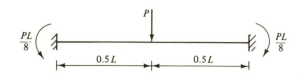

$$\frac{PL}{8}$$

P

$$\frac{PL}{8}$$

$0.5L$ $0.5L$

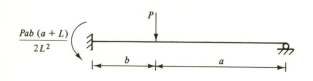

$$\frac{Pab\,(a+L)}{2L^2}$$

P

b a

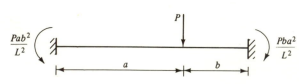

$$\frac{Pab^2}{L^2}$$

P

$$\frac{Pba^2}{L^2}$$

a b

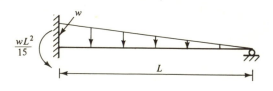

$$\frac{wL^2}{15}$$

w

L

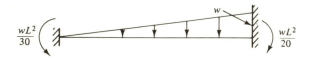

$$\frac{wL^2}{30}$$

w

$$\frac{wL^2}{20}$$

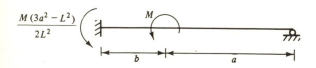

$$\frac{M\,(3a^2-L^2)}{2L^2}$$

M

b a

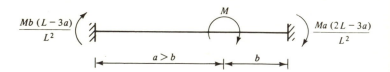

$$\frac{Mb\,(L-3a)}{L^2}$$

M

$$\frac{Ma\,(2L-3a)}{L^2}$$

$a>b$ b